GEOMORPHOLOGY

A Systematic Analysis
of Late Cenozoic Landforms

GEOMORPHOLOGY

A Systematic Analysis
of Late Cenozoic Landforms

SECOND EDITION

Arthur L. Bloom

Department of Geological Sciences
Cornell University

PRENTICE HALL, Englewood Cliffs, New Jersey 07632

Library of Congress Cataloging-in-Publication Data

BLOOM, ARTHUR L. (Arthur Leroy) (date)
Geomorphology : a systematic analysis of late cenozoic landforms /
Arthur L. Bloom.—2nd ed.
p.   cm.
Includes bibliographical references and index.
ISBN 0-13-351560-5
1. Geomorphology.   2. Geology, Stratigraphic—Quaternary.
I. Title.
GB401.5.B55   1991
551.4′1—dc20

90-49389
CIP

To Donna, with love.

Cover: Portion of Landsat Thematic Mapper image no. 50213-1342 of the Subandean Belt
in southern Bolivia along the Pilcomayo River. North is toward the left. Image covers an
area of about 70 × 100 km. See discussion of Fig. 11–21 for details.

Editorial/production supervision: Merrill Peterson
Interior design: Joan Stone
Cover design: Bruce Kenselaar
Prepress buyer: Paula Massenaro
Manufacturing buyer: Lori Bulwin
Acquisitions editor: David Brake

© 1991 by Prentice-Hall, Inc.
a Simon & Schuster Company
Englewood Cliffs, New Jersey 07632

Printed in the United States of America
10 9 8 7 6 5 4 3 2 1

ISBN 0-13-351560-5

Prentice-Hall International (UK) Limited, *London*
Prentice-Hall of Australia Pty. Limited, *Sydney*
Prentice-Hall Canada Inc., *Toronto*
Prentice-Hall Hispanoamericana, S.A., *Mexico*
Prentice-Hall of India Private Limited, *New Delhi*
Prentice-Hall of Japan, Inc., *Tokyo*
Simon & Schuster Asia Pte. Ltd., *Singapore*
Editora Prentice-Hall do Brasil, Ltda., *Rio de Janeiro*

Contents

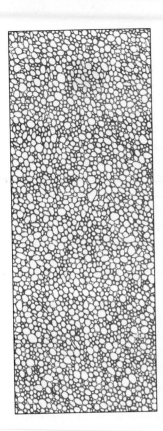

Preface

A useful textbook should be as timeless as possible, but ideas, like landscapes, change. In the thirteen years since the first edition was published, many advances have been made in the field of geomorphology toward understanding the processes that shape landscapes and the rates of change. Older ideas are being tested with a quantitative rigor not available even a decade ago. Mathematical modeling of landscapes and geomorphic processes provide new and testable hypotheses of landscape evolution.

Another trend has been the study of landscapes on a regional scale, aided by photographs and digital images from orbiting spacecraft. *Megageomorphology* is a new aspect of the subject that concerns the origins and evolution of mountain ranges rather than mountains, and deserts rather than desert landforms. Global tectonics emphasizes constructional landforms and relates denudation history and explanatory description of landscapes to tectonic theories and structural analysis. The new *earth systems science,* which visualizes surface relief as the balance between erosion and tectonism, requires the integration of geomorphology with structural geology, tectonics, geophysics, petrology, and stratigraphy.

Geomorphologists are joining other scientists in studying global change and in integrating geologic, atmospheric, oceanographic, and biologic information in attempts to comprehend the "earth system." Especially in this time of concern over climatic and other environmental changes generated by human activity, we need integrated studies of our entire planet. The hypotheses of climatic geomorphology are especially relevant to studies of global change: What regional landforms and processes are diagnostic of certain climates? When climates change, how soon and in what ways do landscapes respond? We know that mountain belts affect climate, but climate can also affect tectonism by determining erosion rates and thereby heat flow, rock deformation, and isostatic adjustment.

At this writing it seems likely that within the next decade we will have orbiting space platforms monitoring our earth system on an hourly or daily basis, providing a massive data stream to computers and their operators. The next generation of geomorphologists, even those who use this book as an introduction to the subject, will have unprecedented opportunities to study the earth's scenery from space and to comprehend the land surface as the product of competing internal and external processes.

In writing and revising this book, I have assumed that the student who uses it has had at least one introductory course in geology or physical geography. Elementary familiarity with climatic zones, planetary winds and ocean currents, the geologic time scale, common rocks and minerals, and topographic maps is assumed. The primary role of geomorphologists is to teach and demonstrate that landscapes change in response to energy supplied from a variety of sources. Among ourselves, we can engage in learned debates about whether the changes are sequential or random and whether the energy systems are best understood as open or closed, but we all accept the transient nature of the present landscape. Reduced to that level, geomorphology is still close to the natural philosophy from which it and other sciences originated.

This book is organized largely along traditional lines, following an outline I have evolved by teaching a one-semester course in geomorphology. It considers first the constructional processes and landforms that result from tectonic and volcanic activity (Part I). Owing to the dynamic nature of plate-tectonic theory, we must consider constructional and erosional landforms on an equal level, and we must consider the submarine landscape as well as the better-known subaerial landscape.

Next, as an introduction to destructional or erosional landforms (Part II), the energy environment at the surface of the earth is analyzed in some detail. It is my conviction that geomorphologists have been regrettably delinquent in considering the energy supply for the changes they have so sweepingly described. Like all other physical processes, geomorphic change must be controlled ultimately by energy gradients and potentials. The concept of physical systems which have definable boundaries through which matter and energy move at definable rates is especially well illustrated by the energy flow in geomorphic processes. The favorable reaction to the concept of a solar-driven "geomorphology machine" in a brief earlier book (Bloom, 1969) and the first edition of this book encouraged me to expand that topic further.

Erosional processes and the landscapes they produce from tectonic or structural blocks occupy three-fourths of the book. The logical scheme of explanatory description in terms of structure, process, and time is followed not out of blind loyalty to the spirit of W. M. Davis but from my studied conviction that the best way to describe landscapes is to explain their origins. For the practical needs of professional geographers, geologists, pedologists, and hydrographers, as well as for the cultural needs of intelligent tourists who enjoy visiting areas of great scenic beauty, explaining how and when a landscape took its form must be the major emphasis in a book of systematic geomorphology.

Part III is a review of the geomorphic processes and forms that are distinctive of specific climatic regions. Arid and glacial landscapes were once regarded as "climatic accidents," but their extent during the late Cenozoic Era requires that they be treated as more than anomalies. Even so, they and their peripheral regions of semiarid and periglacial landscapes are best studied after the fluvial or "normal" processes are well understood. Much of climatic morphogenesis concerns the relative intensity, rather than the kind, of geomorphic processes. Chapter 18 concludes Part III with a series of regional examples that demonstrate the complexity of the Quaternary climatic oscillations that have made our modern landscape so interesting and unique in geologic history. The advantage of precise radiometric dates for documenting rates of change is illustrated repeatedly.

Coasts deserve separate treatment (Part IV) because the energy input that shapes them is so different from the processes that shape subaerial landscapes. More important, the coastal zone is critical to human activities. Two-thirds of the world's people live in the 10 percent of the land defined as the coastal zone. The transience of sea level during the Quaternary Period has been a major factor in the development of the estuaries, lagoons, deltas, and reefs upon which so much modern commerce and food production depend. Future trends of sea-level movement are critical to coastal-zone activities.

Regional descriptions are included to demonstrate how certain kinds of landscapes formed, but this is not a textbook of regional physiography. Examples and illustrations have been chosen from as wide a range as I have been able to find in my own travel and in published descriptions, in order to counter the natural tendency for provinciality that lessens the value of many geomorphology books.

I regret that the length of the manuscript and the continuing explosive growth of new information have precluded special chapters on the morphology of the ocean floor, the moon, and the other terrestrial planets. It is only in the last decades that we have begun to rectify W. M. Davis's lament that "there are not enough kinds of observed facts on the small earth in the momentary present to match the long list of deduced elements of the scheme [of explanatory description]" (Davis, 1905, p. 156). The morphologists who describe the floor of the ocean and the face of the moon and planets must describe whole new lithologic materials and structures and the entirely exotic processes of weathering and erosion that occur under dense fluid "atmospheres" or in near vacuum. They require new, perhaps vastly expanded, time scales. By contrast, the tectonic relief on the floor of the oceans is very young. Yet I confidently predict that the schemes of explanatory description that are ultimately used for both submarine and planetary landscapes will not be very different from the one developed for explaining subaerial terrestrial scenery.

The references at the end of each chapter have been selected first for their relevance and second for their availability. I hope students will be sufficiently stimulated and curious to read at least some of these outstanding studies. In celebration of its centennial year during 1988, the Geological Society of America published in its Bulletin numerous special review articles on various aspects of geology, many with geomorphic implications. These articles, which appeared during the preparation of this second edition, were enormously helpful both for their depth of analysis and their rich bibliographies. Concurrently, the Geological Society of America has been producing a set of splendid monographs, field guides, and special volumes on topics of North America geology, under the theme of the Decade of North America Geology (DNAG). These volumes have also been of great help in preparing this revision. I also thank my academic neighbors, Donald R. Coates and Marie Morisawa, for their 20 years of organizing and promoting the annual Binghamton Geomorphology Symposia, where geological and geographical geomorphologists meet to discuss and evaluate each other's ideas. The 20 published symposium volumes are another major source of current geomorphic ideas. I especially commend these sources to the readers of this book. A textbook should be the beginning, not the end, of understanding.

Students are always their professor's best critics, and I thank mine for all their comments about the first edition of this book. Numerous colleagues also have offered constructive comments concerning both editions. In particular, I thank my Cornell colleagues of the Andes Group, a loosely structured and convivial group of geologists committed to understanding the genesis of the Andes Cordillera of South America. Their cross-disciplinary approach to understanding a major mountain range has been a model of "megageomorphic" thinking.

An unexpected pleasure of authorship has been experiencing the generous and enthusiastic responses to my requests for illustrations. Many friends provided not only the photograph I requested, but several others from which I could choose. Seeing one's name in a caption is a trivial repayment, so I further acknowledge here my debt of gratitude to all contributors. Photographs not otherwise credited are my own. Most of my line drawings were redrafted by professional artists. Mindful of Sir Charles Cotton's gentle scorn (1955, p. 582) for "careless or delegated drafting," I can only extend my sinistral wrist for a deserved ethereal slap by my respected late teacher.

A final note on dimensions: The SI (metric) system is used almost throughout without conversions into English or obsolete metric dimensions. Geomorphologists in the United States have been slow to convert to the metric notation of all other sciences and most of the world's population. Here, as in the use of regional examples, our comfortable provinciality must be overcome.

Arthur L. Bloom

REFERENCES

BLOOM, A. L., 1969, The surface of the earth: Prentice-Hall, Inc., Englewood Cliffs, N.J., 152 pp.

COTTON, C. A., 1955, Theory of secular marine planation: Am. Jour. Sci., v. 253, pp. 580–89 (reprinted 1974 *in* Bold coasts: A.H. & A.W. Reed, Wellington, New Zealand, pp. 164–74).

DAVIS, W. M., 1905, Complications of the geographical cycle: 8th Internat. Geog. Cong., Washington, 1904, Repts., pp. 150–63 (reprinted 1954 *in* Geographical essays: Dover Publications, Inc., New York, pp. 279–95).

1 The Scope of Geomorphology

THE SYSTEMATIC STUDY OF LANDSCAPES

Every sensitive person, on viewing a mountain or river valley, instinctively wonders how the landscape assumed its form. Deeper thinkers sense that no scene is immutable but that all change with the passage of time. A few scientists of each generation choose *the systematic description, analysis, and understanding of landscapes and the processes that change them* as an area of specialization. These scientists are geomorphologists.

The description, analysis, and understanding of landforms (the component parts of landscapes) is a science only when it is done in a disciplined fashion. Many schemes of organized study can be conceived. **Regional description,** by district, nation, or natural provinces, is one logical approach (Thornbury, 1965; Hunt, 1974; Embleton, 1984; Short and Blair, 1986). The earliest geography books were largely regional descriptions of visible landmarks, passes, river valleys, and other landforms that helped guide the traveler. Regional, descriptive geomorphology remains as one of the accepted subdivisions of the science. A variant of growing importance, although not so much in North America, is the preparation of geomorphic maps that show landform components by distinctive colors and symbols (Barsch and Liedtke, 1980).

At the beginning of the nineteenth century, the science of geology became distinguishable from "natural philosophy." As facts and theories about the origins of rocks and mountain ranges accumulated, they were absorbed into regional or descriptive geomorphology as an aid to understanding the origins of landforms. Thus **explanatory description** evolved, in which a landscape is not simply described by the heights of its hills and the declivities of its slopes but by the reconstructed geologic history of its evolution. Explanatory description is a higher form of analysis than regional description in that it requires inferences about past events. In a sense, a landscape cannot be understood until the entire geologic history of the rocks and slopes is known. Fortunately, for most landscapes it is possible to begin an explanatory description with a phrase like: "As the area emerged from the Cretaceous sea," or "Orogenic uplift in the Miocene Epoch produced a mountainous terrain." Whereas regional geomorphology, or physiography, is often regarded as part of the larger science of geography, explanatory or genetic geomorphology is mostly a geologic subject. Geomorphology is thus a part of two sciences, and geomorphologists may regard themselves as either geographers, or geologists, or preferably both (Dury, 1972; Baker, 1988). To be part of the growing field of "Earth System Science," future geographical geomorphologists will need to learn much more about tectonics, and geologically trained geomorphologists will need to learn more climatology.

The subject of landform analysis can also be organized on the basis of the processes that have operated and now operate to shape the land. Many recent geomorphology books deal with geomorphic processes (for example, Leopold et al., 1964; Ollier, 1969; Embleton and Thornes, 1979; Birkeland, 1984; Drewry, 1986; Ritter, 1986). From the number of process-oriented books published in recent decades, one might think that geomorphology has been reduced to a study of processes, without consideration of the resulting landforms. Process-oriented geomorphology is closely related to climatology, because air temperature, precipitation, winds, and atmospheric humidity largely determine the response of rocks to subaerial exposure. **Climatic morphogenesis** is the approach to geomorphology that emphasizes the shaping of landforms in response to various climatic conditions (Birot, 1960, 1968; Tricart and Cailleux, 1965, 1972; Büdel, 1982). This is an important topic, because many landscapes preserve the record of more than one set of climatic-controlled processes, and the correct explanatory description of such a landscape provides evidence for reconstructing ancient climates.

Much of the information about geomorphic processes is derived from other scientific fields. The elastic and plastic deformation of rocks under pressure and the thermal conductivity of rocks are studied by geophysicists, generally not with the primary intention of describing or analyzing landscapes. Seismologists are properly more concerned with the causes and prediction of earthquakes than with the landforms that earthquakes produce. Geochemists and mineralogists study the chemical reactions of rocks and minerals with hydrous solutions and thereby provide data about weathering processes. Pedologists study soil formation; geohydrologists study the work of flowing water and groundwater; sedimentologists study depositional processes and sediments; glaciologists study the physical properties of ice; and oceanographers describe the submarine topography they see traced on their depth recorders. All of these scientists contribute to an understanding of the processes that shape the earth's surface. A geomorphologist must be at least conversant with these peripheral subjects, as well as with the techniques of remote sensing.

How can regional description, historical reconstruction, and the details of the chemical and physical response of rocks to water and air be unified into a science of geomorphology? The best solution that has evolved is summarized by the trinity of *structure, process,* and *time.* Although generally attributed to the great American geomorphologist, William Morris Davis (1850–1934), the organization of geomorphic information under the three headings is implicit in many earlier writings and is such a natural organization that it was probably unconsciously used many times before Davis codified it with such great force and clarity.

To illustrate the concept of geomorphic description in terms of structure, process, and time, consider the following descriptive paragraphs. The first is empirical; it states clearly *what* and *where.* The second is explanatory; it connotes *how* as well as *what* and *where.*

1.   "The city of Ithaca, New York, is built on low ground at the south end of
     Cayuga Lake. Steep slopes surround the town, above which is a broad,

      rolling upland that becomes gradually more rugged southward into the low mountains of Pennsylvania."

2.   "The city of Ithaca, New York, is built on a delta plain at the south end of the Cayuga Lake trough. Glacially oversteepened valley walls around the city are in contrast to the maturely dissected, cuestaform upland of the glaciated northern Appalachian Plateaus."

In the second paragraph the technical terms "delta plain," "cuestaform," and "plateaus" convey information about structure and genesis. Geomorphic processes are explicit in the terms "trough," "glacially oversteepened," "dissected," and "glaciated." Time is implied in the term "maturely dissected" in that a degree or stage of erosional dissection is specified. The first paragraph would be perfectly suitable for a tourist brochure or travel guide, but the second paragraph, in exactly the same number of words, conveys much more information to a trained reader. True, technical terms are used that must be understood by the reader, but technical terms are a kind of shorthand notation to convey concepts and increase understanding in an efficient manner.

If empirical description were the only goal of geomorphology, a regional landscape description like the two samples just given could be reduced (or expanded?) to a mathematical equation. Only three numerical terms are needed to describe the altitude above sea level of any point on the landscape and locate the point by its latitude and longitude coordinates. An integration of the changes in altitude with latitude and longitude will, in principle, give an equation that completely describes any specified area of land surface. But even if such an equation could be written, no one, not even the most astute mathematician, could envision the land surface that it describes. A radar-scanning computer on board a cruise missile can recognize a landscape by mathematical terms, but human minds require descriptive words to convey images of landscapes. Significantly, the most satisfying way to describe a landscape seems to be by describing its history or genesis. Geomorphic description explains *how* and *when,* as well as *what* and *where.* Only the ultimate question of *why* is left to the philosophers.

RANDOM AND DETERMINISTIC RELIEF

What makes a painting or a computer simulation of a landscape look "real"? A cloud is not a sphere, and most mountains are not cones or pyramids. No one is deceived by the simplified background scenery of animated cartoons, although computer programs are beginning to create extremely realistic, if totally imaginary "landscapes" (Fig. 1–1). The key to naturalistic landscape portrayal is the introduction of random roughness to the scene in a scale-invariant manner, so that whether viewed as an entirety or by microscopic examination of a small part, the scene always resolves into still-smaller detail. For example, photographs of a rocky coastline or a drainage network taken from distances of 10 m, 1 km, or 100 km may be indistinguishable unless a scale bar is provided (Turcotte, 1989). Mandelbrot showed that many, and perhaps most, natural objects such as mountain sides, river drainage networks, and island shorelines, do not resolve into simple geometric shapes on close examination, but maintain constant roughness at all scales. He called such objects *fractals* and developed an alternative to Euclidian geometry, which he called fractal (fractional) geometry, for their analysis. An essential measure of fractal character is the *fractal dimension* value which is commonly not an integer (Mandelbrot, 1983). The fractal dimension of terrestrial relief is remarkably consistent over almost six orders of magnitudes (Fig. 1–2). Submarine abyssal hills,

Figure 1-1. "Gaussian hills that never were" by R. F. Voss. A perspective view of a computer-generated landscape made to look realistic by assigning a Gaussian probability distribution to the difference in altitude between any two prescribed points on a map, with suitable mathematical transformations. (Mandelbrot, 1983, Plate CII.)

a 90 × 90 km terrain in Wales, and longer-wavelength submarine topography all show scale-invariant roughness. If raising each hill against gravity requires a certain amount of mechanical energy, it has been shown that the topography represents a random distribution in which the energy of formation is distributed uniformly over hill sizes. Smaller hills may require less energy of formation, but are more numerous, so as to ensure the uniform division of energy (Bell, 1975).

The shape and size of some landforms are not random but are determined by the forces that built them and the inherent properties of their materials. The widths of fault blocks and grabens, for instance, can be related to the thickness of the brittle layer that has been broken (Chap. 3). Scoria cones can be built to only a certain size, after which granular ejecta can no longer be expelled and they fail by lateral slumping or are "floated" aside by a basal lava flow (Chap. 4). On the largest scale, planet Earth is a spheroid of rotation determined by the density of its materials, its size, and its rotational velocity. Even the randomly rough abyssal hills of the Pacific basin show an elongation by a factor of 3 or 4, parallel to the mid-ocean ridges, that was determined by their origin as fault blocks along a spreading ridge. This shape anisotrophy is not large enough to be seen on a diagram such as Figure 1–2 (Bell, 1975, p. 889).

Time Scales for Landform Development

In general, the larger the landform, the more durable it will be (Ahnert, 1988). The spheroidal shape of the earth was established early in its history and since then has not varied significantly. Continents have accreted over geologic time, but although the pieces have been repeatedly disrupted and reorganized, probably up to 90 percent of the continental crust had formed by 2.5 billion years ago, and the rate of accretion of new continental crust has been very slow since then (Taylor and McClennan, 1985, p. 234).

Progressively smaller landforms can be both created and destroyed relatively more rapidly. Carey (1962, p. 98) suggested that geomorphic and other geologic processes have rates that are size-dependent over many orders of magnitude. Earthquakes and avalanches, measured in linear scales of meters to kilometers, are completed within seconds, or minutes. Glaciers and ice sheets, ranging in size from hundreds of meters to a thousand kilometers across, require many years to many

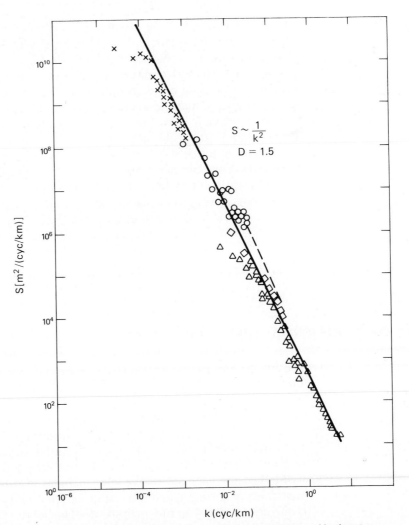

Figure 1-2. Power spectral density S of the earth's topography and bathymetry as a function of wave number k (the reciprocal of length). S is a measure of roughness, and its power law relation to k is defined by the fractal dimension $D = 1.5$. The relation holds over about six orders of magnitude from the spherical harmonics of the earth (x) through several scales of deep-sea topography (o, − −, Δ) to the roughness spectrum of a 90 km × 90 km square in northern Wales (◊). Only at the largest scale of the earth's sphericity ($k \sim 10^{-4}$ to 10^{-5} km^{-1}) does the surface roughness deviate from being a power function of the length of the transect. (Turcotte, 1989, Fig. 14, and references cited therein.)

thousands of years to accumulate, depending on their size. From arguments about finite strain and strain rates in rocks (p. 13) it can be concluded that orogenic deformation intervals typically range in time from 1 to 30 million years, with most events concentrated between 1 and 5 million years. Some periods of intense deformation could be even shorter, perhaps occurring within 100,000 years (Pfiffner and Ramsey, 1982, p. 316). Of all the processes that shape the surface of the earth, only isostasy is more rapidly accomplished for large areas than for small ones, due to both the viscoelastic nature of the mantle and the flexural rigidity of the lithosphere, which can support intense loads on small areas without deflection but deforms rapidly under regional loads of low magnitudes.

From the hypothesis that the rates of geomorphic processes are inversely proportional to the size of the affected area, one can deduce a corollary statement, that the cumulative effect of most geomorphic processes increases progressively slower within either increased time or area. If such changes are exponential, the rate of any process, as defined by the total change/time, would appear progressively more rapid when measured over progressively shorter intervals of time. This is in accord with many personal experiences during floods, avalanches, or earthquakes, when great changes are accomplished in brief intervals. By a similar rule, the total change/area, or *intensity* of change, would be largest when viewed within a very small area, as for instance at the head of an eroding gully (Fig. 9–2). A common experience of geomorphologists is that the net change of a landscape seems to occur at a decreasing rate when we view it over progressively greater area or over longer periods of time. Rates established for any finite time interval, or intensities for a finite area, typically seem excessive when applied to a longer time scale (Gardner et al., 1987) or larger area. For example, sediment yields from small drainage basins always exceed the total sediment yield of the larger combined basin (Chaps. 9 and 10); earth movements during single earthquakes, even in regions where the recurrence interval of comparable earthquakes is known, exceed the net change over long periods of time. The observation that rates and intensities are time- and area-dependent creates problems of geomorphic interpretation that require continued effort to comprehend.

SCALE OF TERRESTRIAL RELIEF FEATURES

A durable concept of both geomorphology and tectonics is that terrestrial relief features can be classified by size. Salisbury (1919) introduced the concept into systematic geomorphology with the following hierarchy: (1) first-order relief features (continents and ocean basins); (2) second-order relief features (tectonic mountain belts, plateaus, and plains); and (3) third-order relief forms (erosion-carved mountains, hills, and valleys, as well as constructional volcanic mountains and equivalent forms). Although difficult to define and of little genetic value, the classification is useful, because each successive level includes features of successively smaller size, greater unity of origin, and more transient durability. Attention is drawn to successively greater detail in the analysis of a landscape. The larger orders of the earth's relief features are not usually analyzed in great detail, but a modernized size classification is introduced here as a prelude to the more complex details of landform analysis and description.

The Shape of the Earth and Other Planets

The best description for the shape of the whole earth is that it is an oblate, or slightly flattened, spheroid (Gaposchkin, 1974; King-Hele, 1976). This reference surface of gravity equipotential, the shape of the earth's mean sea level, and its extension beneath the land, is called the **geoid**. In addition to its fundamentally flattened spheroidal form, the geoid has a number of smaller, unexplained undulations (Fig. 1–3). A boat on the ocean east of Papua New Guinea is 192 m farther from the center of the earth than one south of Sri Lanka, yet by definition, both are at sea level.

The geoid is an important reference surface for geomorphic processes. Rivers, for instance, cease to flow when they reach sea level, unless they drain into depressions below sea level in arid regions (Chap. 13). Thus the geoid is the surface that would result if subaerial processes were ever to level all the continents completely to sea level. The geoidal projection of mean sea level under the subaerial landscape is the best definition of the term **ultimate base level,** which has been the subject of so much debate (Chap. 9).

Figure 1-3. Geoid map of the earth. Contours show deviations from a spheroid of radius 6378.142 km with polar flattening of 1: 298.255. Contour interval: 10 m. (D. L. Turcotte, based on Goddard Earth Model 8.)

Continents and Ocean Basins

The largest units of the earth's rocky *relief* (altitude above or depth below the geoid) are the continental blocks and the abyssal basins of the ocean floors. Continental structure covers about 41 percent of the earth's surface (Chap. 2), but only 29 percent of the surface is now above sea level. On the continents, the average altitude is 823 m above sea level. The average depth of the oceans is about 3700 m. Most of the average depth is formed by abyssal basins, which cover about 30 percent of the earth's surface. Smaller regions of the continents and ocean floors rise much higher and plunge much deeper than the averages, but it is a striking and important characteristic of our planet's primary relief that the *hypsographic curve* is strongly bimodal (Fig. 1–4).

Tectonic theorists now recognize 10 major plates and several smaller ones as the basic components of the earth's lithosphere. In the "new global tectonics," the continents are almost incidental masses that are being rafted along toward the zones where lithospheric plates are sinking back down into the mantle. Because of their low-density rocks, the continental slabs are not drawn down but accumulate on top of the mantle.

Of the inner planets and our moon, only Earth and probably Venus have multiple, laterally mobile, recycling lithospheric plates. The smaller planets (Mercury and Mars) and the earth's moon each have a single, unsegmented lithospheric plate that formed early in the history of the solar system. Venus, approximately the same size as the earth and of similar composition, may recycle lithospheric plates along divergent and convergent margins by processes broadly similar to those that drive terrestrial plate motions (Head and Crumpler, 1987; Head, 1990a, b).

Even if a few slabs or plates are the basic geophysical units of the earth's outer layers, the strong bimodality of the hypsographic curve justifies a primary geomorphic distinction between the continents and abyssal oceanic basins. Whatever the structural cause for their contrasting average altitude, their geomorphic differentiation is very real. How the continents have maintained their altitudes near or above sea level throughout at least the one-half billion years of Paleozoic and later geologic time is one of the most important topics of historical geology. In the face of evidence for erosion to depths of 20 km or more into continental structures (p. 120), the mechanism of long-term continental survival also assumes primary importance as a geomorphic topic.

The oceanic crust, experiencing geologically rapid creation and consumption, has a relatively simple exponential correlation of depth with age (Fig. 1–5). At an accreting plate margin along a mid-ocean ridge, the oceanic lithosphere is hot and has low density. Progressive cooling and thermal contraction of a lithospheric plate as it moves away from an accreting margin accounts well for both the observed geothermal heat flow and the oceanic depth, at least to a depth of about 6 km.

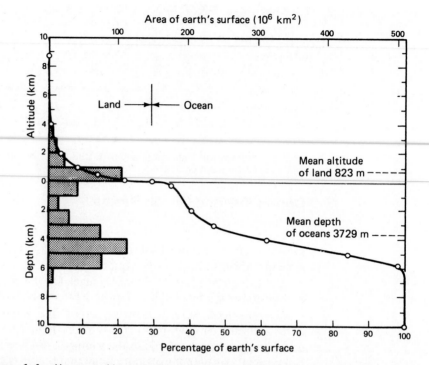

Figure 1-4. Hypsographic curve showing area of earth's solid surface above any given altitude or depth. Bar graph on left shows the distribution of area by 1-km intervals. (From Menard and Smith, 1966; Kossinna *in* Fairbridge, 1968.)

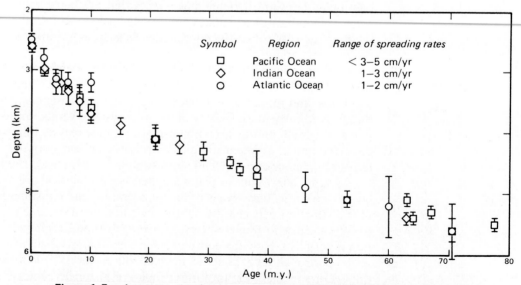

Figure 1-5. Average water depth to smooth oceanic crust as a function of the age of the seafloor. The crust at slowly accreting plate margins is shallower than the average for the first 20 million years. Thereafter, depth is an accurate predictor of crustal age. (Data from Sclater et al., 1971.)

Geomorphic (Physiographic) Provinces

Continental structures are broadly subdivided into cratonic massifs, platforms (thin, nearly flat-lying strata), and orogenic belts of several ages. Similarly, oceanic structures include mid-ocean ridges, abyssal plains, island arcs, back-arc basins, and epicontinental seas. Superimposed on the continental and oceanic structures are major relief units that are recognizable by their size, continuity, uniformity of structure, and geologic durability. The familiar physiographic provinces of the United States (Fig. 1–6), as organized by Fenneman (1917, revised 1928), are examples of such relief features. Each province was defined as a geomorphic (or physiographic) entity that has resulted from a set of processes or a succession of processes acting on a specific structure (in the broad sense of a regional geologic terrane, as noted on page 12) for a certain period of time or to a certain stage of completion. Fenneman's approach was thoroughly Davisian, as the preceding sentence emphasizes. Nevertheless, the boundaries that he drew formally defined regions that had long been informally recognized as "provinces" in loosely defined terms as an aid in identifying the regional setting for local areas of study. Fenneman's provinces have been generally accepted as the geomorphic divisions of the United States. The provinces, subdivided into sections and grouped into larger divisions, are easily recognized as entities, even though their exact boundaries may be matters for scholarly debate. Words such as "Colorado Plateau" and "Appalachian Valley and Ridge province" communicate immediate understanding to the informed reader.

The relief features commonly identified as geomorphic or physiographic provinces seem to be about the optimal size for regional or descriptive geomorphology. Books and maps of the regional geomorphology of the United States (Fenneman, 1931, 1938; Hammond, 1963; Thornbury, 1965; Hunt, 1974), Canada (Bostock, 1964), Alaska (Wahrhaftig, 1965), and the floor of the Atlantic Ocean (Heezen et al., 1959) are but a few examples of the use of geomorphic provinces to organize regional geomorphic description. Although the concept of geomorphic provinces was intended to facilitate regional descriptions of nations or large continental areas, it has been adopted, even by scientists not necessarily aware of the geomorphic tradition, as the framework for describing the moon and other planets. For example, Head (1990a, p. 100) organized a description of a terrane on Venus into two large tectonic and geomorphic units, a *Plains Group* consisting of *Smooth Plains, Ridged Plains,* and *Mottled Plains,* and a *Linear Ridge Group* consisting of a *Banded Unit,* a *Ridged and Domed Unit,* and a *Linear Ridge and Groove Unit.* The divisions were based on structure, topography, and surface roughness as interpreted from radar images, and while neither the formative processes nor the relative age of the geomorphic units on Venus are known with certainty, relative surface roughness and stratigraphic relationships to probable volcanic flows permitted reasonable inferences about both.

Morphotectonic Regions

Among structural geologists and tectonists, the terms *morphotectonic region* and *morphostructure* are sometimes used to define regions or landscapes characterized primarily by their tectonic or structural unity. Although they are not clearly defined, these terms emphasize the origins of topography by tectonic movements rather than by erosion. Morphostructures are large topographic features such as submarine trenches or mid-ocean ridges, formed directly by tectonic activity and little modified by subsequent erosion. The definition of geomorphic provinces includes tectonic origins, so a good geomorphic description is more comprehensive than a purely morphotectonic one.

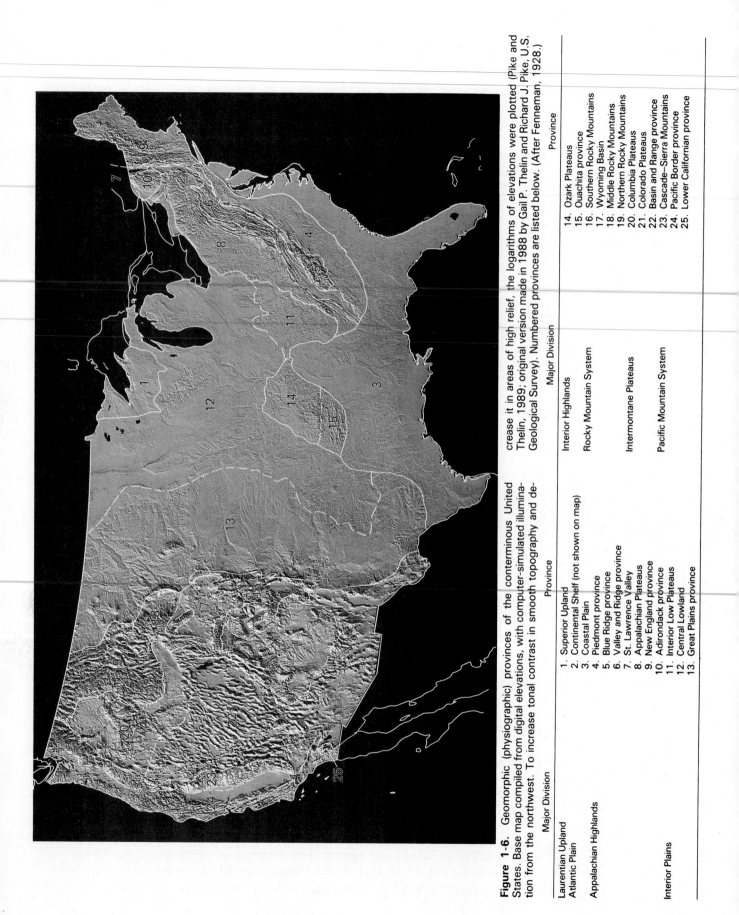

Figure 1-6. Geomorphic (physiographic) provinces of the conterminous United States. Base map compiled from digital elevations, with computer-simulated illumination from the northwest. To increase tonal contrast in smooth topography and de-crease it in areas of high relief, the logarithms of elevations were plotted (Pike and Thelin, 1989; original version made in 1988 by Gail P. Thelin and Richard J. Pike, U.S. Geological Survey). Numbered provinces are listed below. (After Fenneman, 1928.)

Major Division	Province		Major Division	Province
Laurentian Upland	1. Superior Upland		Interior Highlands	14. Ozark Plateaus
Atlantic Plain	2. Continental Shelf (not shown on map)			15. Ouachita province
	3. Coastal Plain		Rocky Mountain System	16. Southern Rocky Mountains
Appalachian Highlands	4. Piedmont province			17. Wyoming Basin
	5. Blue Ridge province			18. Middle Rocky Mountains
	6. Valley and Ridge province			19. Northern Rocky Mountains
	7. St. Lawrence Valley		Intermontane Plateaus	20. Columbia Plateaus
	8. Appalachian Plateaus			21. Colorado Plateaus
	9. New England province			22. Basin and Range province
	10. Adirondack province		Pacific Mountain System	23. Cascade–Sierra Mountains
	11. Interior Low Plateaus			24. Pacific Border province
	12. Central Lowland			25. Lower Californian province
	13. Great Plains province			

Scenery

Von Engeln (1942) elegantly compared the first-order relief features to theaters: "permanent" structures "in which geologic dramas are enacted." The second-order relief features were compared to theatrical stages, which are rebuilt or altered from time to time. The third-order relief features are the scenery, stage settings, and backdrops, which are mobile and temporary. **Scenery** is the assemblage of landforms that can be viewed from a single vantage point, the pleasure of the tourist and the raw material of the geomorphologist. Almost all of the following chapters concern landscapes at this scale. Most scenery, excluding such notable views as isolated volcanic cones (Fig. 4–14), is erosional rather than constructional.

THE LANDFORM AS THE UNIT OF SYSTEMATIC ANALYSIS

Landscapes are surfaces composed of an assemblage of subjectively defined, irregular, lesser surfaces. Each element of the landscape that can be observed in its entirety, and has consistence of form or regular change of form, is defined as a **landform.** The definition is necessarily subjective (Howard and Spock, 1940). To one geomorphologist, a hillside is a landform, continuous and homogeneous. Another observer sees gullies, ridges, and dozens of other small topographic features that either break the continuity of the hillside or establish its continuity, depending on the mental concept. The mathematical concept of the **fractal character** of natural objects (p. 3) is based heavily on geomorphic examples. There is no limit to the scale of observation. For example, the simple question "how long is the coast of Britain?" becomes an issue of great mathematical complexity when it is realized that at each closer scale of observation, new irregularities appear (Mandelbrot, 1967, 1983). In principle, to get the length of the British coast, the wetted perimeter of every pebble and sand grain on every beach at one instant would have to be summed! At another extreme, currently available radar images of Venus only allow landforms to be resolved at a scale of 1 to 3 km, and Landsat Thematic Mapper images of the earth (Figs. 4–19, 11–19, 11–21, book cover) have a resolution of 30 m. Of necessity, a landform will be defined differently by an observer on the ground or one in an orbiting spacecraft or studying a computer-processed digital image.

Landforms may be *constructional,* as in the examples of fault scarps, volcanic cones, glacial moraines, or river deltas. However, because of the peculiar intensity of atmospheric weathering and erosion on our planet, most subaerial landforms are *destructional,* or *erosional.* Properly, the subaerial landscape is to be viewed as an assemblage of residual landforms, each in the process of being consumed or transformed by erosion. One of the fundamental goals of geomorphologic training is to be able to see landscapes not as an assemblage of slopes, ridges, and mountain peaks, but of erosional valleys, gullies, and hollows, between which the residual hills and mountains stand. The eye must be trained to see that the missing rock material is the significant feature, not the material that remains.

From these introductory comments, it can be appreciated how landforms came to be systematically evaluated and explained in terms of the Davisian trinity of *structure, process,* and *time* (or *stage*). Each landform can be visualized as enclosing or covering a mass of rock with specific physical and chemical properties and geometrically disposed discontinuities (bedding planes, joints, faults, etc.). All of these lithologic and mineralogic factors are collected in the geomorphic concept of *structure.* Also included are active and former tectonic movements.

We observe that a complex series of reactions takes place when rocks are exposed to water and air in our planet's gravitational field. Certain phenomena, such as downward movement of detached fragments, affect all matter. Other processes are uniquely effective only on specific structures. Thus landforms are the result of constructive or destructive *processes* acting on structures.

In closed systems, reactants tend toward stable equilibrium states. Some geomorphic processes, such as landslides or earthquakes, illustrate nearly instantaneous response by rocks to some overwhelming stress. The resulting piles of broken rock (and houses) represent a more stable configuration. Most geomorphic processes act much more slowly. A major topic of geomorphic study is whether all stream valleys are always *graded* (fully adjusted to the erosional processes that are actively shaping them) or whether the graded condition develops slowly and progressively. We see, then, that an account of the interaction of surface processes with geologic structures does not give a complete explanation of landforms unless we include in the explanation the length of time the process has been operating or the degree of equilibrium (*stage*) that has been reached. Relatively few landforms show purely constructional processes.

A significant number of present-day landforms were shaped by processes that are no longer active. For example, many landforms in Canada, the northern United States, and northern Europe were shaped by former ice sheets. These are *relict* landforms, some of which are surprisingly pure records of former conditions. Other relict landforms have been buried for long periods of time and have only recently been *exhumed* (Fig. 3–13). These landforms also resulted from the interaction of geologic structures and processes through an interval of time, but that time did not continue to the present.

The systematic analysis of landforms in terms of structure, process, and time seems to be intuitive. Acknowledgment that landscapes are not immutable leads to the immediate conclusion that if change is in progress, something (structure) is being altered by something (process) to a definable extent (stage) or for a definite interval (time). Absolute time is difficult to measure on the geologic scale, and rates of geomorphic change are variable because of the wide range of processes and structures. For both of these reasons, the concept of a relative stage of development, rather than the absolute length of time a process has been acting, has been prevalent. New techniques that radiometrically date the exposure of land surfaces by measuring the accumulation in surface rocks of isotopes produced by cosmic radiation promise to alter fundamentally the geomorphic concept of time. Future trends of geomorphic research will be toward even better defining absolute rates of change and the true elapsed time during which specified processes have acted on specified structures (Jennings, 1973, p. 123).

FURTHER CONSIDERATIONS OF STRUCTURE, PROCESS, AND TIME

Structure

All of the physical and chemical properties of rocks and minerals are loosely gathered into this general heading by geomorphologists. A useful distinction is made between the **terrane**, an area of a certain structure or rock type, as in the expressions "granitic terrane" or "limestone terrane," and the **terrain**, a landscape or "the lay of the land." A military tactician must appreciate the terrain; he need not be concerned with the terrane. However, one of the first things a geomorphologist wants to know about a region is: "What is the structure?" By this question, he or she inquires about the minerals that form the rocks, the structural and stratigraphic arrangement of rock layers and masses, the previous tectonic displacement of the rocks, and the present state of deformation.

Structure is usually considered the passive result of previous depositional or deformational processes. Rocks at the earth's surface are normally regarded as rigid solids, not likely to flow under their own weight to produce landforms of smooth, rounded domes. Yet it is well known that rocks have finite yield strengths, which when exceeded will result in permanent deformation. Under sufficient shear stress, rocks "flow" or "creep." The property of a substance that determines whether it will behave as a solid or fluid has been defined as *rheidity,* and Carey (1954) arbitrarily defined the rheidity of rock as the amount of time necessary for viscous flow to exceed by 1000 times the elastic (or recoverable) deformation. The concept of rheidity is more rigorously quantified by defining the conventional linear strain rate $\dot{e} = e/t$, where elongation e = change of length/original length and t = duration of strain in seconds (Pfiffner and Ramsay, 1982). The dimensions of length cancel each other, so the dimension of strain rate ($\dot{e}$) is the reciprocal of time (in seconds, s^{-1}). For example, measured strain rates for rock salt vary over eight orders of magnitude from 10^{-8} to 10^{-16} s^{-1}. At such rates, under specified conditions of temperature, confining pressure, and shear stress, a mass of salt would deformationally elongate by an amount equal to its original length in 10^8 to 10^{16} s (1 year = 3×10^7 s). On the time scale of tectonic deformation (1 million to 10 million years, or 10^{13} to 10^{14} s), the rheidity of deforming masses of salt is so low that their deformation is approximately viscous (Jackson and Talbot, 1986). *Salt domes* or their upwardly arched cover strata form significant landforms on the Gulf coast of the United States, in northern Germany, and in Iran. In the arid climate of Iran, the salt breaks through to the surface and flows as *salt glaciers* down the flanks of salt domes (Ala, 1974; Jackson and Talbot, 1986). When moistened with groundwater or precipitation, salt flows even more rapidly. Extruded salt at the heads of some Iranian salt glaciers flows at 2 m/yr.

Glacier ice deforms even more rapidly than salt; under low shear stress, it will deform by an amount equal to its original length within one year ($\dot{e} = 10^{-7}$ s^{-1}) (Fig. 16–3). It deforms readily to fill mountain valleys, and the profile of an ice sheet is an elliptical or parabolic curve, steepest at the edges and flatter inland (Fig. 16–10). It also fractures as a brittle substance to form open crevasses as deep as 50 m. Thus the surface morphology of an ice sheet is largely the result of plastic, or rheid, flow but also the result of fracturing and partial melting. Glacier ice is the second most common rock type at the surface of the earth (greatly exceeded by submarine basalt) forming a layer 1 to 2 km in thickness over one entire continent and the world's largest island.

Many volcanic landforms also have their shape determined by the rheidity of molten lava (Chap. 4). Even though the lava solidifies and becomes rigid on cooling, the landforms record the mobility of the former molten state.

Rocks in young orogenic zones, where intense deformation occurred within an interval of 10 million years or less, must have conventional strain rates of between 10^{-13} and 10^{-15} s^{-1} (Pfiffner and Ramsey, 1982). Within the cores of orogenic belts, these rocks flow, as proved by the intricate contortion of gneissic bands or the great folds exposed on eroding mountain sides (Fig. 11–13). On the largest scale of geomorphic units, the earth's surface is an approximate spheroid of rotation on which continents stand a few kilometers higher than ocean floors, because the continents are composed of somewhat lower-density rocks. The strain rate of matter in the earth's interior must be very rapid in order for isostatic equilibrium between large areas of the surface to be so very nearly perfect. The shapes of other planetary masses are also fundamentally related to their masses, temperatures, and the strengths of their materials. Small icy satellites in our solar system, such as some of the moons of Jupiter, Saturn, and Uranus, are irregular in shape if their radii are less than 200 km. Larger icy satellites are more nearly spherical. Rocky satellites show less definite trends, but in general, if they are larger than 500 km in radius, they also are spherical (Thomas, 1989). All the larger objects in the solar system are spherical. Whereas a small icy satellite may have surface roughness equal to a few percent of its mean radius, the maximum surface relief on the earth is only about 10 km, a trivial part of the planet's radius of 6370 km.

In summary, for most geomorphic considerations, rocks are to be regarded as rigid solids, brought near the surface of the earth by tectonic uplift and erosion but not likely to assume distinctive landforms as a result of rheid flow. Only in special cases, such as salt domes, glaciers, and lava flows, is the rheidity of rocks a factor in the description of landforms.

Process

A host of internal and external processes contribute to produce a landscape. Tectonic forces crumple rocks and push up continents and mountain ranges. Air and water react with rocks to cause the familiar results of weathering. Moving water, air, and ice erode rocks and transport the eroded debris to depositional sites. Organisms absorb mineral nutrients from the rocks on which they live. Gravity tends to pull down structures that rise above the general ground level. The rates of geomorphic processes vary enormously, from a few centimeters per thousand years for the surface weathering of ancient monuments to 50 m/s or more for an avalanche roaring down a mountainside.

Geomorphic processes also vary in intensity from one region to another, depending on climate, vegetation, and altitude above or depth beneath the ocean. A large portion of any geomorphology book must be devoted to analyzing the processes that act on geologic structures. The vital feature of geomorphic processes is that like all chemical and physical processes, they are assumed to be time-independent. That is, we assume that under the same set of environmental conditions, processes have acted in the past and will act in the future exactly as they now can be observed to act. Hutton's great principle of uniformity of process, often called "Uniformitarianism," is a keystone of all scientific method.

Rigid uniformitarianism is not useful when studying relict or exhumed landscapes, however. Throughout the decipherable history of the earth, life has been evolving, and present living things are not the same as those that lived formerly. Many geomorphic processes are controlled by organisms, either directly, as when the steepness of hill slopes is determined by the holding capacity of tree or grass roots, or indirectly, as when soil chemical reactions are determined by the microorganisms in the soil water. It is hard to envision a pre-Cenozoic landscape, for example, on which no grasses had yet evolved to anchor the sod. "One of the most remarkable transformations of terrestrial ecosystems during the Tertiary was the spread of grasslands on all major continents of the world" (Retallack, 1982). Open grassland is first recorded in Eocene and Oligocene sediments in Argentina, and by late Oligocene time grasslands were widespread. The implications for geomorphology are so profound that this book has been given a restrictive subtitle. It is only since mid-Miocene time, about 25 million years ago, that landscapes have looked "modern" and have been evolving in response to modern processes. Paleogeomorphologic reconstructions of landscapes older than the late Cenozoic require a modified set of assumptions.

Mammals also had an explosive mid-Cenozoic evolution, based largely on grazing and running on the new open, grassy plains. Even within the Quaternary Period, when most of our modern mammals had already evolved, faunal migrations have influenced landscape development. Grazing and browsing by domesticated animals may be responsible for huge areas of semiarid landscapes. *Homo sapiens,* a late Cenozoic mammal, has reshaped many landscapes, especially in the past 200 years. It has been asserted that human activity now involves an annual flux of earth materials equal to that of plate tectonics (Chap. 5).

These examples illustrate the dangers of regional or temporal extrapolations of the processes that are assumed to shape landforms. Although physical and chemical laws are indeed uniform through geologic time, the oscillatory climatic changes of the late Cenozoic Era have caused repeated transpositions of climate-dominated processes on many regions (Chap. 18). Glaciated terrains in many mid-latitude regions, and loess and sand sheets that record former eolian processes in areas that are now humid and vegetated, remind us that few contemporary landforms can be assumed to have developed under any single set of processes.

Time

One of the fundamental principles of geomorphology is that the processes that act on geologic structures at the earth's surface change the appearance of landscapes with the passage of time. Anyone who has seen or felt the weather-beaten surface of an ancient stone monument understands that rocks react with water and atmosphere, and only a slight extension of reasoning is necessary to deduce that entire landscapes are shaped by the collection of processes that we call weathering and erosion. The key word in the relationship between structure and process is "progressive." If structures are affected by surface processes, are the changes progressive and sequential? Do landscapes pass through predictable evolutionary stages with the passing of time? These questions are at the heart of the science, because in order to be a valid science, geomorphology must generate hypotheses (predictions) that can be further tested and either falsified, in which case they must be modified, or further corroborated, in which case they may rise to the level of theories, or even "laws of nature."

Practical or applied geomorphology emphatically needs prediction. Will an excavation wall remain stable, or will it collapse as it adjusts to groundwater seepage or subaerial exposure? Will the soil on a hillside remain in place when cultivated with row crops, or should it be seeded with grass and used as pasture? These are practical geomorphic questions in which the time scale is in days or years.

In the theoretical analysis of landscapes, time is measured in thousands or even millions of years. How have landscapes evolved when exposed to the agents of weathering and erosion for a geologically significant interval of time such as the Quaternary Period? Traditionally, geomorphologists spoke of *relative* time and compared the development of landscapes only in terms of the relative stage of erosional evolution. The basis for this convention was that some landscapes are on weak or easily eroded terranes, have strong degradational processes at work, and evolve rapidly. Others, forming on resistant structures or experiencing only weak processes, change very slowly. Little value was placed on specifying the rate of development of landscapes in terms of actual time.

With the perfection of methods for determining the age of rocks and sediments, and even of the exposure ages of rock surfaces, by radioactive disintegration of natural isotopes, the determination of real time, and therefore real rates of change, has become possible. When applied to geomorphic problems, radiometric ages can answer such questions as when an ice sheet uncovered a region, when rivers began to dissect a volcanic ash bed, when sand dunes invaded a former river valley, or when a coral reef grew up to the ocean surface. Time is now specified to a precision that was not conceivable only a few decades ago, and the future of geomorphology lies in converting relative ages to real rates of change.

A corollary of the principle that processes of change alter structures through time is that *stages,* or degrees of completeness, of alteration can be defined. W. M. Davis conceived a geomorphic cycle,[1] in which specific stages could be recognized. His dangerously oversimplified terms "youth," "maturity," and "old age" are familiar to every beginning student of geology and geography. They need careful re-evaluation (Chap. 12).

Time has another special significance to geomorphology in that most of the subaerial landscape has evolved in the late Cenozoic Era, a time of great tectonic activity and climatic change on earth (Chaps. 2 and 18). Probably the average geologic status of continental regions is one of low relief and slow erosion, or of shallow marine submergence and slow accumulation of sediment. Our present relief is fantastically diverse by comparison. Geomorphology is a necessary branch of geology precisely because in late Cenozoic time the continental blocks are unusually high above sea level and are undergoing rapid erosion. The latitudinal zonation of climates is extreme.

SUMMARY

At present, agents of geomorphic changes are retouching landscapes that are the result of a complex series of changing environments. Most subaerial landscapes are **palimpsests,** a word originally applied to parchment manuscripts that had been partially scraped clean and reused for later texts. As in a manuscript one or more older texts can sometimes be read beneath the latest words, so in the description of landscapes, previous processes and stages of development can be recognized beneath or within the landforms that are now being shaped. The highest skill of the geomorphologist is in deciphering those fragments of ancient development almost obscured by present processes. Sometimes the task is easy, as when the striking landforms of glacial erosion and deposition are found in regions of temperate climate, or when empty branching systems of river valleys are found in deserts where no rain falls today. However, many landscapes do not record such obvious changes, and we must always apply the principle of uniformity of process with caution, because the present dominant processes may not be the key to past landform development. The science of geomorphology has two major goals. One is to organize and systematize the description of landscapes by intellectually acceptable schemes of classification, and the other is to recognize the deviations from these schemes that are evidence for changes in the environment that could not be inferred if normal conditions had not first been specified.

REFERENCES

AHNERT, F., 1988 Modelling landform change, *in* Anderson, M. G., ed., Modelling geomorphological systems: John Wiley & Sons Ltd., Chichester, England, pp. 375–400.

ALA, M. A., 1974, Salt diapirism in southern Iran: Am. Assoc. Petroleum Geologists Bull., v. 58, pp. 1758–70.

BAKER, V. R., 1988, Geological fluvial geomorphology: Geol. Soc. America Bull., v. 100, pp. 1157–67.

BARSCH, D, and LIEDTKE, H., 1980, Principles, scientific value and practical applicability of the geomorphological map of the Federal Republic of Germany at the scale of 1:25,000

---

[1] His term was "geographic cycle," but the growth of geography into humanistic studies soon rendered the term inappropriate.

(GMK25) and 1:100,000 (GMK100): Zietschr. für Geomorph., Supp. v. 36, pp. 296–313.

BELL, T. H., JR., 1975, Statistical features of sea-floor topography: Deep-Sea Res., v. 22, pp. 883–92.

BIRKELAND, P. W., 1984, Soils and geomorphology: Oxford Univ. Press, New York, 372 pp.

BIROT, P., 1960, Le cycle d'érosion sous les différents climats: Univ. do Brasil Centro de Pesquisas de Geografia do Brasil, Rio de Janeiro, 137 pp.

———, 1968, Cycle of erosion in different climates (transl. by C. I. Jackson and K. M. Clayton): Univ. of California Press, Berkeley, 144 pp.

BOSTOCK, H. S., 1964, Provisional physiographic map of Canada: Canada Geol. Survey Paper 64–35, 24 pp.

BÜDEL, J., 1982, Climatic geomorphology (transl. from German by Fischer, L., and Busche, D.): Princeton Univ. Press, Princeton, N.J., 443 pp.

CAREY, S. W., 1954, The rheid concept in geotectonics: Geol. Soc. Australia Jour., v. 1, pp. 67–117.

———, 1962, Scale of geotectonic phenoma: Jour. Geol. Soc. India, v. 3, pp. 97–105.

DREWRY, D., 1986, Glacial geologic processes: Edward Arnold (Publishers) Ltd., London, 276 pp.

DURY, G. H., 1972, Some recent views on the nature, location, needs, and potential of geomorphology: Professional Geographer, v. 24, pp. 199–202.

EMBLETON, C., ed., 1984, Geomorphology of Europe: The Macmillan Press Ltd. (published in the United States by Wiley-Interscience, a Division of John Wiley & Sons, Inc., New York), 465 pp.

———, and THORNES, J., eds., 1979, Process in geomorphology: Edward Arnold (Publishers) Ltd. (published in the United States by the Halsted Press, a Division of John Wiley & Sons, Inc., New York), 436 pp.

VON ENGELN, O. D., 1942, Geomorphology: The Macmillan Company, New York, 655 pp.

FAIRBRIDGE, R. W., 1968, Continents and oceans—statistics of area, volume and relief, *in* Fairbridge, R. W., ed., Encyclopedia of geomorphology: Reinhold Book Corp., New York, pp. 177–86.

FENNEMAN, N. M., 1928, Physiographic divisions of the United States: Assoc. Am. Geographers Annals, v. 18, pp. 261–353.

———, 1931, Physiography of western United States: McGraw-Hill Book Company, New York, 534 pp.

———, 1938, Physiography of eastern United States: McGraw Hill Book Company, New York, 714 pp.

GAPOSCHKIN, E. M., 1974, Earth's gravity field to the eighteenth degree and geocentric coordinates for 104 stations from satellite and terrestrial data: Jour. Geophys. Res., v. 79, pp. 5377–5411.

GARDNER, T. W., JORGENSEN, D. W., SHUMAN, C., and LEMIEUX, C. R., 1987, Geomorphic and tectonic process rates: Effects of measured time intervals: Geology, v. 15, pp. 259–61.

HAMMOND, E. H., 1963, Classes of land-surface form in the forty-eight states, U.S.A: Map Supp. no. 4, Assoc. Am. Geographers Annals, v. 54 [1964].

HEAD, J. W., 1990a, Formation of mountain belts on Venus: Evidence for large-scale convergence, underthrusting, and crustal imbrication in Freyja Montes, Ishtar Terra: Geology, v. 18, pp. 99–102.

———, 1990b, Venus trough-and-ridge tessera: Analog to Earth oceanic crust formed at spreading centers?: Jour. Geophys. Res., v. 95, pp. 7119–32.

———, and CRUMPLER, L. S., 1987, Evidence for divergent plate-boundary characteristics and crustal spreading on Venus: Science, v. 238, pp. 1380–85.

HEEZEN, B. C., THARP, M., and EWING, M., 1959, Floors of the oceans, I. The North Atlantic: Geol. Soc. America Spec. Paper 65, 122 pp.

HOWARD, A. D., and SPOCK, L. E., 1940, Classification of landforms: Jour. Geomorphology, v. 3, pp. 332–45.

HUNT, C. B., 1974, Natural regions of the United States and Canada: W.H. Freeman and Company, San Francisco, 725 pp.

JACKSON, M. P. A., and TALBOT, C. J., 1986, External shapes, strain rates, and dynamics of salt structures: Geol. Soc. America Bull., v. 97, pp. 305–23

JENNINGS, J. N., 1973, 'Any millenniums today, lady?' The geomorphic bandwaggon parade: Australian Geog. Studies, v. 11, pp. 115–33.

KING-HELE, D., 1976, The shape of the earth: Science, v. 192, pp. 1293–1300.

LEOPOLD, L. B., WOLMAN, M. G., and MILLER, J. P., 1964, Fluvial processes in geomorphology: W.H. Freeman and Company, San Francisco, 522 pp.

MANDELBROT, B., 1967, How long is the coast of Britain? Statistical self-similarity and fractional dimension: Science, v. 156, pp. 636–38.

———, 1983, The fractal geometry of nature: W.H. Freeman and Company, New York, 468 pp.

MENARD, H. W., and SMITH, S. M., 1966, Hypsometry of ocean basin provinces: Jour. Geophys. Res., v. 71, pp. 4305–25.

OLLIER, C. D., 1969, Weathering: American Elsevier Publishing Company, Inc., New York, 304 pp.

PFIFFNER, O. A., and RAMSAY, J. G., 1982, Constraints on geological strain rates: Arguments from finite strain states of naturally deformed rocks: Jour. Geophys. Res., v. 87, no. B1, pp. 311–21.

PIKE, R. J., and THELIN, G. P., 1989, Shaded relief map of U.S. topography from digital elevations: Eos, v. 70, pp. 843 and 853.

RETALLACK, G. R., 1982, Paleopedological perspectives on the development of grasslands during the Tertiary: Third North American Paleontological Convention, Proc., v. 2, pp. 417–21.

RITTER, D. F., 1986, Process geomorphology, 2nd ed.: Wm. C. Brown Publishers, Dubuque, Iowa, 579 pp.

SALISBURY, R. D., 1919, Physiography, 3rd ed., rev.: Henry Holt & Co., New York, 676 pp.

SCLATER, J. G., ANDERSON, R. N., and BELL, M. L., 1971, Elevation of ridges and evolution of the central eastern Pacific: Jour. Geophys. Res., v. 76, pp. 7888–7915.

SHORT, N. M., and BLAIR, R. W., JR., eds., 1986, Geomorphology from space: A global overview of regional landforms: NASA Scientific and Technical Information Branch, Washington, D.C., Publication NASA SP-486, 717 pp.

TAYLOR, S. R. and MCCLENNAN, S. M., 1985, The continental crust: Its composition and evolution: Blackwell Scientific Publications, Oxford, England, 312 pp.

THOMAS, P. C., 1989, The shapes of small satellites: Icarus, v. 77, pp. 248–74.

THORNBURY, W. D., 1965, Regional geomorphology of the United States: John Wiley & Sons, Inc., New York, 609 pp.

TRICART, J., and CAILLEUX, A., 1965, Introduction a la géomorphologie climatique: Soc. d'édition d'enseignement supérieur, Paris, 306 pp.

———, 1972, Introduction to climatic geomorphology (transl. by C. J. K. de Jonge): Longman Group Ltd., London, 295 pp.

TURCOTTE, D. L., 1989, Fractals in geology and geophysics: PAGEOPH, v. 131, pp. 171–96.

WAHRHAFTIG, C., 1965, Physiographic divisions of Alaska: U.S. Geol. Survey Prof. Paper 482, 52 pp.

CONSTRUCTIONAL PROCESSES AND CONSTRUCTIONAL LANDFORMS

Only about 41 percent of the earth's lithosphere is older than 200 million years. The rest has been recycled, perhaps many times, by consumption and accretion at plate margins. The oceanic lithosphere is almost all younger than 200 million years. Continents include rocks with ages of almost 4 billion years, because continental rocks tend to resist subduction and accumulate as thickened crustal slabs. But except for exhumed and other relict landscapes, the continental landscapes we now see are very young, mostly the products of processes that have been at work during late Cenozoic time.

The theory of plate tectonics is based on evidence that large, rigid lithospheric plates move over the surface of the earth at relative velocities of up to 10 cm/yr. Vertical movements measured in millimeters per year are also demonstrated, especially at plate margins. In the face of such evidence for a mobile lithosphere, geomorphologists must constantly reevaluate the traditional concept that rock masses are passive structures from which landscapes are carved by erosion. The next three chapters establish the conceptual framework of landscape evolution during a time of great tectonic and volcanic activity.

Constructional landforms are here defined as all those that have been or are being built (that is, increasing in mass, height, or area), as compared to those that are primarily destructional, or erosional. In Chapters 2, 3, and 4, the tectonic and volcanic constructional processes and landforms are systematically described. Constructional landforms that grow by biologic processes or by accretion of the waste products of erosion are not described until later chapters.

Tectonic and volcanic relief results from dynamic stresses and density changes, all of which are ultimately caused by thermal gradients within the earth. Nothing is ever really "constructed" on any planet, because each total planetary mass remains very nearly constant. All constructional landforms are built by reprocessing or by redistributing some other mass.

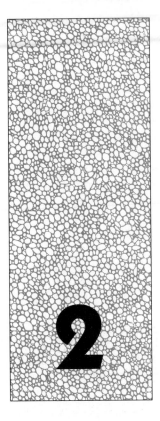

2 Cenozoic Diastrophism and Constructional Processes

The Cenozoic Era, especially that part of it defined as the Quaternary Period, has seen the modern continents generally emergent, with especially intense orogenic activity around the Pacific basin, across southern Europe and Asia in the Alpine-Himalayan zone, and around the Caribbean Sea. Much of our subaerial landscape is developed on relatively young tectonic structures, although some of the rocks exposed by recent uplift and erosion are among the oldest known. Epeirogenic and orogenic activity in the late Cenozoic Era have given us a landscape of exceptional diversity, complexity, and beauty. It may well be true that at any moment of geologic time orogenic belts have been actively rising on some part of the earth. However, if there have been other geologic intervals when extensive continental emergence and high mountain ranges were as notable as those of the late Cenozoic, they are not well documented in the stratigraphic record.

CENOZOIC PLATE MOTIONS, RELIEF, AND CLIMATE

The processes of sea-floor spreading, whereby new oceanic lithosphere is created at mid-ocean ridges along diverging plate margins, generates the dominant relief pattern of the ocean floors, in which depth is primarily determined by age. On the scale of millions of years, the oceanic lithosphere cools and contracts as it moves away from the mid-ocean ridges, and the ocean water over it slowly deepens (Fig. 1–5). Increases in spreading rates are believed to increase the volume of the mid-ocean ridges, decrease the volume of the ocean basins, and cause marine transgressions over low-lying continental areas. Slower spreading rates should produce marine regressions from the continents. Tectonic changes in the volume of the ocean basins and the late Cenozoic growth of the Antarctic ice sheet have combined to lower sea level and expose extensive Miocene and younger coastal plains on the trailing continental margins.

Even the strong latitudinal gradients of our modern climates, and the corresponding modern diversity of climate-determined erosional processes, are in part determined by tectonism. Cenozoic plate motions have moved large continental regions into middle to high latitudes of the northern hemisphere, preparing them for continental glaciation. Cenozoic mountain ranges have become powerful orographic barriers for global atmospheric circulation. Large areas of contemporary mid-latitude continental shields and platforms with low relief are moisture deficient and have become arid or semiarid plains as a result of their size, flatness, and distance from moisture sources. In the last 55 million years, for instance, Antarctica and Australia are believed to have separated by the present 30° of latitude. Probably the Antarctic ice cap and the dominantly arid landscape of Australia are both Cenozoic phenomena that are best explained by plate tectonics (pp. 436, 444).

RATES OF CONSTRUCTIONAL PROCESSES

Rates of geomorphic and other geologic processes cover many orders of magnitude, from less than 1 mm/1000 yr for deep-sea sedimentation rates to more than 50 m/s for avalanches. Tectonic and volcanic landscapes can be built at rates at least an order of magnitude more rapidly than erosion can destroy them (Chap. 12). If this were not the case, our terrestrial landscape would be very dull indeed. To describe rates of processes, the "Bubnoff Unit" (B) has been proposed (Fig. 2–1). For example, the sedimentation rate noted above can be written as 1 B, whereas the avalanche rate exceeds 1.5×10^{12} B. The value of the "Bubnoff Unit" is that it allows us to visualize the relative rates of geomorphic processes and the rates of other geologic processes on a common scale. For example, regional denudation rates, by which continental-size areas are reduced toward sea level, are near the middle of Figure 2–1, and are no more than one-tenth as rapid as mountain uplift. A disadvantage of the "Bubnoff Unit" is that it gives no hint of the duration of a process. As noted in

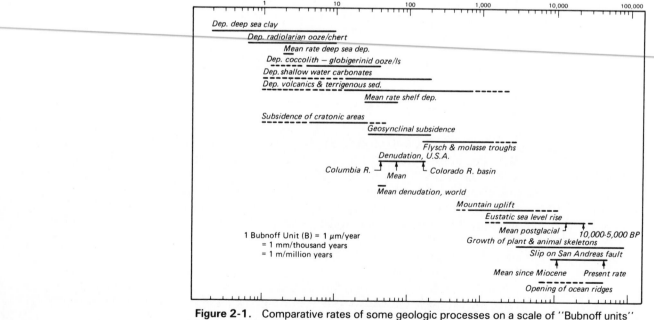

Figure 2-1. Comparative rates of some geologic processes on a scale of "Bubnoff units" (Fischer, 1969). Regional denudation rates range from 30 to 150 B (30 to 150 mm/1000 yr), whereas tectonic uplift rates range upward from 300 B (300 mm/1000 yr).

Chapter 1, the rates of most geomorphic processes are some complex reciprocal function of their size and duration, with small, brief events being the most rapid. The denominators in the rate scales used in this book are chosen to represent the typical duration of the process, whether it be seconds or minutes for avalanches, hours or days for intense precipitation, or thousands of years for tectonic uplift. For both constructional and destructional geomorphic processes, comparisons between rates that have been measured over variable time intervals are not valid. A scaling factor must be calculated in order to compare rates measured over years or decades with those measured over centuries and longer periods of time (Gardner et al., 1987).

Some landforms appear abruptly, literally overnight. In 1943 a Mexican farmer detected volcanic fumes escaping from cracks in the ground of his cornfield, and within a few days an active scoria cone had buried his farm, eventually to become a new volcano, Parícutin. During the 1964 Alaskan earthquake, a sizable region of nearshore ocean floor was raised above tide level to become new land (Fig. 2–2). The duration of the movement was probably only a few minutes or even seconds, although no one witnessed the actual upheaval.

Figure 2-2.   Tectonic uplift and subsidence in south-central Alaska resulting from the earthquake of March 27, 1964. Changes in land level are shown by contours (in meters) that are dashed where approximate or inferred. Edge of continental shelf (−200 m) is shown by dotted line. Active and dormant volcanoes are identified by asterisks. (Plafker, 1965; © 1965 by the American Association for the Advancement of Science.)

At somewhat slower rates, but still rapid by the ordinary scale of geologic change, the earth's surface is displaced along faults or is arched upward into anticlines and domes (Fig. 3–21). Areas that have been deglaciated within the last 10,000 years are still actively rising in isostatic response to the removed load of former ice sheets (Figs. 17–13 and 17–15).

In coastal regions, vertical movements are relatively easy to measure, because emerged or submerged former shore-zone landforms and sediments can often be dated by radiometric techniques and their displacement above or below the present shoreline can be measured. Much of our information about tectonic rates comes from coastal studies. Quaternary tectonic movements can be conveniently categorized as slow, in the range of 0.1 to 1 m/1000 yr, or rapid, 1.0 to 10 m/1000 yr, based on the interaction of tectonic movements with glacially controlled sea-level oscillations (Bloom, 1980) (Chaps. 17 and 20). Quaternary sea level has repeatedly fluctuated through a range of at least 100 m on a time scale of about 100,000 years, and we are now in an interglacial time of relatively high sea level. On a coast that is either rising or subsiding at a slow rate of 0.1 m/1000 yr, the shoreline landforms of the previous interglacial will have risen or subsided about 10 m, and will still be within the range of active coastal erosion or deposition. Such coasts have complex stratigraphic records of repeated shoreline reoccupation at nearly the same level, and complex relict or exhumed landforms (Figs. 20–1 and 20–2).

With rapid tectonic uplift at a rate of 1 m/1000 yr, at least 100 m of uplift during each 100,000-year glacial cycle lifts the coastal erosional and depositional landforms of that cycle up above the landforms of the next cycle (Figs. 20–3 and 20–4). Great flights of marine terraces result (Fig. 2–3). The most rapid known Quaternary uplift rates approach 10 m/1000 yr in active orogenic belts. If continued for only 1 million years, such uplift could produce Himalayan-height mountains. No geomorphic examples of comparably rapid tectonic subsidence are known, probably because former coastal landforms would be deeply submerged in a relatively brief time.

Several new high-precision surveying techniques use portable radio receivers to determine ground position relative to the orbital positions of a group of navigational satellites. The Global Positioning System (GPS) can now measure the horizontal distance between two points to within 0.1 parts per million, or 0.1 mm/km. Vertical measurements are several times less precise. By such methods, it is now possible to measure tectonic movements directly by resurveying points at time intervals of a year or less. A GPS survey across 40 km of the Santa Maria fold and thrust belt northwest of Santa Barbara, California, supplemented by other survey methods, established crustal shortening of 7 ± 1 mm/yr between 1986 and 1987. The shortening axis trends N 03° E, and can be resolved into 6 ± 2 mm/yr shortening perpendicular to the fold belt at N 30° E, and 3 ± 1 mm/yr of right-lateral shear parallel to the fold belt (Feigl et al., 1990). Within a few years, relative plate motions of 1 to 10 cm/yr will be routinely measured by hand-carried equipment, providing the ultimate test of the plate tectonic theory. With slightly better equipment or longer time intervals, soon it will also be possible to measure active vertical tectonic movements.

CENOZOIC EPEIROGENIC UPLIFT AND SUBSIDENCE

The Greek root "epeiros" refers to a continent. Therefore, by literal translation, **epeirogeny** would refer only to the genesis of continental relief, and would not be the appropriate term to refer to movements of the ocean floor. However, the term can be extended to include all continent-sized warping, including thermal contraction of the oceanic lithosphere (Fig. 1–5). No simple correlation of age and

Figure 2-3. Emerged Pleistocene coral-reef terraces on the Huon Peninsula, Papua New Guinea. Terraces are on a fault block that has been rising at a rate of 1 to 3 m/1000 yr for at least 200,000 years. Figure 17-16 shows the terrace chronology for the last 140,000 years.

altitude can be demonstrated for continental landscapes. Some geomorphic provinces seem to have oscillated within a few hundred meters above and below sea level since early Paleozoic time. Precambrian rocks are exposed both in the cores of mountain ranges and in shield areas at sea level. Some tectonists suspect that contemporary compressive stress and seismicity in eastern North America may be related to Cenozoic plate motions (Sbar and Sykes, 1973). Possibly the moving plates ride over sublithospheric "bumps" or "hollows" that cause vertical epeirogenic movements. This is an area of active research, in which geomorphic evidence is as important as seismology, rock mechanics, and geodesy.

Although 41 percent of the total earth's surface is underlain by continental structure, only about 29 percent of the total area is presently above sea level (Fig. 1–4) (Cogley, 1985). The remaining 12 percent of continental structure is submerged to form the continental shelves, the inner halves of which also were exposed during glacial intervals of lower sea level. Thus essentially all of the continental structure of the earth has a surface form determined by subaerial erosion or deposition, and most of it is presently exposed to view. Throughout Phanerozoic geologic history, prevailing isostatic equilibrium has maintained the average continental "freeboard" within less than 100 m of present sea level (Wise, 1974). By analogy with a boat at rest in water, the "freeboard" of a continent is the amount exposed above sea level, supported by the isostatic buoyancy of the low density continental crust in the more dense mantle. With a crustal thickness of 30 km, a continental region will be at isostatic equilibrium just at sea level (Kinsman, 1975, p. 94). That is why relatively minor changes of sea level have been able to episodically flood wide continental areas and cover them with layers of sediments. No "lost continents" lie submerged beneath the oceans.

Epeirogeny ("continent formation") is broad regional movement, with little folding or other local deformation. Except for mountain ranges, much of the continental landscape is eroded from sedimentary rocks that lie in their initial depositional attitude. Broad structural domes and basins with deformation-produced relief of a few hundred meters over distances of hundreds of kilometers are common in epeirogenic regions, but the resulting dips of the strata are usually less than 1 percent and are barely detectable in a single exposure. The cuestaform erosional landscape that develops on the broad domes and basins of epeirogenic regions is analyzed in detail in Chapter 11. Epeirogenic uplift is not considered likely to produce small-scale constructional landforms. More commonly, epeirogenic domes are breached by erosion and their central regions become topographic basins. Uplifted former structural basins become low plateaus surrounded by outward-facing escarpments. The resulting *topographic inversion* by erosion on epeirogenic structures implies that the tectonic movements are either old or extremely slow relative to the rate of erosion (pp. 281–84).

Epeirogenic Rates

Long-term epeirogenic subsidence rates derived from the thickness of sediments that have accumulated in epeirogenic basins or on subsiding continental margins are generally slow. For example, 6 km of Tertiary sediment in the gulf coast geosyncline, deposited over an interval of some 60 million years, imply a subsidence rate of 0.1 m/1000 yr. This is a rather rapid epeirogenic rate, due to the rapid sedimentary loading of the northern Gulf of Mexico. Epeirogenic basins such as the Paleozoic Michigan Basin typically subsided at a rate of about 0.01 m/1000 yr. Rates such as these are comparable to the rates of ocean-floor subsidence, and are probably caused by the same process, thermal contraction (Sleep et al., 1980).

Long-term epeirogenic uplift is harder to document than subsidence, for much of the evidence is removed by erosion. Pliocene and younger marine sediments now emerged on the Atlantic coastal plain of the United States record several million years of net uplift at rates averaging 0.01 to 0.03 m/1000 yr, comparable to the rate of subsidence farther offshore on the Atlantic continental shelf during Cenozoic sediment loading (Cronin, 1981).

Neotectonic Epeirogeny

Geologists in the Soviet Union and eastern Europe have evolved a new theory of epeirogenesis called **neotectonism** (the "newest" tectonics—the study of earth movements of the Quaternary Period and the later part of the Tertiary Period, including those in progress at the present time). According to this theory (Mescherikov, 1968; Fairbridge, 1981), the ancient continental shields and the continental crust in general are subject to repeated late Cenozoic uplifts or depressions of 10 to 100 m. Contemporary wavelike motions with vertical components of as much as 10 mm/yr have been reported. Most of the data for the contemporary motions are derived from repeated precise leveling surveys. From releveling of the geodetic network in England and Wales between 1912 and 1956, uplifts of up to 150 mm and subsidences of as much as 46 mm during the 44-year interval were inferred (Chorley, 1963, p. 964). Recent resurveys of the eastern United States show the Appalachians rising at 6 mm/yr relative to the Atlantic coast. Other leveling lines and tide-gage studies show comparable rates of relative vertical motions (Fig. 2–4). Some of the differences between such successive surveys are probably due to unrecognized instrumental errors, but the possibility of large-scale but episodic or oscillatory warping as a continuing feature of epeirogenic tectonics must be considered as a possible factor in geomorphic analysis. Many scientists are now attempting to verify these subtle vertical movements by the new satellite positioning systems such as GPS. Probably within the next decade the reality of vertical epeirogenic movements of mm/yr will be confirmed or denied.

Figure 2-4. Vertical movements (mm/yr) in the eastern United States, compiled from various sources. Tide gauges all show subsidence relative to sea level, but the cause could be either rising sea level or sinking land, or both. (Brown and Oliver, 1976, Fig. 15.)

Vertical displacements of the scale suggested by the neotectonic theory, while yet to be confirmed, might influence drainage networks, river terraces, lake shorelines, and many other geomorphic features. One example of the effects of a river system crossing an uplift is the Bogue Homo Creek across the Wiggins anticline, a neotectonic uplift on the southern border of Mississippi and Alabama (Fig. 2–4). The river is being affected by uplift of nearly 5 mm/yr across its path. It is responding by straightening its main channel upstream of the ridge in order to shorten its channel length and thereby maintain its downstream gradient against the rising structure. On the downstream flank of the anticline, it has incised to form a low terrace, increased its sinuosity to offset the steepened gradient, and divided into braided segments as a result of the increased sediment loads from the incision across the axis of uplift (Burnett and Schumm, 1983). The river is said to be *antecedent* to the uplift because its route was established prior to, and has persisted in spite of, the uplift. The tectonic significance of antecedent rivers is further discussed in Chapter 11.

Glacial Isostasy

One special form of epeirogenic movement, of particular significance in the Quaternary Period, is **glacial isostasy.** The theory of isostasy holds that the pressure at the base of all vertical sections of the earth, above some deep level of compensation, is equal. Nature's "great experiments in isostasy" (Daly, 1940, p. 309; Peltier, 1987) were the alternate loading and unloading of large continental areas with ice sheets during late Cenozoic time (Chap. 17). Seawater must evaporate and accumulate on land as ice in order for ice sheets to grow. Thus each glaciation involved the mass transfer of a layer of water perhaps 100 m in thickness from the 71 percent of the earth's surface that is ocean, and the concentration of that water mass, as glacier ice as much as 2 to 3 km in thickness, on about 5 percent of the earth's surface. We assume but have not yet proved that the ocean floor responded to the removal of water load by isostatically rising. It is much easier to prove that the glaciated areas were basined beneath their loads of ice, because when the ice sheets were melting, lakes and seas often formed shorelines near the ice margins. The shorelines of the former ice–marginal water bodies show regular tilting, always concentric upward toward the center of former maximum ice thickness. Because these shorelines must have been horizontal initially, their present slope is a measure of the differential postglacial isostatic uplift. Uplift of 300 m is known from areas on the eastern side of Hudson Bay, Canada, in a total time of less than 8000 years. Postglacial isostatic uplift at the northern end of the Baltic Sea totals 520 m, and an estimated 210 m of uplift remains to be accomplished (Flint, 1971, p. 352).

The rate of postglacial uplift is rapid at first, then progressively slower. It approximates an exponential decay curve with a *relaxation time* (the time required for the deformation to decrease to $1/e$, or 37 percent, of the original value) of from 1000 to 5000 years. Initial uplift rates of 9 to 10 m/100 yr have been reported for Greenland (Washburn and Stuiver, 1962; Ten Brink, 1974). Where uplift rates are so rapid, landforms such as storm beaches are raised in a few decades sufficiently high to be unreached by the next major storm, and wide, descending flights of beach ridges mark the margins of seas and lakes in former glaciated regions (Fig. 17–13).

In northern United States, Canada, and Scandinavia, the postglacial uplift is still in progress, although at a greatly decreased rate. Northern Sweden is presently emerging from the Baltic Sea at a rate of 9 mm/yr (Fig. 17–15); Finland annually gains 7 km² of new land by the uplift.

CENOZOIC OROGENY

The preceding paragraphs clearly demonstrate that rapid crustal movements are not confined to mobile orogenic belts that surround or separate "stable" cratonic shields at the cores of the continents. Nevertheless, it is improbable that neotectonic epeirogenic movements continue in one direction for more than a few centuries or a few thousand years. Otherwise, places like the Wiggins anticline on the Gulf coastal plain of the United States would have grown into mountains within the Quaternary Period. (Note that a rate of 1 mm/yr = 1 m/1000 yr = 1 km/million yr.) However, there are many regions of the earth where vertical crustal movements measured in millimeters per year have been cumulative throughout late Cenozoic time, and erosion has been unable to keep pace with tectonic uplift. The developing subdiscipline of **tectonic geomorphology** is concerned with the constructional landscapes of such regions, and of erosional landscapes in which the tectonic origins are still obvious. This chapter adopts the "megageomorphic" approach (see Preface) of describing tectonic landscapes on the scale of entire mountain ranges. Individual tectonic landforms are analyzed in more detail in Chapter 3.

Orogeny clearly refers to mountain formation. Unfortunately, in geologic usage, the term has much more complex and contradictory implications. Gilbert (1890, p. 340) formalized the term: "Displacements of the earth's crust which produce mountain ridges are called orogenic.—The process of mountain formation is orogeny, the process of continent formation is epeirogeny, and the two collectively are diastrophism." This definition ignored Gilbert's earlier discoveries that the orogeny that deformed the rocks of the basin ranges was much older than the block faulting that created their modern relief (p. 46). In fact, many mountain ranges are the result of "postorogenic" or epeirogenic movements from which erosion has etched into mountainous relief the most resistant of the folded and metamorphosed rocks. Structural connotations have almost completely divorced the word from its obvious geomorphic ancestry, so that orogeny is now used for: "the processes by which rock structures within the mountain chains or foldbelts are created" (King, 1969, p. 45; but see Cebull, 1973, 1976). As such, it has little geomorphic significance, although for many general geomorphic descriptions the tectonic niceties of the term can safely be ignored.

Mountains and Orogenic Style

In plate-tectonic theory, all orogenic belts are on or near converging plate margins. Mountain ranges that are now imbedded within continents, such as the Himalaya Mountains and the Ural Mountains of Asia, have been shown to have been near plate margins at their time of formation but have had new continental terranes accreted to them later. Orogenic vertical movements are not necessarily more rapid than epeirogenic ones. However, they seem to be more continuous through time, less likely to stop or reverse on a time scale of less than 1 to 10 million years, and are restricted to relatively narrow, elongate regions.

Cenozoic mountain ranges are not randomly distributed but form distinctive linear or arcuate belts across the earth's surface. Not all earthquake-prone regions are mountainous, but a map of recent earthquakes (Fig. 2–5) outlines very well the most orogenic regions of the late Cenozoic Era, especially the circum-Pacific belt associated with active volcanoes (Fig. 4–2) along subduction zones. Three distinct types of orogenic belts can be identified on converging plate margins: (1) Island arcs and trenches, (2) Cordilleran-type mountain ranges, and (3) Collisional mountain belts.

Island Arcs and Trenches. These form distinctive morphotectonic assemblages (Fig. 2–6) that extend laterally for thousands of kilometers along the north-

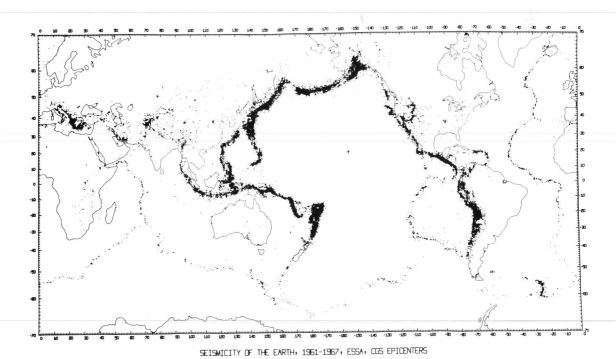

SEISMICITY OF THE EARTH, 1961-1967, ESSA, CGS EPICENTERS

DEPTHS 000-700 KM.

Figure 2-5. Seismicity of the earth, 1961–1967. Compare Fig. 4-2. Each dot is the epicenter of a recorded earthquake. Depths 0 to 700 km. (Barazangi and Dorman, 1969.)

Figure 2-6. Generalized geologic cross section of a typical island-arc system. Horizontal distances are measured from the trench axis; vertical exaggeration is 5:1. (Karig, 1974.)

ern and western Pacific margin, through Indonesia, in the Caribbean Sea, and wherever subduction zones are between two portions of oceanic lithosphere. The downgoing lithospheric plate flexes upward in a broad *outer swell* seaward of the *trench,* into which it descends beneath the overriding plate. Sediments that accumulate in the trench and are scraped off the downgoing plate form a submerged *accretionary prism* of highly deformed mudstones and sandstones on the inner trench wall. Older and similarly deformed sedimentary rocks may rise above sea level as the *frontal arc* of mountainous islands. In many tropical areas, the islands of the frontal arc are capped with coral limestone terraces that record oscillating Quaternary sea levels (Fig. 2–3; Chaps. 17, 20).

Close behind the frontal arc may be a *volcanic arc* of andesitic composite cones. Landward of the volcanic chain may be older, inactive *remnant arcs* and *back arc* or *marginal basins* where the upper plate is spreading in response to the buoyancy of the subducted plate.

Cordilleran-type Mountain Ranges. Where oceanic lithosphere subducts under thicker continental lithosphere, more complicated mountain ranges evolve. On the *Cordilleran type* of converging plate margin, exemplified by the Andes Mountains of South America (Fig. 2–7) (Jordan et al., 1983), there may be narrow *coastal ranges* on the edge of the continent, related to the accretionary prism of trench sediments or to faulting and uplift of continental crustal blocks. Behind the coastal ranges, typically there is an elongate *longitudinal valley,* parallel to the coast. Next is a steep rise to the crest of the *magmatic or volcanic arc* surmounted by a line of andesitic composite cones. Because of compression and thickening of the continental crust in the upper plate, the volcanic peaks on a Cordilleran type of mountain belt rise above a broad plateau or ridge of granitic batholiths and older sedimentary and metamorphic rocks which forms the axial part of the orogenic belt. This central region of the mountain range is sometimes called the *hinterland.* On the continental side of the central ranges is a *foreland* region, where sediments eroded from the central ranges accumulate on the continental platform and may later be folded and thrusted toward the continental interior.

Collisional Mountain Belts. These are found where convergence and subduction have brought an island arc or another plate of continental lithosphere into contact with the overlying plate of the subduction zone. Because the lower density of the converging island-arc or continental rocks prevents their being subducted back into the mantle, they crumple against, thrust over, or are thrust under the

Figure 2-7. Generalized geologic model of the Andes Mountains of South America between 18° and 24°S latitude. This is a typical segment of a cordilleran-type mountain range. Width of the orogenic belt is about 600 km; vertical scale is imprecise. (Revised from Jordan et al., 1983, Fig. 3.)

rocks of the continental orogenic belt, creating high mountains such as the Himalayas or Alps, with intensely deformed rocks of contrasting lithology in contorted or parallel ranges. A sedimentary foreland on a collisional plate boundary may form on the subduction side, rather than on the continental interior side, of the hinterland.

Deformation of orogenic forelands can be subdivided into two main categories: *thin-skinned,* in which stratified sedimentary cover strata of the foreland basin and thin slices of basement rocks are folded or faulted above a detachment zone that separates their deformation from the deeper continental basement, and *thick-skinned,* in which the deeper basement rocks are massively included in the deformation. As the continental basement rocks are usually igneous and metamorphic rocks that lack the "layer cake" contrasts in strength and erodibility of sedimentary cover strata, thick-skinned deformation usually takes the form of basement block faulting, often along preexisting zones of structure weakness, and with or without the draping of cover strata over the displaced basement blocks.

The geomorphic expression of thin-skinned foreland orogeny is by linear anticlines and synclines or thrust slices, from which erosion carves a well-known set of structurally controlled landforms (Chap. 11). Thick-skinned deformation that includes resistant basement rocks is more likely to form bold fault-block mountain ranges arrayed in irregular chains, extending hundreds of kilometers into the continent from the actual orogenic belt (Rodgers, 1987). The Front Ranges of the central Rocky Mountains in the United States and the Sierras Pampeanas of Argentina (Fig. 3–13) are good examples of thick-skinned deformation. Their geomorphic expression contrasts strongly to the landforms of foreland thin-skinned fold and thrust belts, such as the eastern Canadian Rocky Mountains, the Appalachian Valley and Ridge geomorphic province, of the Subandean Ranges of the Andes (Fig. 11–21).

Extensional Mountains. The Basin and Range province of the western United States (Fig. 1–6) is a classic example of a late Cenozoic extensional, rather than compressional, tectonic landscape (Wernicke et al., 1988; Levy and Christie-Blick, 1989). After a long interval of Paleozoic and Mesozoic orogeny, crustal shortening, and accretion of exotic continental terranes, for the past 37 million years the entire Basin and Range province between the Sierra Nevada and the Colorado Plateau has been undergoing lithospheric extension (Levy and Christie-Blick, 1989, p. 1455). Most of the extension has been in the past 15 million years. Near Las Vegas, Nevada, the present width of 360 km of the Basin and Range province has at least doubled, and perhaps increased six-fold, during late Cenozoic extension. The Sierra Nevada has moved 200 to 300 km to the west-northwest relative to the Colorado Plateau, at earlier rates of 20 to 30 mm/yr and at a current rate of less than 10 mm/yr (Wernicke et al., 1988, p. 1755). Extensional tectonics produces an assemblage of fault-bounded landforms that are described in more detail in Chapter 3.

Orogenic Rates

Many examples of rapid orogenic uplift on the converging margins of the Pacific Ocean are cited in Chapter 3. Much of Japan has been uplifted as much as 2000 m during the Quaternary; the amount of tectonic uplift within the last 2 million years is commonly one-half to two-thirds of the highest altitude in any district (Research Group for Quaternary Tectonic Map, 1973).

All the major landforms of New Zealand are tectonic except for depositional forms. Numerous published reports confirm that the late Cenozoic Kaikoura Orogeny is still in progress there. Late Cenozoic uplift along the Alpine fault on the South Island of New Zealand of 14.5 to 21.5 km has been inferred from fission track dating (Kamp et al., 1989). The uplift rate seems to have been accelerating. Between 5.5 million and 1 million years ago, the reported rate was 2.3 to 2.5 mm/yr; during the last million years the average rate was 6.2 mm/yr; in late Quaternary time (<150,000 yr) the reported rate was 10 to 20 mm/yr. However, the common tendency noted in Chapter 1 for rates to appear more rapid when measured over short time intervals should again be considered when evaluating these rates. At Mount Cook (3764 m), as much as 3 to 4 km of rock have been removed in late Cenozoic time, considered by Kamp et al. (1989, p. 191) to cover about the past 5 to 7 million years. The regional uplift rate of the North Island of New Zealand during at least the last 100,000 years has been somewhat slower than on the South Island, reportedly as much as 4 mm/yr over the main axial mountain ranges, with other areas of slower uplift or subsidence (Pillans, 1986).

Similar deformational rates are demonstrated for many other orogenic regions. Emerged late Quaternary reefs in Papua New Guinea (Fig. 2–3) required uplift rates of 1 to 3 m/1000 yr to carry them above the range of glacial-controlled sea-level oscillations (Bloom et al., 1974).

Orogenic movements measured in millimeters per year, if continued for the nearly 2 million years of the Quaternary Period, would produce total uplift measured in kilometers, quite adequate to exceed erosion and create substantial mountain ranges. Along most converging plate margins, such movements are in progress today. They produce not only tectonic relief but also strong zonal climatic gradients, rain-shadow deserts, and other orographic climatic regions. Climate-controlled processes in turn have important feedback effects on the course of tectonic development (pp. 98, 314). Late Cenozoic orogeny has also disrupted ocean currents, intensified latitudinal temperature gradients, and set the stage for the multiple Pleistocene glaciations that have shaped large areas of continents. A complete theory of Cenozoic plate tectonics will be a powerful tool in explaining present-day landforms, not only in orogenic zones but over the entire earth.

REFERENCES

BARAZANGI, M., and DORMAN, J., 1969, World seismicity maps compiled from ESSA, Coast and Geodetic Survey, epicenter data, 1961–1967: Seismol. Soc. America Bull., v. 59, pp 369–80.

BLOOM, A. L., 1980, Late Quaternary sea level change on South Pacific Coasts: A study in tectonic diversity: *in* Morner, N.-A., ed., Earth rheology, isostasy and eustasy, John Wiley & Sons Ltd., Chichester, England, pp. 505–21.

———, BROECKER, W. S., CHAPPELL, J. M. A., MATHEWS, R. K., and MESOLELLA, K. J., 1974, Quaternary sea level fluctuations on a tectonic coast: New ^{230}Th/^{234}U dates from the Huon Peninsula, New Guinea: Quaternary Res., v. 4, pp. 185–205.

BROWN, L. D., and OLIVER, J. E., 1976, Vertical crustal movements from leveling data and their relation to geologic structure in the eastern United States: Rev. Geophys. and Space Phys., v. 14, pp. 13–35.

BURNETT, A. W., and SCHUMM, S. A., 1983, Alluvial-river response to neotectonic deformation in Louisiana and Mississippi: Science, v. 222, pp. 49–50.

CEBULL, S. E., 1973, Concept of orogeny: Geology, v. 1, pp. 101–2.

———, 1976, Concept of orogeny: Reply: Geology, v. 4, pp. 388–89.

CHORLEY, R. J., 1963, Diastrophic background to twentieth-century geomorphological thought: Geol. Soc. America Bull., v. 74, pp. 953–70.

COGLEY, J. G., 1985, Hypsometry of the continents: Zeitschr. für Geomorph., Supp. v. 53, 48 pp.

CRONIN, T. M., 1981, Rates and possible causes of neotectonic vertical crustal movements of the emerged southeastern United States Atlantic Coastal Plain: Geol. Soc. America Bull., v. 92, pp. 812–33.

DALY, R. A., 1940, Strength and structure of the earth: Prentice-Hall, Inc., Englewood Cliffs, N.J., 434 pp.

FAIRBRIDGE, R. W., 1981, The concept of neotectonics: An introduction, *in* Fairbridge, R. W., ed., Neotectonics: Zeitschr. für Geomorph., Supp. v. 40, pp. vii–xii.

FEIGL, K. L., KING, R. W., and JORDAN, T. H., 1990, Geodetic measurement of tectonic deformation in the Santa Maria fold and thrust belt, California: Jour. Geophys. Res., v. 95, pp. 2679–99.

FISCHER, A. G., 1969, Geologic time-distance rates: The Bubnoff unit: Geol. Soc. America Bull., v. 80, pp. 549–51.

FLINT, R. F., 1971, Glacial and Quaternary geology: John Wiley & Sons, Inc., New York, 892 pp.

GARDNER, T. W., JORGENSEN, D. W., SHUMAN, C., and LEMIEUX, C. R., 1987, Geomorphic and tectonic process rates: Effects of measured time interval: Geology, v. 15, pp. 259–61.

GILBERT, G. K., 1890, Lake Bonneville: U.S. Geol. Survey Mon. 1, 438 pp.

JORDAN, T. E., ISACKS, B. L., RAMOS, V. A., and ALLMENDINGER, R. W., 1983, Mountain building in the central Andes: Episodes, v. 1983, no. 3, pp. 20–26.

KAMP, P. J. J., GREEN, P. F., and WHITE, S. H., 1989, Fission track analysis reveals character of collisional tectonics in New Zealand: Tectonics, v. 8, pp. 169–95.

KARIG, D. E., 1974, Evolution of arc systems in the western Pacific: Ann. Rev. Earth and Planetary Sci., v. 2, pp. 51–75.

KING, P. B., 1969, Tectonics of North America—a discussion to accompany the tectonic map of North America, scale 1:5,000,000: U.S. Geol. Survey Prof. Paper 628, 94 pp.

KINSMAN, D. J. J., 1975, Rift valley basins and sedimentary history of trailing continental margins, *in* Fischer, A. G., and Judson, S., eds., Petroleum and global tectonics: Princeton Univ. Press, Princeton, N.J. pp. 83–126.

LEVY, M., and CHRISTIE-BLICK, N., 1989, Pre-Mesozoic palinspastic reconstruction of the eastern Great Basin (western United States): Science, v. 245, pp. 1454–62.

MESCHERIKOV, Y. A., 1968, Neotectonics, *in* Fairbridge, R. W., ed., Encyclopedia of geomorphology: Reinhold Book Corp., New York, pp. 768–73 (see also pp. 223–27).

PELTIER, W. R., 1987, Glacial isostasy, mantle viscosity, and Pleistocene climatic change, *in* Ruddiman, W. F., and Wright, H. E., Jr., eds., North America and adjacent oceans during the last deglaciation: Geol. Soc. America, The Geology of North America, v. K-3, pp. 155–82.

PILLANS, B., 1986, A late Quaternary uplift map for North Island, New Zealand: Royal Soc. New Zealand Bull. 24, pp. 409–17.

PLAFKER, G., 1965, Tectonic deformation associated with the 1964 Alaskan earthquake: Science, v. 148, pp. 1675–87.

RESEARCH GROUP FOR QUATERNARY TECTONIC MAP, 1973, Explanatory text of the Quaternary tectonic map of Japan: Natl. Research Center for Disaster Prevention, Tokyo, 167 pp.

RODGERS, J., 1987, Chains of basement uplifts within cratons marginal to orogenic belts: Am. Jour. Sci., v. 287, pp. 661–92.

SBAR, M. L., and SYKES, L. R., 1973, Contemporary compressive stress and seismicity in eastern North America: An example of intraplate tectonics: Geol. Soc. America Bull., v. 84, pp. 1861–81.

SLEEP, N. H., NUNN, J. H., and CHOU, LEI, 1980 Platform basins: Ann. Rev. Earth Planet. Sci., v. 8, pp. 17–34.

TEN BRINK, N. W., 1974, Glacio-isostasy: New data from West Greenland and geophysical implications: Geol. Soc. America Bull., v. 85, pp. 219–28.

WASHBURN, A. L., and STUIVER, M., 1962, Radiocarbon-dated postglacial delevelling in northeast Greenland and its implications: Arctic, v. 15, pp. 66–73.

WERNICKE, B., AXEN, G. J., and SNOW, J. K., 1988, Basin and Range extensional tectonics at the latitude of Las Vegas, Nevada: Geol. Soc. America Bull., v. 100, pp. 1738–57.

WISE, D. U., 1974, Continental margins, freeboard and the volumes of continents and oceans through time, *in* Burk, C. A., and Drake, C. L., eds., The geology of continental margins: Springer-Verlag New York, Inc., New York, pp. 45–58.

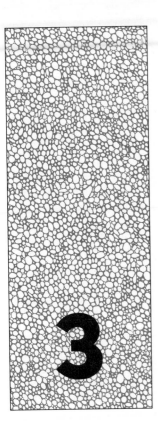

3 Tectonic Landforms

It has become the convention to describe landforms that result from crustal movements as *tectonic*. Some geologists have used "structural" in the same sense, but as was noted in Chapter 1, "structure" in the geomorphic sense has a much broader connotation. Structurally controlled landforms, those that are shaped by differential erosion in response to rock structure, are analyzed in Chapter 11.

It could be argued that no subaerial relief can occur until isostatic or tectonic uplift has raised land above sea level and that therefore all subaerial landscapes are "tectonic" unless they are constructed by depositional (volcanic or sedimentational) processes. However, it is convenient to restrict the term to those landforms that are sufficiently undissected by erosion so that the shape of the fractured or deformed surface can be discerned. All degrees of transition are found between purely tectonic and totally erosional landscapes. The distinction is nevertheless one of the important topics of geomorphology, because if a landform is classed as tectonic, we assume that the deformation has been either recent, or rapid, or both, when compared to the erosion rate. The geomorphic recognition of tectonic landforms is of major value in working out the geologic history of a region. Most tectonic landforms are the result of Quaternary, or at the oldest, late Cenozoic land movements. Older tectonic relief is likely to have been degraded by erosion until its tectonic origin is no longer obvious. In most regions of tectonic landforms, movement continues today.

As noted in Chapter 2, tectonic landforms abound in orogenic belts but are rarely discernible in epeirogenic regions. They occur in well-defined assemblages, appropriate to the style of tectonic deformation in their geomorphic province. Their scale ranges from global or continental, such as mid-ocean ridges, transform faults, and rift-valley systems, to barely measurable offsets in stream channels or rock surfaces.

Whereas Chapter 2 reviews the regional and temporal contexts of tectonic landscapes, this chapter systematically explains individual tectonic landforms or groups of closely related forms. The origin and history of an orogenic belt can only be interpreted by careful analysis and integration of the assemblage of smaller landforms and structures.

TECTONIC SCARPS

A **scarp** (shortened form of "escarpment"; any steep, abrupt slope or cliff along the margin of a plateau, terrace, or other topographic bench) may be formed by a variety of geologic causes, either tectonic, erosional, or depositional. It is well to retain a few purely descriptive topographic terms such as "scarp" and "bench" in the totally nongenetic sense and to add modifying adjectives to them as the need arises. Thus a **tectonic scarp** is a steep slope that results from differential movement of the earth's surface.

Fault Scarps

If a fault displaces the surface of the ground so that one side is higher or lower than the other, a **fault scarp** results. Very recent, high-angle faults in resistant rock may preserve the actual surface of faulting on the scarp complete with slickensides or fault breccia. Such fresh details of fault movement are rare, however. More commonly, fault scarps expose a surface layer of weathered, broken rock and soil. The

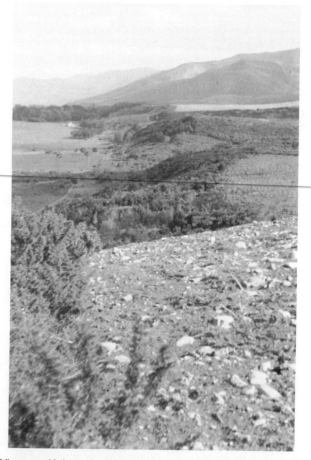

Figure 3-1. View south along the Wairarapa fault near Featherston, New Zealand. Parts of this scarp increased 2 m in height during the 1855 Wellington earthquake.

resulting scarp quickly slumps or collapses to a slope of gravitational stability and gives no clue to the attitude of the fault surface at depth. Even though the actual fault surface is normally not preserved, if the relief on a scarp is the direct result of faulting, the name *fault scarp* is appropriate. This is true even when subsequent erosion on the face of the raised block causes the scarp to retreat laterally some distance from the surface of fault movement, and when the scarp assumes a slope quite different from the dip angle or becomes gullied and dissected. Fault scarps may include segments with varying degrees of degradation, if movement has been intermittent. The typical angle of repose (Fig. 8–1) for the surface sediments on scarps of all origins is between about 25 and 40°. Even historical fault scarps in humid regions quickly become grassy or brush-covered slopes (Fig. 3–1).

The more nearly vertical the dip of the fault surface, the more linear the resulting fault scarp will be across the preexisting landscape. In the case of a low-angle or nearly horizontal thrust fault, even without erosional dissection, the scarp will be along a contour; if the landscape is irregular, the scarp will be as sinuous or crenulate as a contour line.

The absolute direction and amount of motion is not relevant to the formation of a fault scarp. Only the relative motion between adjacent blocks determines the way the scarp will face and its height. Many of the great scarps, or *Pali,* of the Hawaiian islands are fault scarps, facing seaward away from the crests of the volcanic domes (Fig. 3–2). Some authors suggest that the central volcanic areas have been raised up along these faults (Eaton and Murata, 1960; Macdonald, 1972, p. 278). Others hypothesize that flanks of the great volcanic shields are slumping seaward as large, arcuate slices (Cotton, 1944, p. 86). The fault scarp shown in Figure 3–2, like other normal faults, might be considered as a gigantic slump scar (Fig. 8–10).

A fault scarp may die out laterally, break into a series of *splinters,* or become a monoclinal scarp (Fig. 3–3). A major fault or narrow fault zone at depth may break the land surface as a series of closely spaced, *en echelon* scarps, none of which has any conspicuous continuity.

Fault scarps, like all other high, steep landforms, are prone to rapid erosional

Figure 3-2. View southwest along a coastal scarp of the Hilina fault system, south flank of Kilauea Volcano near Puu Kapukapu, Hawaii. The main Hilina Pali (scarp) forms the skyline. (Photo: R. I. Tilling, Hawaiian Volcano Observatory, U.S. Geological Survey.)

(a) Fault scarp dying out laterally

(b) Fault scarp dividing into a series of splinters laterally

(c) Fault scarp passing laterally into a monocline

Figure 3-3. Varieties of lateral termination of fault scarps.

dissection. Streams that formerly flowed across the faulted landscape either will be rejuvenated if they flow from the upthrown to the downthrown block, or they will be impounded against the scarp if they flow in the opposite direction. As most dissected scarps represent repeated late Cenozoic uplift, the first case is much more common. Initially, then, streams cascade down a fault scarp but quickly cut steep gorges that erode progressively headward from the scarp into the uplifted block (Figs. 3–4 and 3–5). The resulting valleys are narrow at their lower ends and widen upstream into the upthrown block (Fig. 3–5). These distinctively shaped scarp valleys have been called *bottle-necked,* or *wine-glass, valleys* (von Engeln, 1942, p. 377; Cotton, 1948, pp. 87, 396, 402). They are not restricted to fault scarps but form on any previously undissected steep slope or scarp.

Dissection by many gullies or valleys causes a scarp to be segmented into a series of **triangular facets,** planar surfaces at an appropriate angle of repose with their bases aligned on or parallel to the fault trace (Fig. 3–6). Excessive debate has been generated about the significance of triangular facets on fault scarps. If misinterpreted as remnants of the original fault surface, the facets imply that the fault was a low-angle normal fault, because the angle of repose is rarely greater than about 30°. By contrast, low scarplets of recent movement at the base of the same facets may show by their straightness in plan or map view to be the trace of a high-angle fault. Thus *two* episodes of faulting, on two faults that are unrelated but coincide at the ground surface, are incorrectly postulated.

In the advanced stage of erosional dissection, a fault scarp may look no different from the face of any other dissected hill or mountain. If the dissecting streams discharge their alluvium into a desert basin, the aggradation may extend headward across the fault trace into embayments in the uplifted block. If a river flows along the foot of the scarp, the alluvium is more likely to be carried away and the tectonic relief preserved, but stream erosion may undercut the fault scarp and make its recognition difficult.

Fault scarps may be buried by alluvium or other sedimentary cover. The tectonic relief of the scarp is then eliminated. If, at a later time, the covering sediments are eroded from the scarp, it is an *exhumed* landform. The relief, then, even though identical to the original fault scarp, is no longer tectonic, but erosional. A scarp produced by structurally controlled erosion at an ancient fault is known as a **fault-line scarp** and is one of many kinds of structurally controlled erosional landforms (p. 276). The distinction between a fault scarp, of tectonic origin, and a fault-line scarp, formed by structurally controlled erosion, is not just a semantic matter. It is "the indispensable prerequisite to a correct interpretation of geological history" (Johnson, 1929, p. 356). The U.S. Nuclear Regulatory Commission stipulates that no nuclear powerplant site should be located in the vicinity of a "capable" fault. A "capable" fault is one that shows movement at or near the ground surface at least once within the past 35,000 years or movement of a recurring nature within the past 500,000 years. Obviously, the correct interpretation of scarps in tectonic regions now has economic and political as well as theoretical significance.

Compound scarps can result from various combinations of tectonic and erosional processes. If a fault scarp were increased in height by a river-cut bank at its base, the upper part of the scarp would be tectonic but the lower part erosional. Conversely, an exhumed fault-line scarp might be the locus of renewed faulting, and its lower part could thus become a true fault scarp (Cotton, 1950b).

Earthquake Scarps and Related Landforms

In seismic regions, many small-scale scarps or "scarplets" are the direct result of displacement during earthquakes (U.S. Geological Survey Staff, 1971; Berberian, 1979; Philip and Meghraoui, 1983). They may be the actual fault scarp (Fig. 3–7) or secondary effects of differential compaction and slumping. Open *fissures* form during some earthquakes, probably due to consolidation of porous surface materials. Soil creep and erosion are likely to obliterate such features quickly. At the foot of recent fault scarps, closed basins or *sag ponds* may develop. The rate at which soil creep and other mass-wasting processes modify fresh scarps can be modeled by a diffusion equation (Chap. 8), and some success has been achieved at estimating the age of prehistoric fault scarps (Colman and Watson, 1983; Nash, 1984).

A variety of other small landforms are commonly observed along recent fault scarps. If the stress field on the fault is compressional, *pressure ridges* (low elongate mounds) may parallel the surface break. Secondary scarps facing the main scarp across a shallow trench have been called *earthquake rents, reverse scarplets,* or *cicatrices* (Cotton, 1948, 1950b). Their significance is not clear, but they may record local or shallow buckling of one of the fault blocks adjacent to the major fault surface.

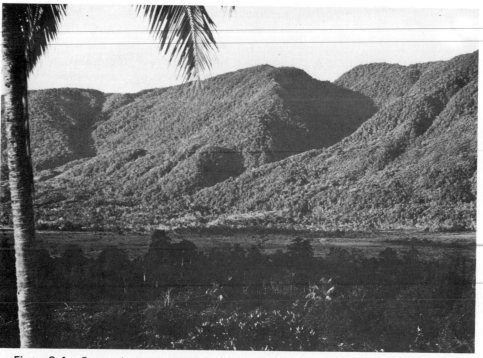

Figure 3-4. Steep, short stream valleys dissecting a fault scarp, north flank of the Cromwell Range, Huon Peninsula, Papua New Guinea.

Landforms Associated with Strike-Slip Faulting

A **strike-slip fault** is a fault of any scale along which movement is parallel to the strike of the fault. A **transform fault** is a strike-slip fault that cuts the entire lithosphere on the boundary between two plates (p. 43); a **transcurrent fault** is a strike-slip fault that is confined to the crust (Sylvester, 1988, p. 1666). Very large strike-slip faults such as the San Andreas fault in California do not move on a single surface, but are defined by a "rift" zone up to 10 km in width in which small surface landforms collectively define the surface trace of the fault. Neither side of the "rift" zone need be higher than the other, and rather than forming a single continuous scarp, the surface expression of a strike-slip fault may be an assemblage of the landforms described in the two previous paragraphs, arranged parallel to or *en echelon* across the width of the fault zone. However, if a series of gullies is offset by a transcurrent fault, many low scarps will result. Typically, if the fault crosses a group of ridges, all the spur ends or ridge crests will be displaced in the same direction (Fig. 3–8). Such half-displaced ridges were named *shutter ridges* by Buwalda (1936). Continued or intermittent strike-slip faulting, especially if it developed a zone of crushed, erodible rock, could be expected to encourage the erosion of a valley along the fault that would intersect and collect the drainage from a number of scarp gullies (Fig. 3–9).

The San Andreas fault in California was the first place where strike-slip fault movement was convincingly demonstrated (Hill, 1981; Sylvester, 1988, p. 1669). It was studied extensively following the 1906 San Francisco earthquake, and is now being closely monitored for precursory evidence of the next damaging earthquake. Some segments moved almost 5 m during the 1906 San Francisco earthquake.

Stratified, organic rich Holocene sediments were deposited in the marshy alluvial fan of Pallett Creek where it crosses the San Andreas fault 55 km north of Los Angeles. By measuring the deformation of the sediments and dating the interbed-

Figure 3-5. Continuation of the view in Figure 3-4. Face-on view of a bottle-necked scarp valley.

Figure 3-6. Triangular facets aligned on the fault scarp of Spanish Peak, 15 km south of Provo, Utah. View east. (Photo: H. J. Bissell.)

ded peat layers by radiocarbon, Sieh (1978, 1984) documented 12 earthquakes between A.D. 260 and 1857 that caused lateral displacement on this segment of the San Andreas fault. At least five of the earthquakes caused movement comparable to the 1857 displacement of 3 to 4 m. He estimated the average recurrence interval of such large-magnitude earthquakes to range from 145 to 200 years, with a suggestion

Figure 3-7. View west along the scarp of a near-vertical fault that resulted from the magnitude 7.3 El Asnam, Algeria, earthquake of October 10, 1980. The fault scarp is on a bedding plane of folded Pliocene strata (B in line rendering) and is offset in a direction opposite to the main earthquake-producing thrust fault (A in line rendering). (Philip and Meghraoui, 1983; photo courtesy M. Meghraoui.)

Figure 3-8. Shutter ridges caused by right-lateral movement on a strike-slip fault across ridge-and-gully topography. Streams are deflected along the fault trace or may be captured by the downstream segment of an adjacent gully.

of shorter intervals during the past 900 years. Farther north, in the Carrizo Plain of central California, similar techniques demonstrated 128 m of right-lateral offset in the modern channel of Wallace Creek, which flows eastward across the San Andreas fault between a series of shutter ridges (Sieh and Jahns, 1984). The rate of right-lateral strike-slip movement (the west side of the fault moving north relative to the east side) has averaged 33.9 ± 2.9 mm/yr for the past 3700 years and 35.8 ± 5 mm/yr for the past 13,250 years. During each of the last three great earthquakes, slip ranged from 9.5 to 12.3 m, giving a recurrence interval of 240 to 450 years for such events. The long-term slip rate at Wallace Creek, although consisting of a discrete major event every few centuries, is indistinguishable from the slip rate on a segment of the fault farther north where continuous creep of a few centimeters per year is measured by deformed buildings and offset rows of orange trees.

Some intermontane basins in New Zealand have been attributed to gaps that opened in the zone of tension where a major strike-slip fault makes an abrupt bend (Clayton, 1966). A similar explanation has been offered for the origin of Death Valley in California and the Dead Sea basin in the Middle East (Fig. 3–18; p. 54). Such tectonic depressions are called *pull-apart basins* (Burchfiel and Stewart, 1966; Sylvester, 1988, pp. 1683–84). They are similar to grabens, but their regional control is strike-slip rather than dip-slip faulting (Fig. 3–9). Anticlines or horsts are forced up in comparable circumstances where the local stress is compressional rather than tensional.

In 1965 a new type of strike-slip fault was devised to explain the offset of submarine topography and paleomagnetic anomalies (Wilson, 1965). The topographic offset of a spreading ridge across a *transform fault* is in the opposite sense from the lateral movement (Fig. 3–10). The transform fault was an entirely new concept and has been one of the major contributions to the theory of plate tectonics. Initially, the transform fault was a structure thought to be unique to the thin, basaltic oceanic crust. Subsequently, several major continental strike-slip faults such as the San Andreas fault have been mapped as plate boundary transform faults (Sylvester, 1988). The oceanic lithosphere has been displaced hundreds of kilometers laterally along great transform faults. Some of them form submarine scarps or ridges several thousand kilometers long and 1 to 2 km high (Menard, 1964). Their height and continuity is not due to active seismicity, but to the fact that the oceanic lithosphere on opposite sides of the transform fault is of different ages, and therefore has cooled and contracted to different depths (Fig. 1–5). Transform faults are typically seismic only on the segments between two spreading ridges (Fig. 3–10). Where an aseismic ocean-floor ridge is subducted, the irregular relief on the subducting plate can produce differential uplift on the overlying plate margin (Taylor et al., 1980).

(a)

(b)

Figure 3-9.   (a) Small-scale (<1 km) landforms developed along recently active strike-slip faults. (Wesson et al., 1975.) (b) Large scale (10 to 100 km) extensional (pull-apart) basins and compressional thrusts and anticlines between *en echelon* segments of strike slip faults.

Figure 3-10. Transform faults that offset a spreading ridge in a left-lateral sense although the seismic portions of the faults are zones of right-lateral slip.

FAULT VALLEYS AND FAULT-BLOCK MOUNTAINS

By a peculiar accident of geography and history, the earliest- and most-studied examples of block-faulted landscapes are in an arid climatic region, where vegetation is sparse and geologic structures are boldly exposed. The regional assemblage of fault-block mountains in the Great Basin section of the Basin and Range province of the western United States (Fig. 1–6; pp. 32, 46) led to the original name "basin ranges" for all the mountains of the region. Subsequently, "basin-range structure" became the general term for block faulting. Many deductions about the erosional evolution of landscapes on faulted structures are really based on an assumption of aridity. Although it is true that few places in the world have such an array of well-exposed fault-block landforms as the U.S. Basin and Range province, the systematic description of these tectonic landforms must not be overly conditioned by the climatically controlled erosional processes that act on them.

Horsts and Grabens

A **horst** is a fault block that has been uplifted relative to the blocks on either side; a **graben** is a depression produced by subsidence along faults. Most authors regard a graben as synonymous with a rift valley, but some (Willis, 1928) have attempted to define grabens as depressions between normal faults, and rift valleys as basins enclosed by escarpments along high-angle thrust faults. The distinction is not geomorphically useful, and it became unfortunately involved with the erroneous concept (p. 39) that facets on fault scarps are remnants of actual fault surfaces. Horsts do not seem to be particularly common landforms. Grabens, on the other hand, are well-known spectacular landforms in many tectonic regions (Fig. 3–11).

Tilted-Block Mountains and Half Grabens (Fault-Angle Depressions)

Rather than being horsts, uplifted fault blocks are commonly tilted, and bounded on only one of their sides by fault scarps (Fig. 3–12). The asymmetric valley or basin between adjacent elongate blocks that are tilted in the same direction is a *half gra-*

Figure 3-11.   The central graben of Iceland. (Photo: S. Thorarinsson.)

ben, or *fault-angle depression.* An assembly of tilted blocks, half grabens, grabens, and horsts is called simply a *fault-block landscape.*

Fault-block landscapes composed of originally flat-lying rocks are called the *Oregon type* (Davis, 1903; Cotton, 1950a, 1950b), named for the block-faulted basalt terrane around Klamath Lake in southern Oregon. According to Davis (1903, p. 130): "There was good evidence for regarding the Oregon lava-block ranges as types of the youngest, most elementary mountain forms known to geographers." The faulting is so recent, the erosional dissection so slight, and the displaced flow sheets so strongly in control of surface slopes that photos of the region approach block diagrams in their simplicity (Donath, 1962). The relief is entirely tectonic. Small alluvial fans and lakes bury or drown some of the grabens and half grabens, but horsts and tilted-block mountains stand in bold profile. Oregon-type landscapes are also found in Iceland (Fig. 3–11) and in the Afar region, near the Red Sea (Tazieff, 1970; Bannert, 1972).

The *basin-range type* of fault-block landscape is characterized by tilted, lifted, or dropped blocks of complexly deformed sedimentary, metamorphic, and igneous rocks in an arid climate, where drainage is perpendicular to the ranges and their bounding faults. Rivers rise near the summits of the ranges and terminate in closed basins between them. Very early in the exploration of the Great Basin, Gilbert (1874, 1928) recognized that the faulting that produced the range-and-basin topography was younger than the orogeny that had intensely deformed and metamorphosed the rocks within the ranges. As noted in Chapter 2, the Basin and Range region experienced multiple and intense orogenies in the Paleozoic, Mesozoic and early Cenozoic Eras, which strongly deformed the rocks of the region. In the late Cenozoic Era, especially in the last 15 million years, the region has undergone east-west extension that has created the present fault block landscape. This is the reason

Figure 3-12.   Block diagram of the Ruby and East Humboldt fault blocks. To show geometric relations, the pre-faulting surface is assumed flat and the post-faulting erosion of the two mountain ranges is ignored. The near edge of the diagram is about 25 km wide; vertical exaggeration is about twofold. (Sharp, 1939, Fig. 5.)

that the word "orogeny" is no longer used for the creation of topography, but is restricted to the deformation of rocks (p. 29). Louderback (1904, 1923) demonstrated that some tilted blocks are capped by lava flows that are neither folded nor thrusted but simply rest in sharp unconformity on erosional surfaces of low relief and are tilted because of the latest episode of extensional block faulting. In western Nevada, an interval of tectonic quiescence between 7.5 and 3.5 million years ago produced the widespread erosion surfaces across earlier structures, which were then covered with latest Pliocene and Pleistocene volcanic rocks and broken by Quaternary faulting into the modern topography (Gilbert and Reynolds, 1973).

The Ruby–East Humboldt Range of northeast Nevada is a typical basin range (Sharp, 1939, 1942). The complex internal structures are of pre-Miocene age; the less deformed Miocene alluvium and lacustrine beds either are faulted against or unconformably overlie the older structures; and the Quaternary alluvium unconformably overlies all older sediment and rock. An imposing fault scarp on the eastern side is 1200 to 1500 m in height, but the westward slopes are more gentle. The western base of the range is linear, recent scarplets there are abundant, and the Miocene beds are displaced as much as 70 m, but the present tectonic form of the range is more like a pair of tilted blocks than a horst (Fig. 3–12). The blocks are bounded by normal faults that dip 60 to 70°, where exposed in canyons.

The basin-range type of fault-block landscape has been described in some detail because of its significance to the historical development of American geology, and because drawings and photographs of basin ranges are featured in almost every American text. Lacking personal familiarity with fault-block landscapes, many geologists and geographers mentally associate them with deserts, forgetting that the arid or semiarid climate and internal drainage of most of the U.S. Basin and Range province impose special geomorphic processes on the tectonic landforms. In particular, almost all deductions about the sequential erosional development of fault-block mountains, half grabens, and grabens started from the assumption that alluvium eroded from the ranges would accumulate in the adjacent basins and progressively bury the base of the range. For instance, Sharp (1942, p. 683) demonstrated that on the east scarp of the tilted fault block in the southern Ruby Mountains, almost one-third of the tectonic relief is buried.

The *Sierra Pampeanas,* or Pampean Ranges, of northwestern Argentina are an interesting variant of the basin-range type of fault-block mountains (Jordan and Allmendinger, 1986). They are composed of Paleozoic and older igneous and metamorphic rocks, which are truncated by an erosion surface of low local relief

(Fig. 3–13). Within the last 10 million years, this ancient continental terrane has been caught up in the east-west compressional stresses of the Andean orogeny. Basement blocks have failed, commonly along inherited zones of weakness or fractures, and the block mountains have risen along one or more bounding reverse faults, which may dip between 30 and 80°. The cover strata are mostly continental arkosic sandstone and conglomerates that are highly erodible. The great contrast in erodibility between the basement blocks and the cover strata has resulted in extensive exhumation of the ancient erosion surface, with inherited structures being etched out in the semiarid climate (Fig. 3–13). Structural relief between the crests of the blocks and the floors of the adjacent sediment-filled basins is as much as 8 km. Many of the range-bounding faults have been active in the last 3 million years, and recent earthquake scarps have been identified across alluvial fans at the bases of some ranges. Although they are bounded by reverse faults, the Pampean Ranges superfi-

Figure 3-13. Sierra de Velasco, a tilted fault-block mountain of crystalline basement rocks in the Sierras Pampeanas (Pampean Ranges) of Argentina. The western slope is an ancient erosion surface exhumed and broken by minor faults. The east-facing scarp is a tectonic scarp along a reverse fault. The range reaches a maximum altitude of 4500 m. (NASA ERTS E-2147-13383.)

cially look like the basin ranges of the western United States because the scarps in both provinces have eroded to the appropriate stable angle for similar rocks in similar climates. Tectonically, the Pampean blocks are a modern analog of the early Cenozoic Laramide fault blocks in the Rocky Mountains of the United States.

To correct the climatic bias toward arid and semiarid regions in geomorphic descriptions of fault-block mountains, we can follow Cotton (1948, 1950a, 1950b, 1958, 1968) and designate a third type of fault-block landscape as the *New Zealand type.* The Awatere fault scarp (Fig. 3–14), one of several northeast-trending fault scarps in eastern Marlborough, is representative of fault scarps in New Zealand, Venezuela, Japan, and other humid regions, where through-flowing rivers dominate the erosional processes. The Awatere fault scarp "differs in one important respect from those of the basin-range type. A vigorous consequent river has been present parallel to the scarp, at or near its base, during much of its long erosional development—a river that has been capable of carrying away the debris furnished by smaller streams dissecting the scarp" (Cotton, 1950a, p. 196). "Distinction of the Awatere from the basin-range type [of fault scarp] depends on the absence of a bahada fringe, all the debris derived from rapid early wasting has long ago been removed, and newer debris, though in abundant supply, continues to be carried off progressively by active river transportation" (Cotton, 1950b, p. 727).

New Zealand–type fault-block landscapes are characterized, then, by half grabens and grabens that are river valleys or stream-dissected lowlands rather than alluvium-filled basins (Stevens, 1974). Where near the sea, New Zealand–type half grabens may be drowned to form excellent harbors, such as Port Nicholson, at Wellington, New Zealand (Fig. 3–15), Spencer and St. Vincent Gulfs near Adelaide, Australia, and the "Port Phillip Sunkland," the outer harbor of Melbourne, Australia (Twidale, 1968).

Early and extensive stream dissection of fault scarps and back slopes is the other diagnostic feature of the New Zealand type of fault-block landscape. Instead of planar triangular facets, rounded or blunt-ended ridges descend to the fault trace, where they are sharply cut off (Figs. 3–14 and 3–15). The dissecting valleys between

Figure 3-14. View southwest along the Awatere fault scarp, South Island, New Zealand. Mountain block on the right has moved up and toward the camera relative to the valley along a high-angle reverse fault. (Photo: D. L. Homer, N. Z. Geological Survey.)

the scarp segments are large and are occupied by vigorous perennial streams that have substantial catchment areas in the interior of the upthrown block.

The unique features of fault-block landscapes of the New Zealand type are not tectonic but erosional. Many major New Zealand faults are high-angle reverse faults, but as in the case of all fault scarps, the dip of the fault surface is not significant to the scarp angle. The fault scarp shown in Figures 3–4 and 3–5 is presumably of the New Zealand type, because it stands in a humid tropical climate, is heavily forested, and lacks alluvial fans at the mouths of the dissecting creeks.

Figure 3-15. View northeast over Port Nicholson (Wellington Harbor), the drowned portion of a half graben (fault-angle depression). Wellington, New Zealand, in foreground. (Photo: Royal New Zealand Air Force. Crown Copyright reserved.)

Other fault-block landscapes in the humid tropics, such as the great graben that forms the Markham Valley westward from Lae, Papua New Guinea, also are of the New Zealand type. Wherever precipitation exceeds evaporation, the excess water must drain along the tectonic depressions to the sea, creating river valleys along the lowest axes of half grabens or on the floors of grabens. Continued uplift on a scarp of the New Zealand type might create flights of uplifted river terraces on the scarp face, in contrast to the multiple earthquake scarplets across alluvial fans at the base of a basin-range scarp. Considering that most of the world's landscapes are drained to the sea by rivers (Chap. 9), it is tempting to conclude that the New Zealand type of fault-block landscape is more common than the basin-range type. Perhaps we have been educated to expect the basin-range type of fault-block landforms and have therefore overlooked or ignored fault-block landforms in humid regions. Highly tectonic regions in southeast Asia, Japan, Alaska, and the Caribbean can be expected to provide many additional examples of the New Zealand type of tectonic landscape.

Rift Valleys

The great *rift valleys* of the Rhine, Lake Baikal, and the Afro-Arabian rift system so completely dominate their tectonic and geomorphic settings that they deserve special treatment as tectonic landforms, although descriptively they are little more than large, branching, complex grabens. Most authorities have emphasized the association of rift valleys with plateaus produced by broad epeirogenic updoming (Holmes, 1965, pp. 1044–78; Illies, 1981). Experiments (Fig. 3–16) show that excellent models of rift-valley morphology and tectonics, including the high plateaus that rise toward the rims of the depression and other details, result if moist clay layers are arched or updomed from beneath. Inward-facing normal faults form the boundary scarps, the model rift valley has a surface width approximating the thickness of the deformed layers, and no isostatic or mechanical inconsistencies develop. Instead of resulting from simple lateral tension, rift valleys are now interpreted as "keystones" subsiding along the crests of thermally uparched plateaus that are underlain by unusually thin continental lithosphere.

The Rhine graben is a small rift valley, compared to those of Africa and the Middle East. It is 30 to 40 km wide (Holmes, 1965, p. 1047; Illies, 1981) and, including several branching segments, about 300 km long. It cuts crystalline rocks or older folded structures, and Mesozoic and Cenozoic formations within the graben are strongly deformed and broken by secondary faults of Miocene and younger age. The fault scarps of the Rhine graben are of the New Zealand type: dissected, rounded spurs end abruptly at the alluvial plain of the Rhine River. Movement continued into the Pleistocene. Hot springs and health spas mark the boundary faults today.

The Afro-Arabian rift system extends along a 60° arc of the earth's crust, for a length of 6500 km from southeastern Turkey to Mozambique. It includes the Dead Sea, Red Sea, Gulf of Aden, and eastern African rifts. Along various segments it exhibits combinations of epeirogenic doming, strike-slip faulting, and normal faulting that have been in progress for at least the Cenozoic Era. The rift valleys of eastern Africa are a branching, *en echelon* series of grabens and half grabens that divide into two major groups, the eastern or Kenya rift and the longer western rift system (Fig. 3–17). The total length of the system southward from the Red Sea approaches 3000 km. Rift valleys within the system are uniformly 30 to 50 km wide, approximately the thickness of the continental crust (Holmes, 1965, p. 1061; Rosendahl et al., 1986). Seismic profiles show that only about 20 percent of the Lake Tanganyika

Figure 3-16.   Experimental production of a rift valley by slowly arching layers of moist clay. (Redrawn from Holmes, 1965; experiments by Hans Cloos.)

sector of the western rift is a true or full graben. The remainder of the rift is an alternating series of half grabens 50 to 70 km in length and 20 to 40 km in width. The half grabens are arcuate or crescent shaped in plan view, so that the rift has a sinusoidal shape (Rosendahl et al., 1986). Four major half graben segments are separated by a variety of transverse or *en echelon* structures that include both grabens and horsts. Most of the major basins in the eastern or Kenya rift are also half grabens (Baker, 1986, p. 52), as are the 32 or more basins that form the western rift system (Ebinger, 1989). Opposed or overlapping half grabens probably are the dominant structural form of all rift systems. A geomorphic consequence is that one rim of the rift will usually be higher than the facing one. In a region of extensional tectonics, half grabens require less energy to form than full grabens (Rosendahl et al., 1986, p. 32). Groshong (1989, Fig. 10) demonstrated that the simple rift-valley structures experimentally produced in clay models (Fig. 3–16) can better be explained by a dominant fault on one side and lesser faults and tilted layers on the other, effectively a half-graben structure. Recent volcanism and transverse defor-

Figure 3-17. Tectonic setting of Lake Victoria and Lake Kyoga between the eastern and western rift valleys in eastern Africa. Former west-flowing streams were defeated by the uplifted shoulder of the western rift and reversed. Their drowned headwaters now form the lakes and drain north to the Nile. (After Holmes, 1965; Bishop and Trendall, 1967.)

mation of the downdropped half grabens has created divides within the rift valleys so that lakes in the southern part of the western rift system drain to the Congo, lakes in the northern part of the western rift drain to the Nile, Lake Nyasa (Malawi) drains south to the Indian Ocean via the Zambezi River, and most of the eastern rifts are areas of internal drainage into highly saline, alkaline lakes (Baker, 1986). Lake Tanganyika, part of the Congo drainage system, is 1400 m deep, its sediment bottom 650 m below sea level. As do many of the other rift lakes, it fills the floor of the rift valley from one side to the other.

In the midst of the western rift valley stands the Ruwenzori horst, the highest nonvolcanic mountain in Africa. Its summit at 5100 m is nearly 4 km higher than the plateau rims on each side of the rift valleys, which in turn are 400 to 500 m above the floors of the rifts. Clearly, the Ruwenzori massif is an uplifted horst, not simply a fault block that did not subside. It is said to be topped by the same mid-Cenozoic erosion surface that is arched upward toward the rift valleys across the adjacent plateaus (McConnell, 1972).

The plateau surface around Lake Victoria, between the eastern and western rift valleys, has been strongly upwarped in the Quaternary (Bishop and Trendall, 1967). By Miocene time, a widespread erosion surface of low relief had beveled the metamorphic terrane of the plateau region. Miocene volcanic rocks buried this erosion surface, although much of it has subsequently been exhumed. Pre-Miocene rivers may have drained westward across Africa, perhaps via the Congo River, but by Miocene time the western rift had developed and the plateau drainage entered it. As the half grabens progressively deepened in mid-Pleistocene time, the plateau erosion surface was lifted and flexed into a sharp western anticline and a broad central basin. An uplifted "shoulder" just east of the rim of the western rift valley defeated and reversed streams that formerly crossed it. The headwaters of these streams ponded to form Lakes Victoria and Kyoga (Fig. 3–17), which deepened against the uplifted plateau rim until they spilled around it to the north. The present Victoria Nile River resulted, draining into Lake Albert (Mobutu) over an exhumed fault-line scarp at Murchison Falls. Lake Kyoga is a drowned dendritic drainage system on the warped early Pleistocene surface. The warping continues: 90 m of relative uplift across the defeated drainage systems may have occurred between 50,000 and 20,000 years ago (Bishop, 1965, p. 317). The implied tectonic differential uplift rate of 3 m/1000 yr approaches that of strongly compressional orogenic regions (p. 24), although the plateau movements are certainly extensional and of epeirogenic scale.

The Dead Sea occupies a pull-apart basin on a rift valley 600 km long and only 10 to 20 km wide that extends northward from the Red Sea to Lebanon (Neev and Emery, 1967) (Fig. 3–18). The eastern scarp is straight and continuous along a single border fault. The western side of the rift is more complex, with several step faults or fault splinters involved. Fault scarps along the Dead Sea rift valley are predominantly of the basin-range type. The region is one of extreme aridity.

The Dead Sea rift has a more complex structure than the Rhine graben and the eastern African rift systems (Quennell, 1958; Hatcher et al., 1981; Gerson et al., 1985). Left-lateral movement totaling 100 to 120 km along a crooked strike-slip fault is believed to have opened a quadrilateral pull-apart basin in the vicinity of the Dead Sea (Fig. 3–18). Pliocene and Pleistocene vertical movements along the boundary faults and several oblique transverse faults have created the present basins. The shore of the Dead Sea is the lowest subaerial place on the earth's surface at −392 m.

Lake Baikal, the world's deepest freshwater lake (1600 m), lies in a rift valley north of Mongolia along the crest of the Baikal arch. The tectonic setting is similar

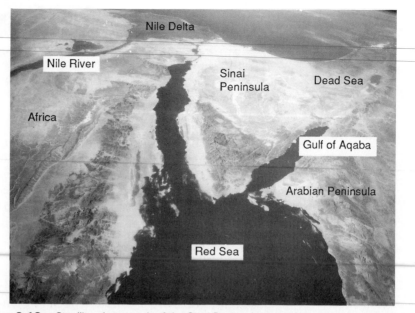

Figure 3-18. Satellite photograph of the Sinai Peninsula viewed toward the northwest. The Dead Sea rift extends northward from the Gulf of Aqaba past the Dead Sea to Lake Tiberias. The eastern side of the rift has moved north relative to the western side. (Space Shuttle handheld photograph 41G-120-180, NASA.)

to that of the eastern African rift lakes. Fault scarps plunge directly into the lake, with only small deltas. Most of the regional drainage is away from the lake along radii of the uplifted Baikal arch.

A median rift valley is a characteristic feature of many submarine mid-ocean ridges or rises (Figs. 3–10 and 3–19). The rift valleys of mid-ocean ridges are bounded by a series of parallel normal faults, each with a throw of several hundred meters. The fault blocks that parallel the central rift valley are commonly lifted and rotated so that their upper surfaces slope away from the rift axis. The morphology of median rift valleys is consistent with their tectonic setting on the crests of expanding ridges at accreting plate margins (Ballard and van Andel, 1977).

LANDFORMS MADE BY FOLDING

Fault scarps are bold, obvious landforms that attract the eyes of geomorphologists and sightseers. Much less has been written about tectonic landforms that result from warping or flexing of the earth's surface without fracture and displacement on faults. A few examples can be cited from regions where earth movements are obvious. Perhaps by calling attention to known examples, this review will stimulate readers to consider other possibilities, especially in regions of humid climate, heavy vegetation cover, active soil-forming processes, and mild diastrophism.

A major difficulty in recognizing tectonic landforms produced by folding is that a visible reference surface is not generally available. Whereas every structure and the landscape itself is sharply displaced across a fault scarp, warping or flexing of an irregular landscape or homogeneous structure is extremely difficult to recognize. Only where approximately planar erosional and depositional surfaces, such as shore platforms or river terraces are exposed, is warping likely to be visible. Even then, surveyed profiles may need to be taken along the surfaces to prove that they are warped.

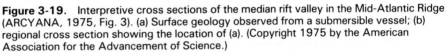

Figure 3-19. Interpretive cross sections of the median rift valley in the Mid-Atlantic Ridge (ARCYANA, 1975, Fig. 3). (a) Surface geology observed from a submersible vessel; (b) regional cross section showing the location of (a). (Copyright 1975 by the American Association for the Advancement of Science.)

Monoclinal Scarps

Tectonic scarps with height, steepness, and lateral continuity comparable to fault scarps are formed by steep monoclines (Fig. 3–20). As is the case with fault scarps, monoclinal scarps are regarded as tectonic landforms only if their relief can be shown to be due to tectonic movement and if substantial parts of the original surface of deformation can be recognized. Monoclines, as other folds, are described by the attitude of the deformed strata, not by the shape of the displaced landscape.

Monoclines are the most distinctive structure of the Colorado Plateau (Kelley, 1955). The longest is over 400 km in length; strata are displaced vertically as much as 4300 m. In the Colorado Plateau many land surfaces coincide with particularly resistant formations, so structural benches are flexed sharply along the monoclines

Figure 3-20. First illustration of a monoclinal fold. (Powell, 1975, Fig. 67.)

to produce imposing scarps. The major period of deformation in the region was the late Cretaceous–Eocene Laramide orogeny (Kelley, 1955, p. 797). It is not known whether monoclinal flexing has continued to the present.

No systematic distinction of monoclinal landforms has ever been made analogous to the distinction between fault scarps and fault-line (erosional) scarps. It would seem that the great monoclines of the Colorado Plateau were folded at some depth. Tertiary strata that postdate the folding have been partly stripped away. Such scarps are better considered as examples of structurally controlled erosion (Chap. 11) than as tectonic landforms.

Topographic Domes

A variety of processes can produce domal tectonic landforms. Shallow intrusions of ice, salt, mud, or molten rock can uparch the overlying rock or sediment. Shallow intrusions may also extrude to create new landforms. Active anticlines in rock structures normally produce topographic domes in the landscape above. All these landforms are tectonic in the sense that they are built by earth movements, even if the movements that are involved are at a very shallow depth.

Salt Tectonics.   More than 400 domal or piercement structures with salt in their cores are known from the northern part of the Gulf of Mexico (Halbouty, 1967; Jackson and Talbot, 1986). The salt is presumed to have risen hydrostatically or under tectonic pressure from source beds of Jurassic or older age that are now buried under many kilometers of rapidly accumulated overburden. A thousand meters is regarded as the minimum thickness of a buried salt bed that will flow into domes (Murray, 1961, p. 260). Some of the U.S. gulf coast salt domes have risen through 7500 m or more of covering strata. They are circular or ridgelike in plan view and may be up to 10 km in diameter. Similar structures are known or suspected in the Colorado Plateau, the Canadian Arctic, many parts of the continental shelves, Saudi Arabia, France, Germany, Iran, and the Soviet Union. Because of the potential for petroleum production from structural and stratigraphic traps on their flanks, salt domes are actively sought by exploration geologists.

Only five salt domes on the U.S. gulf coast have positive topographic relief (Gould and Morgan, 1962, p. 314). Although the domes penetrate Pleistocene and even Holocene sediments, the region has been one of deltaic progradation and fluviatile deposition, and lateral migration of coastal-plain rivers has obliterated most of the salt-dome relief. Avery Island, Louisiana, has a highest point of 46 m and forms a dome about 2.3 km in diameter. Salt is found at a depth of only about 4 m below its surface. Several salt springs and ponds drain radially off the dissected dome.

Many gulf coast salt domes are marked by central depressions ringed by low cuestas of the upwarped overlying rocks. *Annular* drainage patterns (Fig. 11–17) are common (Muehlberger et al., 1962, p. 266). Such morphology is structurally controlled, but not tectonic, and is considered further in Chapter 11.

In the Canadian Arctic archipelago, Tertiary or older folding forced Paleozoic anhydrite and gypsum into anticlinal cores and domes. In contrast to the rapid solution of evaporites in humid regions, in the polar deserts these massive, unjointed, monomineralic rocks are resistant ridge- and dome-formers (St. Onge, 1959). Whether the highlands of gypsum in the Sverdrup Islands of northernmost Canada are tectonic or structural has not been determined (Stott, 1969, p. 37), but aerial reconnaissance shows that many of the higher hills are underlain by massive white gypsum, intrusive into darker sedimentary rocks that form surrounding lowlands.

Mud Lumps. A minor but interesting form of tectonic domal landform has been described from the Mississippi delta region. Rapid forward growth of distributary channels may cause local loading of unstable prodelta clay by 100 m of deltaic sand, mud, and organic sediment. Diapiric intrusions through the overlying sandy sediment by plastic clay causes updoming or even extrusion as low islands near the mouths of distributaries (Morgan et al., 1968). Near the mouth of South Pass, 45 to 50 different mudlump islands have been observed and mapped since 1867. Some reappear at intervals after being eroded away. Surface exposures show mud with marine faunas that must have been initially deposited in water depths of 100 to 120 m. Their appearance and disappearance greatly complicates the definition of state and federal boundaries in offshore petroleum leases.

Igneous Domes. Shallow igneous sills may swell to a planar-convex lens shape and arch overlying sedimentary strata to produce topographic domes. The name **laccolith** (lens-shaped intrusion) was given for such forms in the Henry Mountains of Utah in one of the classic works of early American geology (Gilbert, 1877). Some of the larger type examples subsequently have been claimed to be igneous stocks (Hunt et al., 1953) and not laccoliths, but the debate continues (Jackson and Pollard, 1988).

The laccoliths of the Colorado Plateau were intruded in late Eocene or Miocene time, probably at a depth of at least 1600 m (Hunt et al., 1953, p. 147). Because some of the laccoliths are thicker than that, surface arching must have resulted. The regional drainage pattern suggests that major rivers, flowing on an erosion surface 1000 m or more above the present landscape, were diverted by the rising laccoliths and migrated laterally off the domes (Hunt et al., 1953, p. 212). The subsequent geomorphic history of the Henry Mountains laccoliths, as is true of the monoclines of the region, is one of structurally controlled erosion (Chap. 11).

Numerous instrumental records demonstrate that volcanoes bulge in a domal shape prior to an eruption. The movements are undetectable except by precise tiltmeters or geodetic surveys, but are a useful indication of an impending eruption. The doming is attributed to magma rising into shallow reservoirs within or beneath the volcano. On a larger scale, the Yellowstone Plateau, within Yellowstone National Park, has updomed almost 1 m in 50 years (Fig. 3–21). As is the case with neotectonic movements (Chap. 2) such rapid uplift rates cannot continue for very long without major geomorphic consequences. The domal uplift of the Yellowstone Plateau is attributed to a recent influx of magma into the upper crust within an ancient volcanic caldera (Pelton and Smith, 1982; Dzurisin and Yamashita, 1987).

Rising or Live Anticlines. Orogenic regions such as the margin of the Pacific Ocean have domes or arches in the landscape over the crests of structural anticlines, or at compressional segments of curved strike-slip faults (Fig. 3–9). Warped coastal and river terraces (Figs. 3–22, 3–23) are a common form of evidence. The Ventura River in southern California flows south across the active Ventura Avenue anticline (Rockwell et al., 1984, 1988). For the past 200,000 years, based on radiocarbon dating, soil stratigraphy, and correlation with dated marine terraces, the anticline has been rising across the path of the river. As terraces were cut, they were progressively arched upward (Fig. 3–23). The rate of uplift along the anticlinal axis has decreased from about 20 m/1000 yr for the first 100,000 years to about 5 m/1000 yr at the present time.

In New Zealand, several active anticlines warp a coastal-plain landscape (Te Punga, 1957). One of them, the Pohangina anticline can be traced for 30 km. Recent alluvium and Pleistocene coastal-plain formations dip 2 to 4° off the west flank of the anticline, but on the steeper eastern flank, beds dip as steeply as 60°.

Figure 3-21. Vertical changes in surface levels in Yellowstone National Park from surveys done in 1923 and 1975. Contour interval is 100 mm with corresponding velocity value given in parentheses in mm/yr. Resurgent domes shaded. Caldera rim and outer edge of ring-fracture zone are also shown. (Pelton and Smith, 1982, Fig. 7.)

Figure 3-22. Longitudinal profile of the Oguni River and its terraces, Yamagata Prefecture, northeastern Japan. Vertical scale is exaggerated tenfold. Quaternary terraces are progressively upwarped over the anticlines in the Miocene and Pliocene strata shown on the geologic cross section beneath the profile. A fault near the eastern end of the profile displaces older terraces. (Sugimura, 1967, Fig. 2.)

On this rising anticline in a humid climate, vigorous gullies migrate rapidly headward on the steeper flank, and stream divides have shifted laterally beyond the crest of the structure onto the less-steep dip slope. Many drainage diversions have been noticed in the century since the land was cleared and fenced for agriculture. Seismic surveys and the regional geologic setting suggest that active anticlines in both New Zealand and Japan (Ota, 1975) are surface manifestations of rising fault blocks of indurated basement rock. Each anticline is thought to be in sedimentary cover strata that are draped over the highest edge of an upfaulted basement block.

Numerous examples of rising anticlines can be cited from other orogenic re-

Figure 3-23. Ventura River terrace profiles over the Ventura Avenue anticline and Red Mountain fault. For location, see inset map. Age control is from ¹⁴C dates (terrace B, D, E, F) and correlation to a marine terrace east of the Venture River (terrace H) dated by amino-acid racemization. YBP = years before present. (Rockwell et al., 1988, Figs. 3 and 4.)

gions. Lees (1955) described a Persian *Qanat*, or horizontal tube well, that had been arched by a rise of 18 m in the middle part of its 4-km course across an anticline in the Khuzistan Plain at Shaur, Iran. The tunnel was dug 1700 years ago, giving a differential uplift rate across the structure 10 m/1000 yr. Prehistoric irrigation canals in Peru, built between about A.D. 900 and 1400, show successive repairs and rerouting that were at least in part required by tectonic tilting and faulting (Ortloff et al., 1985). In both Japan and California, Holocene deformation of river terraces corresponds well with continuing deformation documented by geodetic releveling (Sugimura, 1967; Castle et al., 1974).

REFERENCES

ARCYANA, 1975, Transform fault and rift valley from bathyscaph and diving saucer: Science, v. 190, pp. 108–16.

BAKER, B. H., 1986, Tectonics and volcanism of the southern Kenya rift valley and its influence on rift sedimentation, in Frostick, L. E., Renaut, R. W., Reid, I., and Tiercelin, J. J., eds., Sedimentation in the African rifts: Geol. Soc. (London) Spec. Pub. no. 25, pp. 45–57.

BALLARD, R. D., and VAN ANDEL, T. H., 1977, Morphology and tectonics of the inner rift valley at lat. 36°50′ N on the Mid-Atlantic Ridge: Geol. Soc. America Bull., v. 88, pp. 507–30.

BANNERT, D., 1972, Afar tectonics from space photographs: Am. Assoc. Petroleum Geologists Bull., v. 56, pp. 903–15.

BERBERIAN, M., 1979, Earthquake faulting and bedding thrust associated with the Tabas-e-Golshan (Iran) earthquake of September 16, 1978: Bull. Seismological Soc. America, v. 69, pp. 1861–87.

BISHOP, W. W., 1965, Quaternary geology and geomorphology in the Albertine Rift Valley, Uganda: Geol. Soc. America Spec. Paper 84, pp. 293–321.

———, and TRENDALL, A. F., 1967, Erosion-surfaces, tectonics and volcanic activity in Uganda: Geol. Soc. London Quart. Jour., v. 122, pp. 385–420.

BURCHFIEL, B. C., and STEWART, J. H., 1966, "Pull-apart" origin of the central segment of Death Valley, California: Geol. Soc. America Bull., v. 77, pp. 439–41.

BUWALDA, J. P., 1936, Shutterridges, characteristic physiographic features of active faults [abst.]: Geol. Soc. America, Proc. 1936, p. 307 [1937].

CASTLE, R. O., ALT, J. N., SAVAGE, J. C., and BALAZS, E. I., 1974, Elevation changes preceding the San Fernando earthquake of February 9, 1971: Geology, v. 2, pp. 61–66.

CLAYTON, L., 1966, Tectonic depressions along the Hope fault, a transcurrent fault in North Canterbury, New Zealand: New Zealand Jour. Geol. Geophys., v. 9, pp. 95–104.

COLMAN, S. M., and WATSON, K., 1983, Age estimated from a diffusion equation model for scarp degradation: Science, v. 221, pp. 263–5.

COTTON, C. A., 1944, Volcanoes as landscape forms. Whitcombe and Tombs Ltd., Christchurch, New Zealand, 416 pp. (republished with minor corrections, 1952).

———,1948, Landscape as developed by the processes of normal erosion, 2nd ed: Whitcombe and Tombs Ltd., Wellington, New Zealand, 509 pp.

———, 1950a, Tectonic scarps and fault valleys: Internat. Geog. Cong., 16th, Lisbon 1949, Comptes rendus, pp. 191–200.

———, 1950b, Tectonic scarps and fault valleys: Geol. Soc. America Bull., v. 61, pp. 717–58.

———, 1958, Geomorphology, 7th ed: Whitcombe and Tombs Ltd., Christchurch, New Zealand, 505 pp.

———,1968, Tectonic landscapes, in Fairbridge, R. W., ed., Encyclopedia of geomorphology: Reinhold Book Corp., New York, pp. 1109–16.

DAVIS, W. M., 1903, Mountain ranges of the Great Basin: Mus. Comp. Zool. Harvard Bull. 42, pp. 129–77 (reprinted 1954 in Geographical essays: Dover Publications, Inc., New York, pp. 725–72).

DONATH, F. A., 1962, Analysis of basin-range structure, south-central Oregon: Geol. Soc. America Bull., v. 73, pp. 1–15.

DZURISIN, D., and YAMASHITA, K. M., 1987, Vertical surface displacements at Yellowstone Caldera, Wyoming, 1976–1986: Jour. Geophy. Res., v. 92, pp. 13,753–66.

EATON, J. P., and MURATA, K. J., 1960, How volcanoes grow: Science, v. 132, pp. 925–38.

EBINGER, C. J., 1989, Tectonic development of the western branch of the East African rift system: Geol. Soc. America Bull., v. 101, pp. 885–903.

VON ENGELN, O. D., 1942, Geomorphology: The Macmillan Company, New York, 655 pp.

GERSON, R., GROSSMAN, S., and BOWMAN, D., 1985, Stages in creation of a large rift valley-geomorphic evolution along the southern Dead Sea rift, in Morisawa, M., and Hack, J. T., eds., Tectonics Geomorphology: Proc., 15th Binghamton Geomorphology Symp., 1984, Allen & Unwin, Inc. Boston, pp. 53–73.

GILBERT, C. M., and REYNOLDS, M. W., 1973, Character and chronology of basin development, western margin of the Basin and Range province: Geol. Soc. America Bull., v. 84, pp. 2489–2509.

GILBERT, G. K., 1874, Preliminary geological report, expedition of 1872: U. S. Geog. Geol. Survey W. 100th Mer. (Wheeler), Progress Rept., pp. 48–52.

———, 1877, Report on the geology of the Henry Mountains [Utah]: U.S. Geog. Geol. Survey Rocky Mtn. Region (Powell), pp. 18–98.

———, 1928, Studies of basin-range structure: U. S. Geol. Survey Prof. Paper 153, 92 pp.

GOULD, H. R., and MORGAN, J. P., 1962, Coastal Louisiana swamps and marshlands, Field Trip no. 9, *in* Rainwater, E. H., and Zingula, R. P., eds., Geology of the Gulf coast and central Texas and guidebook of excursions: Houston Geological Soc., Houston, Texas, pp. 287–341.

GROSHONG, R. H., JR., 1989, Half-graben structures: Balanced models of extensional fault-bend folds: Geol. Soc. America Bull., v. 101, pp. 96–105.

HALBOUTY, M. T., 1967, Salt domes, Gulf region, United States and Mexico: Gulf Publishing Co., Houston, Texas, 425 pp.

HATCHER, R. D., JR., ZIETZ, I., REGAN, R. D., and ABU-AJAMIEH, M., 1981, Sinistral strike-slip motion on the Dead Sea Rift: Confirmation from new magnetic data: Geology, v. 9, pp. 458–62.

HILL, M. L., 1981, San Andreas fault: History of concepts: Geol. Soc. America Bull., v. 92, pp. 112–31.

HOLMES, A., 1965, Principles of physical geology, 2nd ed: Ronald Press, New York, 1288 pp.

HUNT, C. B., AVERITT, P., and MILLER, R. L., 1953, Geology and geography of the Henry Mountains region, Utah: U. S. Geol. Survey Prof. Paper 228, 234 pp.

ILLIES, J. H., 1981, Epilogue: Mechanism of graben formation, *in* Illies, J. H., ed., Special Issue: Mechanism of graben formation: *Tectonophysics,* v. 73, pp. 249–66.

JACKSON, M. D., and POLLARD, D. D., 1988, The laccolith-stock controversy: New results from the southern Henry Mountains, Utah: Geol. Soc. America, v. 100, pp. 117–39.

JACKSON, M. P. A., and TALBOT, C. J., 1986, External shapes, strain rates, and dynamics of salt structures: Geol. Soc. America Bull., v. 97, pp. 305–23.

JOHNSON, D. W., 1929, Geomorphic aspects of rift valleys: Internat. Geol. Cong., 15th, South Africa, 1929, Comptes rendus, v. 2, pp. 354–73.

JORDAN, T. E., and ALLMENDINGER, R. W., 1986, The Sierras Pampeanas of Argentina: A modern analogue of Rocky Mountain foreland deformation: Am. Jour. Sci., v. 286, pp. 737–64.

KELLEY, V. C., 1955, Monoclines of the Colorado Plateau: Geol. Soc. America Bull., v. 66, pp. 789–804.

LEES, G. M., 1955, Recent earth movements in the Middle East: Geol. Rundschau, v. 43, pp. 221–26.

LOUDERBACK, G. D., 1904, Basin range structure of the Humboldt region: Geol. Soc. America Bull., v. 15, pp. 289–346.

———, 1923, Basin range structure in the Great Basin: Calif. Univ. Pub. Geol. Sci., v. 14, pp. 329–76.

MACDONALD, G. A., 1972, Volcanoes: Prentice-Hall, Inc., Englewood Cliffs, N.J., 510 pp.

McCONNELL, R. B., 1972, Geological development of the rift system of east Africa: Geol. Soc. America Bull., v. 83, pp. 2549–72.

MENARD, H. W., 1964, Marine geology of the Pacific: McGraw-Hill Book Company, New York, 271 pp.

MORGAN, J. P., COLEMAN, J. M., and GAGLIANO, S. M., 1968, Mudlumps: Diapiric structures in Mississippi delta sediments: Am. Assoc. Petroleum Geologists Mem., no. 8, pp. 145–61.

MUEHLBERGER, W. R., CLABOUGH, P. S., and HIGHTOWER, M. L., 1962, Palestine and Grand Saline salt domes, eastern Texas, Field Trip No. 6, *in* Rainwater, E. H., and Zingula, R. P., eds., Geology of the Gulf coast and central Texas and guide book of excursions: Houston Geological Soc., Houston, Texas, pp. 266–77.

MURRAY, G. E., 1961, Geology of the Atlantic and Gulf Coastal Province of North America: Harper and Brothers, New York, 692 pp.

NASH, D. B., 1984 Morphologic dating of fluvial terrace scarps and fault scarps near West Yellowstone, Montana: Geol. Soc. America Bull., v. 95, pp. 1413–24.

NEEV, D., and EMERY, K. O., 1967, The Dead Sea: Israel Geol. Survey Bull. 41, 147 pp.

ORTLOFF, C. R., FELDMAN, R. A., and MOSELEY, M. E., 1985, Hydraulic engineering and historical aspects of the pre-Columbian intravalley canal system of the Moche Valley, Peru: Jour. Field Archeol., v. 12, 77–98.

OTA, Y., 1975, Late Quaternary vertical movement in Japan estimated from deformed shore-lines: Royal Soc. New Zealand Bull. 13, pp. 231–39.

PELTON, J. R., and SMITH, R. B., 1982, Contemporary vertical surface displacements in Yellowstone National Park: Jour. Geophysical Res., v. 87, pp. 2745–61.

PHILIP, H., and MEGHRAOUI, M., 1983, Structural analysis and interpretation of the surface deformations of the El Asnam earthquake of October 10, 1980: Tectonics, v. 2, pp. 17–49.

POWELL, J. W., 1875, Exploration of the Colorado River of the west and its tributaries. U. S. Govt. Printing Office, Washington, D.C., 291 pp.

QUENNELL, A. M., 1958, Structural and geomorphic evolution of the Dead Sea Rift: Geol. Soc. London Quart. Jour., v. 114, pp. 1–24.

ROCKWELL, T. K., KELLER, E. A., CLARK, M. N., and JOHNSON, D. L., 1984, Chronology and rates of faulting of Ventura River terraces, California: Geol. Soc. America Bull., v. 95, pp. 1466–74.

ROCKWELL, T. K., KELLER, E. A., and DEMBROFF, G. R., 1988, Quaternary rate of folding of the Ventura Avenue anticline, western Transverse Ranges, southern California: Geol. Soc. America Bull., v. 100, pp. 850–58.

ROSENDAHL, B. R., REYNOLDS, D. J., LORBER, P. M., BURGESS, C. F., MCGILL, J., SCOTT, D., LAMBIASE, J. J., and DERKSEN, S. J., 1986, Structural expressions of rifting: Lessons from Lake Tanganyika, Africa, *in* Frostick, L. E., Renaut, R. W., Reid, I., and Tiercelin, J. J., eds., Sedimentation in the African rifts: Geol. Soc. (London) Spec. Pub. no. 25, pp. 29–43.

ST. ONGE, D. A., 1959, Note sur l'érosion du gypse en climat périglaciaire: Rev. Canadienne de Geog., v. 13, pp. 155–62.

SHARP, R. P., 1939, Basin-range structure of the Ruby–East Humboldt Range, northeastern Nevada: Geol. Soc. America Bull., v. 50, pp. 881–920.

————, 1942, Stratigraphy and structure of the southern Ruby Mountains, Nevada: Geol. Soc. America Bull., v. 53, pp. 647–90.

SIEH, K. E., 1978, Prehistoric large earthquakes produced by slip on the San Andreas fault at Pallett Creek, California: Jour. Geophys. Res., v. 83, pp. 3907–39.

————, 1984, Lateral offsets and revised dates of large prehistoric earthquakes at Pallett Creek, southern California: Jour. Geophys. Res., v. 89, pp. 7641–70.

————, and JAHNS, R. H., 1984, Holocene activity of the San Andreas fault at Wallace Creek, California: Geol. Soc. America Bull., v. 95, pp. 883–96.

STEVENS, G. R., 1974, Rugged landscapes: The geology of central New Zealand: A. H. & A. W. Reed Ltd., Wellington, New Zealand, 286 pp.

STOTT, D. F., 1969, Ellef Ringnes Island, Canadian Arctic Archipelago: Canada Geol. Survey Paper 68-16, 44 pp.

SUGIMURA, A., 1967, Uniform rates and duration period of Quaternary earth movements in Japan: Osaka City Univ. Jour. Geosci., v. 10, pp. 25–35.

SYLVESTER, D. G., 1988, Strike-slip faults: Geol. Soc. America Bull., v. 100, pp. 1666–1703.

TAYLOR, F. W., ISACKS, B. L., JOUANNIC, C., BLOOM, A. L., and DUBOIS, J., 1980, Coseismic and Quaternary vertical tectonic movements, Santo and Malekula Islands, New Hebrides island arc: Jour. Geophysical Res., v. 85, pp. 5367–81.

TAZIEFF, H., 1970, The Afar triangle: Sci. American, v. 222, no. 2, pp. 32–40.

TE PUNGA, M. T., 1957, Live anticlines in western Wellington: New Zealand Jour. Sci. and Technology, sec. B., v. 38, pp. 433–46.

TWIDALE, C. R., 1968, Geomorphology with special reference to Australia: Thomas Nelson (Australia) Ltd., Melbourne, Australia, 406 pp.

U. S. GEOLOGICAL SURVEY STAFF, 1971, Surface faulting: The San Fernando, California, earthquake of February 9, 1971: U. S. Geol. Survey Prof. Paper 733, pp. 55–76.

WESSON, R. L., HELLEY, E. J., LAJOIE, K. R., and WENTWORTH, C. M., 1975, Faults and future earthquakes, in Borcherdt, R. D., ed., Studies for seismic zonation of the San Francisco Bay region: U. S. Geol. Survey Prof. Paper 941A, pp. 5–30.

WILLIS, B., 1928, The Dead Sea problem: Rift valley or ramp valley: Geol. Soc. America Bull., v. 39, pp. 490–542.

WILSON, J. T., 1965, A new class of faults and their bearing on continental drift: Nature, v. 207, pp. 343–47.

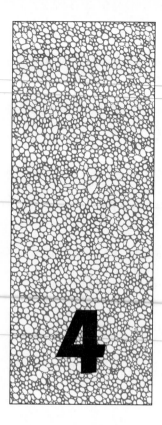

4 Volcanoes

Just as tectonic processes and landscapes can be systematically studied either in the temporal and spacial context of plate tectonics (Chap. 2) or as discrete landforms created by tectonic deformation (Chap. 3), so volcanoes can be studied either on a "megageomorphic" regional or planetary scale or as a special group of landforms built by a set of localized, dramatic, and often dangerous processes. In this chapter, the plate tectonic setting of volcanoes is described and explained. However, the emphasis is on volcanoes as constructional landforms: parts of the modern landscape that either are being built or were built recently enough for their constructional form to be recognized in spite of dissection. Volcanic activity ("eruptions," "volcanicity," and "volcanism" are synonyms) must be described, as must the products of eruptions, in order to understand the structure and genesis of volcanic landforms. Post-eruptive erosional development on volcanic structures is deferred and considered with other forms of structurally controlled erosion (Chap. 11). By this arrangement, full attention can here be given to volcanoes as regional assemblages of dramatic, often strikingly beautiful, constructional landforms, built by processes that have had legendary impact on human senses.

Volcanoes have always had an uncertain place in systematic geomorphology. W. M. Davis (1905) treated volcanism as an "accident" that occurs so arbitrarily in time and place and is so disruptive to the erosional development of landscapes that the landforms cannot be treated in a systematic manner. A common arrangement of modern geomorphology textbooks is to review both constructional and erosional landforms on volcanic structures in a single chapter late in the book. Some books define volcanoes as the vents or fissures from which hot subsurface materials are transferred to above the former ground surface. Others define volcanoes as the mountains or hills of erupted rock around the vents. Literally, then, there is no general agreement as to whether volcanoes should be defined as holes or hills! Cotton (1944, preface) summarized the issue: "An approach to volcanism may be made through either petrology or geomorphology, a volcano and its mechanism being regarded either as a geological crucible or as a builder of landforms." Information on volcanoes, their activity, and their products can be found in petrology books as well as in geomorphology books and descriptive monographs (Cotton, 1944; Rittmann, 1962; Simkin et al., 1981; Blong, 1984; Fisher and Schmincke, 1984; Short, 1986; Ollier, 1988; Decker and Decker, 1989).

It is generally understood that volcanoes are characterized by eruptive activity. That is, hot gases, liquids, molten rock, and shattered rock fragments are forcibly or rapidly ejected from openings in the earth's surface. Classification of volcanoes beyond that level of understanding has been based on a wide variety of criteria, such as (1) chemical composition and temperature of the ejecta; (2) physical state of the ejecta—whether a gas, liquid, or solid phase is dominant; (3) historic record of volcanicity—active, dormant, or extinct; (4) shape of the aperture— whether a central vent or linear fissure, for instance; (5) nature of the volcanic activity—exhalative gas discharge, effusive lava flows, explosive ejection of fragments, frothing of gas-charged liquid, fire fountains, and so on; or (6) shape of the resulting landforms—mounds, rings, craters, domes, cones, shields, or plateaus. For geomorphic analysis, both volcanic processes and landforms must be understood. For evaluating volcanic hazards, understanding the nature and variability of volcanic activity is absolutely essential.

GEOLOGIC AND GEOGRAPHIC SIGNIFICANCE OF VOLCANOES

Planetary Volcanology

More than twenty years of exploration of Mercury, Venus, Mars, and the Earth's moon has stimulated much contemporaneous study of terrestrial volcanoes. The smaller planets (Mercury, Mars) and our moon are probably "single plate" objects whose surfaces were shaped by meteorite impact and voluminous lava flows in their early history. The earth and probably Venus have active plate motions that recycle their crustal layers (Chap. 1). The entire oceanic crust is volcanic basalt and its intrusive equivalents, and is less than 250 million years old. Continental plateaus of flood basalts (p. 85) have been claimed to be terrestrial *maria* of solidified lava lakes that were formed by meteorite impact, in the same manner as the lunar maria (Alt et al., 1988). Another current hypothesis is that meteorite impact has caused periodic fracturing of the continental lithosphere by creating "hot spots" of upwelling magma at the impact sites (Rampino and Stothers, 1988). These provocative hypotheses imply that even as late as the Mesozoic and Cenozoic eras of earth history, terrestrial volcanism has been caused in part by the same processes that we infer for the other terrestrial planets and the moon (p. 103).

Plate Tectonics and Volcanism

According to less catastrophic hypotheses of terrestrial volcanism, the location of volcanoes is determined by large-scale motions in the earth's interior, involving both the crust and mantle. Cylindrical rising segments of convection cells or upwelling plumes, perhaps from deep within the mantle adjacent to the core–mantle boundary, probably determine the location of "hot spots" beneath the lithosphere (Loper et al., 1988; Richards et al., 1989; White, 1989). A chain of basalt volcanic islands and seamounts such as the Hawaiian Islands may be built as an oceanic lithospheric plate moves over the "hot spot," or a continent may be stretched and rifted with an outpouring of flood basalt. The downward or sinking limbs of mantle convection cells are more sheetlike or slablike, and are found beneath subduction

Figure 4-1. Size comparisons for some subaerial volcanoes, volcanic islands, and **guyots** (flat-topped, volcanic seamounts). No vertical exaggeration. (Menard, 1964, Fig. 4.8.)

zones on converging plate margins (Bercovici et al., 1989). Where slabs of relatively cool, dense, oceanic lithosphere sink to a depth of about 125 km in a subduction zone, andesitic magmas are generated to form volcanic arcs, either as part of an island arc on oceanic lithosphere (Fig. 2–6) or as part of a cordilleran or collisional orogenic belt (Fig. 2–7). The most abundant effusive volcanism, which accretes new oceanic basalt lithosphere at diverging plate margins along mid-ocean spreading ridges, is probably not generated by upwelling of deep mantle, but is the result of shallow separation of oceanic lithospheric plates under the pull of their sinking margins in subduction zones.

The annual production rate of volcanic ejecta is estimated at 6 to 8 km^3, of which 4 to 6 km^3 is basalt accreted at mid-ocean spreading ridges. Only minor amounts of this rock appear above sea level in islands such as Iceland and the Azores. Another 1 km^3 per year may be added to the submarine relief over oceanic hotspots, and lesser amounts are extruded in volcanic arcs, basalt plateaus, and continental rifts (Fisher and Schmincke, 1984, Table 1–4).

Abundance and Geographic Distribution of Subaerial Volcanoes

Even though only 10 to 20 percent of the annual volcanic output appears above sea level, and the volcanoes that we call mountains are small by comparison to their submarine counterparts (Fig. 4–1), volcanic contribution to the terrestrial landscape is significant. Entire geomorphic provinces are volcanic in origin. The great Cenozoic basalt plateaus of Iceland, India, and the northwestern United States are among the largest constructional relief features of the subaerial landscape (although exceeded greatly by the ice caps of Greenland and Antarctica). An estimated 2 million km^2 of land surface have been covered to depths of 500 to 3000 m by lava flows in geologically recent time (Rittmann, 1962, p. xiii). Probably the entire earth has been covered by layers of fragmental volcanic ejecta (ash and dust), as recorded by their frequent occurrence in deep-sea sediment cores. More than 500 volcanoes have historical records of eruptions, and over 1300 have probably erupted in Holocene time (Simkin et al., 1981). Tens of thousands of extinct volcanoes can be recognized by their form or structure. About 62 percent of the active subaerial volcanoes are on the subduction zones around the Pacific basin, in the circum-Pacific "ring of fire" (Fig. 4–2). Another 22 percent are in Indonesia, 10 percent are in the Atlantic Ocean (including the Caribbean region), and a few percent each are in Africa and the Mediterranean–Middle East. The remaining few form the Hawaiian Islands and other mid-ocean islands.

Volcanoes have attracted renewed geomorphic interest in recent years because new techniques of geochronometry permit precise dating of eruptive rocks. It is now possible, by measuring the volume of rock eroded from a dated volcanic landscape, to measure the absolute rates of landscape modification (Chap. 12). Because many volcanic landforms have symmetrical geometry, their original constructional shape can be restored with much greater confidence than is possible with tectonic constructional landforms. Furthermore, key marker beds of volcanic ash, dated by radioactive isotopes and chemically identified as to place of origin, can blanket millions of square kilometers of landscapes formed by other processes. It is hard to overstate the scientific value of these distinctively colored or otherwise identifiable ash beds that were inserted at an instant of geologic time into a stratigraphic sequence or onto an evolving landscape (pp. 78, 437).

A final point about the geomorphic importance of volcanoes concerns their wide latitudinal and climatic distribution. Volcanic landforms are constructed independently of any climatically controlled processes. Volcanic edifices are built in or on the Antarctic ice cap, in the tropical forests of Melanesia and Indonesia,

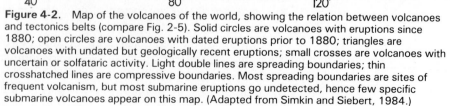

Figure 4-2. Map of the volcanoes of the world, showing the relation between volcanoes and tectonics belts (compare Fig. 2-5). Solid circles are volcanoes with eruptions since 1880; open circles are volcanoes with dated eruptions prior to 1880; triangles are volcanoes with undated but geologically recent eruptions; small crosses are volcanoes with uncertain or solfataric activity. Light double lines are spreading boundaries; thin crosshatched lines are compressive boundaries. Most spreading boundaries are sites of frequent volcanism, but most submarine eruptions go undetected, hence few specific submarine volcanoes appear on this map. (Adapted from Simkin and Siebert, 1984.)

in deserts, and in every other geomorphically significant climate. In each instance, the initial structure and form of the constructed landforms are similar. A great opportunity is thus available for geomorphologists to study contrasting erosional forms and rates in many climatic regions on constructed volcanic landforms that are not only similar but datable. This aspect of geomorphic volcanology alone justifies an extended systematic treatment of volcanic constructional landforms.

VOLCANIC ACTIVITY AND PRODUCTS

The products of eruption can be described as the results of certain types of activity, such as *exhalative* (gas), *effusive* (lava), or *explosive* (tephra). Alternatively, characteristic modes of activity and products can be defined by describing active volcanoes that behave in certain ways. Both devices are used in the following pages.

Exhalative Activity and Mud Volcanoes

Hot springs, geysers, fumaroles, solfateras, and many other terms are used to describe vents that continuously, intermittently, or cyclically discharge hot water, steam, and other gases but rarely any solid or molten rock. Such vents are not generally considered volcanoes even though their activity is similar to volcanic eruption. They build minor landforms such as *sinter mounds* and *cones* of precipitated minerals.

Mud volcanoes make impressive small landforms over exhalative vents. Several in the Copper River Basin of Alaska were described by Nichols and Yehle (1961). They are between 45 and 95 m in height and discharge mineralized warm water and gas, including light hydrocarbons probably derived from the decay of buried peat beds or coal. One of the largest cones contains abundant fragments of lava and volcanic glass that are typical of shallow hydrothermal eruptions but are unlike the glacial drift that forms the surficial cover of the region. Travertine crusts intermingle with the hardened mud and angular rock fragments of the crater rims. Such cones are probably the minimum accepted "volcanic" landform to most geomorphologists.

In 1951 a spectacular mud volcano in northeastern California suddenly erupted steam, gases, and mud to an estimated height of 1000 m (White, 1955). Fine debris, including pellets of mud, fell on a town and farms 7 km downwind from the eruption. The entire episode ended within four days. After a few months of quiescence, probably with some ground subsidence, the area was flooded with boiling water and mud springs. This kind of mud volcano draws its energy from volcanic heat, but most of the water seems to be of near-surface origin. If pressure is suddenly reduced at the top, by chance overflow of a pool, for instance, the confining pressure is decreased throughout the entire water column. A feedback cycle of steam generation ensues exactly as in a geyser eruption, except that overlying fine-grained sediment is liquefied and erupted as hot mud, along with blocks of indurated bedrock.

Identical mud-cone landforms are built by an entirely nonvolcanic process. In orogenic regions, where methane-rich, plastic, organic muds accumulate rapidly and are rapidly buried, gas and water may liquefy the mud and cause it to flow under tectonic pressure from vents or fissures along faults (Fig. 4–3). The best known examples are on the Makran coast of Pakistan, 100 km west of Karachi (Snead, 1964). These cool-mud volcanoes, and others in comparable areas of hydrocarbon-rich sediment accumulation, are genetically more related to the mud lumps of the Mississippi delta (p. 57) than to volcanic landforms.

Figure 4-3. Dormant mud volcano, about 50 m high, on the south flank of the larger Chandragup mud volcano on the Arabian Sea coast of Pakistan. (Photo: R. E. Snead.)

Effusive Activity: Lava

Lava is the name given to erupted molten rock and to the subsequently cooled, solid equivalent, provided that it has an obvious flowlike surface form. More often, solidified lava is simply called "volcanic rock," along with solidified ejecta, because the fine grain sizes or glassy textures of eruptive rocks may not permit a clear conclusion about their origin. **Effusive** (as contrasted to explosive) activity refers to copious outpourings of lava from a vent or fissure.

Most lavas are molten silicates, although silica-free carbonate melts are known, especially in the rift systems of eastern Africa (Krafft and Keller, 1989). A lava flow of elemental sulfur in Japan has been commercially mined. The ordinary range of SiO_2 in volcanic rocks is from about 35 percent to about 75 percent. The viscosity of natural silicate melts ranges through 13 orders of magnitude. At any given temperature, acidic (silica-rich) lavas are more viscous than basic (silica-poor) lavas (Fisher and Schmincke, 1984, p. 53; Ryan and Blevins, 1987). The composition of the molten phase is a major factor in landform development because of the control it exerts on viscosity. Low-silica Hawaiian basalt freely flows down slopes of only a few degrees. The resulting landforms are likely to be low compared to their lateral extent, with smooth, domal form. On the other extreme, silica-rich dacite and rhyolite may not flow from a vent but form a protruding plug or spine.

Temperature and dissolved volatiles are the two other factors that determine the viscosity of lava. Confining pressure alone normally elevates the melting temperature of solids, but when water and other volatile components such as CO_2 and SO_2 are dissolved in a silicate melt, the melting temperature is lowered. Although acidic magmas can dissolve more volatile compounds than basic magmas, and thereby be highly mobile at depth, they become highly viscous at surface pressure, when the volatiles have boiled off. In the process of degassing, they become violently explosive.

Surface pools and flows of Hawaiian basalt have a surprisingly narrow range of temperatures, between 1100° and 1200° C. The viscosity increases by a factor of 10 with cooling of only 100° C (Ryan and Blevins, 1987, pp. 458–9). On cooling to a temperature of below 1150° C, some of the minerals in a silica-poor melt begin to crystallize, and the viscosity increases rapidly with only slightly lower temperature. The heat of fusion is probably a factor that keeps most lava flows in the relatively narrow temperature range noted. These considerations of temperature, volatile pressure, and composition help to clarify why effusive activity is almost totally confined to silica-poor, or basic, melts. By far the most characteristic effusive lava is *basalt,* although intermediate-composition lavas such as andesite and dacite also form stiff, rough, slow-moving flows.

Lava emerges from vents or fissures as white-hot lobes or tongues that flow downhill (Fig. 4–4). Each lobe or tongue is called a **lava flow.** Hawaiian lava flows generally advance at slower than human walking speed (3 to 5 km/h), but double or triple that rate on steep slopes. Lava streams in open channels on steep slopes may reach speeds of 55 km/h. They range in thickness from less than 1 m to 20 m (Mullineaux et al., 1987, p. 601).

Lava flows follow preexisting topography downhill and may completely fill river valleys. On more open slopes, they can be diverted by walls or dikes. A flow front may congeal and stop when a lateral lobe breaks out higher upslope and follows an easier path.

The upper surface of a basalt lava flow is usually highly *vesicular* or porous due to gas bubbles that escape during the final stages of cooling. If sufficiently "clinker-like" in texture, the rock is called *scoria.* Two contrasting surface textures develop on flows, both commonly known by the traditional Hawaiian terms: *aa* for angular, blocky, scoriaceous surfaces, and *pahoehoe* (or ropy) for smoothly twisted, convolute surfaces (Fig. 4–5). Pahoehoe surfaces develop on hotter, more fluid, flows. An extreme variant of pahoehoe or ropy lava occurs when hot, fluid lava flows into or erupts under water. Blobs or lobes of lava up to a meter in diameter form steam-jacketed, tough, flexible skins and pile up like pillows or sandbags while still molten in their interiors. The result is *pillow lava* (Fig. 4–6), a diagnostic feature of subaqueous eruption and because much of the oceanic crust is composed of it, probably the most abundant rock type on earth (Moore, 1975).

Repeated observations confirm that pahoehoe flows remain hot and fluid inside even though their top and bottom surfaces and lateral margins have solidified. One result of fluid lava becoming encased in cooler margins is the formation of *lava tubes.* These caves, many of which are several kilometers in length and large enough to be entered, mark the draining out of the molten lava from within a solidified surface. As they drain, air may enter at their lower end and, escaping somewhere

Figure 4-4.   Active lava flows, Mauna Loa, Hawaii. (Photo: Hawaiian Volcano Observatory, U.S. Geological Survey.)

Figure 4-5. Pahoehoe lava draped over a former sea cliff south of Kilauea Volcano, Hawaii. (Photo: D. A. Swanson, U.S. Geological Survey.)

Figure 4-6. Submarine lava pillows in the median valley of the Mid-Atlantic Ridge. Pillows show effects of repeated surface quenching during expansion. (Photo: R. D. Ballard.)

higher upstream, set up a convective "blast furnace" condition that glazes the interior walls and creates stalactites of melted rock (Greeley and Hyde, 1972).

Rough aa surfaces on flows result from the cooled and broken surface layer being carried forward by laminar flow and then being "plowed under" at the front of the advancing lava. The motion is analogous to that of the endless tread on a tracked vehicle. *Blocky lava* is sometimes distinguished from aa surfaces on the basis of a slablike or blocky character rather than the extremely scoriaceous, spinose, or "clinkery" aa texture.

Lava flows, in common with intrusive dikes and sills and thick sheets of fused tephra, develop distinct sets of intersecting contraction joints upon cooling that result in the familiar *hexagonal* or *columnar* jointing (Fig. 11–1). These primary structures exert powerful controls on the landforms eroded from volcanic terranes (Chap. 11).

Explosive Activity: Tephra

Explosive volcanism along converging plate boundaries supplies only 10 to 13 percent of the magma that reaches the earth's surface annually, but includes 63 percent of all known Holocene volcanoes and 84 percent of all historical eruptions (Simkin and Siebert, 1984, p. 110). To classify the "bigness" of historic volcanic eruptions, a *volcanic explosivity index* (VEI) was devised (Simkin et al., 1981, pp. 20–21; Newhall and Self, 1982). The VEI combines the volume of ejecta, height of the ejecta and gas cloud above the volcano, descriptive terms, and other criteria on a simple scale of 0 to 8 of increasing explosivity (Table 4–1). More than 4800 historical eruptions were used to compile the index. A voluminous lava flow on Hawaii might be locally important, yet have an explosivity index of 0. Nevertheless, the VEI scale is the best available measure of the relative size of 80 percent of the world's known eruptions.

Obviously, explosive volcanism has a human impact far out of proportion to its role in shaping the earth's surface. Explosive eruptions are potentially capable of killing millions of people and devastating thousands of square kilometers. They may give little warning. Of 205 well-documented historical eruptions with a VEI ≥ 3, selected from the list of 4800 known events (Table 4–1), 45 percent experienced the major eruption within one day of the first indication of activity, and half of the first-day eruptions occurred within the first hour (Simkin and Siebert, 1984, p. 112). Active magma chambers under large calderas such as Yellowstone, Wyoming and Long Valley, California are now monitored for advance warnings of possible eruptions. Neotectonic movements (Fig. 3–21) are important clues to possible eruptions, as are certain kinds of rhythmic seismicity, increased heat flow, and fumarolic activity. Unfortunately, the largest explosive volcanoes have the longest intervals between eruptions, typically hundreds to thousands of years, and people may forget or be unaware that they are living in a hazardous region.

In an explosive eruption, rock may be fragmented by a sudden, foaming loss of volatiles, by abrupt contact with groundwater or surface water, by grinding abrasion in the volcanic neck, or by dispersion of molten droplets. Volcanic ejecta that settle out of air or water are called *pyroclastic* or *volcaniclastic* sediments or rocks. **Tephra** is a less cumbersome collective term for all fragmental ejecta from volcanoes. It is derived from the Greek word for ash and is intended to be functionally equivalent to lava and magma (Thorarinsson, 1951, p. 5; Fisher and Schmincke, 1984, p. 89). No connotation of grain size is intended or implied, although the related subscience of **tephrochronology** is often thought of as the use of ash beds in stratigraphic correlation and dating, ignoring the coarser grades of ejecta. Tephrochronology has many applications in geomorphology, stratigraphy, and archeology.

Tephra consists of fragments with a wide range of grain sizes and shapes. *Ash* is the most common term for sand-size and finer tephra. Larger fragments or aggregates are called *lapilli* (gravel size; ejected either molten or solid), *blocks* (cobble or boulder-sized solid ejecta), or *bombs* (twisted, air-cooled masses ejected in a molten

TABLE 4-1

Criteria for Estimation of the Volcanic Explosivity Index (VEI)

Criteria	VEI: 0	1	2	3	4	5	6	7	8
Description	non-explosive	small	moderate	mod-large	large	very large →			
Volume of ejecta (m³)	$<10^4$	$10^4–10^6$	$10^6–10^7$	$10^7–10^8$	$10^8–10^9$	$10^9–10^{10}$	$10^{10}–10^{11}$	$10^{11}–10^{12}$	$>10^{12}$
Column height (km)*	<0.1	$0.1–1$	$1–5$	$3–15$	$10–25$	>25 →			
Qualitative description	← "gentle, effusive" →		← "explosive" →		← "severe, violent, terrific" → "cataclysmic, paroxysmal, colossal" →				
Classification	← "Hawaiian" →		"Strombolian" ← "Vulcanian" →		← "Plinian" →		"Ultraplinian" →		
Duration (hours) of continuous blast	← <1 →		← $1–6$ →		← $6–12$ → >12 →				
Tropospheric injection	negligible	minor	moderate	substantial →					
Stratospheric injection	none	none	none	possible	definite	significant →			
Eruptions (total in file)	443	361	3108	720	131	35	16	1	0

Source: simplified from Newhall and Self, 1982, Table 1.

* For VEI's 0–2, uses km above crater; for VEI's 3–8, uses km above sea level.

state) (Fisher and Schmincke, 1984, pp. 89–96). The terminology is elaborate and extensive, because explosive volcanic eruptions are among the most feared and discussed phenomena on earth.

Tephra is strongly size-sorted during air transport. Bombs and blocks fall only as far from the vent or fissure as the explosive force is able to hurl them, rarely more than 5 km from the vent (Blong, 1984, p. 25). Lapilli and ash are airborne for many kilometers, the grain size and thickness of the resulting tephra layer decreasing with increasing distance from the source (Porter, 1972; Fisher and Schmincke, 1984, pp. 125–62).

A **nuée ardente,** or glowing cloud, is the most terrifying form of explosive volcanic activity. In May 1902 a nuée ardente from Mt. Pelée, Martinique, killed 30,000 people in the city of St. Pierre. The eruption, the first of its kind to be scientifically studied, gave the name Peléan to such peculiarly explosive activity (p. 80). The phenomenon can be produced either by a lateral, or directed, eruption of superheated steam and glowing hot tephra or by "boiling over" of foaming lava. Beneath the towering, glowing cloud of hot, fine tephra and gasses, the leading edge of a nuée ardente may speed downhill at 50 m/s. The base of the fluidized mass is heavier than air, so it remains in contact with the ground and flows in a turbulent fashion as a pyroclastic flow.

Volcanologists emphasize that pyroclastic flows beneath nuées ardentes are soundless, evidence that despite their forward velocity comparable to the winds of a hurricane, the glowing fragments in them are cushioned by hot gas and are not abraded by the rapid, turbulent motion. The temperature in the nuée ardente that devastated St. Pierre is estimated to have been between 650 and 1060° C.

Most of the volume of a nuée ardente is hot gas, primarily superheated steam. The tephra accumulation over the ruins of St. Pierre was less than a meter in thickness, generally only about 30 cm. A nuée ardente in the Philippines charred trees to a height of 4.5 m, clearly marking the thickness of the flowing layer, yet left only 13 mm of ash on the ground (Macdonald, 1972, p. 5). On cooling, the steam condenses to make scalding hot mud of the tephra. Although the cloud of airborne tephra that towers above and obscures the fast-moving ground layer is technically the nuée ardente, the basal pyroclastic flow is responsible for the mass of the resulting deposit (Fisher and Schmincke, 1984, p. 187).

Pyroclastic flows are believed to be responsible for some of the largest constructional volcanic landforms. The ejecta is hot enough to fuse into a layer of porous volcanic rock. These are the *ignimbrite* or *welded tuff* sheets that cover thousands of square kilometers in New Zealand, the Yellowstone Plateau, Katmai National Monument in Alaska, the Lake Toba region of northern Sumatra, the Altiplano Plateau of the central Andes, and elsewhere (p. 91). The ignimbrite may be hundreds of meters thick.

A variety of explosive volcanic activity has been recorded by direct observations of relatively small eruptions. In the eruption of Mount St. Helens, Washington, on May 18, 1980, the north flank of the volcano swelled, oversteepened and collapsed as an avalanche. The sudden pressure drop in the central vent led to a *directed blast,* a violent lateral explosion that flattened forests as far as 25 km from the vent (Fig. 4–7) (Lipman and Mullineaux, 1981). A much larger flank collapse about 7500 years ago on Socompa volcano in northern Chile covered about 600 km² with avalanche debris (Francis and Self, 1987). By contrast, an eruption of Cotapaxi volcano in Ecuador involved a gas-charged pumice froth, like the "foam from a boiling-over rice pot" (Fisher and Schmincke, 1984, p. 187). The eruption of Nevada del Ruiz, Colombia, on November 13, 1985 was the culmination of almost a year of continuing explosive activity. In the final explosion, relatively small hot pyroclastic flows on the glaciated summit melted part of the summit glacier, which

Figure 4-7. Major kinds of volcaniclastic deposits, excluding airfall ash, from the May 18, 1980 eruption of Mount St. Helens, Washington. (After Lipman and Mullineaux, 1981, Plate 1 and Fig. 219.)

generated rapidly moving **lahars,** or volcanic mudflows, that destroyed numerous villages along lower river valleys with the loss of 25,000 lives (Naranjo et al., 1986).

The largest historical eruption, and the largest known eruption during the past 10,000 years, was the explosion of Tambora volcano in Indonesia in A.D. 1815 (Self et al., 1984; Stothers, 1984). The volcanic cone more than 1400 m in height was totally destroyed. The earth-circling dust cloud caused spectacular sunsets for months after the eruption and may have been responsible for the colder-than-normal temperatures recorded in the northern hemisphere during the following two years. Another large explosive eruption occurred when the Greek island of Thera (Santorini) exploded in about 1450 B.C., covering the eastern Mediterranean basin with pumice and ash (Stanley and Sheng, 1986; Newhall and Dzurisin, 1988, pp. 129–34). The event has been blamed for the decline of the Minoan culture, the darkness and plagues in Egypt during the exodus of the Jewish people, and the legends of lost Atlantis.

Less thick than pyroclastic flows and avalanche deposits but even more widespread are *ash showers* that result from some explosive eruptions. Explosive force cannot propel tephra more than a few kilometers, so wind must be the transporting

Figure 4-8.   Thickness map of air-fall ejecta from Mount St. Helens on May 18, 1980. Lines represent uncompacted thickness in mm. +, light dusting of ash; x, no ash observed; circles, observation sites (Modified from Sarna-Wojcicki et al., 1981.)

force for widespread ash showers. They are cold when they reach the ground. When a tephra sheet is mapped in detail (Fig. 4–8), the direction of prevalent wind during the eruption is obvious.

Widespread ash showers are the great tools of tephrochronology. Individual ash beds can be identified by a variety of analytical methods, including the refractive index of the glass shards, chemical analyses, especially of the rare earth elements, and mineralogy. Ash from a huge eruption of Toba volcano in Sumatra about 75,000 years ago has been identified in numerous submarine cores in the eastern Indian Ocean at distances of 500 to 2100 km from the volcano, and in a sample from India, 3100 km from the volcano (Rose and Chesner, 1987).

Tephrochronology has become an important tool for measuring rates of constructional and erosional landscape development, especially in the volcanic terranes of Japan, New Zealand, Iceland, and western North America. One example from Japan suggests the potential of the method. West of Tokyo, beds of lapilli and ash totaling 100 m in maximum thickness comprise the Older Fuji Tephra, which was deposited by hundreds of ash falls over a long period of time, probably between 25,000 and 10,000 years ago (Machida, 1967). It contains preceramic artifacts in its upper part. The Younger Fuji Tephra, which also accumulated during multiple ash falls, is 6 m thick in the vicinity of Mt. Fuji but less than 1 m thick near Tokyo, 80 km to the east (Fig. 4–9). It overlies archeologic sites with pottery, indicating a young age. The uppermost tephra layer is known to date from historical eruptions since A.D. 781. The weathered zone, fine ash, and soil between the Older and Younger Fuji Tephras shows that the latest historic activity of Fuji began after a long period of quiescence. These widespread tephra layers are used to measure tectonic displacement on faults, warping of coastal terraces, subsidence in the metropolitan Tokyo area, and many other tectonic movements of geomorphic interest. They have also been used to study rates of soil formation and for dating archeological sites. In Arctic Canada, floating pumice 5000 years old marks a distinctive emerged beach from which postglacial isostatic uplift can be accurately measured.

Figure 4-9. Isopach map of the Younger Fuji Tephra Formation (solid contours) and the Older Fuji Tephra Formation (dashed contours). Locations of measured sections are shown by numbered crosses. (Machida, 1967.)

Subaqueous Volcanism: Hyaloclastite

Pillow lavas, an extreme form of pahoehoe surface texture, have been noted as one diagnostic feature of subaqueous eruptions of fluid lava. They are especially common on the deep ocean floor, where basalt is the only eruptive rock, and where the high hydrostatic pressure prevents degassing of the lava and minimizes fracturing (Fig. 4–6).

More viscous lavas, and those erupted at lesser depths, develop shattered glassy margins on pillows and flow surfaces. The related volcanic product is **hyaloclastite** (literally: glassy-fragment rock). Ancient hyaloclastites are common in association with pillow lavas and deep-sea sedimentary rocks and are presumed common on the seafloor today. Hyaloclastites are also common in Iceland, where basalt eruptions under glaciers or through glaciers have been frequent. In Marie Byrd Land of Antarctica, hyaloclastites make up significant proportions of several volcanic peaks that protrude through the ice sheet. The volcanic rocks range in age from 36 million to 0.5 million years (Oligocene to Pleistocene). The abundance of hyaloclastites in the volcanic rocks demonstrates that a thick ice sheet has been present in West Antarctica during most of the late Cenozoic Era (LeMasurier, 1972).

Characteristic Eruptive Types

It has been the tradition of volcanologists to use the names of certain well-known volcanoes to classify characteristic modes of eruptive activity (Table 4–1). As no two volcanoes are alike, such a classification could become nothing less than a description of all known eruptions. A summary of eruptive types, based on about 8 named volcanoes or other widely recognized modes of activity, is widely used for descriptions (Short, 1986; Ollier, 1988). Only the four most traditional "typical" eruptive types and a few of their major variants are listed here to illustrate this alternative method of classifying volcanic activity. None of the named volcanoes erupt with only the activity described, however.

Hawaiian Eruption. The simplest, most predictable, and least dangerous type of eruption to study is the effusive outpouring of basalt lava from craters, lava lakes, or fissures. The VEI is 0 to 1 (Table 4–1). Individual flows are on the order of

10 m thick, and either spread widely over open slopes or pour down valleys as lava rivers. Little gas or tephra is produced. The great basalt plateaus of the Columbia Plateau and Iceland are large-scale examples of Hawaiian activity.

Strombolian Eruption. The persistent lava-fountain activity of Stromboli, reflected at night as a red glow on the underside of a towering steam plume, has caused the volcano to be known for milleniums as the "lighthouse of the Mediterranean." Strombolian activity is similar to Hawaiian, except that the somewhat more viscous lava is ejected upward in fountains from a lava lake in the crater at regular, rhythmic intervals of about 15 minutes, and flows are unusual. The VEI ranges from 1 to 3 (Table 4–1). The continued source of heat for maintaining the liquid pool of lava seems to be bubbles of burning gas that rise freely through the melt. The vapor cloud over the volcano is white, showing that no tephra is generated.

Vulcanian Eruption. Vulcano, an island north of Sicily, was named for the Roman god of fire and smelting and in turn gave its name (with modified spelling) to the entire range of volcanic eruptive activity and landforms. Vulcanian activity, based specifically on observations of the 1888–1890 eruption there, is explosive, with a VEI of 2 to 4. The crater fills with viscous lava, but rather than remaining molten as in nearby Stromboli, the lava solidifies and is explosively ejected as a great cauliflower cloud of dark tephra. Bombs, blocks, and lapilli shower the surrounding area. Pumice is a common component of the ejecta. Only minor lava flows result. Vulcano is known to have erupted explosively at least 12 times between the fifth century B.C. and the 1888 eruption. After each eruptive cycle, the volcano was dormant for decades or centuries.

An awesome feature of Vulcanian eruptions is the brilliant lightning flashes that play in the cloud or between it and the ground. The friction of tephra particles generates great electrical discharges. Thunderstorms are generated by the heat, and torrential muddy rainfall is an added geomorphic feature of this type of activity. Some volcanologists consider Vesuvius a better example of Vulcanian activity than Vulcano. A narrative account of the Vesuvius eruption of A.D. 79 by the Roman scholar Pliny is widely quoted and an extreme Vulcanian or Vesuvian eruption is often called a **Plinian** eruption. The popular image of a volcanic eruption is surely the Vulcanian or Plinian type. The VEI for Plinian eruptions has no defined upper limit.

Peléan Eruption. The nuées ardentes of Mount Pelée, Martinique; Mount St. Helens, Washington; and elsewhere were described in a preceding section. The activity is violently explosive, the result of very viscous, gas-rich, silicic lava plugging the vent and either frothing violently over the crater rim or breaking out laterally. An explosion of the Peléan type differs from a Vulcanian eruption in that the very hot gas and lava mixture is not carried skyward by the updraft to become cold tephra but collapses and spreads downslope as a nuée ardente, continuing to evolve gas that cushions the flowing fragments. In a closely related process, a stiff lava plug may choke the vent and force a lateral, or directed, explosion of the gas and lava emulsion.

Another variant of Peléan activity was displayed by a 1912 ash eruption near Katmai volcano in Alaska, which buried the Valley of Ten Thousand Smokes. In that eruption, glowing ash flows welled out of numerous fissures nearly simultaneously. The fluidized ash filled the valley floor, burying water-saturated alluvium. The numerous fumaroles and hot springs that subsequently formed gave the valley its new name. Some authors give a unique name, *Katmaian,* to such massive explosions of tephra. The widespread occurrence of ignimbrite sheets testifies to the importance of Peléan, or Katmaian, activity.

It will be noted that the fourfold classification above can be consolidated to a twofold classification of effusive, or lava, volcanoes, and explosive, or tephra, volcanoes (Rittmann, 1962, p. 117). Not included in the fourfold classification are violent gas or steam explosions that erupt only shattered old rock, with little or no inclusion of new lava or tephra. These explosions are called either *Ultra-Plinian* (Table 4–1) because their violence exceeds the Plinian type, or *phreatic,* because they are caused by groundwater or seawater entering the magma chamber below a vent. The enormous 1883 eruption of Krakatoa, in Indonesia, was a major phreatic eruption, as was part of the 1815 Tambora eruption.

VOLCANIC LANDFORMS

Having extensively described and classified the types of volcanic activity and the products of volcanic eruptions, we now turn to the explanatory description of constructional volcanic landforms. The viscosity of the magma is the major factor that determines the type of eruptive activity and the products that are erupted. (Viscosity, of course, is a parameter that is determined by the several more fundamental properties of composition, temperatures, and dissolved-gas content of the magma rising within the feeder conduits.) Viscosity and the size of the various edifices can be combined as the basis for a classification of volcanic landforms (Table 4–2; Fig. 4–10).

The vertical component of Table 4–2 is the viscosity of the magma, or as an equivalent, the type of activity (effusive, mixed, or explosive). For a horizontal component, the table uses the volume of erupted matter. Increasing quantities of lava and tephra do not simply produce larger constructional landforms of the same shape. The intriguing concept of *morphological capacity* was introduced by Rittmann (1962, p. 113) to account for changes in landform with increasing size. *Scoria cones* (Fig. 4–11), for instance, can only grow to a certain size before they can no longer support the column of lava rising in the throat of the volcano. Then flows break out on the flank or even from beneath the base of the cone (Fig. 4–12) and subsequent growth of the cone follows a different trend. The change in the relative importance of specific volcanic landforms as the total edifice grows larger is similar to the biologic concept of *allometry,* the study of changing body proportions as an organism grows.

In the following paragraphs, the forms that are more common and more significant to a general explanatory–descriptive scheme are described in a sequence beginning with the forms built by the least viscous, effusive lava eruptions, and continuing to consideration of the forms built by extremely viscous lava and explosive tephra eruptions. Within each eruptive type, a sequence of forms from small to large is reviewed. Most of the small landforms result from a single, or **monogenetic,** eruption.

Basalt Flows, Domes, and Shield Volcanoes

The basic geomorphic unit of a lava eruption is the *flow,* described previously. Flows may reduce the preexisting relief by filling valleys or may increase it by building elongate, flat-topped lobate sheets and tongues of lava on top of the previous landscape (Fig. 4–4). Their edges and upper surfaces may have pahoehoe or aa texture or may be a jumble of lava blocks. On the flow, a variety of small landforms may be found, such as *tumuli,* formed by small-scale laccolithic swellings on a pahoehoe surface, *pressure ridges, trenches* from collapsed lava tubes, and *spatter cones* or *hornitos,* built where secondary gas eruptions from within or beneath the

Figure 4-10.   Schematic block diagrams of simple central volcanoes. The blocks are arranged according to the classification scheme of Table 4-2 except that the largest volcanic structures are not represented. (From Rittmann, 1962, Fig. 57.)

TABLE 4–2

Classification of Volcanic Landforms

Quality of Magma	Type of Activity	Quantity of Eruptive Material Small ← → Great				
Fluid, very hot, basic	Effusive	Lava flows	Exogenous domes	Basalt plateaus and shield volcanoes		
					Icelandic	Hawaiian
Increasing viscosity, gas content, and silica percentage	Mixed	Scoria cones with flows	Composite cones or Strato-volcanoes		Volcanic fields with mutiple cones	
		Loose tephra cones and thick flows				
		Endogenous domes (plug domes, tholoids, spines)	Ruptured endogenous domes with thick lava flows			
Viscous, relatively cool, acidic	Explosive	Maars of tephra	Maars with ramparts	Collapse and explosion calderas	Ignimbrite sheets	
Extremely viscous, abundant crystals	Explosive, mostly gas	Gas maars	Explosion craters			

Source: Simplified from Rittmann (1962), Tables 4 and 5.

flow throws up small cones of lava clots (Fig. 4–12). *Lava blisters* are created by steam pockets beneath thin flows that have spread over marshy ground. Thorarinsson (1951, pp. 63–73) reviewed the origins of these minor landforms and introduced a new term, *pseudocraters,* for lines or clusters of explosion craters along valley-filling lava flows. He demonstrated that the crater fields and lines of Iceland are not vents along the fissures that fed the flows but result from steam eruptions when flows spread over water-saturated alluvium.

Effusive flows from a cental vent in flat terrain may build a low dome over the vent. A dome will develop only if slopes are nearly flat because of the low viscosity of basalt flows. These domes built of successive flows piled around eruptive vents are called **exogenous domes,** to distinguish them from **endogenous domes,** or cumulodomes of viscous lava that swell within a skin or crumbly encrustation of obsidian and pumice by slow, continued extrusion of highly viscous lava (Cotton, 1944, pp. 158–60). The latter forms are described on pages 87–90.

Exogenous domes grade upward in size to become **shield volcanoes.** No definitive size criteria have been suggested. Shield volcanoes of the Icelandic type are between 100 and 1000 m in elevation and have a base diameter about 20 times their height. Surface slopes on shield volcanoes average only a few degrees, although slopes of 20° may surround the summit crater. Late-stage caldera subsidence may flatten the summit profile. A genetic sequence for the Icelandic-type shield volcanoes in the Galápagos Islands has been deduced by Nordlie (1973).

Figure 4-11.   Scoria cones partly burying a lava flow, Sheeptrail Butte, Craters of the Moon National Monument, Idaho. (Photo: U.S. National Park Service.)

Shield volcanoes of the Hawaiian type rank with basalt plateaus and ignimbrite sheets as the largest volcanic landforms. They are not simply scaled-up Icelandic shield volcanoes or exogenous domes; again, the morphologic capacity of the domal form is involved. When a lava dome has grown to a critical size by central eruption, it is no longer strong enough to support a central lava lake or carry a central conduit up to the summit crater. Flank eruptions break out, sometimes along the normal faults that encircle the growing mass (Fig. 3–2), and the dome grows peripherally with only slow increase in height. Subaerial slopes range from only 3 to

Figure 4-12.   Small spatter cone (hornito) over a partly collapsed lava tube, breached by a lava flow that emerged from its base. North flank of Mauna Loa Volcano, Hawaii. (Photo: J. P. Lockwood, Hawaiian Volcano Observatory, U.S. Geological Survey.)

Figure 4-13. View northward across the summit of Mauna Loa shield volcano, including Mokuaweoweo Caldera and South Pit (foreground). (Photo: U.S. Navy, courtesy D. P. Cruikshank.)

10°; somewhat steeper submarine slopes range up to 20° (Peterson and Moore, 1987, p. 161). The crest of the dome becomes a broad, convex-upward plateau around a rimless *pit crater,* or a *summit caldera* (Fig. 4–13). Because of hydrostatic balance in the internal magma chambers, a pit crater may temporarily hold a lava lake. When the lake drains down due to a flank eruption, the crater rim subsides as giant slump blocks along concentric normal faults.

Hawaiian-type shield volcanoes are enormous in size (Decker et al., 1987). Mauna Kea and Mauna Loa, the two major shields of the five that merge to form the island of Hawaii, have summit elevations of 4206 m and 4175 m above sea level. More significantly, they rise smoothly from a submarine base about 400 km in diameter, and the total height of the multiple-shield volcano is about 10 km (Fig. 4–1). Recent submarine surveys have demonstrated that very large submarine landslides have occurred throughout the submarine and subaerial phases of Hawaiian shield volcano growth (Moore et al., 1989). Slumps and debris avalanches (Chap. 8) covering hundreds of square kilometers surrounded the growing shields and were incorporated into their structure.

Basalt Plateaus

During various intervals of geologic time, continental regions measuring from 0.2×10^6 to 2×10^6 km² have been engulfed by massive eruptions of fluid basalt, some exceeding 1 km in thickness. These *flood basalts* are probably caused by mantle "hot spots" beneath continents, although meteorite impacts have also been invoked (p. 65). Many of them are now near the margins of continents, and were emplaced at the time of a continental breakup (Macdougall, 1988, p. 388; Richards et al., 1989; White, 1989). Central vents are not responsible; rather, *fissure eruptions* of regional extent provide the lava. Each fissure eruption is monogenetic. Dikes can be seen that cut individual flows but rarely are continuous through the entire thickness of multiple flows. The former land surface bows down beneath the accumulating sheets of basalt, either because of isostatic compensation or because of the enormous volume of magma that is erupted.

The result of such massive lava effusions are plateaus of considerable size and monotonous structure. The largest such area in the United States is the 164,000 km² of the Columbia Plateaus in the states of Washington and Oregon. The Columbia Plateau flood basalts define an entire geomorphic province of the United States (Fig. 1–6). A total volume of 170,000 km³ of basalt erupted during the Miocene Epoch, between 17 million and 11 million years ago. The most voluminous eruptions were within a brief interval of about 2 million years and accounted for 85 percent of the total volume (Mangan et al., 1986; Hooper, 1988). Adjacent to the south is the smaller Snake River Plain, with an area of 52,000 km² extending across southern Idaho and merging into the Columbia Plateau. The basalt flows of the Snake River Plain are of Quaternary age, although no historic eruptions are known. These basalts overlie 1000 m of Pliocene ignimbrite along an elongate structural depression called the Snake River downwarp (Malde and Powers, 1962; Eaton et al., 1975).

The largest known flood basalt region is the Deccan district of northwestern India, which now covers 500,000 km² and originally may have been three times as large (Mahoney, 1988, p. 152). The Deccan flood basalts were emplaced within an interval of 2 million years between 65 and 69 million years ago, and most of the eruptions may have occurred within an even shorter interval of 0.5 million years (Richards et al., 1989, p. 103). This huge volcanic event was at the time of the Cretaceous-Tertiary geologic boundary when widespread extinctions occurred. It is unusually well dated and figures strongly in debates about dinosaur extinctions, meteorite impact, and the mechanisms of plate tectonics (Hallam, 1987; White, 1989). Similar basalt plateaus of recent enough age to form constructional landscapes are present in southern Brazil, Manchuria, and central Siberia, and in fragments around the North Atlantic basin on Greenland, Iceland, Ireland, the Faeroe Islands, and Jan Mayen.

Except for exogenous domes or minor scoria mounds, the constructional morphology of a basalt plateau is extraordinarily flat. Flows have initial slopes of 1° or less, barely detectable to the eye. The impression is created that the entire regional landscape is the surface of a single lava flow, which is, of course, not the case. However, individual flows within the Deccan Plateau have been mapped over several thousand square kilometers and traced laterally for more than 160 km without showing signs of dying out (Choubey, 1973, 1974; Hooper, 1988, p. 155). In central Oregon, some individual lava flows have volumes two or three times as great as those of large andesitic cones (McBirney et al., 1974).

Erosional dissection of basalt plateaus is strongly controlled by the sheetlike continuity and columnar jointing of the flows. Flat benches and vertical cliffs dominate the landscape (Fig. 11–2). Hills are commonly flat-topped. The Germanic name for the steplike landscape (*treppen*) has survived in English not as a geomorphic term but as the petrographic term "trap" or "traprock" for all dense, dark igneous rocks that look like basalt.

Tephra Cones and Composite Cones (or Strato-Volcanoes)

Most subaerial historic volcanoes have been characterized by explosive rather than effusive activity (Table 4–1). The percentage of tephra in the total ejecta (tephra plus lava) for historic circum-Pacific, Indonesian, and Caribbean volcanoes generally has exceeded 90 percent (Rittmann, 1962, p. 153). None of these estimates include the enormous amount of quiet extrusive basalt volcanism along the mid-ocean ridges, however.

Mixtures of tephra and lava, erupted by the explosive Vulcanian or Peléan activity, build cones with radial slopes at or less than the angle of repose of the erupted

Figure 4-14.   Mount Mayon (height: 2400 m) in southern Luzon, Philippines; reputedly the most perfectly symmetric volcanic cone. (Photo: Philippine Department of Tourism.)

material. All proportions of tephra and lava are found in the composite cones that form the earth's most perfectly symmetrical landforms (Fig. 4–14).

Small *tephra cones* have straight, steep sides at the angle of repose of tephra and scoria (Porter, 1972) (Fig. 4–15). They are permeable enough to resist erosional dissection by streams, but straight gullies may descend radially from their rims. Larger cones cannot be built entirely of scoria or tephra, because the material is not strong enough to support continued summit eruptions. Lava flows that mantle the slopes help to support the cone. More important, several sets of systematic intrusive dikes and sills (Fig. 4–16) are integral parts of large volcanic edifices, and these, more than the eruptive lava flows, provide a strong framework to the cone. The intermingling of tephra, lava flows, intrusive pipes, dikes, and sills makes the structure of large volcanic cones too complex for any simple designation of lava or tephra cone, so they are called *composite cones,* or less commonly, *strato-volcanoes.* Their complex internal structure plays an important controlling role in their subsequent erosional dissection (Chap.11).

The magnificent symmetry of composite cones has been the subject of much scientific and esthetic inquiry. Their smooth, concave-up profiles (Fig. 4–14) are best fitted by logarithmic or exponential curves and are attributed to summit lava flows, slumping, or angle-of-repose conditions (Cotton, 1944, pp. 235–36). Complexities of tephra accumulation by size sorting, fallout rates, and shifting wind patterns complicate the pattern (Fisher and Schmincke, 1984, pp. 125–62). The symmetry of composite cones has also been modeled by assuming that the surface of the volcano is a surface of uniform hydraulic potential on a uniformly porous medium. Theories of groundwater flow adapted to magma intrusions show that as the volcano becomes too high, flank eruptions will restore the shape (Lacey et al., 1981). It is remarkable that the simplifying assumptions of fluid flow in an unconfined, uniform aquifer can reproduce the shape of a complex volcanic cone.

Plug Domes, Cumulo-Domes, Tholoids, and Spines

Very viscous lava of highly silicic composition may surge upward into a crater but be too stiff to flow and too degassed to explode. The result is an **endogenous dome** (Table 4–2), a "skinned-over," bulbous protrusion of lava from the vent. These domes vary greatly in size and shape from the awesome "shark-fin" spine that grad-

Figure 4-15. Four major stages in the development of tephra cones. Stage 1: Tephra ring with shape controlled by ballistic trajectories of particles. Stage 2: Angle-of-repose talus forms a straight midslope segment. Stage 3: As cone grows, talus extends upward to crater rim. Stage 4: Talus extends from rim to beyond ballistic limit. From this stage until the morphological capacity of the cone is exceeded, the slope angle remains constant. (Redrawn from McGetchin et al., 1974, Fig. 14; © 1974 by the American Geophysical Union.)

ually extruded from Mount Pelée for nine months after the St. Pierre disaster and eventually reached a height of more than 300 m above the crater, to the mass of Mount Lassen, which is a dome about 1.6 km in diameter and over 600 m high. After the Mount St. Helens eruption of May 1980, a series of plug domes formed in the breached crater, the largest of which had grown to over 90 m in height by the end of that year (Moore et al., 1981). **Tholoid** (dome-shaped) is the term generally used for a flattened hemispheroidal dome within a crater rim (Fig. 4–17). **Cumulo-domes** or **plug domes** are more irregular or cylindrical masses of rigid lava that form

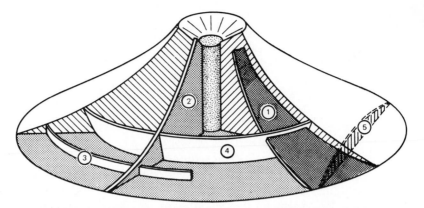

Figure 4-16. Systems of dikes and sills combine with lava flows to strengthen large composite cones. (1) Mantle sill or buried flow, (2) radial dike, (3) ring dike, (4) cone sheet, (5) Peripheral dike. (Redrawn from Rittmann, 1962, Fig. 73.)

Figure 4-17. Wilson Butte, a tholoid or plug dome at Mono Craters, California. Diameter of base: 1 km. Surface features include explosion craters, spines, and an encircling talus. (Photo: E. I. Smith.)

hills or low domal mountains (Smith, 1973; deSilva et al., 1988). Sometimes their surfaces are cratered, or covered with small flows and spinelike viscous extrusions, proving that they are not entirely endogenous.

The surface of an endogenous dome is constantly stretched by continued inflation, so blocks or breccia accumulate around its base or on its flanks and may nearly bury the dome. The breccia is usually of obsidian glass and forms straight taluses radiating from the sides of the dome. Rarely, the striated, polished surface of an extruded plug will be visible above its talus, and may solidify, preserving the marks of extrusion on its flanks.

Plug domes should not be confused with **volcanic necks,** which are the result of long-continued erosion of a composite cone to expose the solidified lava-filled conduit or neck. A plug dome is a constructional landform; a volcanic neck is a structurally controlled erosional remnant. Both may be present in the same volcanic region, as in the Auvergne region of France. There, the same term, *Puy,* is given to domes and eroded necks.

Maars, Pumice Rings, and Tuff Rings

Extremely explosive monogenetic eruptions blast out broad, shallow craters below general ground level with pumice or tuff rings around them (Ollier, 1967). The Eifel district of Germany, at the northern end of the Rhine graben (p. 51), is the type locality for these landforms. The name **maar** (not to be confused with the lunar *mare*) refers to the shallow lake that fills the crater in a humid climate such as that of western Europe. Maars (Germanic plural: *maare*) and their enclosing ring walls are surrounded by widespread thin sheets of tephra, especially in the prevailing downwind direction. If there is no central lake, such an explosive eruption results in no geomorphic expression other than a low rampart or ring wall, grading radially outward into thin tephra, and inward into a pumice-covered shallow crater underlain by shattered country rock (Fig. 4–18).

Purely gas explosions produce craters surrounded by little or no ejecta. A lake may form in the resulting crater (a *gas maar*), or the geologic evidence for the eruption may be nothing more than an area of shattered rock. If volcanic breccia can be identified, the dike or stocklike intrusive mass is called a *diatreme.* Some large circular areas of shattered sedimentary rocks in the central United States were once thought to have resulted from gas eruptions, but because of the total lack of volcanic rocks, they were called *cryptovolcanoes.* Most of them are now interpreted as ancient meteorite impact scars.

Large Explosion Craters and Calderas: Ignimbrite Sheets

Plinian or Katmaian types of explosive activity, or the even more explosive ultra-Plinian and phreatic types, may create a "constructional" landscape of catastrophically destructive appearance. A number of large composite cones have had either

Figure 4-18.   Crater Elegante, a maar in the Pinacate volcanic field, Sonora, Mexico. (Arvidson and Mutch, 1974.) Pre-explosion tephra and lava beds are exposed on the crater walls, overlain by a tephra ring. (Photo: R. E. Arvidson.)

their entire summits blown off or subside inward following a major explosion. The subsidence hypothesis is favored, because the mass of ejecta is substantially less than the volume of the resulting explosion crater or caldera, and also because as much as 95 percent of the tephra traceable to a source in an explosion caldera is derived from new magma rather than older volcanic or other kinds of rock (Macdonald, 1972, p. 305).

The distinction between craters and calderas was arbitrarily set at a diameter of 1 mile (1.6 km) by Macdonald (1972, p. 290). Although arbitrary, this dimensional distinction is related to the morphological capacity concept mentioned earlier in this chapter, in that summit craters up to a mile in diameter are normal components of the constructional morphology of composite cones, whereas calderas larger than that are probably always due to subsidence during or following a Katmaian-type terminal eruption.

Large explosion craters and calderas should be distinguished from the pit craters and summit calderas of basalt shields (Fig. 4–13) that are the site of nonexplosive lava lakes or lava fountains. Explosion craters are surrounded by tephra sheets that thicken and become coarser toward the crater. The inward-facing walls of explosion craters are often slump scars or fault scarps of blocks that subsided into the funnel-shaped opening prepared by an explosion, so the resulting crater is substantially broader than the actual explosive vent.

Calderas are huge basins, ranging in diameter from the defined minimum of 1.6 km to one of the largest yet recognized, 70 km long and 45 km wide, largely buried beneath the thick ignimbrite sheets of the Yellowstone Plateau (Eaton et al., 1975). At least 138 modern calderas are defined as "large" ($\geq$ 5 km in diameter) (Newhall and Dzurisin, 1988). Collectively they have experienced 1299 known episodes of unrest in historic time, although not all of the episodes of activity have been eruptions. Calderas may be on the summits of composite cones or in volcanic plateaus. The site of Crater Lake, Oregon, is a caldera 8 to 10 km in diameter and 1100 m deep that was formed by the Mount Mazama tephra eruption 6700 years ago. Other calderas are so large that the area of subsidence includes entire clusters of volcanic vents, or grabenlike elongate trenches across ignimbrite plateaus. Calderas are bounded by inward-facing fault scarps, and tilted fault blocks or horsts may rise above the basin floors. Late-stage scoria cones or basalt flows may build minor volcanic landforms on the floor of a caldera. In humid regions, calderas are frequently the sites of deep lakes.

The largest calderas are those associated with massive ignimbrite sheets (Marshall, 1932; Francis, 1983). The best hypothesis for this relationship is that a large subsurface reservoir of magma persists long enough to become compositionally stratified, with the most silicic, viscous, and gas-rich fraction near the top. When the magma begins to subside within the chamber (by a cause and to a place unknown), the pressure drop causes a massive effusion of pumice and glowing hot tephra along fissures that are associated with the subsidence of the caprock over the magma chamber. In a geyserlike self-perpetuating sequence, the explosive eruption lowers the pressure still more and further eruptions occur. In some of the great ignimbrite sheets, the later eruptives are more basic than the earlier, showing that deeper levels of more basic magmas were progressively expelled.

The quantities of tephra ejected from calderas are staggering. Whereas the 1980 eruption of Mount St. Helens ejected an estimated 0.6 km^3 of magma and left a crater 2 km in diameter, the most recent of the Yellowstone caldera, 600,000 years ago, erupted 10^3 km^3 of tephra. Toba caldera, in western Sumatra, Indonesia, erupted an estimated 2800 km^3 of tephra in its most recent eruption about 75,000 years ago. The Toba caldera is about 100 km long and 30 km wide, and is the youngest and largest caldera known (Francis, 1983; Rose and Chesner, 1987). It now contains a large lake.

Following eruption and collapse, many of the large calderas then slowly rise, probably caused by the intrusion of fresh magma into the chamber beneath. The result is a central broad dome, or an island in the lake of a flooded caldera, known as a *resurgent caldera* (Smith and Bailey, 1969). A large example is Cerro Galán, a volcanic complex in the northwestern Argentine Andes (Francis, 1983). The broad central resurgent dome rises 1 km above the surrounding caldera floor, and occupies almost half of the elliptical caldera, which is 34 km long from north to south (Fig. 4–19). Because of the extreme aridity of the high Andean plateau, only a small lake remains in the caldera. Ignimbrite sheets from the Cerro Galán eruption 2.5 million years ago are up to 500 m in thickness and can be traced 70 km from the caldera.

Figure 4-19.   Cerro Galán caldera, Catamarca Province, northwestern Argentina. See text for description. (Landsat Thematic Mapper image no. 50451-13553.)

The Yellowstone Plateau was built during three great cycles of Katmaian volcanism, each of which culminated in the eruption of massive ignimbrite sheets. They occurred approximately 1.9, 1.2, and 0.6 million years ago. The huge caldera was filled by several thousand meters of tephra and rhyolite flows (Eaton et al., 1975). Individual ash beds totaling hundreds of meters in thickness cooled as single units, with columnar joints as well developed as in lava flows. Tephra from the multiple Yellowstone eruptions is well known in the older Pleistocene deposits of the Great Plains from Texas to Kansas (p. 437). Two resurgent domes are developing at the present time (Fig. 3–21).

REFERENCES

ALT, D., SEARS, J. M., and HYNDMAN, D. W., 1988, Terrestrial maria: The origins of large basalt plateaus, hotspot tracks and spreading ridges: Jour. Geology, v. 96, pp. 647–62.

ARVIDSON, R. E., and MUTCH, T. A., 1974, Sedimentary patterns in and around craters from the Pinacate volcanic field, Sonora, Mexico: Some comparisons with Mars: Geol. Soc. America Bull., v. 85, pp. 99–104.

BERCOVICI, D., SCHUBERT, G., and GLATZMAIER, G. A., 1989, Three-dimensional spherical models of convection in the earth's mantle: Science, v. 244, pp. 950–55.

BLONG, R. J., 1984, Volcanic hazards: Academic Press Inc., Orlando, Florida, 424 pp.

CHOUBEY, V. D., 1973, Long-distance correlation of Deccan basalt flows, central India: Geol. Soc. America Bull., v. 84, pp. 2785–90.

———, 1974, Long-distance correlation of Deccan basalt flows, central India: Reply: Geol. Soc. America Bull., v. 85, pp. 1008–10.

COTTON, C. A., 1944, Volcanoes as landscape forms: Whitcombe and Tombs Ltd., Christchurch, New Zealand, 416 pp. (republished with minor corrections, 1952).

DAVIS, W. M., 1905, Complications of the geographical cycle: Internat. Geog. Cong., 8th, Washington, 1904, Repts., pp. 150–63 (reprinted 1954 in Geographical essays: Dover Publications, Inc., New York, pp. 279–95).

DECKER, R., and DECKER, B., 1989, Volcanoes: Revised and updated ed: W. H. Freeman and Co., New York, 285 pp.

DECKER, R. W., WRIGHT, T. L., and STAUFFER, P. H., 1987, eds., Volcanism in Hawaii: U. S. Geol. Survey Prof. Paper 1350, 1667 pp.

EATON, G. P., CHRISTIANSEN, R. L., IYER, H. M., PITT, A. M., MABEY, D. R., BLANK, H. R., ZIETZ, I., and GETTINGS, M. E., 1975, Magma beneath Yellowstone National Park: Science, v. 188, pp. 787–96.

FISHER, R. V., and SCHMINCKE, H.-U., 1984, Pyroclastic rocks: Springer-Verlag, Berlin, 472 pp.

FRANCIS, P., 1983, Giant volcanic calderas: Sci. American, v. 248, no. 6, pp. 60–70.

———, and SELF, S., 1987, Collapsing volcanoes: Sci. American, v. 256, no. 6, pp. 90–97.

GREELEY, R., and HYDE, J. H., 1972, Lava tubes of the cave basalt, Mount St. Helens, Washington: Geol. Soc. America Bull., v. 83, pp. 2397–2418.

HALLAM, A., 1987, End-Cretaceous mass extinction event: Argument for terrestrial causation: Science, v. 238, pp. 1237–42.

HOOPER, P. R., 1988, The Columbia River basalt, in Macdougall, J. D., ed., Continental flood basalts, Kluwer Academic Publishers, Dordrecht, The Netherlands, pp. 331–41.

KRAFFT, M., and KELLER, J., 1989, Temperature measurements in carbonatite lava lakes and flows from Oldoinyo Lengai, Tanzania: Science, v. 245, pp. 168–70.

LACEY, A., OCKENDON, J. R., and TURCOTTE, D. L., 1981, On the geometrical form of volcanoes: Earth and Planetary Science Letters, v. 54, pp. 139–43.

LEMASURIER, W. E., 1972, Volcanic record of Antarctic glacial history: Implications with regard to Cenozoic sea levels, in Price, R. J., and Sugden, D. E., eds., Polar geomorphology: Instit. Brit. Geographers Spec. Pub. 4, pp. 59–74.

LIPMAN, P. W., and MULLINEAUX, D. R., eds., 1981, The 1980 eruptions of Mount St. Helens, Washington: U.S. Geol. Survey Prof. Paper 1250, 844 pp.

LOPER, D. E., McCARTNEY, K., and BUZYNA, G., 1988, A model of correlated episodicity in magnetic-field reversals, climate, and mass extinctions: Jour. Geology, v. 96, pp. 1–15.

MACDONALD, G. A., 1972, Volcanoes: Prentice-Hall, Inc., Englewood Cliffs, N.J., 510 pp.

MACDOUGALL, J. D., 1988, Continental flood basalts and MORB: A brief discussion of similarities and differences in their petrogenesis, in Macdougall, J. D., ed., Continental flood basalts: Kluwer Academic Publishers, Dordrecht, The Netherlands, pp. 331–41.

MACHIDA, H., 1967, The recent development of Fuji Volcano, Japan: Tokyo Metropolitan Univ. Geog. Repts., no. 2, pp. 11–20.

MAHONEY, J. J., 1988, The Deccan traps, in Macdougall, J. D., ed., Continental flood basalts: Kluwer Academic Publishers, Dordrecht, The Netherlands, pp. 151–94.

MALDE, H. E., and POWERS, H. A., 1962, Upper Cenozoic stratigraphy of western Snake River plain, Idaho: Geol. Soc. America Bull., v. 73, pp. 1197–1220.

MANGAN, M. T., WRIGHT, T. L., SWANSON, D. A., and BYERLY, G. R., 1986, Regional correlation of Grande Ronde basalt flows, Columbia River Basalt Group, Washington, Oregon, and Idaho: Geol. Soc. America Bull., v. 97, pp. 1300–1318.

MARSHALL, P., 1932, Notes on some volcanic rocks of the North Island of New Zealand: New Zealand Jour. Sci. and Technology, v. 13, pp. 198–200.

MCBIRNEY, A. R., SUTTER, J. F., NASLUND, H. R., SUTTON, K. G., and WHITE, C. M., 1974, Episodic volcanism in the central Oregon Cascade Range: Geology, v. 2, pp. 585–89.

MCGETCHIN, T. R., SETTLE, M., and CHOUET, B. A., 1974, Cinder cone growth modeled after northeast crater, Mount Etna, Sicily: Jour. Geophys. Res., v. 79, pp. 3257–72.

MENARD, H. W., 1964, Marine geology of the Pacific: McGraw-Hill Book Company, New York, 271 pp.

MOORE, J. G., 1975, Mechanism of formation of pillow lava: American Scientist, v. 63, pp. 269–77.

——, CLAGUE, D. A., HOLCOMB, R. T., LIPMAN, P. W., NORMARK, W. R., and TORRE-SAN, M. E., 1989, Prodigious submarine landslides on the Hawaiian Ridge: Jour. Geophs. Res., v. 94, pp. 17465–84.

MOORE, J. G., LIPMAN, P. W., SWANSON, D. A., and ALPHA, T. R., 1981, Growth of lava domes in the crater, June 1980–January 1981, in Lippman, P. W., and Mullineaux, D. R., eds., The 1980 eruptions of Mount St. Helens, Washington: U.S. Geol. Survey Prof. Paper 1250, pp. 541–47.

MULLINEAUX, D. R., PETERSON, D. W., and CRANDELL, D. R., 1987, Volcanic hazards in the Hawaiian Islands, in Decker, R. W., Wright, T. L., and Stauffer, P. H., eds., Volcanism in Hawaii: U.S. Geol. Survey Prof. Paper 1350, pp. 599–621.

NARANJO, J. L., SIGURDSSON, H., CAREY, S. N., and FRITZ, W., 1986, Eruption of the Ne-vado del Ruiz volcano, Colombia, on November 13, 1985: Tephra fall and lahars: Science, v. 233, pp. 961–63.

NEWHALL, C. G., and DZURISIN, D., 1988, Historical unrest at large calderas of the world: U.S. Geol. Survey Bull. 1855, 1108 pp.

NEWHALL, C. G., and SELF, S., 1982, The volcanic explosivity index (VEI): An estimate of explosive magnitude for historical volcanism: Jour. Geophys. Res., v. 87, pp. 1231–38.

NICHOLS, D. R., and YEHLE, L. A., 1961, Mud volcanoes in the Copper River Basin, Alaska, in Raasch, G. O., ed., Geology of the Arctic: Univ. of Toronto Press, Toronto, pp. 1063–87.

NORDLIE, B. E., 1973, Morphology and structure of the western Galápagos volcanoes and a model for their origin: Geol. Soc. America Bull., v. 84, pp. 2931–55.

OLLIER, C. D., 1967, Maars, their characteristics, varieties, and definition: Bull. Volcanol., v. 31, pp. 45–73.

——, 1988, Volcanoes: Basil Blackwell Ltd., Oxford, England, 228 pp.

PETERSON, D. W., and MOORE, R. B., 1987, Geologic history and evolution of geologic con-cepts, island of Hawaii, in Decker, R. W., Wright, T. L., and Stauffer, P. H., eds., Volca-nism in Hawaii: U.S. Geol. Survey Prof. Paper 1350, pp. 149–89.

PORTER, S. C., 1972, Distribution, morphology, and size frequency of cinder cones on Mauna Kea volcano, Hawaii: Geol. Soc. America Bull., v. 83, pp. 3607–12.

RAMPINO, M. R., and STOTHERS, R. B., 1988, Flood basalt volcanism during the past 250 million years: Science, v. 241, pp. 663–68.

RICHARDS, M. A., DUNCAN, R. A., and COURTILLOT, V. E., 1989, Flood basalts and hot-spot tracks: Plume heads and tails: Science, v. 246, pp. 103–07.

RITTMANN, A., 1962, Volcanoes and their activity (transl. by E. A. Vincent): Wiley-Inter-science, New York, 305 pp.

ROSE, W. I., and CHESNER, C. A., 1987, Dispersal of ash in the great Toba eruption, 75 ka: Geology, v. 15, pp. 913–17.

RYAN, M. P., and BLEVINS, J. Y. K., 1987, Viscosity of synthetic and natural silicate melts and glasses at high temperatures and 1 bar (10^5 Pascals) pressure and at higher pressures: U.S. Geol. Survey Bull. 1764, 563 pp.

SARNA-WOJCICKI, A. M., SHIPLEY, S., WAITT, R. B., JR., DZURISIN, D., and WOOD, S. H., 1981: Areal distribution, thickness, mass volume, and grain size of air-fall ash from the six major eruptions of 1980, *in* Lippman, P. W., and Mullineaux, D. R., eds., The 1980 eruptions of Mount St. Helens, Washington: U.S. Geol. Survey Prof. Paper 1250, pp. 577–600.

SELF, S., RAMPINO, M. R., NEWTON, M. S., and WOLFF, J. A., 1984, Volcanological study of the great Tambora eruption of 1815: Geology, v. 12, pp. 659–63.

SHORT, N. M., 1986, Volcanic landforms, *in* Short, N. M., and Blair, R. W., Jr., eds., Geomorphology from space: NASA Spec. Pub. 486, pp. 185–253.

deSILVA, S. L., SELF, S., and FRANCIS, P. W., 1988, The Chao dacite revisited (abst.): EOS, v. 69, no. 44, pp. 1487–88.

SIMKIN, T., and SIEBERT, L., 1984, Explosive eruptions in space and time: Durations, intervals, and a comparison of the world's active volcanic belts, *in* National Research Council Geophysics Study Committee, Explosive volcanism: Inception, evolution, and hazards: National Academy Press, Washington, D.C., pp. 110–21.

———, McCLELLAND, L., BRIDGE, D., NEWHALL, C., and LATTER, J. H., 1981, Volcanoes of the world: Hutchinson Ross Publishing Co., Stroudsburg, Pa., 232 pp.

SMITH, E. I., 1973, Mono Craters, California: A new interpretation of the eruptive sequence: Geol. Soc. America Bull., v. 84, pp. 2685–90.

SMITH, R. L., and BAILEY, R. A., 1969, Resurgent cauldrons, *in* Coats, R. L., Hay, R. L., and Anderson, C. A., eds., Studies in volcanology: Geol. Soc. America Mem. 116, pp. 613–62.

SNEAD, R. E., 1964, Active mud volcanoes of Baluchistan, West Pakistan: Geog. Rev., v. 54, pp. 546–60.

STANLEY, D. J., and SHENG, H., 1986, Volcanic shards from Santorini (Upper Minoan ash) in the Nile Delta, Egypt: Nature, v. 320, pp.733–35.

STOTHERS, R. B., 1984, The great Tambora eruption in 1815 and its aftermath: Science, v. 224, pp. 1191–98.

THORARINSSON, S., 1951, Laxárgljúfur and Laxárhraun: A tephrochronological study: Geograf. Annaler, v. 33, pp. 1–88.

WHITE, D. E., 1955, Violent mud-volcano eruption of Lake City Hot Springs, northeastern California: Geol. Soc. America Bull., v. 66, pp. 1109–30.

WHITE, R. S., 1989, Igneous outbursts and mass extinctions: EOS, v. 70, pp. 1480, 1490–91.

PART II

SUBAERIAL DESTRUCTIONAL (EROSIONAL) PROCESSES AND EROSIONAL LANDFORMS

However they may be constructed, subaerial landforms cannot stand immutable. The rocks that make them were formed in physical and chemical environments that are grossly out of equilibrium with the oxygen-rich, low-pressure, hydrous atmosphere into which they are thrust or erupted by constructional processes, and they are sure to crumble or dissolve. Landscapes are carved from structures by agents of chemical and physical weathering and erosion. Water, which occurs naturally as both a chemically reactive fluid and an erosive solid at earthly temperatures and pressures, is the primary agent. The energy drive is provided by solar radiation, converted into heat at the earth's surface and eventually reradiated into space. Gravity and internal heat are secondary sources of energy for geomorphic change. We call these processes *destructional, degradational,* or *erosional,* because they tend to destroy the edifices that are constructed by diastrophism and volcanism. The eroded debris accumulates in tectonic basins on the continents or is washed to the sea by rivers. Distinctive landforms are built by the accumulated waste, but the far greater proportion of the subaerial landscape consists of erosional remnants, still in the process of reduction.

Present-day processes conform to universal thermodynamic laws but are also influenced by the steady evolution of the earth's lithosphere, hydrosphere, atmosphere, and biosphere. We can see a geologic record of at least 1 billion years of weathering and erosional processes similar to those now operating, although different to the degree that different forms of life were involved in the processes.

The certainty of geomorphic change by subaerial processes acting on geologic structures has been the subject of philosophical comment since the beginning of written history, at least. A Chinese scholar, in 240 B.C., observed: ''Nothing under heaven is softer or more yielding than water; but when it attacks things hard and resistant there is not one of

97

them that can prevail'' (Waley, 1935, p. 238).[1] One of the oldest geomorphic deductions is: ''Every valley shall be exalted and every mountain and hill shall be made low; and the crooked shall be made straight, and the rough places plain'' (Isaiah, 40:4).

This philosophical tradition of skepticism about the dubious durability of the ''everlasting'' hills was summarized in a single statement by J. W. Powell, one of the great geomorphologists who learned his trade while exploring the mountains of the western United States. *Powell's dictum* is: ''We may now conclude that the higher the mountain, the more rapid its degradation; that high mountains cannot live much longer than low mountains, and that mountains cannot remain long as mountains: they are ephemeral topographic forms. Geologically all existing mountains are recent; the ancient mountains are gone'' (Powell, 1876, p. 193).[2]

New knowledge of the absolute rates of erosion make it clear how true Powell's dictum really is. All mountains except volcanoes are carved from either late Cenozoic orogenic belts or areas of more ancient terranes that were uplifted by late Cenozoic epeirogeny. The sediment loads carried in modern rivers demonstrate that the watersheds ought to be reduced to near sea level in 10 to 12 million years. The great issue before geomorphologists today is not whether erosional processes can accomplish the work they are alleged to have done but how continents have managed to survive despite such enormous denudation. With this outlook, geomorphologists have joined sedimentologists, volcanologists, and tectonists in a reexamination of the Cenozoic history of continents and ocean basins, using the inferred motions of lithospheric plates as a guiding theory. The explanatory description of major landforms becomes an account of complexly interrelated or alternating episodes of erosion and diastrophism, with continental ice sheets and other climate-controlled processes adding fascinating complexity.

Orogenic belts create special climates of cold highlands, regions of heavy precipitation on windward slopes, and rainshadow deserts. It is not so obvious, but equally important, that erosion can exert powerful controls on orogeny. A mountain belt of erodible sedimentary rocks from accretionary trench sediments in a humid tropical zone, as for example in eastern Taiwan, may never be more than 1500 m high despite rapid uplift rates, because erosion rates are also high (Suppe, 1981).[3] In contrast, the late Cenozoic central Andes Mountains rose by crustal shortening and thickening at a converging plate margin in a hyperarid climate (Isacks, 1988).[4] Rapid uplift and aridity permitted the high Altiplano plateau to be created without significant erosion by externally draining rivers. Perhaps the lack of erosion will, through isostatic equilibrium, prevent the Altiplano from rising further and stop the orogeny, or widen the orogenic belt (p. 314). The Tibetan plateau may be a similar example of climate imposing limits on orogenic landscape evolution.

In the following eight chapters we consider the processes and landforms of subaerial weathering and fluvial erosion and make deductions about sequential changes in these forms through time. Although climatic influence on fluvial landforms must be considered, the distinctively climate-controlled processes and landforms are reserved for Part III.

[1] See References, Chapter 5.

[2] See note 1.

[3] See note 1.

[4] See note 1.

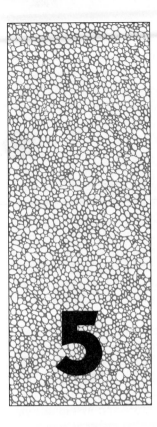

Energy Flow in Geomorphic Systems

THE SYSTEMS CONCEPT IN GEOMORPHOLOGY

All physical phenomena may be conceived as being organized into **systems,** a system being defined as a collection of objects and the relationships between those objects. **Closed systems** are those that have boundaries across which no energy or matter moves; **open systems** have a flow of energy and matter through their boundaries. A closed system must change toward a time-independent equilibrium state, according to the second law of thermodynamics, in which the ratios of various physical phases remain constant, the amount of free energy (such as chemical potential, potential energy of position, etc.) is minimal, and the entropy (unavailable energy, often visualized as the diffuse heat of random molecular motion) is maximal. An open system, on the other hand, may achieve a steady state, wherein the system and its phases remain in a constant condition even though matter and energy enter and leave it (von Bertalanffy, 1950).

Only the most simple geomorphic processes may be studied as closed systems. Certain weathering reactions, such as solution of limestone by acidified rainwater, can be duplicated by adding measured quantities of the reactants to a test tube and observing that after a period of time, predicted amounts of new compounds have formed and are in stable equilibrium with each other. Hot, newly erupted lava radiates and conducts heat energy to its surroundings until the temperature gradient is reduced to zero and the mass has solidified. Then it ceases to give off heat. Even in these simple examples of physical and thermal systems, it is obvious that the natural phenomena are much more complex than these brief statements imply.

The entire subaerial portion of the earth's surface can be profitably regarded as an open system (Chorley, 1962). Matter is supplied by diastrophism or volcanism, with an additional trivial net increase of mass from meteoritic infall. Energy is derived from solar radiation, gravitation, rotational inertia, and internal heat. Surface processes erode the rock structures, and the eroded debris is deposited elsewhere, usually below sea level. It is conceivable that a steady state could be reached in such a complex open system so that the form of the landscape would not change even though the mass that composes it were constantly changing and energy constantly being expended. One of the great issues of geomorphology is whether landforms change sequentially through time or whether they achieve steady states and then change only slowly or not at all.

Compare a landscape (a "geomorphic system") with a rotary kiln used to manufacture portland cement (Fig. 5–1). Simple raw materials, primarily limestone and shale, are fed into the upper end of a rotating, heated drum. The fragments tumble and are pulverized within the drum as they move toward the lower end, but in the presence of heat they react chemically to form the desired end product, a powder that reacts with water to form calcium silicate minerals of great adhesive strength. The heat that is not stored in the newly created chemical compounds is liberated as wasted energy, along with the mechanical frictional losses. Mechanical energy is provided by a gravity potential and also by the kinetics of rotation. It can be seen that if the supply of raw materials is adjusted to the supply of mechanical and thermal energy, a desired product is produced at a desired rate. If material is supplied at a more rapid rate, the energy supply also has to be increased to maintain the flow of the desired product. A steady state at various levels of net flow can be obtained by mutually adjusting the several variables of material and energy supply in this simple open system.

A landscape above sea level (or above the geoid) is similar to the rotary kiln model. Rainfall sweeps weathered rock fragments downhill to the sea in rivers. The rate of flow is a function of the climatic variables of temperature and precipitation, the gradient is established primarily by diastrophism, and the intensity of chemical and physical change of the transported debris is a function of the time of travel through the system. As long as the energy and material flow are adjusted, the system stays in a steady state.

Suppose, however, that the intake end of a rotary kiln had been unwisely built on the supply dump, which is finite. As the supply of raw material (the constructional landform) is consumed, the longitudinal gradient of the kiln is reduced. To maintain quality control during a longer residence time in the kiln, the heat must be reduced. The rate of cement production gradually declines. By proper control, even such a poorly designed system would appear to be in a steady state over any brief period of observation, although on an extended time scale the potential energy of the system would be declining. This is analogous with erosional landscapes that are not being renewed by constructional processes. Although rivers may appear to be *graded,* that is, nicely adjusted open systems in steady states doing just the work they are required to do, on a longer time scale they are seen to be open systems of declining energy, apparently in a steady state only during short periods of observation (Chap. 9).

We assume that the mass of the earth is constant, ignoring the continuous small increment of meteoritic dust and the extremely rare impact of large extraterrestrial fragments. In principle, it should be possible to define the geoid (Chap. 2) as the boundary of an open system and to monitor the movement of rock material upward through it by diastrophism and volcanism and back downward through it by diastrophism and erosion (Fig. 5–2). Because the geoid is in equilibrium, the masses transported through this reference surface must balance, and lateral transport systems must also operate above and below the geoid to maintain isostatic equilibrium. The study of the geomorphic system has not, however, progressed to the point where such mass balances can be calculated.

The energy balance in the geomorphic system is becoming quite well known, owing to related research in biology, meteorology, astronomy, and geophysics. The earth's surface has not grown significantly colder or warmer in geologic time, so it must dissipate to space as much heat as it receives from various sources. In the uni-

Figure 5-1. The rotary kiln analog to an open geomorphic system.

verse as a whole, gravity is the dominant form of energy (Dyson, 1971), greatly exceeding heat, light, and nuclear energy. Within our geomorphic system, solar radiation (in itself a result of the gravitational force that sustains thermonuclear fusion in the sun) dominates all other energy sources. A diagram of the energy flow at the surface of the earth (Fig. 5–2) shows the major sources of energy for geomorphic change.

It is important for the next generation of geomorphologists to understand better the energy potentials available for any system they wish to study. Many poor generalizations have been made in the past that were not consistent with energy considerations. Trends toward quantification require that energy as well as material rates and balances be calculated. The remainder of this chapter is an evaluation of the energy flow in the geomorphic system, essentially a detailed review and discussion of Figure 5–2. A few rare or trivial forms of energy that might produce landforms, such as lightning discharge, cosmic radiation, and magnetic fields, are not considered further. Human beings, who have become significant geomorphic agents during the last few thousand years, are now using energy on a scale comparable to that released by wind, tides, and geothermal heat. We are processing as much rock annually as is being recycled by plate tectonic processes. If we can learn how to harness safely the energy of nuclear fusion, which drives solar radiation, and nuclear fission, which provides most of the geothermal heat, our role in the future modification of this and other planets could be almost limitless.

THE POWER AVAILABLE FOR GEOMORPHIC CHANGE

Figure 5–2 outlines the rate of energy conversion (power) near the surface of the earth. Solar radiation, mostly in the visible and near-infrared wavelengths, supplies about 99.98 percent of the energy entering the geomorphic system. About 30 percent of the solar radiation is directly reflected, mostly by clouds (the planetary **albedo,** or brightness, is the measure of this reflected energy). Most of the nonreflected energy is converted to thermal vibration of air, water, and mineral molecules as sensible heat, activating chemical and physical reactions and possibly even heating rocks until they fracture. A very significant fraction is absorbed as latent heat of vaporization by the abundant water on the surface of the earth and is

Figure 5-2. Energy and matter flux in the geomorphic system. System boundaries are the geoid and the air and water interfaces. Energy enters the system as solar radiation, rotational energy, and radiogenic heat. The rate of outgoing infrared radiation balances all incoming energy. Matter enters the system by tectonic uplift above the geoid and leaves by tectonic subsidence and erosion. Trivial amounts of matter enter as meteorites and escape as molecular hydrogen and helium (Flohn, 1969; Hubbert, 1971; Inman and Brush, 1973; other references cited in text).

transported by water vapor in the atmosphere until it is released by condensation in clouds. Much smaller fractions of the energy are converted directly to kinetic motion of air and water masses, driving convective and advective currents in the atmosphere and oceans, or are absorbed by photosynthetic plants. A very small amount of the energy drives geomorphic processes. All of this energy is ultimately returned to outer space by reradiation from the moderately warmed earth.

Solar radiation provides energy for geomorphic change at a rate 4000 times faster than all other energy sources combined, yet these other sources cannot be ignored. They include the gravitational and inertial forces associated with the mass and motion of the earth, moon, sun, and other objects of the solar system, expressed at the earth's surface by tides in both the lithosphere and the hydrosphere. Also included is the heat that flows outward through the surface from the interior of the earth. Only about 2 percent of the earth's internal heat is released by volcanic activity. A comparable amount is released during earthquakes. The rest is conducted to the surface along the geothermal gradient from the hotter interior to the cooler surface rocks and is a small but worldwide phenomenon. In ways not yet understood, heat energy and geothermal gradients in the interior of the earth, aided by kinetic rotational forces and gravity, drive all of the diastrophic activity in the lithosphere. Any comparison of internal and external energy sources for geomorphic processes must consider that the power of the latter is 4000 times greater than the former, although internal energy is much more localized.

Not shown in Figure 5–2 is the energy released by meteorite impact (Melosh, 1989). More than 120 impact craters have now been identified on the earth, mostly in the continental shields where erosion is slow and the oldest parts of the continental crust are preserved. Half of the craters larger than 10 km in diameter have ages of less than 200 million years (Grieve, 1987; 1990). The energy of the iron meteorite 60 m in diameter that created the Barringer Crater in Arizona about 30,000 years ago was comparable to that of one of the most powerful modern nuclear weapons, and it created a "simple" crater 1.2 km in diameter and 170 m deep. Larger, complex craters with a central area of uplift and down-faulted margins range from 4 to 100 km or more in diameter. The largest were caused by impacts of kilometer-size meteorites. Lunar and terrestrial crater abundances indicate that from one to three craters wider than 20 km in diameter are created on earth every million years (Grieve, 1990, p. 68). More frequent or giant impacts in the early history of the earth may have been important in creating the moon, tilting the earth's rotational axis, and separating the earth's core and mantle (Stevenson, 1987). The possible role of a 10-km diameter meteorite impact in causing the mass extinctions that define the Cretaceous-Tertiary boundary (Alvarez et al., 1980), and possibly caused the eruption of the Deccan flood basalts (p. 85), was debated at a major conference in 1981 (Silver and Schultz, 1982) and has continued to be a major controversial topic of earth history. While the enormous energy of such infrequent but catastrophic events cannot be calculated, their role in both creating individual landforms and affecting global geomorphic processes cannot be ignored.

One difficulty with evaluating the power of the subaerial geomorphic system is that there is no simple way to measure the *power density,* or localized concentration of energy expenditure. For example, a meteorite impact could produce a 100-km diameter crater in a few seconds. The small proportion of geothermal heat flow that is concentrated at volcanoes and faults creates spectacular landforms, whereas the other 90 to 95 percent drives tectonic processes but makes no specific landforms.

Biological systems, especially human beings, have an exceptional ability to store and concentrate diffuse energy at the earth's surface (Häfele, 1974; Holdren and Ehrlich, 1974). The geomorphic impact of human beings is unique because we can concentrate power so intensely. Sunlight, when fixed by photosynthesis, accumulated as fossil fuel, burned in an engine, and focused at the tip of a drill bit, can alter the landscape along a highway right-of-way at a rate that millions of years of normal incident radiant energy could not accomplish. Industrial gas lasers used for cutting, welding, and heat-treating metals can focus incredible power densities of 10^4 to 10^9 W/cm^2 in bursts of energy of microsecond duration. Holes can even be drilled through diamonds by such concentrated energy.

SOLAR RADIATION

A major goal of geomorphology is to understand how the immense but diffuse energy of solar radiation is converted into mechanical work that shapes landscapes. We can visualize the process in terms of a "geomorphology machine" (Fig. 5–3) in which a steam engine fired by the sun powers a series of fans, saws, files, grinders, and hydraulic jets to reduce the landscape. The analogy with a steam engine is a good one, because it is the phase changes in the abundant water on the earth's surface that convert solar energy into geomorphically significant mechanical work. Of even greater significance is the fact that solar heat keeps the surface of our planet in the temperature range in which water naturally occurs in solid, liquid, and vapor phases. In a sense, we are inside the boiler and condenser of the machine rather than witnessing the final mechanical conversion of the solar energy.

The sun radiates energy over an electromagnetic spectrum of wavelengths that is closely approximated by a perfectly radiating blackbody at an absolute temperature of 5785K (Goody and Walker, 1972). The radiant energy intercepted by a sur-

Figure 5-3.   The "geomorphology machine." (Bloom, 1969.)

face at the earth's mean orbital radius from the sun is about 1.4 kW/m². This value is called the **solar constant**. The earth intercepts solar energy at a rate equal to the product of the solar constant and its cross-sectional area. However, the earth is not a disk but approximately a rotating sphere, with a surface area of $4\pi r^2$ that is four times as large as the equivalent disk cross section. Thus the average input of solar radiation on each square meter of the upper atmosphere is only one-fourth of the solar constant, or 349 W/m² (Flohn, 1969, p. 10). The total power of solar radiation shown in Figure 5–2 is equal to this average energy expenditure on the total area of the earth (510×10^{12} m²).

In addition to reflecting nearly one-third of the total incoming energy, the atmosphere acts as a selective filter to restrict further the wavelengths of solar radiation that reach the earth's surface. In the lower atmosphere, water vapor (H_2O) and carbon dioxide (CO_2) strongly absorb incoming solar radiation in specific infrared wavelengths. It is one of the remarkable balances of nature that the earth is warmed by the sun to just the right temperature so that it reradiates energy out through the atmosphere into space at wavelengths that are in spectral "windows" between the incoming wavelengths that are so strongly absorbed by water and by carbon dioxide. The mean terrestrial surface temperature of 288K (15°C) is a result of the balance between both the wavelengths and the intensity of incoming and outgoing radiation. As long as the earth has had its present atmospheric composition, the mean surface temperature cannot have fluctuated more than a few degrees. However, future changes in the amount of atmospheric CO_2 can have a dramatic effect on surface temperatures.

Terrestrial Thermal Gradients

Solar energy is absorbed and reradiated unequally by the lower atmosphere, hydrosphere, and lithosphere. The result is a complex of thermal gradients that cause atmospheric and oceanic circulation and produce climatic regions on the earth's surface. Because climate is such an important aspect of geomorphic processes, the major thermal energy gradients need specific discussion.

Latitudinal Gradients. For several reasons, most of the solar energy is received by the earth within the tropics. First, the sun is high in the tropical sky at all seasons, and the reflectance is least when the incoming radiation is at a high angle to the surface. Second, the tropics contain large oceanic areas, favoring absorption of heat by water. Third, the albedo of tropical regions is much lower than the planetary average. Fourth, the atmosphere is cloudy and humid, absorbing reradiated earth heat. Fifth, the surface area between successive parallels of latitude is larger near the equator than near the poles. Almost 40 percent of the earth's surface is within the tropics. For all these reasons, in the lower latitudes more energy is received from the sun than is reradiated back into space (Vonder Haar and Suomi, 1969).

The polar regions receive very little of the total solar radiation. The low incident angle of the sunlight, even during the time of 24-hour daylight in high latitudes, promotes reflection. Both the Arctic Ocean and the Antarctic Continent are covered with ice and snow, which further aid in reflecting the incoming radiation. It is true that for more than a month of midsummer when the sun shines continuously, each polar region in turn receives more radiation per day than is ever received per day in the tropics, but the season is very brief, and for the rest of the year the energy input at the poles is far lower than the output. Figure 5–4 summarizes the latitudinal distribution of incoming and outgoing radiation in the northern hemisphere and shows that at a latitude of about 38° north, the annual heat budget of the earth is balanced. The southern hemisphere is assumed to have a similar

Figure 5-4. Radiation balance of the northern hemisphere. Within 38° north of the equator, energy is received at a faster rate per unit area than it is reradiated; north of 38° each unit area radiates at a faster rate than it receives energy from the sun. The southern hemisphere has a similar, but not identical, balance. (After Houghton, 1954.)

budget. The excess heat received in the tropics is carried poleward by winds and ocean currents, and it balances the net deficit between incoming and outgoing radiation at high latitudes.

Altitudinal Gradients. The atmosphere is reasonably transparent to incoming solar radiation of visible and near-infrared wavelengths but is strongly absorbent in certain wavelengths of the outgoing longer-wave infrared radiation that leaves the solid earth. Thus the atmosphere is heated from beneath, and a *normal temperature lapse rate* of about −6.4°C/1000 m is recorded during ascent from sea level. With increasing altitude, the mean annual temperature on a mountainside steadily decreases until it is below freezing. If precipitation is adequate, a permanent snowline is reached. The altitude of the regional snowline varies from nearly 6 km above sea level in the subtropical dry belts to sea level in the polar regions. Near the equator, the snowline dips to about 5000 m because of the greater cloudiness and precipitation there than in the adjacent clear, dry subtropical belts.

If a mountain is high enough to intersect the regional *snowline,* the altitude above which snow cover persists throughout the year, it will carry a year-round snow cap and possibly glaciers (Chaps. 16 to 18). The sediment loads in streams draining glaciated and nonglaciated mountain peaks suggest that the presence of glaciers in its headwaters increases a stream's erosion rate by a factor of 5 to 100 (Table 12–2). Thus altitude alone can be a powerful factor in determining the effectiveness and even the kind of geomorphic processes acting on a landform. Powell's dictum (p. 98) easily comes to mind.

Although mountain peaks are cold, they may nevertheless be subject to a very high energy flux. Unimpeded by most clouds, haze, and atmospheric dust, the incoming radiation intensity on a mountain greatly exceeds the average at sea level. Reradiation is also rapid. Sunlit rock surfaces at high altitudes may be much warmer than the air only a few centimeters away, but as soon as the sun sets, the rocks reradiate into cold, dry space and cool rapidly. Freeze and thaw cycles are frequent and intense (Chap. 15).

Seasonal Gradients; Land–Sea Contrasts. Because the earth's equatorial plane of daily rotation is now inclined 23½° to the orbital plane, the sun is vertically overhead across a wide intratropical belt during some part of the year. The well-known phenomena of seasonality result. In extratropical regions, the names "winter" and "summer" express the dominant role of temperature change. Within the tropics, where temperatures are generally high, seasonality is defined by precipitation maxima and minima, so that rainy seasons, or wet monsoons, alternate annually with dry seasons of several months' duration. In both high and low latitudes, the earth's climate slowly changes because of cyclical variations in the tilt of the axis between 22° and 24.5° on a time scale of 40,000 years, as well as other cyclical orbital variations (Fig. 18–4).

Many geomorphic processes are associated directly with seasonality. The annual alternation of torrential rain and prolonged drought creates a distinctive savanna landscape. Subtropical desert landforms are shaped by flash floods and wind. Tundra regions have long, smooth slopes formed by soil creep and earth flow in a seasonally thawed, water-saturated surface layer that overlies perennially frozen ground. Climate-controlled geomorphic processes give such a distinctive shape to landscapes that climatic regions can be used to organize an alternative approach to systematic geomorphology (Part III).

Associated with seasonal climatic changes are the temperature contrasts and gradients between land and sea. Maritime climates are characterized by moderate daily and seasonal temperature contrasts, because the abundance of water and water vapor provides a "buffer" of latent heat. The continental interiors, on the other hand, have extreme seasonal temperature contrasts. The greatest seasonal contrasts are not in the subtropical deserts but in high-latitude continental interiors. The greatest range of mean annual air temperature on earth is 62°C, in Siberia. In contrast, both the annual and daily temperature range on small tropical islands is commonly less than 3°C. There, temperature change is totally irrelevant to geomorphic processes. Despite their moderate temperatures, the general high humidity of maritime climates promotes chemical weathering, and in cold maritime regions, glaciers are likely to be significant geomorphic features of mountains and plateaus.

Daily Temperature Fluctuations. Daily temperature changes are especially marked in tropical highlands and in subtropical deserts. In the Sahara, for instance, the daily temperature fluctuation in a shady location is over 60°C. The daily temperature range of exposed rock surfaces exceeds 120°C. These fluctuations are reliably reported to shatter rocks (p. 123). In nondesert regions, night also is normally cooler than day, but reradiation back to earth from the humid atmosphere keeps the diurnal range small. Geomorphic processes related to condensation of dew on rocks, or freezing, may be dominated by a daily cycle.

The Hydrologic Cycle

A very large proportion of the solar radiation intercepted by the earth is isothermally absorbed as *latent heat of vaporization* by water (Fig. 5–2). Ours is truly the hydrous planet, and water is as anomalous a compound in its thermal properties as in its chemical activity. About 2.4 kJ is required to evaporate 1 g of water at 20°C. The large area of oceans in tropical regions ensures that much of the incoming radiation is absorbed not as an increase in sensible temperature but as latent, or insensible, heat by the change in state of water from liquid to vapor. When the water vapor condenses to form clouds of droplets, the latent heat is liberated to the surrounding atmosphere. The largest single component in the horizontal transfer of heat over the surface of the earth from equatorial to polar regions is undoubtedly this latent heat of vaporization and condensation.

On the most awesome scale, local intense evaporation and precipitation over tropical oceans spawn the funnel-like spiraling convective cells variously called cyclones, hurricanes, or typhoons. These storms draw huge quantities of heat from the surface waters of the ocean and transfer it to higher altitudes and latitudes. In the process, they generate winds and waves that cause as much geomorphic change and human tragedy as any other form of catastrophe. Tropical cyclones are an extreme example of highly concentrated solar energy.

The **hydrologic cycle,** or interchange of water between oceans, atmosphere, land, glaciers, and organisms, is summarized in Figure 5–5 and Tables 5–1 and 5–2. Two kinds of information are compiled: Table 5–1 is an estimate of the amounts of water that are present in various parts of the outer layers of the earth; Table 5–2 is the calculated annual exchange of water among the various reservoirs. The figure ideally portrays a steady-state open system. The energy inflow equals the outflow, and, with negligible exceptions, the amount of water in the system is constant.

First, examine the amounts of water in the system. In their order of importance, the reservoirs of water are: (1) oceans, (2) glaciers, (3) groundwater, (4) lakes and rivers, (5) atmosphere, and (6) biomass (all living matter). Actually, over 97 percent of all the water near the surface of the earth is in the oceans, and most of the remainder is in glaciers. The interior of the earth must contain some water, chemically combined or dissolved in solid or molten rock, but for at least the last billion years, the amount of water at or near the surface of the earth has been approximately constant. Only a small amount of new water might be added annually by condensation of volcanic gases, and an equally small amount of water is probably lost from the earth annually by the photochemical dissociation of water vapor by solar radiation.

The hydrologic cycle is usually shown as a rate of transfer or *flux,* commonly as cubic kilometers of water per year. It can be converted into energy or power units, too. Over the past several decades, various estimates of the annual water budget have increased by about 15 percent, primarily because of increases in the estimated radiation budget of the earth. However, no high degree of accuracy is claimed even for the latest estimates.

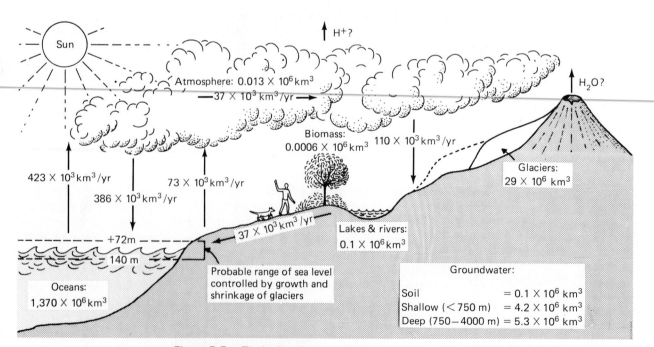

Figure 5-5.   The hydrosphere, or the water on or near the surface of the lithosphere. For easy comparisons, all the amounts of water in the reservoirs are given in millions of cubic kilometers ($10^6 km^3$); all the annual exchanges among the reservoirs (the hydrologic cycle) are given in thousands of cubic kilometers per year ($10^3 km^3$/yr). The accuracy of the various estimates varies widely. (Compiled from various sources; see Tables 5-1 and 5-2.)

TABLE 5–1

Inventory of Water at the Earth's Surface

Reservoir	Volume (10⁶ km³)	Percent of Total
Oceans	1370	97.25
Ice caps and glaciers	29	2.05
Deep groundwater (750–4000 m)	5.3	0.38
Shallow groundwater (<750 m)	4.2	0.30
Lakes	0.125	0.01
Soil moisture	0.065	0.005
Atmosphere[a]	0.013	0.001
Rivers	0.0017	0.0001
Biosphere	0.0006	0.00004
Total	1408.7	100

Source: Compiled from various sources by Berner and Berner (1987).
[a] As liquid equivalent of water vapor.

We see from Figure 5–5 that 423,000 km³ of water is evaporated from the sea annually, representing a layer 117 cm thick from the entire ocean. Most of the water precipitates back onto the ocean, but the excess falls on the land. Because more snow and rain fall on land every year than can be evaporated, 37,000 km³ of water annually drips, seeps, and flows from the land back into the sea.

Much of the water evaporated from the land is not simply evaporated from the exposed surfaces of lakes and streams but is used by plants and animals. A field of crops may annually use an amount of water equivalent to a layer 45 to 60 cm deep over the field. A forest of Douglas fir may annually pump into the atmosphere the equivalent of a layer of water 120 cm deep over its area. Over most of the land, rainwater falls on vegetation. Some evaporates, and some is absorbed by the plant roots and transpired through leaves. Because the variables cannot be clearly distinguished, the term *evapotranspiration* is used for the collective effect. The water requirements of typical plants on the parts of the land that are neither desert nor glacier could evapotranspire all of the 73,000 km³ of water that annually evaporates

TABLE 5–2

Fluxes in the Water Cycle with Methods of Determination

Process	Flux km³/yr	Flux cm/yr[a]	Source
Precipitation on land	110,300	74	Lvovitch (1973)
Evaporation from land	72,900	49	(Precipitation on land minus runoff)
Runoff from land (river runoff and direct groundwater discharge to the ocean of ~ 6% of total)	37,400	25	Baumgartner and Reichel (1975); groundwater discharge, Meybeck (1987)
Precipitation on oceans	385,700	107	(Total precipitation minus precipitation on land)
Evaporation from oceans	423,100	117	(Precipitation on oceans minus runoff)
Total precipitation on earth	496,000	97	Baumgartner and Reichel (1975)
Total evaporation on earth	496,000	97	Baumgartner and Reichel (1975) (equal to total precipitation)

Source: Berner and Berner (1987).
Note: Because of use of different areas, values in cm/yr do not balance between land and oceans.

[a] Fluxes in cm/yr calculated on the following basis: area of earth = 510 × 10⁶ km² (total evaporation and precipitation); area of oceans = 362 × 10⁶ km² (precipitation and evaporation over oceans); and area of land = 148 × 10⁶ km² (runoff, precipitation, and evaporation over land).

from the land, so it is likely that each year a significant amount of water passes through a biologic cycle as a part of the larger hydrologic cycle. Deserts are hot not because they receive more solar radiation than vegetated regions at the same latitude, but because without evapotranspiration, more of the incoming radiation is converted into heat. Only a small part of the evapotranspired water is held at any time as living matter, and the turnover is continuous. The energy required for photosynthesis is only about 0.1 percent of the total received from the sun (Woodwell, 1970), but is almost 1 percent of the total absorbed by plants, most of which is used to warm the plant (Miller, 1981, p. 182). By combining water and carbon dioxide in photosynthesis, an estimated 2×10^{11} tons of dry organic matter is produced annually. Some of it builds landforms such as peat bogs. Photosynthesis is vital to geomorphology because it permits plants and soil to cover most land surfaces, but also because the use of fossil fuels permits human beings to achieve tremendous power densities (p. 103).

Glaciers store large amounts of water on land, temporarily removing it from the hydrologic cycle. If the present glaciers were to melt, sea level would rise 60 to 70 m and submerge the most heavily populated areas of the earth. In the last 2 million years, continental ice sheets have repeatedly developed and melted again, each time temporarily upsetting the hydrologic cycle. During maximum glaciation (Chap. 17), the amount of glacier ice on land may have been three times as great as the present ice volume, and sea level was lowered as much as 120 m, exposing most of the continental shelves (Figs. 17–16 and 18–1).

All of these aspects of the hydrologic cycle relate to geomorphology, but one aspect in particular must be emphasized. The average continental height is 823 m above sea level (Fig. 1–4). If we assume that the 37,000 km³ of annual runoff flows downhill an average of 823 m, the potential mechanical power of the system can be calculated. Potentially, the runoff from all lands would continuously generate almost 9×10^9 kW. If all this power were used to erode the land, it would be comparable to having (in the English units of power and area) one horse-drawn scraper or scoop at work on each 3-acre piece of land, day and night, year around. Of course, a large part of the potential energy of the runoff is wasted as frictional heat by the turbulent flow and splashing of water. The "geomorphology machine" has a low but definable efficiency, eroding and transporting rock debris down to the oceans or into arid tectonic basins (Caine, 1976).

ROTATIONAL ENERGY OF THE EARTH-MOON SYSTEM

Terrestrial Gravity

Every mass has an inherent attraction for every other mass. This weak force, called *gravitation,* is an intrinsic property of mass. Although its nature is yet to be defined, we know something about its operation. We can assume, for one thing, that the gravitational force between two large objects operates as though the entire mass of each object were concentrated at the center of the mass, or the center of gravity. Because all objects at the surface have small masses compared to that of the planet, we usually regard the force of gravitation as a tendency for an object at the surface to fall or accelerate toward the center of the earth.

The gravitational force is reduced by the tendency of objects on this rotating spheroid to maintain straight-line motion and hence to be lifted from the surface (the centrifugal effect), especially near the equator. Objects moving horizontally appear to deviate to the right in the northern hemisphere and to the left in the southern hemisphere when viewed by an observer who is being carried in a circular path by the earth's rotation. This *Coriolis effect* is proportional to the sine of latitude and the velocity of the object. It has been claimed to cause northern-hemisphere rivers to erode their right banks preferentially, but the effect is actually much too weak. However, the Coriolis effect is of great importance to atmospheric and oceanic circulation.

Gravity, as we commonly use the word, refers to the net force of gravitation as reduced by the centrifugal and other lesser effects. The geoid is controlled purely by terrestrial gravitation and rotational forces (Fig. 1–3). It is worth noting that the word "level," as in "sea level," does not connote flatness, or planarity, on earth but rather a line or surface parallel to the geoid. A billiard table must be level, but if it were truly flat or planar, it would be tangent to the earth at only one point and all the balls would, in principle, roll to that point and stop. We commonly diagram sea level as a straight line, as in showing the concave long profile of rivers (Fig. 9–8). In reality, of course, the bed of a major river has a convex-skyward longitudinal profile, just as has any line on the geoid. The Mississippi River, for instance, originates in Lake Itasca, Minnesota, at an altitude of 450 m above sea level. In flowing 3800 km southward to sea level, it follows a convex-skyward meridianal arc over 15° in length, more than one-sixth of a quadrant. No reasonably sized scale drawing could show the difference in altitude of head and mouth of the river along the convex arc of its southward course. Furthermore, because of the oblate spheroidal shape of the earth, the mouth of the Mississippi is actually several kilometers "higher" (with respect to the center of the earth) than its source. These amusing but real considerations are easily overlooked in most geomorphic thinking but should at least be mentioned once so that words like "up," "down," "horizontal," and "level" can be seen in their true sense.

All water that falls as rain or snow on the 29 percent of the earth's surface that projects above the sea tends to move downhill by gravity, back to the ocean. Thus every raindrop that strikes the ground has potential energy proportional to the product of its mass and the altitude above sea level at which it strikes. The few localities where dry land extends well below sea level (Death Valley, California, at −86 m; the Dead Sea at −390 m; and others) are exceptions to the rule that sea level is the limit for the downhill flow of water. However, each basin that extends below sea level is in an arid region where few raindrops ever fall. In fact, such basins never last very long in humid regions, because they are soon filled with water to overflowing, and sea level again becomes the ultimate downward limit of water flow from the basins. The tendency of water to flow downhill is the cause of most erosional landforms.

Rock particles as well as raindrops are attracted toward the center of the earth by the force of gravity. When rock material moves downslope under the influence of gravity, but without a transporting agent such as flowing water or glacier ice, the process is called *mass wasting* (Chap. 8). Mass wasting includes not only such spectacular events as landslides and avalanches but also the slow, barely perceptible process of soil creep.

The small contribution of rotational forces to the massive gravity field of the earth may in some way be related to diastrophism. A correlation between earthquake frequency, volcanism, and a "wobble" of the rotational axis of the earth has been reported (Myerson, 1970; Anderson, 1974).

Extraterrestrial Gravity: Tides

The gravitational forces of the moon, sun, and nearby planets are expended as frictional energy in the earth's hydrosphere and lithosphere. The force of extraterrestrial gravity causes variations in the surface gravity field of only about 1 part in 5 million. Most of the energy generated by extraterrestrial gravitation is expended in the oceans, much of it in the turbulent friction of ebbing and flowing tides in shallow marginal seas (Hendershott, 1973; Wahr, 1988). Because of the orbital motions of the planets and our moon, tides vary cyclically, but complexly. According to the estimates used in compiling Figure 5–2, the tidal energy expended on the lithosphere contributes about 10 percent of the geothermal heat flow. It is hard to account for the 5.7 (estimates range from 3 to 6) terawatts (1 TW = 10^{12} W) of power consumed by tidal drag. Perhaps internal waves on boundary layers within the oceans contribute to the consumption. A related problem is that shallow oceanic basins have varied greatly in size and distribution, even within the late Cenozoic Era. Tidal friction in the oceans may be larger now than it has been during most of earth history (Wahr, 1988, p. 240).

Tidal energy is an important factor in coastal processes (Chap. 19), and for most purposes, tides are best studied in the context of the marine environment. However, *earth tides,* with a regular semidiurnal rhythm similar in period to ocean tides but smaller in amplitude, can be measured by sensitive strain meters and gravity meters. The lithosphere apparently has two high and two low tides, with an amplitude of about 30 cm, that follow the moon in its orbit. As the earth rotates daily through the tidal "bulges," rocks are constantly but minutely flexed. The combined effect of oceanic and lithospheric tidal friction is to decrease the number of daily rotations of the earth per year by about 1 day every 10 million years (Wells, 1963). Our moon has long since "locked on" the earth with one rotation per month for the same reason, so that we never see the far side of the moon (Goldreich, 1972).

The role of lunar gravity as a geomorphic process (excluding the important tides in the hydrosphere) is not well known. Tidal friction causes the earth's axis to precess and cyclically changes the intensity of seasonal climates (Fig. 18–4). Some seismologists suspect that earthquakes occur with greater frequency during the full and new phases of the moon, when the sun's gravitational force is additive to the moon's. The hypothesis was strengthened by the discovery that "moonquakes" were recorded by seismometers on the moon with greater frequency every 28.5 days when the moon was at perigee, its closest orbital approach to the earth (Latham et al., 1971).

One of the longer cyclic components of tides is the 18.6-year regression of the lunar nodes, a shift of the lunar orbit that causes a 10 percent variation in the average tide-producing force. Geyser activity in Yellowstone Park has been correlated with the 18.6-year cycle (Rhinehart, 1972). Attempts also have been made to correlate volcanic activity and diastrophism with tide-producing forces (Shaw et al., 1971; Hamilton, 1973). On a very long time scale, it is possible that the slight flexing of crustal rocks by earth tides could lead to "fatigue" or brittle fracture. Joint patterns in young rocks may be propagated upward from older patterns beneath by such weak, but persistent, forces.

INTERNAL HEAT

Geothermal Gradient

Direct observation proves that the interior of the earth is hotter than the exterior. The increase in temperature with depth is the **geothermal gradient,** a typical near-surface gradient being about 20°C/km. Only one well penetrates more than 12 km into the earth, and beyond that depth, the gradient can only be inferred from indirect evidence (Kozlovsky, 1984). Clearly, the near-surface geothermal gradients must decrease sharply at depth; otherwise, the mantle could not behave as a solid and transmit seismic shear waves.

A steep geothermal gradient is obviously important for volcanic processes, including exhalative activity. However, any rock that solidifies or lithifies at depth and is later exposed by erosion has moved upward through the geothermal gradient. Cooler near-surface temperatures should cause thermal contraction of rocks. Fractures caused by elastic stresses in rocks are at least in part the result of their passage upward along the geothermal gradient. A complicating factor is the pressure gradient in the earth that counteracts the geothermal gradient (p. 120).

Heat Flow

The energetics of geomorphic processes are better understood in terms of the energy that flows to the surface from the interior of the earth than from the temperature gradient within the earth. Heat flow is the product of the geothermal gradient and the thermal conductivity of the rocks through which the gradient was measured, plus the heat transported by convective fluid transport. The rate of energy receipt at the earth's surface from internal heat is about 40×10^{12} watts, calculated from the measured average heat flow on continents and in ocean sediments, and with an additional estimate for high heat flow at accreting plate margins (Williams and Von Herzen, 1974; Chapman and Pollack, 1975; Verhoogen, 1980, p. 3). According to Figure 5–2, volcanoes and hot springs carry about 2 percent of the total internal heat flow to the surface by convection, the remainder being conducted to the surface of ocean floors and continents, especially along midocean rises. High heat flow is observed in volcanic regions, but cause and effect are not clear. Other sources estimate that as much as 2 or 3 percent of the total heat flow is dissipated by seismicity.

Most of the geothermal heat flow is generated by radioactive decay of natural isotopes of uranium, thorium, and potassium, with a contribution from rotational or tidal friction, and an uncertain amount of primordial heat from the origin of the earth. Although small in comparison to the solar energy received by the earth, this internal heat must provide nearly the entire energy drive for diastrophism and volcanism. It is odd that the heat flow in volcanic and orogenic areas is rarely more than two or three times greater than the global average. One of the puzzles of geophysics is the small range of heat-flow values. There is no problem with potential sources of heat. Most estimates of radioactivity, rotational energy, and original heat provide an "embarrassment of riches," with radioactivity alone accounting for 60 to 100 percent of the observed heat flow. This contribution would have been even greater during the earlier history of the earth, as would the original heat and the additional release of gravitational and rotational energy associated with core–mantle separation (Verhoogen, 1980, p. 28).

Orogenic Energy

To uplift a mass of sedimentary rocks from a depositional site below sea level to form a mountain range, work must be done against gravity. However, if tectonic uplift were simply to replace the annual erosional mass, only about 7×10^9 W of power would be required (Verhoogen, 1980, p. 8). This is several orders of magnitude less than the power of volcanoes and earthquakes (Fig. 5–2). Along the 200-km length of fold belt in eastern Taiwan, only about 10 percent of the power input does useful work against gravity, a measure of the efficiency of orogeny (Barr and Dahlen, 1989). Most of the tectonic energy is consumed by mechanical deformation of the rock mass and by friction on the subducting lithospheric slab.

Lithospheric plate motions (Chap. 2) are driven almost entirely by thermal and density gradients in the lithosphere and asthenosphere. Therefore, they require no energy not included in the average geothermal heat flow.

Lattice Energy of Minerals; Metamorphic Heat

A difficult minor factor to evaluate in the energetics of geomorphic processes is the highly ordered atomic lattices of mineral phases that crystallize at the high temperatures within the earth. When exposed to weathering reactions, especially oxidation and hydrolysis, these minerals normally react exothermally and produce new phases that are more stable under surface conditions. In a fashion better explained by a physical chemist than by a geomorphologist, the free energy of the original crystal lattice is degraded to lost heat at the earth's surface. When the debris of weathering and erosion is buried at great depths in geosynclines and recrystallized, internal heat is again absorbed. The process has been repeated countless times in earth history. The rate of energy conversion is probably of the same order as that of volcanic heat.

Degradation of Geothermal Energy

At a future distant time, the radiogenic energy source for diastrophism will be expended, and when the surface is worn down, no internal force will be able to renew it again. The half-lives of the radioactive isotopes that supply the heat are on the order of 10^9 or 10^{10} years, however, so in the total geologic history of the earth, only a part of the internal energy supply has been consumed. This is another grand example of an energy system that is slowly declining, although it appears essentially in a steady state even when measured by the scale of geologic time.

REFERENCES

ALVAREZ, L. W., ALVAREZ, W., ASARO, F., and MICHEL, H. V., 1980, Extraterrestrial cause for the Cretaceous-Tertiary extinction: Science, v. 208, pp. 1095–1108.

ANDERSON, D. L., 1974, Earthquakes and the rotation of the earth: Science, v. 186, pp. 49–50.

BARR, T. D., and DAHLEN, F. A., 1989, Brittle frictional mountain building 2. Thermal structure and heat budget: Jour. Geophys. Res., v. 94, pp. 3923–47.

BAUMGARTNER, A., and REICHEL, E., 1975, The world water balance: R. Olenburg Verlag GmbH, Munich, 179 pp.

BERNER, E. K., and BERNER, R. A., 1987, The global water cycle: Prentice-Hall, Inc., Englewood Cliffs, N.J., 397 pp.

VON BERTALANFFY, L., 1950, Theory of open systems in physics and biology: Science, v. 111, pp. 23–29

BLOOM, A. L., 1969, The surface of the earth: Prentice-Hall, Inc., Englewood Cliffs, N.J., 152 pp.

CAINE, N., 1976, A uniform measure of subaerial erosion: Geol. Soc. America Bull., v. 87, pp. 137–40.

CHAPMAN, D. S., and POLLACK, H. N., 1975, Global heat flow: A new look: Earth and Planetary Sci. Letters, v. 28, pp. 23–32.

CHORLEY, R. J., 1962, Geomorphology and general systems theory: U.S. Geol. Survey Prof. Paper 500-B, 10 pp.

DYSON, F. J., 1971, Energy in the universe: Sci. American, v. 225, no. 3, pp. 50–59.

FLOHN, H., 1969, Climate and weather (transl. by B. V. de G. Walden): McGraw-Hill Book Company, New York, 253 pp.

GOLDREICH, P., 1972, Tides and the earth–moon system: Sci. American, v. 226, no. 4, pp. 42–52.

GOODY, R. M., and WALKER, J. C. G., 1972, Atmospheres: Prentice-Hall, Inc., Englewood Cliffs, N.J., 150 pp.

GRIEVE, R. A. F., 1987, Terrestrial impact structures: Ann. Rev. Earth Planet. Sci., v. 15, pp. 245–70.

————, 1990, Impact cratering on the earth: Sci. American, v. 262, no. 4, pp. 66–73.

HÄFELE, W., 1974, A systems approach to energy: Am. Scientist, v. 62, pp. 438–47.

HAMILTON, W. L., 1973, Tidal cycles of volcanic eruptions: Fortnightly to 19 yearly periods: Jour. Geophys. Res., v. 78, pp. 3363–75.

HENDERSHOTT, M. C., 1973, Ocean tides: EOS (Am. Geophys. Un., Trans.), v. 54, pp. 76–86.

HOLDREN, J. P., and EHRLICH, P. R., 1974, Human population and the global environment: Am. Scientist, v. 62, pp. 282–92.

HOUGHTON, H. G., 1954, On the annual heat balance of the northern hemisphere: Jour. Meteorology, v. 11, pp. 1–9.

HUBBERT, M. K., 1971, The energy resources of the earth: Sci. American, v. 225, no. 3, pp. 60–70.

INMAN, D. L., and BRUSH, B. M., 1973, The coastal challenge: Science, v. 181, pp. 20–32.

ISACKS, B. L., 1988, Uplift of the central Andean plateau and bending of the Bolivian orocline: Jour. Geophys. Res., v. 93, no. B4, pp. 3211–31.

KOZLOVSKY, Y. A., 1984, The world's deepest well: Sci. American, v. 251, no. 6, pp. 98–104.

LATHAM, G., EWING, M., DORMAN, J., LAMMLEIN, D., PRESS, F., TOKSOZ, N., SUTTON, G., DUENNEBIER, F., and NAKAMURA, Y., 1971, Moonquakes: Science, v. 174, pp. 687–92.

LVOVITCH, M. I., 1973, Global water balance: EOS (Am. Geophys. Un., Trans.), v. 54, no. 1, pp. 28–42.

MELOSH, H. J., 1989, Impact cratering: A geologic process: Oxford monographs on geology and geophysics no. 11, Oxford Univ. Press, Inc., New York, 245 pp.

MEYBECK, M., 1987, Global chemical weathering of surficial rocks estimated from river dissolved loads: Am. Jour. Sci., v. 287, pp. 401–28.

MILLER, D. H., 1981, Energy at the surface of the earth: Academic Press, Inc., New York, 516 pp.

MYERSON, R. J., 1970, Long-term evidence for the association of earthquakes with the excitation of the Chandler wobble: Jour. Geophys. Res., v. 75, pp. 6612–17.

POWELL, J. W., 1876, Report on the geology of the eastern portion of the Uinta Mountains: U.S. Geol. and Geog. Survey Terr. (Powell), Washington, D.C., 218 pp.

RHINEHART, J. S., 1972, 18.6-Year earth tide regulates geyser activity: Science, v. 177, pp. 346–47.

SHAW, H. R., KISTLER, R. W., and EVERNDEN, J. F., 1971, Sierra Nevada plutonic cycle: Part II, tidal energy and a hypothesis for orogenic–epeirogenic periodicities: Geol. Soc. America Bull., v. 82, pp. 869–96.

SILVER, L. T., and SCHULTZ, P. H., eds., 1982, Geologic implications of impacts of large asteriods and comets on the earth: Geol. Soc. America Spec. Paper 190, 528 pp.

STEVENSON, D. J., 1987, Origin of the moon—the collision hypothesis: Ann. Rev. Earth Planet, Sci., v. 15, pp. 271–315.

SUPPE, J., 1981, Mechanics of mountain building and metamorphism in Taiwan: Mem. Geol. Soc. China, no. 4, pp. 67–89.

VERHOOGEN, J., 1980, Energetics of the earth: National Academy of Sciences, Washington, D.C., 139 pp.

VONDER HAAR, T. H., and SUOMI, V. E., 1969, Satellite observations of the earth's radiation budget: Science, v. 163, pp. 667–69.

WAHR, J. M., 1988, The earth's rotation: Ann. Rev. Earth Planet Sci., v. 16, pp. 231–49.

WALEY, A., 1935, The way and its power: Houghton Mifflin Company, Boston, 262 pp.

WELLS, J. W., 1963, Coral growth and geochronometry: Nature, v. 197, pp. 948–50.

WILLIAMS, D. L., and VON HERZEN, R. P., 1974, Heat loss from the earth: New estimate: Geology, v. 2, pp. 327–28.

WOODWELL, G. M., 1970, Energy cycle of the biosphere: Sci. American, v. 223, no. 3, pp. 64–74.

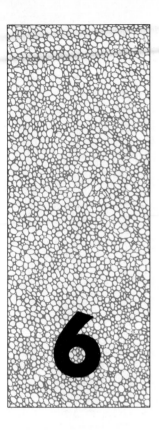

6 Rock Weathering

The processes that alter the physical or chemical state of rocks at or near the surface of the earth, without necessarily eroding or transporting the products of alteration, are collectively called **rock weathering.** Weathering does not make landforms but only altered or broken rock from which landforms are shaped. Rock weathering must always be studied in detail as a set of mineralogical or geochemical processes, because rocks are assemblages of minerals, and every mineral has specific chemical and physical responses to the near-surface environment. However, for many geomorphic generalizations, it is sufficient to speak of weathering at the scale in which an entire mass of rock is the unit of alteration.

Weathering is the assemblage of rock-altering processes that are powered by an exogenic, essentially solar, energy supply. The depth of weathering is thereby restricted by the depth to which exogenic-powered processes can operate.

Silicate minerals are strong and brittle solids in the environmental of the shallow continental crust as long as aqueous solutions are not present. However, numerous observations and experiments verify that in the presence of water, rock-forming minerals lose strength and are much more easily broken and deformed (Kirby, 1984, and references therein). These mineralogical studies become geomorphically significant with the demonstrations that *meteoric water* (derived from the hydrologic cycle) circulates to depths of 10 to 20 km. A deep well drilled in the Soviet Union encountered copious water to a depth of 12 km, although this hot, heavily mineralized water may have been derived from metamorphic processes (Kozlovsky, 1984). Fracture permeability permits groundwater to circulate to depths of 10 km or even 20 km along fault zones (Costain et al., 1987; Nesbitt and Muehlenbachs, 1989). Circulation to depths greater than 3 km is a widespread phenomenon in the Appalachian provinces of the eastern United States (Tillman, 1980), and water of meteoric isotopic composition circulated around Eocene-age plutons in Idaho to depths of 5 to 7 km (Criss and Taylor, 1983). Thus rocks are altered, at least in part by exogenic processes, at depths of many kilometers and for millions of years before they are exposed by uplift and erosion to create a landscape.

Weathering is distinguished from other destructive processes by the inclusion in its definition of the concept of in situ, or nontransported, alteration. Mass wasting and erosion, considered in subsequent chapters, always involve translocation or transportation of material. Of course, when a mineral grain reacts with water, the soluble products are carried away in solution. Yet the basic structures of the rock mass, including bedding and other primary layered structures, granularity, and intrusive dikes or veins, are not obliterated until extremely advanced stages of weathering (Fig. 6–1). Weathering is the precursor of mass wasting and erosion, although weathering continues even after a rock fragment has been dislodged from a hillside ledge or pried loose from the bed of a stream. Indeed, weathering processes do not end until a rock fragment has been finally buried or submerged beyond contact with the atmosphere or circulating groundwater.

A rational classification of weathering processes must be arbitrary. Traditionally, if artificially, weathering processes have been subdivided into a mechanical, or physical, group and a chemical group (Reiche, 1950). The recently recognized importance of organic processes in weathering now requires the additional consideration of biological weathering. Mechanical weathering is also called **disintegration,** implying that pieces of rock are taken apart or disaggregated without alteration. In contrast, chemical weathering is also called **decomposition,** emphasizing the breakdown of the chemical composition of the mineral grains that make a rock. The more that is learned about weathering processes, the less sharp becomes the distinction among the mechanical, chemical, and biological categories because the same thermodynamic principles govern all three. The most rational geomorphic approach is to consider mechanical weathering processes first, because mechanical fracturing of rocks usually is necessary before air, water, and organisms can begin their largely chemical attack.

Figure 6-1. Deeply weathered syenite under glacial drift in southwestern Maine. A narrow basalt dike cuts the syenite vertically, left of the shovel. Horizontal foliation and the dike have not been disrupted, although the entire exposure can be dug with a shovel.

MECHANICAL WEATHERING PROCESSES

Whenever a brittle mass such as rock is fractured, whether by internal or external stresses, the resulting fragments have sizes with a statistical distribution that follows *Rosin's law* (Bennett, 1936; Blatt et al., 1972, pp. 37, 58). Rosin distribution is not Gaussian, or Normal; instead, fracturing produces an excess of fine fragments. The theory of crushing has important applications to mining and quarrying operations, but it is also instructive as a rigorous mathematical explanation of why rocks broken by any mechanical process yield many small fragments with a large total surface area.

Mechanical fracturing is controlled by preexisting cracks in or between mineral grains. Cracks form from thermal, mechanical, and chemical stresses that few rocks can escape. Microfractures and micropores as small as 1 to 10 μm have been photographed (Brace et al., 1972) (Fig. 6-2). The fundamental elastic properties of rocks and minerals are probably controlled by cracks, so a consideration of mechanical weathering can be built upon an assumption of their presence (Stout, 1974; Simmons et al., 1975). Tensile microcracks grow much more rapidly and easily in ceramics, glasses, and minerals if moisture is present, again emphasizing the merging of physical and chemical processes in weathering (Kirby, 1984, p. 3992, and references therein; Michalske and Bunker, 1987).

The most important processes by which rocks are mechanically broken or disintegrated are: (1) differential expansion with pressure release; (2) thermal expansion and contraction, including fire damage; and (3) growth of foreign crystals in cracks and pores. The mechanical pressures generated by growing and moving organisms might be important enough to rate a place on this list, as well. Each of these processes affects different rock types in different ways, and the second and third processes are strongly dependent on climatic conditions.

Figure 6-2. Photomicrograph of the boundary between grains of feldspar (upper left) and quartz (lower right) in Rutland quartzite. Feldspar grain is full of pores or tubes about 1 μm in diameter. On the grain boundary is a nearly continuous crack with larger triangular cavities aligned on it. (Photo: W. F. Brace.)

A common effect of weathering is the detachment of slabs, sheets, spalls, or chips from rock surfaces. **Exfoliation** is the general term for the loosening or separation of concentric surface shells or layers of rock. Exfoliation is caused by chemical as well as thermal and physical processes, but only the physical processes and results are described in the following section. Other examples of exfoliation are described in subsequent sections.

Pressure Release on Unloading: Sheeting

Several lines of evidence can be used to prove that some rocks now at the earth's surface were formerly buried at depths of 20 km or more. The reconstructed geometry of structures in the Valley and Ridge province of the Appalachian Mountains requires that at least 8 km of folded sedimentary rocks have been removed since late Paleozoic time. From the metamorphic grades of certain mineral assemblages, 24 km of rock is estimated to have been eroded from the southeastern Adirondacks (Whitney and McLelland, 1973). Post-Jurassic erosion to a depth of 8 to 16 km has been required to expose the igneous rocks of the Sierra Nevada (Bateman and Eaton, 1967); 16 to 24 km of post-Triassic erosion has exposed the tightly folded sedimentary rocks of the Wellington District, New Zealand (Stevens, 1974, p. 26). The Alps in Switzerland have suffered 30 km of denudation in the last 30 million years, one of the largest values and most rapid rates to be found in the literature (Clark and Jager, 1969).

The confining pressure that is typically released by erosion of the magnitudes just listed is on the order of 1.5 to 8×10^5 kPa (1.5 to 8 kilobar). The coefficients of elastic expansion of typical rocks are such that expansion of 0.1 to 0.8 percent must result (Birch, 1966). However, thermal contraction due to cooling and the increased area of an uplifted segment of spherical crust subtended by a constant solid angle should more than compensate for the expansion due to unloading of crystalline rocks, and the predicted state of stress in uplifted, cooled, and unloaded rock masses should be strongly tensile (Haxby and Turcotte, 1976; Bruner, 1984). As most surface rocks are in a state of strong horizontal compression, some other factor must be negating the predicted thermal contraction. That factor seems to be the growth of microcracks within and between mineral grains due to their anisotropic contraction during cooling (Bruner, 1984). The net result is that most rocks are under compressive stresses measured as high as 41 MPa (0.4 kilobar) in surface exposures. Newly quarried blocks of Stone Mountain granite (Fig. 6–3) expand 0.1 percent in the direction of greatest confinement (Hopson, 1958, p. 70). If a massive rock is confined on all sides but one, corresponding to the land surface, the expansion must be toward the free face.

Quarry workers and miners are well aware of dangers of *rock bursts* on a freshly excavated quarry or mine floor or wall. The rapid pressure drop at the surface can cause slabs of rocks to bulge or burst into the mined-out space, sometimes with explosive force. It is common practice in mining and quarrying operations to detonate explosive charges at the end of a work shift, allowing time for the immediate and most dangerous expansive stresses to decay before the next crew enters the area.

Sheeting joints or simply **sheeting,** the separation of rock into layers parallel to the land surface, is a near-surface phenomenon, most common in massive rocks such as igneous intrusives, quartzites, and thickly bedded sandstones (Fig. 6–3). Rocks with other types of fractures, such as columnar joints or bedding planes, are able to expand upon unloading without developing sheeting, even though they may develop minor sets of *relief joints* parallel to gorge walls and excavations. Granite in quarries near Boston, Massachusetts, has an average thickness of joint-bounded sheets of 10 cm to 1 m near the surface, but the average sheet thickness increases

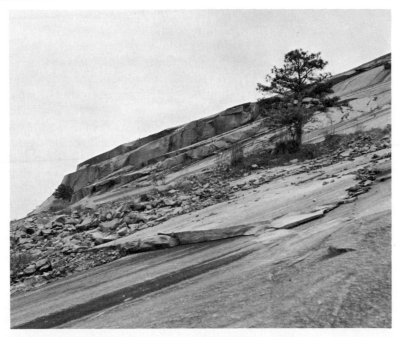

Figure 6-3. Sheeting joints on the southeast flank of Stone Mountain, a granite dome near Atlanta, Georgia. An estimated 12 to 15 km of overlying rock have been removed from Stone Mountain in 300 million years (Whitney et al., 1976). (Photo: C. A. Hopson.)

rapidly to about 5 m at a depth of 20 m, and to more than 10 m at depths of 30 to 40 m. At greater depths, the sheeting joints become so widely spaced as to be insignificant (Jahns, 1943, p. 97). Sheeting joints on granitic monoliths in Yosemite National Park, California, are spaced from a few centimeters apart at the surface to a few meters apart at depth. The aggregate thickness of the sheets rarely exceeds 30 m (Matthes, 1930, p. 116). In tunnels through sandstone at the Glen Canyon dam site on the Colorado Plateau, all joints parallel to the canyon wall were encountered within 10 m of the surface (Bradley, 1963, p. 522).

Few rock structures have been attributed to so many causes as sheeting. Jahns (1943, p. 78) reviewed seven processes that deserve serious consideration as the specific cause of sheeting:

1. Tensional or contractional strains set up during cooling of an igneous mass.
2. Local or regional compressional stresses (tectonic).
3. Insolation, with attendant daily and secular temperature changes.
4. Progressive hydration and formation of chemical alteration products in susceptible minerals.
5. Mechanical action of fire, frost, and vegetation.
6. Diminution of primary confining pressure by removal of superincumbent load.
7. Combinations of the foregoing causes.

Of the various proposed causes of sheeting, only number 6 generally satisfies the observed field relations. A minority opinion favors number 2, however (Twidale, 1973). Sheeting cuts across primary structures and intrusive contacts within igneous rocks; the curvature and orientation are much too local to be related to tectonic stresses; it is similar in climatic regions of great insolation variation, such as polar and tropical, wet and dry regions; the minerals within the sheets show negligible chemical alteration; and sheeting invariably conforms to the present or a previous land surface (Dale, 1923; Hopson, 1958).

Sheeting joints develop only near the land surface probably because the geothermal gradient decreases to much less than normal in the outer 200 m of the crust and the temperature of the rock becomes nearly constant. The overburden pressure, however, continues to decrease at a rate proportional to the rate at which overlying rock is peeled off by erosion. No compensating contraction due to decreased temperature is possible at shallow depths, so the expanding outer layers of rock are bowed upward or outward toward the land surface, and sheeting joints develop. Preexisting microcracks determine the final fracturing, and because they are so abundant in rocks, slight directional stresses can cause large, continuous joints to propagate.

Massive igneous rocks and thick-bedded sandstones typically have low geothermal gradients. Their "tight" structure permits efficient heat conduction. These are the same rocks that exhibit the best sheeting joints, not only because no other fractures are available to absorb the expansion or release of pressure, but because the low geothermal gradient provides a minimal compensatory contraction upon cooling.

A variety of systematic and non-systematic joints other than sheeting may also develop in rocks because of thermal and tectonic stresses. One of the identifying characteristics of all rock formations is the spacing and degree of development of jointing. Joints are important structural controls for differential erosion (Chap. 11). The size of monolithic domes is determined largely by spacing of faults and joints other than sheeting, but their shape is determined by sheeting. If a rock mass is irregularly fractured at depth, the most fractured portions adjust readily as the superincumbent load is removed, but portions with few or widely spaced fractures store the stress of expansion until the overburden pressure is so low that failure occurs. Such failure is parallel to the land surface at a depth of 100 m or less, so the sheeting joints conform closely to local topography. In glaciated mountains, sheeting on cirque walls and floors is concave skyward, parallel to the bowl shape of the cirques, whereas the sheeting joints are convex on the adjacent convex summits (Ollier, 1969, p. 6). At the rims of cirques, the upland and cirque-wall joint sets intersect.

A combination of relief joints and expansion toward a free face upon unloading produces structures that are controlled by erosion but that subsequently control erosion. A notable correspondence of valleys with small-scale anticlines in Northampton, England, was explained by a process called *cambering,* whereby brittle sandstone beds fractured along vertical joints and moved apart, while underlying plastic claystones flowed or bulged upward between the separated blocks (Hollingsworth et al., 1944). A sinuous anticline in the Muav Limestone along the floor of the Grand Canyon has been similarly attributed to elastic rebound or plastic flow caused by erosion-induced pressure release (Sturgul and Grinshpan, 1975). Large-scale circular or arcuate patterns in the crystalline terrane of North Carolina have been attributed to the structural control of sheeting (Hack, 1966). Structurally controlled erosion is examined in greater detail in Chapter 11.

Thermal Expansion and Contraction

For rocks buried at shallow depth, thermal contraction or expansion is minor except during cooling of igneous rocks. However, a dark-colored rock in bright sunlight on a desert or mountain top may heat from below 0°C to 50°C each day (Roth, 1965) and cool a comparable amount each night. Rock is a poor conductor of heat, so the diurnal thermal effects are shallow, extending perhaps only a few centimeters deep. Shattered, angular rock fragments litter the stony plateaus in arid parts of Australia. The fragments can be reassembled into cobbles and boulders up to 30 cm in diameter. Ollier (1963, 1969) suggested that these "gibbers" are heat-cracked, because there is no other plausible weathering process. Thermal "crazing" by a network of fine surface cracks was also thought by Ollier to shatter the surfaces of quartzite boulders to a depth of 10 cm or less and to produce blocky detritus.

Strangely, no one has ever verified experimentally that solar heating is intense enough to break stones. Specimens of granite have been heated dry to 105°C and cooled to −10°C for 2000 cycles with no evidence of deterioration (Kessler et al., 1940). In another often-quoted study, Griggs (1936) alternately heated the surface of a 3-inch cube of dry granite for 5 minutes to about 140°C, then cooled it for 10 minutes with an electric fan to about 30°C, for 89,400 cycles, equal to 244 years of diurnal heating. His experiment lasted over three years, but no change in the rock specimens could be detected, even by microscopic examination. Larger specimens may be more subject to thermal stresses (Rice, 1976). There are also stories of rocks shattering in extreme cold, their fragments strewn about over the snow (Washburn, 1969, pp. 32–33). Even if the process is insignificant on earth, thermal expansion–contraction is likely to be one of the major processes of lunar weathering.

Fire has a notably destructive effect on rocks, especially those with granitic texture. Fire, alternating with sudden water quenching, was probably the principal quarrying tool of primitive people. Generally, limestone (including marble) and sandstone are more resistant than granite to fire damage, and coarse-grained rocks are less resistant than fine-grained rocks to fire damage (Tarr, 1915, p. 349). After a forest fire, granitic boulders are commonly mottled with freshly exposed areas and the ground around them is littered with spalls and flakes. Unequal thermal expansion of the dominant minerals is the controlling factor. For instance, the thermal volumetric expansion of quartz is three times that of feldspar. In addition, quartz, feldspar, and many other common minerals expand under heat in a highly anisotropic fashion, with expansion along certain crystallographic axes being as much as 20 times greater than along other axes.

Growth of Foreign Crystals in Cracks and Pores

Certain minerals, such as pyrite (FeS_2) in shale or slate, oxidize readily on exposure to oxygen-bearing groundwater or a moist atmosphere. The newly formed iron oxides are of lower density and larger volume than the original minerals, and the volume increase can be sufficient to split weaker rocks or cause spalls or "pop-outs" in stronger rocks.

Of greater significance to weathering processes are the numerous hydrated salts that are water-soluble (Evans, 1970). When pore water percolates to the surface of a rock and evaporates, any dissolved salts precipitate. Rainwater may also accumulate salts, especially when it flows over soot-stained city buildings and through masonry joints. The force of crystallization around nuclei of precipitating salts is substantial. Dendritic salt crystals, or even crusts, may form in cracks, along mortar joints, or on grain boundaries within a permeable but otherwise coherent rock.

A particularly troublesome salt is hydrous calcium sulfate (the mineral gypsum). In industrial regions, where the air is polluted by sulfurous smoke from fossil fuels, rainwater becomes a dilute sulfuric acid (Likens and Bormann, 1974; Schwartz, 1989) that corrodes limestone and marble buildings. The reaction product of the corrosion is gypsum, a relatively insoluble salt that crystallizes in rock cracks, flakes off thin pieces of rock, and accelerates chemical attack. The formation of salt crystals in rock openings is regarded as a mechanical weathering process, but the close relationship with chemical processes is obvious.

Tafoni, or honeycomb weathering pits, are a common feature of deserts and semiarid regions, especially near coasts or where dew or fog is frequent (Mustoe, 1982; Young, 1987). Hemispherical pits or small caves form on the undersides of overhanging blocks or in exposed cliffs of a variety of rocks, including granite, sandstone, and volcaniclastic rocks. Some of the pits have overhanging visors or rims that partly enclose the opening. In granite, tafoni grow by flaking (p. 142) of centimeter-sized thin sheets from the inner walls of the opening (Dragovich, 1967). The flakes have been shown to be enriched in gypsum and halite relative to the mineralogy of the fresh granite (Bradley et al., 1978, 1980; Young, 1987). In permeable sandstones, volcaniclastic rocks, and limestone, secondary cementation by precipitates in the outer layers of rock may contribute to the resistant visors or "monks hoods" that partly enclose tafoni (Martini, 1978; Winkler, 1980).

Many water-soluble salts precipitate as hydrates or invert to hydrated compounds within the normal ranges of atmospheric temperature and humidity. Hydration pressure is more effective than simple crystal growth, because as temperature and humidity change seasonally or daily, the salts may hydrate and dehydrate repeatedly (Winkler and Wilhelm, 1970). The hydration pressures of common salts range from 1×10^4 kPa to over 2×10^5 kPa (100 to over 2000 bar). In theory, the pressure should be highest at low temperature and high relative humidity.

Mirabilite (Glauber's salt, $Na_2SO_4 \cdot 10H_2O$) dehydrates to thenardite (Na_2SO_4) readily, quickly, and reversibly. At 39°C, the dehydration is complete in about 20 minutes (Winkler and Wilhelm, 1970, p. 570). The standard test for building-stone durability makes use of this hydrated sodium sulfate salt. Blocks of construction-grade granite that can be subjected to 5000 cycles of alternate freezing (6 hours at -12°C) and thawing (1 hour in water at 20°C) with barely visible disintegration crumble after an average of only 42 cycles when soaked in a saturated solution of sodium sulfate (17 hours at room temperature) and dried (7 hours at 105°C) (Kessler et al., 1940; Evans, 1970).

Until the last century, salt precipitation and hydration were weathering phenomena of concern only in desert regions with a high water table, as, for example, along the Nile River or the Rio Grande. However, many of the same salts are leached out of bricks and mortar and attack stone and concrete masonry in ancient and modern cities. The physical force of hydrating crystals is causing growing concern over ways to maintain and preserve historical monuments in heavily populated industrialized regions (Fig. 6–4) (Winkler, 1973; Committee on Conservation of Historic Stone Buildings and Monuments, 1982).

Frost Cracking and Hydrofracturing

A foreign crystal of peculiar abundance in surface rocks is ice. Water volumetrically expands 9 percent on freezing at 0°C, but the expansion ratio increases to 13.5 percent for confined water at −22°C. The maximum pressure generated is 207 MPa (2070 bar) (Fig. 16–5). At higher pressures, more dense phases of ice crystallize (Tharp, 1987, p. 94). Intuitively, the impressive pressures of freezing that break cast-iron water pipes and soft-drink containers in freezers are assumed to break rocks. However, the actual process of *frost cracking* is much more complicated than the simple total freezing of a fully confined mass. Two popular concepts—(1) that expansion of confined water by freezing breaks rocks, and (2) that the process is enhanced by frequent cycles through the freezing temperature—are both probably wrong.

Few rocks could withstand the high pressure generated by freezing confined water. However, total confinement is required, and there is no natural way for water to be totally confined in rock. A more likely cause of frost cracking, sup-

Figure 6-4. Detail of the entrance vault at Cathedral St. Coretin, Quimper, France. Original carvings of the late fifteenth century are in Caen Stone, a soft Jurassic limestone widely used in western European architecture. Three inner archivolts have been restored recently, with similar stone. Outer carvings have been nearly obliterated by 500 years of weathering.

ported both by observation and thermodynamic considerations, is that a capillary film of water along the ice-rock interface resists freezing, because of the strong molecular adhesion of water to rock. As the temperature is lowered and the freezing front advances into the rock, some of this water freezes, and adjacent pore water migrates to the area of freezing to maintain the thickness of the capillary film and preserve thermodynamic equilibrium. The crack walls are forced apart by the increasing volume of water and ice, but the expansion is limited by the availability of unfrozen water and the permeability of the adjacent rock. The maximum pressure generated is an order of magnitude less than the pressure at the same temperature in a confined system (Tharp, 1987, p. 95). Frost cracking is thus promoted by a rather slow but steady cooling rate of -0.1 to $-0.5°C/hr$, maintained in a temperature range of -4 to $-15°C$. At higher temperatures, thermodynamic limitations prevent significant ice growth; at colder temperatures or with more rapid cooling, the necessary migration of pore water to the growing ice wedge is inhibited (Walder and Hallet, 1985). In such an open system, water is drawn toward the freezing zone, not driven away from it, as the pressure rise in a closed system would cause. Only in the special closed system of a block of permeable rock freezing inward from all surfaces, would water be driven into the unfrozen interior and perhaps cause *hydrofracturing* along pressurized water-filled joints or microcracks. Normally, capillary water will not generate sufficient stresses to propagate cracks (Tharp, 1987).

Many mountain slopes above the treeline are wastelands of shattered rock rubble that is gradually being broken down to particle sizes that can be attacked by other weathering and erosional agents. In humid subpolar regions of perennially frozen ground, freezing and hydrofracturing become the dominant weathering processes (Chap. 15).

Water in shallow soil or at the ground surface freezes at 0°C and freely expands 9 percent because it is unconfined. Small pebbles, garden soil, and grassy turf are "heaved" by unconfined freezing, but no great pressure is involved. During repeated cycles of freezing and thawing, pebbles and cobbles are lifted by the expanding ice and then supported by finer grains washed under them during melting, causing crude sorting.

Thin films of water that form a semicrystalline "ordered" molecular structure on mineral surfaces (especially on clays) can pry apart microfractures and propagate cracks (Dunn and Hudec, 1966; Winkler, 1968). The powerful molecular forces in films of pure water and in saline solutions are analogous to those involved in freezing, drawing more water to the region where the ordered structure is developing (Evans, 1970, p. 167). Studies of "ordered films" of water once more emphasize the disappearing distinction between mechanical and chemical weathering processes.

Plants and Animals as Agents of Mechanical Weathering

The prying or wedging action of growing plant roots, especially of trees, is often described as mechanical weathering. Two-dimensional networks or sheets of interlaced roots can be followed for many meters along bedding planes or joints, deep into fresh rock. It has been supposed that the growing roots exert a pressure on the rock and force cracks to open. However, the efficacy of roots as agents of mechanical weathering probably has been overestimated, just as their importance as agents of chemical weathering has been underestimated. Roots follow the paths of least resistance and conform to each little irregularity of a crack, but they do not exert much force on the rock. Cracks opened by other processes can be maintained by roots, however, and decaying vegetable matter and washed-in dirt can keep rock surfaces wet and chemically active. In addition, when trees sway in a strong wind, their roots pry apart rocks.

Figure 6-5. Stems of *Ficus* encasing prehistoric stone blocks, Kosrae Island, Micronesia. Blocks are 30 to 50 cm high.

The "strangler fig" (*Ficus* sp.) of tropical regions has the ability to envelop other trees, masonry, and bedrock with snakelike, anastomosing stems (Fig. 6–5). As the stems grow and cross each other, they fuse, encasing a host tree with a wooden latticework sufficiently strong to prevent its further growth and cause its eventual death (Argo, 1964). They also pry apart masonry and natural joint blocks, but it is doubtful that they could crush or pry apart sound rock. The mature *Ficus* may grow to huge size, however, and when such a tree is toppled during a storm, large masses of rock may be displaced. Under temperate and boreal forests, soils usually include many mounds and pits formed by tree roots that were torn up when trees fell (Fanning and Fanning, 1989, pp. 54–61). Because translocation is so obviously involved, *root throw* is better treated as a category of erosion than as weathering.

Soils teem with animal as well as plant life. It is claimed that the species diversity of soil animals, even in a small area, equals that of a coral reef. Almost all the animal phyla have representatives that spend at least part of their life within the soil layer (Hole, 1981). They digest, abrade, and mix mineral grains, admit water and air to the soil through their burrows, and supply chemically active humus and excrement to the soil. Most of their activity merges mechanical and chemical weathering with mass wasting and erosion. Larger animals, including human beings, are obvious excavators of soil and rock, and are primarily erosional agents.

CHEMICAL WEATHERING PROCESSES

The general trends of chemical weathering can be predicted from the conditions under which rock-forming minerals crystallize or precipitate. If they formed in an environment of high thermal energy and high pressure, the minerals in rocks at the surface of the earth tend to weather by exothermic chemical reactions that produce new compounds of greater volume and lower density. **Oxidation** is one of the most typical exothermic, volume-increasing reactions between minerals and the wet atmosphere; especially common is the reaction of iron-bearing minerals with oxygen dissolved in water. Other typical weathering reactions are **carbonation,** the reaction of minerals with dissolved CO_2 in water; **hydrolysis,** the decomposition and reaction with water; **hydration,** the addition of water to the molecular structure of a mineral; **base exchange,** the exchange of one cation for another between a solution and a mineral solid; and **chelation,** the incorporation of cations from the mineral into organic compounds. Hydration has been discussed in a preceding section as a primarily mechanical process, as demonstrated by the volumetric expansion of hydrated salts that enter cracks and pores, yet it is certainly a chemical process when the water is absorbed or reacts with the minerals that make the rock.

An important principle of chemical weathering is that mineral dissolution is concentrated along sites of excess surface energy such as crystal dislocations, cleavage planes, microcracks, and other crystal imperfections. In photomicrographs, grains of weathered feldspars, pyroxenes, and amphiboles all show micron-sized etch pits aligned on crystal defects. Chemical weathering is governed by the kinetics of chemical reactions at these activated sites on mineral surfaces, and not by diffusion rates. Contrary to earlier theories, chemical weathering of mineral grains continues at a nearly linear rate rather than being progressively slowed by diffusion through a thickening leached outer zone or a layer of reaction products (Berner and Holdren, 1979; Holdren and Berner, 1979; Berner et al., 1980; Aagaard and Helgeson, 1982).

Water in Chemical Weathering

All chemical weathering reactions on this planet involve water either as a reactant or as the carrier of the reaction products. It is not necessary, therefore, to list *solution* as a unique chemical weathering process, because solutions in water are always a part of chemical weathering.

Water is by far the most abundant material near the surface of the earth. In the outer 5 km of the earth, water is about three times as abundant as the next most common substance, the feldspar group of minerals (Kuenen, 1955). Water is the only compound that occurs naturally at the earth's surface in gaseous, liquid, and solid states. Its general solvent power and its surface tension are greater than those of any other fluid. Its heat of vaporization is the highest of all substances; recall the significance of this property in the conversion of solar energy to geomorphic work (Chap. 5). Its maximum density at 4°C in the liquid phase, with expansion toward the freezing temperature as well as away from it, is a strange and almost unique property. A recitation of the properties of water could go on at length, but this summary is enough to suggest that the outermost few kilometers of the earth are generally saturated with an abundant and chemically active compound, made available in unending supply by the hydrologic cycle, that readily attacks and reacts with rock-forming minerals. The chemical potential of rain is largely expended by various reactions with minerals soon after the drops penetrate the ground. River water has just about the dissolved chemical load that could be predicted from analyzing the rocks through and over which it has flowed.

Oxidation: Weathering of Iron-Bearing Minerals

Iron-bearing minerals commonly contain iron in the ferrous state. This is true for the major iron sulfide (pyrite, FeS_2), iron carbonate (siderite, $FeCO_3$), and various iron silicates. On contact with water that has oxygen in solution, the ferrous iron oxidizes to the ferric state and forms nearly totally insoluble ferric oxides or hydroxides. Only oxygen-free, chemically reducing, alkaline water permits iron to remain in the more soluble ferrous state. Thus, when aerated water reacts with iron-bearing minerals, the reaction is more complex than simple solution. *Oxidation-reduction reactions* are inevitable, with the result that instead of the mineral going into hydrous solution as ionic species, new compounds are formed, some of which are nearly insoluble.

Weathering by oxidation takes place with water as the intermediary. Unprotected iron surfaces stay clean and bright in dry air, as is proved by the condition of tools and equipment abandoned and later found in the polar regions or of aircraft that have been found in the Libyan desert decades after they were abandoned. But iron rusts, or oxidizes, quickly beneath even a dew-thick film of moisture. There is always enough dissolved oxygen in rainwater and circulating groundwater to oxidize metallic iron and to change the ferrous iron in mineral compounds to the more oxidized ferric state. As long as water is in contact with atmospheric molecular oxygen on the one hand and incompletely oxidized iron on the other hand, oxygen dissolves from the air, diffuses through the water, and combines with the iron. When the iron or other elements from a mineral grain combine with oxygen, the original mineral structure is destroyed, and the remaining mineral components are free to participate in other chemical reactions. The oxidation-reduction reactions of water with compounds of aluminum, magnesium, manganese, chromium, and some less common metals are similar to the reactions with iron compounds. The several oxides often occur together.

Iron oxides have intense yellow or red colors. Rocks and soil are stained various shades of red or brown by mere traces of iron oxide. The reactions are so rapid that a brown surface layer is the first indication of weathering on most rocks. Black iron-sulfide muds from the anaerobic conditions of a swamp, for instance, oxidize and develop a brown color in a few minutes after exposure to air. Iron-rich rock-forming minerals such as pyroxenes, amphiboles, and biotite are usually rust-colored on all but the most recent natural exposures.

Iron oxides, and the oxides of related metals, are exceptionally stable chemical compounds, even on the geologic time scale. Precambrian iron-rich rocks are widely distributed on the continents, most of them dating from the time about 2.5 billion years ago or less when our evolving atmosphere began to be oxidizing (Pinto and Holland, 1988). It has been suggested that every Phanerozoic geologic system contains "ironstones," or iron-oxide-rich sedimentary formations, that formed during contemporaneous weathering and have thereafter persisted to the present (Degens, 1965, p. 87).

Hydrolysis and Base Exchange: Reactions of Carbonated Water with Silicates

Hydrolysis, like oxidation, is more than just a mineral dissolving in water; both the mineral and water molecules decompose and react to form new compounds. It is assumed that hydrolysis takes place in a great excess of water, that is, in an aqueous solution. It is the most important chemical weathering reaction of silicate minerals.

Pure water ionizes only slightly, but it does react with some easily weathered silicate minerals in the following fashion:

$$Mg_2SiO_4 + 4H^+ + 4OH^- \rightarrow 2Mg^{2+} + 4OH^- + H_4SiO_4$$

$$\underset{\substack{\text{olivine} \\ \text{(forsterite)}}}{} + \underset{\substack{\text{four ionized} \\ \text{water molecules}}}{} = \underset{\substack{\text{ions} \\ \text{in solution}}}{} + \underset{\substack{\text{silicic acid} \\ \text{in solution}}}{}$$

The result of such complete hydrolysis is that the mineral is entirely dissolved, assuming that a great excess of water is available to carry the ions in solution. Silicic acid, one of the reaction products, is such a weak acid that we can disregard its name and simply think of it as silica (SiO_2) dissolved in water.

The foregoing reaction is deceptively simple, because it implies that water is an ionized compound. Such is not the case. Pure water is known to be a very poor H^+ donor and electrolyte. Its electrical conductivity is extremely low, proving that it has few free ions. Of great importance to weathering processes, therefore, is the fact that carbon dioxide gas dissolves readily in water to form a weak acid, or H^+ donor. Even though the terrestrial atmosphere contains only 0.03 percent CO_2 (by weight), raindrops absorb CO_2 as they fall, so that rainwater is usually slightly acidic, with a pH of 5.7 or less. The reversible reactions between water and carbon dioxide are simplified as follows:

$$CO_2 + H_2O \rightleftharpoons H_2CO_3 \rightleftharpoons H^+ + HCO_3^-$$

$$\underset{\text{gas}}{} + \underset{\text{water}}{} = \underset{\substack{\text{carbonic} \\ \text{acid}}}{} = \underset{\substack{\text{hydrogen} \\ \text{ion}}}{} + \underset{\substack{\text{bicarbonate} \\ \text{ion}}}{}$$

Cold water dissolves more carbon dioxide gas than does warm water; water under pressure also dissolves more gas. For every condition of temperature and pressure, an equilibrium is established among the several compounds and ions so that an increase or decrease of any one of them is offset by shifts among the others to distribute the change evenly. For instance, an increase in available carbon dioxide gas quickly causes an increase in acidity of the equilibrium aqueous solution. Addition of a strong mineral acid such as H_2SO_4 from atmospheric pollution drives the equilibria to the left and decreases the amount of bicarbonate ion in solution (Johnson et al., 1972).

Air that fills pore spaces in soil is greatly enriched in carbon dioxide by the decay of humus. Soil air drawn from the biologically active upper layer of a soil profile may have from 10 to 1000 times more CO_2 than that in the free atmosphere; that is, CO_2 may constitute as much as 30 percent of soil air as compared to the 0.03 percent of dry normal atmosphere (p. 150). Biogenic carbon dioxide in soil air is surely the major source of carbonated groundwater.

Any reaction that increases the H^+ ion concentration in water also increases the effectiveness of hydrolysis. Carbon dioxide dissolving in soil water is the most common and important way that water is provided with H^+ ions, or acidified, for hydrolysis. We can elaborate the chemical equation previously given for the reaction between water and olivine as follows:

$$Mg_2SiO_4 + 4CO_2 + 4H_2O \rightarrow \quad 2Mg^{2+} + 4HCO_3^- \quad + H_4SiO_4$$

olivine (forsterite) + carbon dioxide + water = magnesium and bicarbonate ions in solution + silicic acid in solution

This hydrolysis reaction is much more common than the one with pure water shown previously. Note that the H^+ ions that react to form silicic acid and bicarbonate are derived from water molecules through the carbonation process and the accompanying formation of bicarbonate ions in solution. Carbonic acid is consumed by silicate weathering, and the resulting solutions become more alkaline because of the dissolved bicarbonate. The common cations freed by silicate weathering are quite soluble in solution with bicarbonate ions, and most of the dissolved load of rivers is in the form of bicarbonates.

Quartz (SiO_2) is an extremely common silicate mineral, but it is only slightly soluble in water, and solution reaction is very slow. Silica glass and other polymorphs of quartz are significantly more soluble in water than is quartz, but they are not common except in volcanic terranes. The world average for the dissolved silica concentration in river water is 13 parts per million, about twice the calculated solubility of quartz (Livingston, 1963). Therefore, reactions such as those given previously for olivine and those that follow for feldspars are the most likely sources for silica in solution. Quartz (the oxide of silicon) is chemically inert at the earth's surface, and can be classed with the oxides of iron and aluminum as stable compounds.

Probably the most common weathering reaction on earth is the hydrolysis of feldspars by carbonic acid in water. A typical, although simplified weathering reaction between potassium feldspar (orthoclase) and carbonated water is as follows:

$$2KAlSi_3O_8 + 2H_2CO_3 + 9H_2O \rightarrow Al_2Si_2O_5(OH)_4 + 4H_4SiO_4 + \quad 2K^+ + 2HCO_3^-$$

orthoclase + carbonic acid + water = kaolinite— a clay mineral + silicic acid in solution + potassium and bicarbonate ions in solution

In carbonated water, calcium and sodium feldspars (plagioclase) hydrolyze even more readily than does orthoclase. In the foregoing example, as in many others, the initial mineral and acid are consumed and the end products are: (1) a detrital clay mineral, (2) silica in solution, and (3) bicarbonate of potassium, sodium, or calcium in solution. The process is essentially a surface reaction between the solution and the mineral grain (Berner and Holdren, 1977, 1979). Clay minerals are stable residual solids under all but the most humid and tropical climatic conditions, where they may be desilicified to hydrated aluminum oxides (bauxite).

Soil scientists in recent years have recognized the importance of another chemical weathering reaction, *base exchange*. Base exchange involves a mutual transfer of cations such as Ca^{2+}, Mg^{2+}, Na^+, or K^+ between an aqueous solution rich in one cation and a mineral rich in another. The rate of exchange depends on the chemical activity and the abundance of the various cations as well as the acidity, temperature, and other properties of the solution. In a sense, hydrolysis is a special kind of base exchange in which hydrogen ions (H^+) in solution displace the cations in silicate mineral lattices (Deju, 1971; Holdren and Speyer, 1986, p. 61). The exchange of cations between minerals and groundwater may expand or collapse the mineral structure and free other chemical components. As in other chemical reactions, if one mineral grain in a rock is so destroyed, adjacent grains are detached and exposed to the same and other weathering processes.

The residual clay minerals of hydrolysis and base exchange are hydrated compounds. The closely related hydration and hydrolysis of feldspars to form clays are the major processes in the weathering of granite and related coarse-grained polymineralic rocks. Weathering feldspar grains expand and, combined with the oxidized and hydrolyzed iron-rich minerals such as biotite, cause granitic rocks to crumble into a mass of disaggregated grains called *grus* (Fig. 6-1).

Weatherability of Silicate Minerals

By comparing the rates of chemical reactions between water and various powdered minerals in controlled laboratory experiments, it is possible to predict which minerals will weather most rapidly under natural conditions. Extensive studies of this sort, especially on the silicate minerals that form most of the earth's crust, have led to the recognition of a mineral stability series or **weathering series,** a list of common silicate minerals arranged in order of relative susceptibility to chemical weathering (Fig. 6-6).

The silicate minerals that crystallize at the highest temperatures and that have the lowest silicon/oxygen ratios weather most rapidly. The strength of the silicon-oxygen bond is so much greater than the bonding energies of oxygen with metallic cations that it determines the total bond energy for each mineral molecule. The higher the Si/O ratio, and therefore the higher the bond energy, the more resistant

Forsterite (29,789)   Anorthite (31,935)

Augite (30,728)

Hornblende (31,883)

Biotite (30, 475)

Albite (34,335)

Orthoclase (34,266)

Muscovite (32,494)

Quartz (37,320)

Figure 6-6. Common minerals of igneous rocks arranged as a weathering series, with the least resistant minerals at the top and more resistant minerals progressively lower in the series. Energies of formation in parentheses (kcal/mol) are calculated from the bond strengths between oxygen and the various cations. This same arrangement is more commonly known as Bowen's Reaction Series for the sequence of formation of minerals from a silicate melt. (After Blatt et al., 1972, p. 234.)

the mineral, with few exceptions (Blatt et al., 1972, pp. 233–35). Biotite weathers faster than might be expected from its bond energy, because the iron oxidizes and the layered mica structure expands greatly on hydrolysis. Biotite weathering is a large contributor to the granular disaggregation of granite to grus (Isherwood and Street, 1976; Birkeland, 1984, pp. 85, 176).

Another way to study relative chemical weathering rates is to make a ratio of the concentration of each cation in solution in groundwater or rivers with the abundance of that cation in the regional terrane. Feth et al. (1964) established a **cation mobility series** in this way for a granitic terrane in California; Na, Ca, and Mg are the most mobile elements, K and Si are less mobile, and Al and Fe are by far the least mobile. Colman (1982, p. 36) identified a similar mobility sequence by analyzing the weathering rinds on basalt and andesite cobbles from soils on glacial moraines of various Quaternary ages in the western United States. Based on the abundance of each element in the weathered rinds compared to the abundance in the unaltered rock, the sequence of mobility is:

$$Ca >> Na > Mg > Si > Al \geq K > Fe > Ti.$$

In general, the cations that are lost to solution in greatest proportion are those that are derived from easily weathered plagioclase feldspars and mafic minerals. An exception is iron, which is abundant in the dark mafic silicate minerals that weather most easily, but which oxidizes to an insoluble oxide that does not move away in solution. The iron and titanium in magnetite and titanomagnetite in basic igneous rocks are extremely stable, and also remain in weathered rock as part of the original grains (Colman, 1982, pp. 13–14). Another anomalous result of weathering involves potassium, which is of comparable crustal abundance and chemical behavior to sodium, yet whose chemical mobility is of a much lower order, comparable to that of silica. Potassium is "fixed" in the crystal lattice of many clay minerals and tends to stay behind in residual clays, whereas the other alkalic ions are carried away in solution (Loughnan, 1969, pp. 45–47; Colman, 1982, p. 36).

BIOLOGICAL WEATHERING PROCESSES

Chelation: Chemical Weathering by Plants

Chelation is a complex organic process by which metallic cations are incorporated into hydrocarbon molecules (Lehman, 1963). The word "chelate," which means "clawlike," refers to the tight chemical bonds that hydrocarbons may impose on metallic cations. To function, many organic processes require metallic–organic chelates. The role of iron in hemoglobin to carry oxygen from the lungs is a good example. Synthetic chelating agents are widely used in analytical chemistry laboratories.

Soils in which plants are growing and decaying contain complex organic compounds that can chelate metallic cations (Birkeland, 1984, pp. 74–76). By analogy with laboratory procedures, if natural organic molecules in soil can remove metallic cations directly from mineral grains by chelation, the mineral lattices are disrupted, and possibly fragments of crystal lattices, including otherwise-insoluble elements such as aluminum, can be mobilized.

Vascular plants promote chemical weathering by mutually reinforcing interactions among chelation, hydrolysis, and base-exchange reactions. Each plant rootlet maintains a negative electrical charge on its surface (Keller and Frederickson, 1952) and a surrounding field of H^+ ions. Hydrolysis and base exchange are strongly promoted by the availability of H^+ ions in the vicinity of the root hair, and organic compounds in the roots absorb and chelate the essential metallic cations as they hydrolyze. Through vascular plants, the energy derived from sunlight falling on green leaves is used to weather minerals many meters underground.

An impressive demonstration of the ability of plants to weather rock was provided by Lovering and Engel (1967). In controlled-environment greenhouses, they grew *Equisetum* (scouring rush) and three grasses in pots of freshly crushed, sterilized rock samples. The silica uptake rate by the plants, all of which are known to be silica accumulators, was equivalent to removing all the silica from a basalt substrate 30 cm thick in only 5350 years. This remarkable rate of withdrawal of silica from igneous rocks is exclusive of any chemical weathering or leaching by percolating water, although some of the silica taken up by the plants might have been first dissolved by other weathering reactions.

Less complex, nonphotosynthetic plants also have the ability to dissolve elements from minerals. Fungi (*Penicillium* sp.) grown in a nutrient solution mixed with finely crushed rock greatly increased the quantity of Si, Al, Fe, and Mg entering solution, compared to the amount of elements entering solution in similar, but uninoculated control solutions (Silverman and Munoz, 1970). During the incubation period of the experiment, the pH of the nutrient solution dropped from 6.8 to below 3.5 in 7 days. As the acidity increased, the rate of mineral solution increased correspondingly. Other experiments were conducted with a variety of organic and mineral acid solutions of equivalent concentrations to the fungal growth medium, with essentially similar effects on rocks. Thus fungal weathering was proved to be due not to some special biogenic process but simply to chemical weathering in acidified water.

Laboratory experiments such as those just cited prove that plants are capable of putting into solution or absorbing substantial fractions of the total mass of rock on which they grow, in intervals from a few thousand years to as short as one week. Even though the rocks in the experiments had been mechanically crushed in advance and the plants were growing under optimal conditions, the amounts of translocated mineral matter are impressive. Field study supports the experimental work. Some parts of Iceland today are barren basalt plains, essentially devoid of macroscopic plants even though the climate is humid. Chemical weathering on these barren plains is one-half to one-third as fast as on similar areas that have plant cover, as judged by the dissolved loads in streams (Cawley et al., 1969).

On mainland Antarctica, where no higher plants live, weathering is also very slow. The lack of organic activity is compounded by the fact that liquid water, although available as thin films at ice/rock interfaces (p. 125), does not permeate through rocks and transport dissolved compounds (Ugolini, 1986). Meteorites that were collected from the surface of the Antarctic ice sheet showed slightly less oxidation and hydration than the weathering on chemically similar meteorites that fell in Arizona in 1912 and were collected in 1931 and 1968. However, the Antarctic meteorites probably fell in middle Pleistocene time, so a few hundred thousand years of weathering on the barren Antarctic ice sheet had less effect than 19 to 56 years of weathering in central Arizona (Gooding, 1981, p. 1121). The weathering of the Antarctic specimens was probably restricted to brief, rare intervals when the black meteorites were exposed on the ice surface and absorbed enough solar radiation to melt some nearby ice.

Microbial Weathering

Even in regions devoid of macroscopic plants, algae and fungi may contribute to weathering processes. Cyanobacteria (blue-green algae) have been discovered living in near-surface pores and cracks in sandstone and quartzite in both the Sahara and Antarctica, with no known source of liquid-phase water (Friedmann and Ocampo,

1976). These cyanobacteria are representatives of the oldest known forms of cellular life on our planet. Their well-preserved remains have been found in early Archean rocks that are 3.3 to 3.5 billion years old (Schopf and Packer, 1987). The fossil record of oxygen-producing photosynthetic microorganisms (microbes) demonstrate that biological weathering processes have been active for essentially the entire geologic history of the earth. All of the purely chemical or thermodynamic weathering processes described in this chapter are enormously enhanced or facilitated by microbial activity. For example, the oxidation of ferrous iron, one of the most obvious chemical reactions, is accelerated by a factor of 10^5 to 10^6 in the weathering of pyrite (FeS_2) by *thiobacilli* (colorless sulfur bacteria) (Jorgensen, 1983, p. 117).

Microbes are being cultivated or genetically engineered to leach commercial heavy metals from mine dumps, to immobilize radioactive and toxic wastes from leachates, and to increase food production by higher plants (Lindow et al., 1989). The importance of symbiotic mycorrhizal fungi and associated bacteria that live on the roots of higher plants and facilitate their mineral uptake has lead to the introduction of the term **rhizosphere** for that part of the soil weathering profile in which microbes are active (Berthelin, 1983, p. 249). It is even appropriate to redefine all weathering as "the chemical and mineralogical modifications of the different types of rocks and minerals within the biosphere . . . " (Berthelin, 1983, p. 223), with the demonstration of microbial activity 2 km deep within the earth's crust in salt domes and petroleum reservoirs, at the vents of submarine volcanoes, and in the most hostile earth surface environments known (Berthelin, 1988, p. 34). Life on Venus and Mars, if found, is most likely to be microbial.

ROCK WEATHERING AND SOIL FORMATION

Definitions

The term **soil** is used to describe the rock detritus at the surface of the earth that has been sufficiently weathered by physical, chemical, and biological processes so that it supports the growth of rooted plants. This is an agricultural definition, emphasizing that soil is a biologic as well as a geologic material. Engineers are less specific about their definition of a soil. To them, all the loose, unconsolidated, or broken rock material at the surface of the earth, whether residual from weathering at that place, or transported by rivers, glaciers, or wind, is soil. However, for most of the engineers' "soils," genetic terms such as "alluvium" or "glacial drift" are available. As a general term for the surficial layer of broken rock, *regolith* is also useful. In most contexts, the engineering or agricultural implication of the term is clear enough so that tedious definitions are unnecessary (Legget, 1967).

Soils are characterized by **horizons**: distinctive weathered zones, approximately parallel to the surface of the ground, that are produced by soil-forming processes. A **soil profile** is a vertical cross section through these horizons (Fig. 6–7). The horizons in a soil profile illustrate the principle that the many mechanical, chemical, and biological weathering processes operate simultaneously near the surface of the earth but at different rates and to various depths. By color, chemical analyses, grain size, and other diagnostic criteria, soil profiles are divided into horizons and subhorizons that clearly record the intensity and duration of the various soil-forming, or weathering, processes. Despite the strong genetic implications of horizons, they are intended to be defined by observable facts, not genetic inferences.

The five principal horizons, from the surface down into unaltered rock, are conventionally given the capital-letter symbols O, A, B, C, and R (Fig. 6–7). The **O**

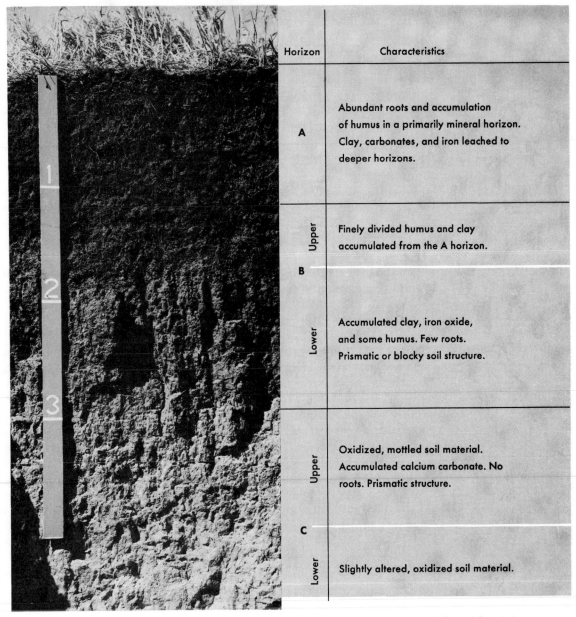

Horizon		Characteristics
A		Abundant roots and accumulation of humus in a primarily mineral horizon. Clay, carbonates, and iron leached to deeper horizons.
B	Upper	Finely divided humus and clay accumulated from the A horizon.
	Lower	Accumulated clay, iron oxide, and some humus. Few roots. Prismatic or blocky soil structure.
C	Upper	Oxidized, mottled soil material. Accumulated calcium carbonate. No roots. Prismatic structure.
	Lower	Slightly altered, oxidized soil material.

Figure 6-7. A Mollisol soil profile in Nebraska showing horizons A, B, and C and the notable characteristics of each. Depth scale is in feet. The R horizon is not exposed in this profile. (Photo: W. M. Johnson, U.S. Department of Agriculture-Soil Conservation Service.)

horizon is the upper organic horizon, dominated by fresh or partly decomposed organic material. The other four horizons are dominantly mineral in nature, defined by the manner and degree in which they differ from the parent material. The **A horizon** is a zone in which organic matter has accumulated and mixed with mineral detritus and from which clay, iron, or aluminum has been transferred downward to underlying horizons. The **B horizon** is characterized by the secondary accumulation or enrichment of clay, iron, or aluminum and by development of distinctive structures not present in the parent material, such as granular, blocky, or prismatic aggregates of partly decomposed minerals, loosely or firmly cemented by the clay or oxides that have moved into the horizon. The B horizon typically has more intense colors than those of the overlying and underlying horizons. The **C horizon** is below the zone of major biologic activity. It retains some characteristics of the parent material but may be mechanically fractured, stained by oxidation, or loosely recemented by calcium carbonate, gypsum, iron oxide, silica, or the more soluble salts. The **R horizon** is the underlying **parent material,** such as bedrock, alluvium, or other material from which the soil has formed. Although not specified in the definition of the soils scientists, most geologists would observe the effects of mechanical weathering on bedrock in the upper part of the R horizon. Its lower limit is not specified by pedologists.

Soil Genesis

Five key factors of soil formation have been long recognized by pedologists: *parent material, climate, organisms, slope,* and *time.* The kind of soil that forms is the result of chemical and mechanical weathering of some rock type ("structure," in the geomorphic sense defined in Chap. 1) under a certain climatic regimen of temperature and precipitation in the presence of plants, animals, and microorganisms (that are at least in part controlled by geologic and climatic factors) on some kind of landform through an interval of time. Under certain conditions, each of the five variables can exert the controlling influence. Three of the factors—climate, organisms, and slope—may change during the progress of soil genesis.

Geologists are inclined to assume that parent material is the dominant soil-forming factor. A residual soil cannot contain minerals that are not present in, or cannot be made by weathering of, the parent material. We assume that a soil on weathered limestone will be calcareous and that one on quartz sand will be siliceous. However, the weathering processes that have been reviewed earlier in this chapter should demonstrate that the residual or new compounds may be in very different proportions from the primary minerals. For instance, the Paleozoic Beekmantown limestone in Tennessee is composed of 90 percent calcium and magnesium carbonate, 7 percent quartz sand, and 3 percent detrital clay. A simple calculation shows that solution of 5 m of this limestone produces a residual parent material 1 m thick if the residuum has one-half the bulk density of the original rock. The residuum is 70 percent sand and 30 percent clay and hydrated iron oxides. Unless we comprehend the relative solubilities of limestone, quartz, and clay minerals, it is hard to believe that a typical red, sandy, clay loam soil on the Beekmantown limestone can result from in situ weathering.

An even more extreme example of the contrast between bedrock and residual soil is the great surficial deposit of lateritic bauxite (hydrated aluminum and iron oxides) at Weipa, Queensland, Australia (Loughnan and Bayliss, 1961). The bedrock is sandstone consisting of 90 percent quartz and 10 percent kaolinite (Fig. 6–8). Under a hot monsoon climate, with strongly seasonal rainfall averaging over 1500 mm/yr, most of the quartz has been leached to a depth of over 8 m, and the upper 5 m is mostly hydrated aluminum oxides stained by iron oxides. Only 5 percent of the thick surface layer is quartz, yet the bedrock is dominantly quartz, which is a relatively insoluble mineral even under the intense heat and rainfall of northern Queensland. A long interval of weathering must be assumed in order to explain the Weipa bauxite ore, because the continuity of the weathering profile shown in Figure 6–8 is clear evidence that the surface bauxite ore is the residuum of the quartz sandstone at depth.

The two examples just cited demonstrate that residual soils can be very different from their parent material, given the right combination of the other four soil-forming factors. Nevertheless, "structural" control of weathering is as important in soil genesis as it is in landscape development.

Climate must compete with parent material for the status of "first among equals" in the list of soil-forming factors. Like geomorphologists, pedologists can

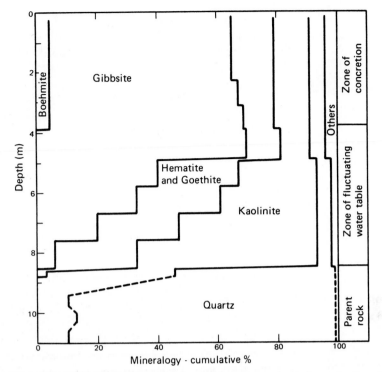

Figure 6-8. Mineralogy in relation to depth at Weipa, Queensland, Australia. (Loughnan and Bayliss, 1961.) See also pp. 142–43.

probably be divided into two groups: those who believe that weathering processes (primarily a function of climate) give the distinctive character to a soil or a landscape despite lithologic or structural differences, and those who believe that structure always shows through despite climate-controlled differences in the intensity of soil-forming and weathering processes. Readers of this book should compare the approach of Part I on constructional landforms with the approach of Part III on geomorphic processes unique to certain climates. Either structure or process can be emphasized, depending on the interest and experience of the writer or reader.

Organisms as a factor in soil formation have been reviewed in their role as chemical and biological weathering agents. Some pedologists view plants as a secondary response to climate and parent material. Others emphasize the role of soil microorganisms and give reduced emphasis to the rooted macroflora. Others see organisms as little more than the creators of a certain chemical soil environment in which essentially chemical weathering reactions form the various horizons of the profile.

Slope, or topographic influence on soil formation, is emphasized more by pedologists than by geologists. The geologist sees a slope as the result of processes acting on structures through time (Chap. 8). The pedologist sees slope as a preexisting factor in the formation of soil. If the slope is steep, runoff is rapid, erosion removes the soil as fast as it forms, little water enters the soil, and profiles are thin and poorly developed. On more level terrain, runoff is inhibited and more water enters the ground to weather minerals and translocate clay and other mobile components of the profile. Yet while a soil is forming, so is the slope around it. Pedologists need the geomorphic appreciation of the origin of slopes, lest they assume that the primary condition of a region is of soil-less but otherwise fully shaped landforms. This cannot be true, because weathering, and therefore soil formation, is as much a factor in slope development as slope is a factor in soil development. No simple cause-and-effect relationship is possible.

The role of the fifth soil-forming factor—time—is at the heart of both geomorphology and pedology. Simply and directly asked: Do soils and landscapes continue to evolve through time at rates appropriate to the energy expended on them? Do soils and landscapes reach some end form in a closed system and cease to evolve; or do they reach some dynamic equilibrium between processes and form and maintain thereafter a steady-state morphology, even though energy and matter continue to move through their open physical systems? These underlying questions of both process-oriented and explanatory geomorphology are evaluated again and again in this book, especially in Chapters 1 and 12. If soils are to be described in terms of the interaction of five soil-forming factors, pedologists must be geomorphologists, just as geomorphologists must be pedologists.

Soil Classification

The very definition of soil as the abode of rooted plants emphasizes the practical reason for studying and classifying soils. Each developed nation has a scheme of soil mapping that enables its agronomists and economists to plan the effective use of available land. Although genetic assumptions are useful and unavoidable, modern soil classifications attempt to classify according to definable, observable chemical and physical properties of the horizons in soil profiles.

In 1965, the United States Department of Agriculture adopted a totally new classification of soils to replace the classification that had been in use since 1938 (Soil Survey Staff, 1975). The new system had been under development since 1951 as a series of "approximations" until it was formally adopted for use by the U.S. Soil Conservation Service and state cooperative programs. Although created to meet the needs of describing and mapping soils in the United States, the new U.S. soil taxonomy was designed for worldwide applicability. Special attention was given to tropical soils, which received woefully inadequate treatment in previous soils classifications. The classification is outlined here in the confident belief that it is the best and most comprehensive classification of soils and soil-forming processes yet devised. Students of geomorphology should become familiar with the new, sometimes curiously compounded terminology. It is regrettable that some recent textbooks of geology and geography have persisted in using the highly useful but now obsolete terminology of the 1938 U.S. classification.

The nomenclature of the U.S. soil taxonomy (Soil Survey Staff, 1975) is systematic. The hierarchy of terms, in descending order of rank, is: order, suborder, great group, subgroup, family, and series. Ten soil orders are defined, based on the presence or degree of development of the various horizons in soil profiles. The name of each soil order (Table 6–1) ends in sol (L. solum, soil) and contains a formative element that is used as the final syllable in the names of taxa in suborders, great groups, and subgroups. Thus the name of each of the 47 suborders consists of two syllables, a formative element from a list of terms (Table 6–2) prefixed to the formative element of an order. The name of each great group, now about 185 in number, consists of three or four syllables, created by prefixing a suitable element from a list (Table 6–3) to the name of a suborder. The names of the 970 subgroups are created by an adjective modifying the name of a great group. At first glance, the resulting names are totally exotic, but all can be decoded quickly and precisely. The formative elements are derived from Greek or Latin roots, are as short as possible, and are designed for use in any modern language that uses the Latin alphabet. By compounding the several ranks of formative elements, any soil type on earth can be classified, and in the definition of the formative elements, the soil is also described.

For example, the Dunkirk Series of central New York State is a well-drained soil formed in silty, calcareous lacustrine deposits of late Pleistocene age. The Dunkirk Series is classified at the subgroup level as a *Glossoboric Hapludalf.* Decoded, using the formative elements from Tables 6–1, 6–2, and 6–3, the Dunkirk Series belongs to a subgroup of the great group *Hapludalf,* the suborder *Udalf,* and the order *Alfisol.* Translating the formative elements of the name, the soil has a gray to brown surface horizon, moderate to high base (cation) saturation, and clay accumulation in a subsurface horizon (formative element *alf,* Table 6–1). It is of the suborder of Alfisol that forms in humid regions (formative element *ud,* Table 6–2). More specifically, as described in the sources cited in the next paragraph, it is one of the alfisols of temperate and tropic regions that are usually moist but may be intermittently dry in some horizons during summer. It is of the great group that has the least complex horizons of the suborder (formative element *hapl,* Table 6–3); specifically, it is described in greater detail as a brownish or reddish Udalf with only a slight loss of clay from the A horizon and a moderate accumulation of clay in the B horizon. It is without a dense, clay-rich crust (fragipan) in the profile. It is in an area characterized by mean summer and mean winter soil temperatures that differ by 5°C or more and a mean annual temperature of 8°C or more (temperate climate).

TABLE 6–1

Names of Soil Orders, with Simplified Definitions

ALFISOL	Soil with gray to brown surface horizon, medium to high base supply, and a subsurface horizon of clay accumulation. Formative element: *alf.*
ARIDISOL	Soil with pedogenic horizons, low in organic matter, usually dry. Formative element: *id.*
ENTISOL	Soil without pedogenic horizons. Formative element: *ent.*
HISTOSOL	Organic (peat and muck) soil. Formative element: *ist.*
INCEPTISOL	Soil with weakly differentiated horizons showing alteration of parentmaterials. Formative element: *ept.*
MOLLISOL	Soil with a nearly black, organic-rich surface horizon and high basesupply. Formative element: *oll.*
OXISOL	Soil that is a mixture principally of kaolin, hydrate oxides, and quartz. Formative element: *ox.*
SPODOSOL	Soil that has an accumulation of amorphous materials in subsurface horizons. Formative element: *od.*
ULTISOL	Soil with a horizon of clay accumulation and low base supply. Formative element: *ult.*
VERTISOL	Cracking clay soil. Formative element: *ert.*

TABLE 6–2

Principal Formative Elements to be Used in Naming Suborders

Names of suborders consist of two syllables, for example, *Aqualf.* Formative elements are as follows:

alb	L. *albus,* white; soils from which clay and iron oxides have been removed.
aqu	L. *aqua,* water; soils that are wet for long periods.
arg	modified from L. *argilla,* clay; soils with a horizon of clay accumulation.
bor	Gr. *boreas,* northern; cool.
fluv	L. *fluvius,* river; soils formed in alluvium.
ochr	Gr. base of *ochros,* pale; soils with little organic matter.
orth	Gr. *orthos,* true; the common or typical.
psamm	Gr. *psammos,* sand; sandy soils.
ud	L. *udus,* humid; of humid climates.
umbr	L. *umbra,* shade; dark colors reflecting much organic matter.
ust	L. *ustus,* burnt; of dry climates with summer rains.
xer	Gr. *xeros,* dry; of dry climates with winter rains.

TABLE 6–3

Principal Formative Elements to be Used in Naming Great Groups

Names of great groups consist of more than two syllables and are formed by adding a prefix to the suborder name, for example, *Cryoboralf.* Some of the formative elements are as follows:

cry	Gr. *kryos,* icy cold; cold soils.
dystr, dys	modified from Gr. *dys,* ill; infertile.
eutr, eu	modified from Gr. *eu,* good; fertile.
frag	modified from L. *fragilis,* brittle; a brittle pan.
gloss	Gr. *glossa,* tongue; deep, wide tongues of albic materials into the argillic horizon.
hapl	Gr. *haplous,* simple; the least advanced horizons.
quartz	Ger. *quarz,* quartz; soils with very high content of quartz.
torr	L. *torridus,* hot and dry; soils of very dry climates.
trop	modified from Gr. *tropikos,* of the solstice; humid and continually warm.

The great group Hapludalf is equivalent to the Gray-Brown Podzolic soils of the 1938 classification. To classify the Dunkirk Series to the subgroup level, the compound adjective *glossoboric* precedes the great group name. *Glossic* refers to minor but characteristic tongues of the clay-depleted A horizon that extend down into the clay-enriched B horizon; if this tonguing were more pronounced, the soil would be a *Glossudalf* instead of a *Hapludalf. Boric* is from the root *boreal;* as defined for this subgroup, the mean annual soil temperature is less than 10°C.

The decoding exercise of the preceding paragraph is tedious and complex for a novice, but it illustrates that approximately a full page of written description about a soil and its climatic environment is contained in the name of the soil at the subgroup level. If additional modifiers concerning texture, mineralogy, grain size, permeability, and soil temperature are prefixed to describe the soil at the family level, a comprehensive impression of the soil, its genesis, and its agricultural potential can be gained just from the name alone. A map of the principal orders, suborders, and great groups of soils found in the United States is published in the U.S. National Atlas (U.S. Geol. Survey, 1970, Plate 38) with a useful accompanying descriptive text. Other than by consulting the definitive document (Soil Survey Staff, 1975) or appropriate textbooks of soil classification and morphology (Wilding et al., 1983; Birkeland, 1984; Buol et al., 1989; Fanning and Fanning, 1989), the Atlas map and text are recommended as a good introduction to the U.S. soil taxonomy.

STRUCTURE, CLIMATE, AND ROCK WEATHERING

To what extent does structure influence or control rock weathering? To what degree are landforms shaped by climate-controlled weathering processes? The remainder of this chapter considers these two questions, in review and in anticipation of expanded treatment in later chapters where structural and climatic control of mass wasting and erosion are also considered.

Structural Influences in Rock Weathering

Lithology, the physical character of a rock, exerts an obvious structural control on weathering. Some rocks are massive; others have strong *fabrics* that preferentially control mechanical disaggregation and chemical disintegration. Some are permeable; others are nearly impermeable. Related to lithology are the *secondary fractures* that have resulted from tectonic deformation or other stresses. Fractures of many kinds, from microscopic pores and cracks to major faults, break rocks and create surfaces that air and water can reach.

Mineralogic factors are part of the structure that controls weathering. Chemical composition is obviously important, as noted by the fact that each example of chemical weathering cited earlier in the chapter used a particular chemical compound (mineral) in the equation. Two other less obvious mineralogic factors also influence rock weathering. One is grain size, and the other can be called "crystallinity," or the amount of crystalline, as opposed to amorphous or glassy, minerals in a rock. For example, different textural types of compositionally identical granite have varying resistance to weathering and erosion. In the humid tropical climate of Hong Kong, coarse porphyritic granite is least resistant, coarse-grained granite is stronger, and medium-grained granite, fine-grained granite, and granite porphyry dikes are the most resistant (Ruxton and Berry, 1957, p. 1273). Obsidian, basalt glass, and amorphous silica (chert and chalcedony) are notably more soluble in water than are their crystalline equivalents (Colman, 1982, p. 10). Hydrated rinds have formed on the surface of obsidian and basalt-glass artifacts only a few thousand years old. Their thickness can be calibrated as an archeologic dating method.

Almost every rock type has a distinctive weathering character. Limestones are often smooth, gray, and lichen-covered, or deeply pitted by solution (Chap. 7). Shales split and crumble. Granites form sandy grus. The list could be extended at length. The weathered appearance of rock is so useful that it is included among the diagnostic criteria for defining formations, the basic units of geologic mapping (Dunbar and Rodgers, 1957, pp. 267, 273).

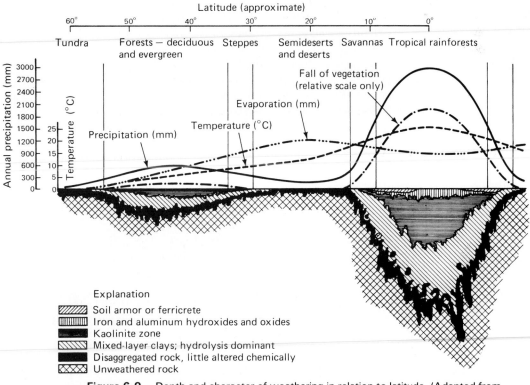

Figure 6-9. Depth and character of weathering in relation to latitude. (Adapted from Strakhov, 1967, Fig. 2.)

Climatic Influence on Rock Weathering

Climate is the integration of weather over a period of years. Weather is related to weathering by more than the identical origin of the terms; the same moisture, temperature, and seasonality factors that define a regional climate also define the weathering processes that operate there.

Strakhov (1967) drew an instructive, if idealized meridianal cross section of the earth's weathered layer (Fig. 6-9). From polar to equatorial latitudes, his figure not only shows the varying depths of weathering but illustrates that certain sets of weathering and soil-forming processes follow each other in prescribed order. Mechanically fractured rocks, but little altered chemically, are at the surface in polar and subtropical deserts. Either because of extreme cold or lack of precipitation, or both, these are latitudinal belts of minimal chemical and biological weathering.

Desert weathering in warm climates proceeds with a minimum of precipitation, but dew and rare rainfalls provide some moisture. Evaporation draws salt-laden pore water toward the surface, where the salts precipitate and cause exfoliation. Surface *flaking* of granular rock under projecting ledges in arid parts of Australia (Dragovich, 1967; Bradley et al., 1978, 1980) occurs because dew and other atmospheric moisture is uniformly absorbed to a depth of 15 mm or less, and small chips or flakes of slightly hydrated and salinized but otherwise fresh rock are spalled off. Weathering and soil formation in dry climates are described in more detail in Chap. 13.

In a humid climate with seasonal freezing, cold weather brings a dormant period for both vegetation and soil microorganisms. Trees shed their leaves or needles, and the humus layer at the surface of the ground accumulates faster than microorganisms, worms, and insects can consume it. Rainwater or snowmelt that soaks through the rotting humus collects organic compounds and carbon dioxide that chelate and hydrolyze metallic cations from the underlying minerals and generally leave a silica-rich clay residue. Iron oxides and clay minerals that are washed from the surface layer accumulate in the upper meter of soil. Chemical weathering processes penetrate only a few meters into rock, but mechanical weathering may extend considerably deeper (Fig. 6-9). Clay minerals in the soil prevent water from freely penetrating the ground and encourage overland runoff. The resulting landscape develops broad, gentle, soil-covered slopes shaped by soil creep and stream erosion. Ridge crests commonly have thin soils or exposed bedrock ledges.

In the humid tropics, hydrolysis and residual clay-mineral formation reach to depths of 100 m or more (Fig. 6-9). The great depth of tropical weathering is due in part to temperature, but abundant precipitation is an even greater climatic factor. Despite the enormous biological productivity of a tropical rainforest, little humus accumulates because of continuous microbial and fungal attack and rapid recycling of nutrients. The heavy rainfall literally flushes the soil free of all soluble compounds.

Chemical weathering in the tropics is so intense that clay minerals, the stable residuals of weathering in cooler latitudes, are desilicified. A typical reaction of kaolinite in a great excess of water is

$$Al_2Si_2O_5(OH)_4 + 5H_2O \rightarrow 2Al(OH)_3 + 2H_4SiO_4$$
$$\text{kaolinite} + \text{water} = \text{gibbsite} + \text{silicic acid}$$
$$\text{in solution}$$

The silica that is removed by leaching is shown as silicic acid in solution, as in earlier illustrations of hydrolysis (p. 130). The actual form of the leached silica is not certain. Some is in amorphous or colloidal form and may reprecipitate at the water table in seasonally dry tropical regions to form nodules, crusts, and layers of chalcedony or opal. Gibbsite, the aluminum hydroxide residual, recrystallizes slowly as insoluble hydrated alumina:

$$2\text{Al(OH)}_3 \rightarrow \text{Al}_2\text{O}_3 \cdot 3\text{H}_2\text{O}$$

Gibbsite is the major mineral in the heterogeneous earthy mixture of hydrated aluminum oxides called *bauxite,* the ore of aluminum. Tropical weathering of iron-poor aluminosilicates produces a surface layer of earthy or nodular bauxite (Fig. 6-8) that can be handled by ordinary earth-moving equipment.

In latitudinal belts peripheral to the tropical rainforests are the regions of tropical savanna climate (Chap. 13). During the wet season, normally when the sun is overhead, the savannas are flooded, but during the dry half of the year, almost no rain may fall. The seasonal wetting promotes deep weathering, but the dry season accelerates generally irreversible oxidation of aluminum and iron hydroxides. Perhaps 25 percent of the African landscape has soils in the Oxisol or Ultisol orders (Lal, 1987, p. 1070). Massive *laterite* crusts may form cap rock that exerts a strong structural control on the landscape (Chap. 11). In some regions, the layer is called *soil armor,* or *ferricrete* (Nahon, 1986).

The chemistry of tropical weathering is complex and as yet relatively unstudied. At least some of the soils stay soft and clay-rich as long as they are under forest cover and remain wet throughout the year. If they are allowed to dry, either seasonally or by exposure through agricultural use or road construction, they irreversibly harden, apparently within a few years. These are the true **laterites,** defined by their tendency to harden. In sub-Saharan western Africa, hardened laterites may cover 2.5 million km², an area equal to 30 percent of the conterminous United States (Lal, 1987, p. 1070). Large areas of laterite in southeast Asia are believed to have been created by traditional slash-and-burn, or *swidden,* agriculture. After being gardened for a few years, a patch of burned forest floor becomes hard and infertile, and a new patch must be cleared. Geographers and geologists have tended to exaggerate the extent of laterite and lateritic soils in the tropics, incorrectly equating the common reddish color of iron-oxide-rich soils with the tendency to harden. Perhaps only 7 percent of the tropical soils are laterites, and many of the other soils can be productively farmed with improved technology (Sanchez and Buol, 1975; Lal, 1987).

Tropical weathering has been discussed at length because of the dramatic intensity and depth of chemical activity. But the contrasts with weathering in cooler and less humid climates are those of degree and not of kind. The latitudinal temperature gradient of the earth is really not very great, especially when considered on the Kelvin scale of absolute temperature, which controls the kinetics of chemical reactions. The intensity of tropical weathering is caused by the continuous biological activity and massive quantities of rainfall that flush through the weathered layer.

In general, the intensity of tropical weathering minimizes the effects of structural control. The depth of the weathered zone is equal to the local relief on hills, escarpments, and other landforms, so that entire landscapes may be within the weathered zone, rather than the weathered zone being a rind on the surface of the landforms. Geomorphologists trained in the cool, humid climates of Europe and North America have always been profoundly impressed with the scale of climate influence in tropical landscapes. As the developing nations of the tropics have become more influential, and as tropical mineral resources such as bauxite and laterite have become primary raw materials of industrial nations, tropical weathering has been given much greater attention.

REFERENCES

AAGAARD, P., and HELGESON, H. C., Thermodynamic and kinematic constraints on reaction rates among minerals and aqueous solutions. I. Theoretical considerations: Am. Jour. Sci., v. 282, pp. 237–85.

ARGO, V. N., 1964, Strangler fig: Native epiphyte: Nat. Hist., v. 73, no. 9, pp. 26–31.

BATEMAN, P. C., and EATON, J. P., 1967, Sierra Nevada batholith: Science, v. 158, pp. 1407–17.

BENNETT, J. G., 1936, Broken coal: Jour. Inst. Fuel, v. 10, no. 49, pp. 22–39.

BERNER, R. A., and HOLDREN, G. R., JR., 1977, Mechanism of feldspar weathering: Some observational evidence: Geology, v. 5, pp. 369–72.

———, Mechanism of feldspar weathering, II. Observations of feldspar in soils: Geochim. Cosmochim. Acta, v. 43, pp. 1173–86.

BERNER, R. A., SJÖBERG, E. L., VELBEL, M. A., and KROM, M. D., 1980, Dissolution of pyroxenes and amphiboles during weathering: Science, v. 207, pp. 1205–6.

BERTHELIN, J., 1983, Microbial weathering processes, in Krumbein, W. E., ed., Microbial geochemistry: Blackwell Scientific Publications, Oxford, England, pp. 223–62.

———, 1988, Microbial weathering processes in natural environments, in Lerman, A., and Meybeck, M., eds., Physical and Chemical Weathering in Geochemical Cycles: NATO Advanced Science Institutes Series C, v. 251, Kluwer Academic Publishers, Dordrecht, The Netherlands, pp. 33–59.

BIRCH, F., 1966, Compressibility; elastic constants, in Clark, S. P., Jr., ed., Handbook of physical constants: Geol. Soc. America Mem. 97, pp. 97–173.

BIRKELAND, P. W., 1984, Soils and geomorphology: Oxford Univ. Press, New York, 372 pp.

BLATT, H., MIDDLETON, G. V., and MURRAY, R. C., 1972, Origin of sedimentary rocks: Prentice-Hall, Inc., Englewood Cliffs, N.J., 634 pp.

BRACE, W. F., SILVER, E., HADLEY, H., and GOETZE, C., 1972, Cracks and pores: A closer look: Science, v. 178, pp. 162–64.

BRADLEY, W. C., 1963, Large-scale exfoliation in massive sandstones of the Colorado Plateau: Geol. Soc. America Bull., v. 74, pp. 519–28.

———, 1980, Role of salts in development of granitic tafoni, South Australia: A reply: Jour. Geology, v. 88, pp. 121–22.

———, HUTTON, J. T., and TWIDALE, C. R., 1978, Role of salts in development of granitic tafoni, South Australia: Jour. Geology, v. 86, pp. 647–54.

BRUNER, W. M., 1984, Crack growth during unroofing of crustal rocks: Effects on thermoelastic behavior and near-surface stresses: Jour. Geophys. Res., v. 89, pp. 4167–84.

BUOL, S. W., HOLE, F. D., and McCRACKEN, R. J., 1989, Soil genesis and classification, 3rd ed.: Iowa State Univ. Press, Ames, Iowa, 446 pp.

CAWLEY, J. L., BURRUSS, R. C., and HOLLAND, H. D., 1969, Chemical weathering in central Iceland: An analog of pre-Silurian weathering: Science, v. 165, pp. 391–92.

CLARK, S. P., JR., and JAGER, E., 1969, Denudation rate in the Alps from geochronology and heat flow: Am. Jour. Sci., v. 267, p. 1143–60.

COLMAN, S. M., 1982, Chemical weathering of basalts and andesites: U.S. Geol. Survey Prof. Paper 1245, 51 pp.

COMMITTEE ON CONSERVATION OF HISTORIC STONE BUILDINGS AND MONUMENTS, 1982, Conservation of historic stone buildings and monuments: National Materials Advisory Board, National Research Council Publication NMAB-397: National Academy Press, Washington, D.C., 365 pp.

COSTAIN, J. K., BOLLINGER, G. A., and SPEER, J. A., 1987, Hydroseismicity—a hypothesis for the role of water in the generation of intraplate seismicity: Geology, v. 15, pp. 618–21.

CRISS, R. E., and TAYLOR, H. P., JR., 1983, An $^{18}O/^{16}O$ and D/H study of Tertiary hydrothermal systems in the southern half of the Idaho batholith: Geol. Soc. America Bull., v. 94, pp. 640–63.

DALE, T. N., 1923, Commercial granites of New England: U.S. Geol. Survey Bull. 738, 488 pp.

DEGENS, E. T., 1965, Geochemistry of sediments: Prentice-Hall, Inc., Englewood Cliffs, N.J., 342 pp.

DEJU, R. A., 1971, A model of chemical weathering of silicate minerals: Geol. Soc. America Bull., v. 82, pp. 1055–62.

DRAGOVICH, D., 1967, Flaking, a weathering process operating on cavernous rock surfaces: Geol. Soc. America Bull., v. 78, pp. 801–4.

DUNBAR, C. O., and RODGERS, J., 1957, Principles of stratigraphy: John Wiley & Sons, Inc., New York, 356 pp.

DUNN, J. R., and HUDEC, P. P., 1966, Water, clay and rock soundness: Ohio Jour. Sci., v. 66, no. 2, pp. 153–67.

EVANS, I. S., 1970, Salt crystallization and rock weathering: A review: Rev. Geomorph. Dynamique, v. 19, no. 4, pp. 153–77.

FANNING, D. S., and FANNING, M. C. B., 1989, Soil: Morphology, genesis, and classification: John Wiley & Sons, Inc., New York, 395 pp.

FETH, J. H., ROBERSON, C. E., and POLZER, W. L., 1964, Sources of mineral constituents in water from granitic rocks, Sierra Nevada, California and Nevada: U.S. Geol. Survey Water-Supply Paper 1525-I, 70 pp.

FRIEDMANN, E. I., and OCAMPO, R., 1976, Endolithic blue-green algae in the dry valleys: Primary producers in the Antarctic desert ecosystem: Science, v. 193, pp. 1247–49.

GOODING, J. L., 1981, Mineralogical aspects of terrestrial weathering effects in chondrites from Allan Hills, Antarctica: Proc., Lunar Planet. Sci., v. 12B, pp. 1105–22.

GRIGGS, D. T., 1936, The factor of fatigue in rock exfoliation: Jour. Geology, v. 44, pp. 783–96.

HACK, J. T., 1966, Circular patterns and exfoliation in crystalline terrane, Grandfather Mountain, North Carolina: Geol. Soc. America Bull., v. 77, pp. 975–86.

HAXBY, W. F., and TURCOTTE, D. L., 1976, Stresses induced by the addition or removal of overburden and associated thermal effects: Geology, v. 4, pp. 181–84.

HOLDREN, G. R., JR., and BERNER, R. A., 1979, Mechanism of feldspar weathering - I. Experimental studies: Geochim. Cosmochim. Acta, v. 43, pp. 1161–71.

HOLDREN, G. R., JR., and SPEYER, P. M., 1986, Stoichiometry of alkali feldspar dissolution at room temperature and various pH values, in Colman, S. M., and Dethier, D. P., eds. Rates of chemical weathering of rocks and minerals: Academic Press, Inc., Orlando, Florida, pp. 61–81.

HOLE, F. D., 1981, Effects of animals on soil: Geoderma, v. 25, pp. 75-112.

HOLLINGWORTH, S. E., TAYLOR, J. H., and KELLAWAY, G. A., 1944, Large-scale superficial structures in the Northampton ironstone field: Geol. Soc. London Quart. Jour., v. 100, pp. 1–44.

HOPSON, C. A., 1958, Exfoliation and weathering at Stone Mountain, Georgia, and their bearing on disfigurement of the Confederate Memorial: Georgia Mineral Newsletter, v. 11, pp. 65–79.

ISHERWOOD, D., and STREET, F. A., 1976, Biotite-induced grussification of the Boulder Creek Granodiorite, Boulder County, Colorado: Geol. Soc. America Bull., v. 87, pp. 366–70.

JAHNS, R. H., 1943, Sheet structure in granite: Its origin and use as a measure of glacial erosion in New England: Jour. Geology, v. 51, pp. 71–98.

JOHNSON, N. M., REYNOLDS, R. C., and LIKENS, G. E., 1972, Atmospheric sulfur: Its effect on the chemical weathering of New England: Science, v. 177, pp. 514–16.

JORGENSEN, B. B., 1983, The microbial sulphur cycle, in Krumbein, W. E., ed., Microbial geochemistry: Blackwell Scientific Publications, Oxford, England, pp. 91–124.

KELLER, W. D., and FREDERICKSON, A. F., 1952, Role of plants and colloidal acids in the mechanism of weathering: Am. Jour. Sci., v. 250, pp. 594–608.

KESSLER, D. W., INSLEY, H., and SLIGH, W. H., 1940, Physical, mineralogical, and durability studies on the building and monumental granites of the United States: U.S. Nat. Bureau of Standards Jour. Res., v. 25, pp. 161–206.

KIRBY, S. H., 1984, Introduction and digest to the special issue on chemical effects of water on the deformation and strengths of rocks: Jour. Geophys. Res., v. 89, pp. 3991–95.

KOZLOVSKY, Y. A., 1984, The world's deepest well: Sci. American, v. 251, no. 6, pp. 98–104.

KUENEN, P. H., 1955, Realms of water: John Wiley & Sons, Inc., New York, 327 pp.

LAL, R., 1987, Managing the soils of sub-Saharan Africa: Science, v. 236, pp. 1069–76.

LEGGET, R. P., 1967, Soil: Its geology and use: Geol. Soc. America Bull., v. 78, pp. 1433–59.

LEHMAN, D. S., 1963, Some principles of chelation chemistry: Soil Sci. Soc. America Proc., v. 27, pp. 167–70.

LIKENS, G. E., and BORMANN, F. H., 1974, Acid rain: A serious regional environmental problem: Science, v. 184, pp. 1176–79.

LINDOW, S. E., PANOPOULOS, N. J., and McFARLAND, B. L., 1989, Genetic engineering of bacteria from managed and natural habitats: Science, v. 244, pp. 1300–7.

LIVINGSTONE, D. A., 1963, Data of geochemistry, Chap. G, Chemical composition of rivers and lakes: U.S. Geol. Survey Prof. Paper 440-G, 64 pp.

LOUGHNAN, F. C., 1969, Chemical weathering of the silicate minerals: American Elsevier Publishing Co., Inc., New York, 154 pp.

————, and BAYLISS, P., 1961, Mineralogy of the bauxite deposits near Weipa, Queensland: Am. Mineralogist, v. 46, pp. 209–17.

LOVERING, T. S., and ENGEL, C., 1967, Translocation of silica and other elements from rock into *Eqisetum* and three grasses: U.S. Geol. Survey Prof. Paper 594-B, 16 pp.

MARTINI, I. P., 1978, Tafoni weathering, with examples from Tuscany, Italy: Zeitschr. für Geomorph., v. 22, pp. 44–67.

MATTHES, F. E., 1930, Geologic history of the Yosemite Valley: U.S. Geol. Survey Prof. Paper 160, 137 pp.

MICHALSKE, T. A., and BUNKER, B. C., 1987, The fracturing of glass: Sci. American, v. 257, no. 6, pp. 122–29.

MUSTOE, G. E., 1982, The origin of honeycomb weathering: Geol. Soc. America Bull., v. 93, pp. 108–15.

NAHON, D. B., 1986, Evolution of iron crusts in tropical landscapes, *in* Colman, S. M., and Dethier, D. P., eds., Rates of chemical weathering of rocks and minerals: Academic Press, Inc., Orlando, Florida, pp. 169–91.

NESBITT, B. E., and MUEHLENBACHS, K., 1989, Origins and movement of fluids during deformation and metamorphism in the Canadian cordillera: Science, v. 245, pp. 733–36.

OLLIER, C. D., 1963, Insolation weathering: Examples from central Australia: Am. Jour. Sci., v. 261, pp. 378–81.

————, 1969, Weathering: American Elsevier Publishing Co., Inc., New York, 304 pp.

PINTO, J. P., and HOLLAND, H. D., 1988, Paleosols and the evolution of the atmosphere; Part II, *in* Reinhardt, J., and Sigleo, W. R., eds., Paleosols and weathering through geologic time: Geol. Soc. America Spec. Paper 216, pp. 21–34.

REICHE, P., 1950, Survey of weathering processes and products, 2nd ed: New Mexico Univ. Press, Albuquerque, N.M., 95 pp.

RICE A., 1976, Insolation warmed over: Geology, v. 4, pp. 61–62.

ROTH, E. S., 1965, Temperature and water content as factors in desert weathering: Jour. Geology, v. 73, pp. 454–68.

RUXTON, B. P., and BERRY, L., 1957, Weathering of granite and associated erosional features in Hong Kong: Geol. Soc. America Bull., v. 68, p. 1263–92.

SANCHEZ, R. A., and BUOL, S. W., 1975, Soils of the tropics and the world food crisis: Science, v. 188, pp. 598–603.

SCHOPF, J. W., and PACKER, B. M., 1987, Early Archean (3.3-billion to 3.5-billion-year-old) microfossils from Warrawoona, Australia: Science, v. 237, pp. 70–73.

SCHWARTZ, S. E., 1989, Acid deposition: Unraveling a regional phenomenon: Science, v. 243, pp. 753–63.

SILVERMAN, M. P., and MUNOZ, E. F., 1970, Fungal attack on rock: Solubilization and altered infrared spectra: Science, v. 169, pp. 985–87.

SIMMONS, G., TODD, T., and BALDRIDGE, W. S., 1975, Toward a quantitative relationship between elastic properties and cracks in low porosity rocks: Am. Jour. Sci., v. 275, pp. 318–45.

SOIL SURVEY STAFF, 1975, Soil taxonomy: A basic system of soil classification for making and interpreting soil surveys: U.S. Dept. Agr. Handbook 436, 754 pp.

STEVENS, G. R., 1974, Rugged landscape: The geology of central New Zealand: A.H. & A.W. Reed, Wellington, New Zealand, 286 pp.

STOUT, J. H., 1974, Helicoidal crack propagation in aluminous orthoamphiboles: Science, v. 185, pp. 251–53.

STRAKHOV, N. M., 1967, Principles of lithogenesis, v. 1 (transl. by J. P. Fitzsimmons): Oliver & Boyd Ltd., Edinburgh, 245 pp.

STURGUL, J. R., and GRINSHPAN, Z., 1975, Finite-element model for possible isostatic rebound in the Grand Canyon: Geology, v. 3, pp. 169–71.

TARR, W. A., 1915, A study of some heating tests, and of the light they throw on the cause of the disaggregation of granite: Econ. Geology, v. 10, pp. 348–67.

THARP, T. M., 1987, Conditions for crack propagation by frost wedging: Geol. Soc. America Bull., v. 99, pp. 94–102.

TILLMAN, J. E., 1980, Eastern geothermal resources: Should we pursue them?: Science, v. 210, pp. 595–600.

TWIDALE, C. R., 1973, On the origin of sheet jointing: R163–87.

UGOLINI, F. C., 1986, Processes and rates of weathering in cold and polar desert environments, in Colman, S. M., and Dethier, D. P., eds., Rates of chemical weathering of rocks and minerals: Academic Press, Inc., Orlando, Florida, pp. 193–235.

U.S. GEOLOGICAL SURVEY, 1970, National atlas of the United States of America: U.S. Dept. of Interior, Geol. Surv., Washington, D.C., 417 pp.

WALDER, J., and HALLET, B., 1985, A theoretical model of the fracture of rock during freezing: Geol. Soc. America Bull., v. 96, pp. 336–46.

WASHBURN, A. L., 1969, Weathering, frost action, and patterned ground in the Mesters Vig district, northeast Greenland: Medd. om Grønland, v. 176, no. 4, 303 pp.

WHITNEY, J. A., JONES, L. M., and WALKER, R. L., 1976, Age and origin of the Stone Mountain Granite, Lithonia district, Georgia: Geol. Soc. America Bull., v. 87, pp. 1067–77.

WHITNEY, P. R., and McCLELLAND, J. M., 1973, Origin of coronas in metagabbros of the Adirondack Mts., N.Y: Contr. Mineral. and Petrol., v. 39, pp. 81–98.

WILDING, L. P., SMECK, N. E., and HALL, G. F., eds., 1983, Pedogenesis and Soil Taxonomy, Developments in soil science 11A and 11B: Elsevier Publishing Co., Amsterdam, The Netherlands, v. 1, 303 pp., v. 2, 410 pp.

WINKLER, E. M., 1968, Frost damage to stone and concrete: Geological considerations: Eng. Geology, v. 2, pp. 315–23.

———, 1973, Stone: Properties, durability in man's environment: Springer-Verlag New York, Inc., New York, 230 pp.

———, 1980, Role of salts in development of granitic tafoni, South Australia: Discussion: Jour. Geology, v. 88, pp. 119–20.

———, and WILHELM, E. J., 1970, Salt burst by hydration pressures in architectural stone in urban atmosphere: Geol. Soc. America Bull., v. 81, pp. 567–72.

YOUNG, A. R. M., 1987, Salt as an agent in the development of cavernous weathering: Geology, v. 15, pp. 962–66.

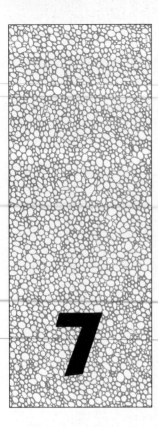

7 Karst

In the preceding chapter, solution (dissolution) was excluded as a specific process of chemical weathering, on the argument that all chemical processes of rock weathering involve aqueous solutions. In one special case, however, solution of massive quantities of abnormally soluble rock gives rise to a terrain so distinctive that almost every dialect has a special set of descriptive terms for the landforms. The Germanized form of an ancient Slavic term for such a region of solution topography in northwestern Yugoslavia is *karst*. The term connotes both an assemblage of landforms and a set of processes. It has become a truly international term, and although an enormous and provincial vocabulary has accumulated to describe karst processes and forms, including caves and their deposits, the one word offers a central and unifying concept for the special landforms on exceptionally soluble rocks.

DEFINITIONS AND HISTORICAL CONCEPTS

Karst is *"terrain with distinctive hydrology and landforms arising from a combination of high rock solubility and well-developed secondary porosity"* (Ford and Williams, 1989, p. 1). This sensible and practical definition stresses geomorphic *form,* but emphasizes *process* in stressing the role of groundwater hydrology, and emphasizes *structural control* in stressing that abnormally soluble rocks with well-developed secondary porosity are involved. Solution becomes a process of both weathering and erosion inasmuch as the weathered products are removed from the surface and carried away in solution.

Attention to the definition is important. For some students, *karst* is synonymous with only limestone landscapes. However, other soluble-rock terranes may be karst, and limestone weathering in both hot and polar deserts is not characterized by solution. Dolomite, gypsum, and salt all show karst landforms in some regions. The term has been deplorably extended (with modifying prefixes, as in "glacier karst," "thermokarst," and "pseudokarst") to a variety of landforms, such as those formed by compaction of fine-grained volcanic tephra, melting of glacier ice and permafrost, and subsidence of desert soils by a lowering of the water table. Under very long intervals of humid tropical weathering, the dissolution of silica from quartzite (Fig. 6–8) has produced karst analogs (Bögli, 1980, p. 1), but the usual concept of karst is restricted to limestone, dolomite, and much less commonly, gypsum and salt.

No branch of geomorphology is so plagued by terminology as "karstology." Part of the problem is psychological in that every language and dialect has special, often highly imaginative, terms to describe caverns and other unusual karst landforms. The other part of the problem is historical. The first systematic studies of karst were carried out in the Dinaric Alps of northwestern coastal Yugoslavia in a region traditionally known as The Karst (Demek et al., 1984, Fig. 16–4). Albrecht Penck (the author of the first textbook of geomorphology) and his students popularized the study of karst landforms (Cvijić, 1893; Penck, 1894). They translated local Serbo-Croatian terms for the forms first into German, then into French, then Italian, and subsequently into other languages. In each country, the words were given slightly different equivalents. Only since World War II has there been international integration of terms and theories of hydrogeologists, speleologists, and geomorphologists to create a body of karst literature.

Even before Penck and his students began their systematic study of the Yugoslavian karst, descriptions of tropical karst had been published that emphasized solution landforms unlike those in the Mediterranean basin and other temperate karst regions. Sawkins (1869) described the "cockpits" of Jamaican karst, a maze of solution depressions separated by steep ridges. Climatic control of karst processes received greater attention in the decades of 1920 and 1930 with studies of tropical karst in south China and in Java (Roglic, 1972). The *tower karst* landscape (p. 165) with isolated, steep- or vertical-sided hills rising from flat plains in the humid tropics, was proposed to be a fundamentally different, climate-controlled karst landscape. The barrel-shaped or conical isolated peaks were judged to be the residuals of marginal corrosion by acid swamp water, a process unique to warm, humid regions.

Hydrologists have long been aware of the unique problems of groundwater movement in soluble rocks. Theories of groundwater movement have had to be modified from models of uniform motion along regional hydrostatic gradients to models more related to flow in pipes or networks of pipes. The very assumption of a water table in karst has been questioned by some hydrologists, even though geomorphologists have debated whether caves form above or below the water table. Although briefly reviewed in this chapter, karst hydrogeology is a field in itself, largely independent of geomorphology (Brahana et al., 1988; Troester and Moore, 1989).

The introduction of hydrologic ideas about subsurface water movement leads to a consideration of the role of caves in karst geomorphology. Topologically, a cave open to the atmosphere is part of the landscape, yet caves are considered "subterranean," unfamiliar, and the objects of fearful superstition by most people. Cave exploration has developed its own heroes and terminology. The literature on cave morphology and genesis must be sought in many journals not normally read by geomorphologists. This chapter concludes with a brief review of the origin of limestone caves and the usefulness of caves in preserving and dating the record of human prehistory and Quaternary climatic changes.

THE KARST PROCESS

Solution, as the term is used in the context of limestone karst, is essentially the hydrolysis of $CaCO_3$ (p. 129). Solution of calcite in groundwater made acid by dissolved CO_2 is summarized by the following reaction:

$$CaCO_3 + H_2O + CO_2 \rightleftharpoons Ca^{2+} + 2HCO_3^-$$

As simplified, the reaction omits the several steps by which atmospheric CO_2 becomes dissolved in water to form carbonic acid, which in turn ionizes to free H^+ ions that convert CO_3^{2-} to the more soluble form of bicarbonate ion HCO_3^-. The Ca^{2+} ion does not form a hydroxide but remains in ionic equilibrium with the bicarbonate. It should be noted that the dissolved ions are those that also enter solution when a calcium-rich silicate rock weathers by hydrolysis. The reversible reaction is forced to the right if carbon dioxide in solution is increased. If carbon dioxide is removed from the left side of the reaction, a saturated bicarbonate solution will become cloudy with finely divided $CaCO_3$ precipitate, which is much less soluble than the bicarbonate. Limestone solubility is controlled primarily by addition or loss of CO_2 gas to a great excess of water. Changes of temperature or pressure, mixing of unlike water masses, and biologic process can all promote solution or deposition.

The solubility of $CaCO_3$ in water over the full range of partial pressures of CO_2 is graphed in Figure 7–1. The partial pressure of CO_2 in an aerated aqueous solution is only 30 Pa and the related solubility of $CaCO_3$ is very low, only about 63 mg/L. However, during decay of humus in a soil the partial pressure of CO_2 may increase to 30 kPa and result in an increase of $CaCO_3$ in solution to 700 mg/L, more than a tenfold increase over the amount in aqueous solution at equilibrium with normal atmosphere. Most groundwater contains dissolved $CaCO_3$ in a concentration of between 200 and 400 mg/L, indicating CO_2 partial pressures well within the range of observed composition of soil air. Limestone solution, and therefore karst, is clearly accentuated by biologically generated CO_2 in decaying humus. The climatic implications of this statement are discussed in a subsequent section.

Biological Effects

In addition to providing the decaying humus that enriches soil air in CO_2, plants and animals may corrode limestone directly. Folk et al. (1973) designated a distinctive minor karst landform on Grand Cayman Island as **phytokarst.** The surface is intricately pitted, with pits intersecting to form a sharp-edged, spongy network of pinnacles over 2 m in height. Algae cover the surface and penetrate the karst surfaces to a depth of 0.1 to 0.2 mm. Presumably, the algae generate acidic solvents that attack limestone.

Subsequent work has shown that *cyanobacteria* (blue-green algae) are the principal microorganisms that live on the surface (epilithic), within pre-existing cavities (chasmolithic), and within the outer millimeter (endolithic) of limestone above sea level on Aldabra Atoll and probably in most other tropical regions (Fig. 7–9) (Viles, 1987). In all three micro-environments, the algae weather limestone, but they also may create biochemical conditions that inhibit solution or promote calcite deposition. In particular, the endolithic cyanobacteria penetrate the outer millimeter of rock with micro-size borings that cover 20 to 80 percent of the rock surface (Viles, 1987, p. 326). The pores are the pathway for water and acids from respiration and decay to enter the rock, and the chemically active surface area is greatly increased. However, relict boreholes are usually filled with fine grained

Figure 7-1. Solubility of $CaCO_3$ in water at various partial pressures of CO_2 (t = 16.1°C). (After Adams and Swinnerton, 1937.)

high-magnesium calcite (micrite) that is resistant to weathering and sometimes forms a crust that protects more soluble coralline limestone. It is also proposed that photosynthesis by the cyanobacteria may actually deplete the CO_2 in pore water and thereby inhibit solution. Like the related cyanobacteria that live within the outer rind of quartz boulders in deserts and in Antarctica (p. 133), the algae in and on limestone produce a complex micro-environment for weathering within 1 mm beneath the surface. The weathering environment for limestone is further complicated by snails and other "grazing herbivores" that eat the endolithic algae and simultaneously erode impressive amounts of limestone (p. 480; Shachak et al., 1987).

On certain tropical limestone coasts, flocks of sea birds have nested for many generations and accumulated thick deposits of droppings. This *guano* is rich in phosphate from the scales and skeletons of the small fish that are the birds' primary food. Decay and leaching of the guano makes phosphoric acid, which strongly corrodes the limestone and reacts with it to form commercial deposits of rock phosphate (earthy calcium phosphate of variable composition). Rock phosphate has been mined extensively, especially on Nauru and Ocean Island in the Pacific (Fig. 7–2). It fills karst cavities in limestone to a depth of 20 m (Power, 1925). On many atolls, the location of former groves of trees in which sea birds roosted can be inferred from areas of phosphatized reef limestone. The acid humus of trees also corrodes peat-filled depressions into limestone at their bases.

Temperature Effect

Water at 10°C dissolves about twice as much atmospheric CO_2 as water at 30°C; water at 0°C dissolves almost three times as much as at 30°C. Cold water saturated with CO_2 should be more "aggressive" than warm water in dissolving soluble rocks. A conclusion is that karst should be well developed in cold, humid climates, especially in snow climates. Snow air is enriched in CO_2 because atmospheric nitrogen and oxygen molecules diffuse out of snow more readily than do the larger CO_2 molecules. However, the theoretical saturation values are not the determining factors in karst solution. Even though cold water can hold more CO_2, if the cold landscape is barren of vegetation and the cold soils are poor in humus and microbial activity, there is no source of dissolved CO_2 except the atmosphere. Karst is not a conspicuous feature of cold regions, especially cold dry regions.

Figure 7-2.   Karst pinnacles on Nauru Island exposed by removing the overlying phosphate rock. (Photo: British Phosphate Commissioners.)

Pressure Effect

In theory, groundwater moving downward under hydrostatic pressure will absorb additional CO_2 in proportion to the partial pressure graph (Fig. 7–1) and will dissolve more $CaCO_3$. The theory is excellent, but once groundwater has moved deeper than the A horizon of the soil profile, little additional CO_2 is available for absorption, so pressure has little additive effect on solution. However, the release of pressure when percolating groundwater under hydrostatic pressure enters the free atmosphere of a cave may be an important factor in dripstone deposition.

Related to the pressure effect is turbulence. Dissolved CO_2 gas may be exsolved by splashing or turbulent flow in an underground stream, and $CaCO_3$ may be deposited. The movement of groundwater through granular materials is not believed to be turbulent.

Mixing Effect

The nonlinear solubility of calcium carbonate in carbonated water gives rise to an interesting phenomenon when two unlike water masses mix. For clarity, a portion of Figure 7–1 is enlarged (Fig. 7–3) to show the *mixing effect*. Suppose that deep groundwater, saturated with $CaCO_3$ at a low CO_2 partial pressure, and water from the aerated zone, also saturated but at a higher CO_2 partial pressure, meet and mix at the water table. Because of the curved solubility graph, mixing of any two saturated but unlike water masses invariably results in an undersaturated solution (Ford and Williams, 1989, Fig. 3–10). This mixing effect is probably the reason that most caves develop just below the water table (Moore and Sullivan, 1977). Mixing is also very important in the coastal zone, where groundwater and seawater mix.

Flow-Rate Effect

Somewhat related to the mixing effect are the solubility changes related to variations in rate of flow of surface and underground water. If water is moving slowly over or through soluble rock, it should approach saturation equilibrium. However,

Figure 7-3.   The result of mixing unlike water masses that are each saturated with respect to calcite is a water mass that is undersaturated. (Diagram courtesy of G. W. Moore.)

during rapid runoff periods, river water is notably undersaturated, because the water is not in contact with mineral soil long enough. If such undersaturated water, normally with a low dissolved CO_2 partial pressure, enters a karst terrane and mixes with saturated groundwater, the extreme difference in their composition greatly enhances the mixing effect. This process may be especially important in enlarging caves beneath plateaus that normally drain outward to a surface river but become backflooded when the river is in flood (Thrailkill, 1968, p. 36).

Climatic Control of Karst Processes

Temperature, precipitation, biological activity, and amount and seasonality of runoff are all climatic factors that affect the intensity of karst processes. Therefore, it might be supposed that a climatic model of karst processes would have been developed, similar to the latitudinal belts of weathering and soil-forming processes (Fig. 6–9). However, the chemical solution of carbonate and other soluble rocks is such a complex process that only a few extreme climatic "end members" are easily deduced (Jennings, 1985, pp. 194–214).

One obvious extreme is the hot dry climate, where low precipitation, rapid evaporation, and lack of vegetation produce a negligible flow of groundwater, in which biogenic carbon dioxide is negligible. The high temperature further reduces the amount of atmospheric CO_2 in solution. Deserts are not karst-prone. Massive limestone and dolomite formations assume the role of ridge- and cliff-formers, perhaps rivaling granite in structural control of landforms. Caves in deserts are known, but at least some are relict from former more humid climatic conditions.

The lack of conspicuous karst in polar regions may be due primarily to the low rate of chemical reactions in cold water, the negligible bacterial decay of humus, the short season of runoff, and permafrost that limits infiltration. Corbel (1959a,b) emphasized the importance of karst solution in cold humid regions, but few other authors have supported him (Jennings, 1985, p. 200; White, 1988, p. 215). In Alpine karst in Switzerland, even though the precipitation is 15 percent greater above the treeline, the denudation rate is only 78 percent of the rate lower on the mountains, under forest cover (Bögli, 1980, p. 48). Ford (1971) made extensive measurements of dissolved components in limestone regions of the Canadian Rockies. He found that glacial meltwater above the treeline was saturated with respect to calcite at only 50 to 90 mg/L $CaCO_3$, indicating that little CO_2 was in solution. Below the treeline, groundwater was not saturated with respect to calcite until it contained 100 to 265 mg/L $CaCO_3$, and even creeks and lake water had enough dissolved CO_2 to carry 100 to 140 mg/L $CaCO_3$ in solution at saturation. Ford concluded that the groundwater on forested mountains in a subarctic climate zone is as capable as the groundwater in temperate or tropical regions of dissolving limestone but that there is no particular dissolving power, or "aggressiveness," associated with glacier or snowbank meltwater. There is no doubt, however, that the large seasonal runoff of meltwater in cold humid climates can promote turbulent solution and abrasion of soluble rocks. Subglacial meltwater is highly charged with CO_2 and is capable of corroding carbonate rocks (p. 417). A well-developed cave system under Castleguard Mountain in the Canadian Rockies extends under the Columbia Icefield and derives water from melting ice (Ford, 1983).

The cool-humid mid-latitude zone of increased weathering (Fig. 6–9) has well-developed karst on suitable rocks. Precipitation is greater than in polar regions, more of the precipitation falls as rain, infiltration is not inhibited by frozen ground, and the seasonal leaf fall and dormant period promote humus accumulation. Groundwater is generally abundant. Valley systems eroded by rivers dominate the landscape, but karst landforms mingle with landforms of surface erosion. This common landscape is called *fluviokarst,* distinguished from *holokarst,* in which the entire regional drainage is underground (Ford et al., 1988, p. 401; White, 1988, p. 112).

The "botanic hothouse" climatic extreme of the humid tropics has the most diverse and dramatic examples of karst landscape, although some have argued that the controls are lithologic and tectonic rather than climatic (Jennings, 1985, p. 201; Ford et. al., 1988, p. 403). The combination of heavy precipitation (in part negated by rapid evapotranspiration), intense biologic decay, groundwater rich in CO_2 and organic compounds, and year-round high temperature all promote karst processes. In a multiple linear regression model, the partial pressure of CO_2 in soil air measured during the growing season at a number of world-wide sites was correlated with temperature, precipitation, and evapotranspiration. The best predictor of CO_2 in soil air was identified as the evapotranspiration measured at the known sites. The study is valuable for karst research because the climatic variables it used are readily available at many weather stations, and on a global map, extrapolated or predicted high values of CO_2 in soil air broadly coincide with subtropical and tropical areas of active karst development such as the southeastern United States, Caribbean Islands, Mexico and Central America, Southeast Asia, and the limestone islands of the southwestern Pacific (Brook et al., 1983; White 1988, pp. 196–99). In contradiction, the calculated karst denudation rates of temperate and tropical humid regions have been shown to be statistically indistinguishable (Fig. 7–6). Whether due to the abundance of geologically recent, highly soluble limestone within the Cenozoic orogenic belts of the tropics (Chap. 19), or the exceptional hydrogeology of humid tropical climates, the karst landforms of the tropics are truly remarkable.

Structural Control of Karst Processes

As noted in Chapter 1, *structure* is used in a broad sense in geomorphology to include all of the mineralogic, lithologic, and deformational properties of rock. The most obvious structural factor in karst is lithology. Most karst is on limestone, which is an extremely diverse rock type. To be defined as a limestone for industrial purposes, a rock should contain more than 80 percent calcium and magnesium carbonate, but many limestones are shaly or sandy. Karst limestones, in general, are quite pure, typically > 70 percent carbonate minerals (Ford and Williams, 1989, p. 30). For example, the Dinaric karst is in limestones that range from 80 to 98 percent $CaCO_3$, with many of the analyses in the upper part of the range (Herak, 1972, p. 28). The Annville Limestone of Ordovician age that underlies a karst area near Hershey, Pennsylvania, is more than 95 percent $CaCO_3$ (Foose, 1953, p. 624). The lower Tertiary White Limestone of the famous "cockpit country" of Jamaica rarely has more than 2 percent noncarbonate minerals and generally has less than 0.5 percent (Versey, 1972, p. 447). By contrast, the Luray Caverns of Virginia are in coarsely crystalline Ordovician dolomite that contains as much as 10 percent quartz, chert, feldspar, and clay, but no calcite (Hack and Durloo, 1962, p. 8).

More than 60 variables have been identified that affect the porosity and permeability of carbonate rocks (Brahana et al., 1988, Table 1). Next to lithology, factors of crystallinity, stratification, and fractures are the structural factors that most control karst processes. Highly permeable limestone such as chalk, or reef limestones that have massive primary voids through which water can freely drain, do not develop good karst. The landscape of southeast England and northwest France, for instance, is underlain by thick beds of Cretaceous chalk, a soluble, fine-grained, soft, and massively permeable limestone. Almost all of the chalk terrain lacks surface streams except during peak runoff intervals, yet the chalk does not form the unique landscape of karst. The uniform permeability precludes it. As noted on p. 148 in the definition of karst, well-developed secondary porosity, or the tendency of solution to create and enlarge voids in the rock, is essential for karst. Fine-grained limestone with moderately thick bedding and widely spaced joints is considered ideal for karst development (Ford and Williams, 1989, pp. 31–37). The intersections of vertical joint sets are conducive to concentrating the downward movement of groundwater and are dissolved to make open *shafts*. Joints and bedding planes in flat-lying rocks localize lateral movement and dissolve lateral *galleries* along their intersections (Deike, 1969). Flow lines in groundwater probably determine the depth and orientation of caves, but most minor features are controlled by structures.

KARST HYDROGEOLOGY

The science of underground water movement in carbonate rocks is closely related to the process of solution and to the formation of caves. Unfortunately, carbonate rocks have the most variable hydrogeologic conditions of all major rock types (Brahana et al., 1988), and groundwater movement in karst is highly irregular and complex. Open solution shafts and galleries are so much larger than the intergranular pore spaces through which water moves in granular rocks that some authors have questioned the applicability of the fundamental concept of a water table beneath karst terrain (Jennings, 1985, p. 66; Ford and Williams, 1989, p. 129, pp. 142–48).

In the classical theories of hydrogeology, precipitation percolates into soil and rock and infiltrates downward through an unsaturated zone of variable and discontinuous intergranular water and air. This zone is known as the **zone of aeration,** or *vadose zone.* At some depth, ranging from zero to hundreds of meters, pore spaces are full of water. This is the **zone of saturation,** or *phreatic zone.* The surface of the zone of saturation is the **water table** (Fig. 7–4) (Freeze and Cherry, 1979, p. 39). Groundwater moves so slowly that it establishes local hydrodynamic gradients, and in humid regions the water table is a subdued replica of the topography, rising beneath hills and intersecting lower parts of the landscape at the surface of rivers, lakes, or springs.

Water beneath the water table moves in accordance with laws of hydraulics along flow lines, perpendicular to surfaces of pressure equipotential (Fig. 7–4).

Figure 7-4.   Flow lines (solid lines) in uniformly permeable material between an intake area at the upland water table and discharge areas on valley floors. Flow lines are orthogonal to surfaces of pressure equipotential (dashed lines). (Hubbert, 1940, Fig. 45: © Univ. Chicago Press.)

From each point on the water table, there is but one unique flow path to an outlet. The general hydrogeologic equation, known as *Darcy's law,* is fully analogous to Ohm's law for electricity, because the amount of flow (amperage) per unit area of cross section is proportional to the pressure-potential difference (voltage) and inversely proportional to the resistance (ohmage) and the length of the conduit.

The fundamental problem of karst hydrogeology is scale. Models of classical groundwater flow (Fig. 7–4) are large compared to the grain size or pore size of the granular medium through which water permeates. Dense, well-crystallized Paleozoic or Mesozoic limestone and marble have extremely low permeability (Brahana et al., 1988, Table 2). *Fracture permeability* is the only way water can move through such rocks. Initial fractures such as bedding planes and joints may be not wider than 10 to 50 μm. Water moves slowly through the fractures as laminar films, so that a large volume of the rock is statistically similar to a uniform porous material of rather low permeability. The slowly moving water is at most only slightly unsaturated.

A definitive criterion of karst-prone rocks is that their permeability increases with time (Ford and Williams, 1989, p. 133). As solution slowly widens fractures, three thresholds are passed, which change the nature of flow (Watson and White, 1985, p. 121). First, in a widened fracture, flow changes from laminar to turbulent. Second, the rate of solution increases rapidly because a larger volume of undersaturated water moves through the fracture. Third, mechanical transport of abrasive clastic particles begins when the flow velocity exceeds a certain level (Fig. 9–10). All three thresholds occur under typical groundwater conditions when fractures are opened to a width of 5 to 10 mm. A time span of 3000 to 10,000 years is suggested for opening an initial tight joint to the critical threshold size. The subsequent opening of selected conduits to meter-sized cave passages may require another 10,000 to 1 million years (Watson and White, 1985, p. 122). The slowness of the initial solution is supported by the opinion of Bögli (1980, p. 58) that shallow surface fractures may require more time to widen to 2 mm than they require subsequently to widen from 2 mm to 200 mm.

Computers can model the flow of karst groundwater as a network of pipes of specified shape and internal friction, with water added at the intersections and draining out at the margins of the network. Figure 7–5 is a map model of the karst plain in the vicinity of Mammoth Cave, Kentucky (Thrailkill, 1972). Water is assumed to move downward through the zone of aeration through open shafts beneath sinkholes. The plain is surrounded by intrenched rivers, into which groundwater emerges as springs and seeps. The computer model reproduces the flow of the large springs and also predicts discharge at other points on the periphery of the network. It correctly models the large flow through the underground river in the lower level of Mammoth Cave. It could be used to predict the sites of wells that would intercept the maximum flows in the zone of saturation. With further refinement and incorporation of solubility factors, a computer model could be used to predict future cavern enlargement.

Figure 7-5. Digital computer two-dimensional pipe model of groundwater movement in the vicinity of Mammoth Cave, Kentucky. (Thrailkill, 1972, Fig. 4.)

KARST LANDFORMS

Regional and Temporal Significance of Karst

Limestones and dolomites constitute between 5 and 15 percent by weight of the total sedimentary rock mass (Garrells and Mackenzie, 1971, p. 40); evaporites constitute an additional 5 percent. About 15 percent of the area of the conterminous United States has karst-prone rock at or near the surface (Davies and LeGrande, 1972, p. 469), and about 10 percent of the earth's land area has exposed carbonate rocks (Ford et al., 1988, p. 401). Karst landforms are worthy of study not only because of their dramatic form and unusual origin but because of their abundance.

The late Cenozoic Era has been a time of unusually extensive karst landscapes. Both the Indonesian–Philippines–New Guinea orogenic belt and the West Indies–Caribbean belt are in climatic zones of submarine carbonate deposition and coral-reef growth. Active tectonism has uplifted coral limestone and other carbonate rocks (Figs. 2–3 and 19–13). The Alpine orogenic belt, extending from the Pyrenees Mountains through southern Europe, Turkey, and Iran, includes thick sections of recrystallized, fractured, and deformed carbonate rocks in which karst processes have been active in late Cenozoic time. The original karst region in Yugoslavia is in the Dinaric part of the Alpine orogenic belt (Demek et al., 1984, Figs. 16–1 and 16–3).

Quaternary glacier-controlled fluctuations of sea level (Fig. 17–16) have alternately exposed and drowned karst landscapes on all carbonate-rock coasts. When sea level is lowered by expanding glaciers, coastal regions develop deeper groundwater circulation, and karst landforms and caves are actively developed. When, as at the present time, the level of the sea is at a high interglacial level, coastal karst terranes are drowned. Dripstone, formed in aerated caves, has been collected by divers from caves 45 m below sea level in the Bahama Islands (Gascoyne et al., 1979), and observed at a depth of 122 m in a "blue hole" in a British Honduras reef (Dill, 1977). Hundreds or thousands of "blue holes" below sea level today are almost certainly dolines (p. 161) that formed during glacial-age low sea levels. The water supply for coastal cities on drowned karst, as in the state of Florida and in Greece, is drawn from karst aquifers below sea level. The shape of atolls also depends to a large extent on karst solution during Pleistocene low sea levels (Chap. 19).

Karst Denudation Rates

The rate of limestone solution is measured by mass balance equations involving the water discharge per unit area and the concentration of dissolved carbonate in the water. Corrections are required for the percentage of the discharge area underlain by noncarbonate rocks (which also are weathered to dissolved carbonate species; Chap. 6), the bulk density of the rocks, the amount of the dissolved load reprecipitated within the drainage area, seasonal variations in discharge and dissolved load, and the underground discharge area, which may not coincide with the surface drainage area. Results are usually reported in volume loss per unit area per year (m^3 km^{-2} yr^{-1}), which is equivalent to a regional *denudation rate,* or average landscape lowering, of mm/1000 yr (Fig. 7–6).

Denudation rates for limestone terranes are deceptive in that much of the volume loss is in the subsurface, but on the scale of geologic time, the net result must be the lowering of the landscape toward sea level. The denudation rates compiled in Figure 7–6 fall well within the range of regional fluvial denudation rates (Chap. 12), not surprising since a significant percentage of all landscapes consist of soluble rocks. Clearly, the most important variable in limestone denudation is runoff, or the excess of precipitation over evapotranspiration (Ford and Williams, 1989, p. 97). This is the water that is available to carry away the dissolved compounds.

Figure 7-6. Limestone removal rate as a function of annual runoff: AUS, Australia; CAN, Canada; ENG, England; GER, West Germany; IRE, Ireland; IT, Italy; JAM, Jamaica; MAD, Madagascar; NZ, New Zealand; NOR, Norway; POL, Poland; PR, Puerto Rico; PYR, Pyrenees; SWE, Sweden; USA, United States; WAL, Wales; YUG, Yugoslavia. (Jennings, 1985, p. 193.) Units of limestone removal rate are equivalent to denudation rate in mm/1000 yr.

Denudation rates are also significantly correlated with both evapotranspiration and precipitation. Much less important is the climatic variable of temperature. More important than temperature is the proportion of bare and soil-covered ground. Although arctic and alpine environments with little or no soil cover show significantly less rapid denudation rates for any runoff than do vegetated and soil-covered terranes, temperate and tropical humid climates have similar denudation rates for any runoff (Jennings, 1985, pp. 192–94; White, 1988, pp. 215–19).

Minor Karst Landforms[1]

Limestone and dolomite, and to a lesser degree other soluble rocks, commonly develop distinctive surface weathering features. Surfaces may be smoothly rounded forms (Figs. 7–7 and 7–8), or hemispherical pits that intersect as razor-sharp points and ridges. In the extreme, a spongelike framework of rock is all that remains (Fig. 7–9). Phytokarst (p. 150) is independent of gravity, with surface pits corroding equidimensionally beneath an algal film, whereas solution forms tend to elongate vertically or downslope.

Karren, Solution Pits, Facets, Flutes, and Runnels. On exposed rock (bare karst, pavement karst), horizontal surfaces develop shallow **solution pits** (rainpits, makatea). Larger features have been named **solution pans** (tinajitas) (Smith and Albritton, 1941). On planar surfaces inclined a few degrees from the horizontal, sheet runoff produces flat **solution facets** (solution bevels). On steeply inclined karst surfaces, elongate **solution flutes** (Rillenkarren, lapies) are common landforms. These are straight, parallel-sided grooves with sharp crests separating them. They follow the steepest slope down a rock face. **Solution runnels** (Rinnenkarren) are branching channels in networks on gently inclined bare surfaces. Many runnels probably first form under soil cover but, having formed, are perpetuated by runoff on bare rock surfaces.

Figure 7-7. Well-developed, but small, clint-and-grike topography on a dolomite pavement. This example is on Ca-rich reefal dolomites in the Bruce Peninsula, Ontario, Canada. (Ford et al., 1988, Fig. 1.)

[1] Because of the multilingual terminology of karst forms, a single boldface term is defined. At the first use, synonyms follow in parentheses but are not defined. The reader is urged to consult glossaries or expanded texts for the detailed shades of meaning (see Monroe, 1970; Sweeting, 1973; Jennings, 1985; White, 1988; Ford and Williams, 1989).

Approximate scale in meters

Figure 7-8. Labeled sketch of minor karst forms on Silurian dolomite near Hamilton, Ontario. (After Pluhar and Ford, 1970, Fig. 2.)

Covered Karst: Grikes, Clints, Solution Pipes, and Solution Notches. If soil or an organic mat covers a limestone surface, the effectiveness of solution increases manyfold. Solution pits are even more common under soil cover than on bare ledges, but the divides between pits are likely to be more rounded and smooth than the knife-edged ridges between intersecting pits on bare karst.

Vertical openings such as joints become widened by solution (p. 155). The British term is **grike** (Kluftkarren, solution slot). Grikes intersect as circular, verti-

Figure 7-9. Spongelike phytokarst on Pleistocene coral limestone, Huon Peninsula, Papua New Guinea. One-meter axe handle is nearly concealed.

cal **solution pipes** (karst wells, potholes). The remnant pavement surfaces between open grikes are called **clints,** on which minor solution features develop (Figs. 7–7 and 7–8) (Pluhar and Ford, 1970).

At the level where bare rock projects above the soil level, a **solution notch** may develop. A distinct undercut measures the greater effectiveness of karst processes below ground level. An extreme example of a solution notch has been named a **swamp slot** (Wilford and Wall, 1965) or foot cave (Jennings, 1985, p. 208). It is probably restricted to the humid tropics. Acid swamp water corrodes a narrow slit a meter or more horizontally into the base of limestone cliffs. Residual clay or an organic detrital layer may prevent downward solution, especially if the water table is at the swamp surface and if percolation is slow. The swamp slot may be a basic component of tropical tower karst (p. 165), characterized by marginal corrosion around swampy plains at the base of steep-sided residual karst hills.

Major Karst Landforms

Karst is essentially a near-surface process, in vegetated and soil-covered landscapes, where most of the solution takes place in or just beneath the soil profile. On bare rock, solution is concentrated at the water table, in the zone of mixing. Thus, there is a continuum of landforms from the minor surface features just described to distinct landforms comparable to the hills and valleys of fluvial landscapes. The forms described in this section are of a scale that can be studied on aerial photographs and topographic maps. It should be noted that portraying karst topography on contour maps poses special problems for the cartographer. Hachured depression contours, contour lines that cross to show overhanging cliffs, and dotted contours in natural bridges are some of the unusual mapping conventions that are required to contour karst landforms.

Dolines (Sinkholes, Sinks, Swallow Holes, Cockpits, Blue Holes, Cenotes). The fundamental genetic component of karst topography is the **doline,** or limestone sink (Figs. 7–10 and 7–11) (Ford and Williams, 1989, p. 396). Whether due to joint control (the most common cause), differential solubility, or random events, localized areas of karst terrain are lowered more rapidly than the surrounding area and form closed depressions. Their sides slope inward at the angle of repose of the adjacent material, typically 20° to 30°. Some have much more gentle side slopes, and others, especially those forms known as *collapse sinks,* have vertical or overhanging cliffs. They are usually circular in plan view and less commonly elongate or oval. They range in size from shallow soil depressions a few meters in diameter and a meter deep to major landforms several kilometers in diameter and hundreds of meters deep.

Doline and its synonyms are genetic terms. They describe a landform produced by the karst process. Identically shaped closed depressions can form by subsidence, vulcanism, wind deflation, or glaciation, in fact, by any process that selectively translocates a mass of rock and permits the surrounding material to slump into the excavation. In areas such as southern Indiana, where glacial moraines cross a region of karst topography, real judgment and careful investigation are required to distinguish dolines from glacial kettles.

Dolines are the karst analog of river valleys. They are the fundamental unit of erosion as well as weathering. As localized sources of surface water, they concentrate and maximize the mixing effect (pp. 152 and 169). As the dissolving water goes underground, the eroding landmass literally disappears "down the hole," and the resulting landform has much in common with the surface forms on dry sand above a hole in the floor of a container (Fig. 7–10).

Figure 7-10. Doline about 120 m in diameter and 45 m deep, which abruptly collapsed on the night of December 2, 1972, near Montevallo, Alabama. (Photo: U.S. Geological Survey.)

The bottom of a doline may open into a subterranean passage, down which water pours during a rainstorm. More commonly, the subsurface passage is partly blocked by rock rubble, soil, or vegetation, and temporary lakes form until the water percolates away. Some karst lakes are notorious for irregularly flushing out a clay plug at depth and draining, only to fill unexpectedly at a later date. The karst lake district of north-central Florida is a real estate lawyer's paradise, full of "shore-front" property that suddenly becomes property at the edge of a sawgrass "prairie" or a mud flat.

Five major classes of dolines are recognized (Fig. 7–11). The two most contrasting types are the funnel-shaped *solution doline* and the steep or cliffed *collapse doline*. In the field, distinction between the two varieties of doline is rarely as obvious as the sketches of Figure 7–11 suggest. *Subsidence dolines* and *subjacent karst collapse dolines* are surface forms in nonsoluble rock, caused by solution of a buried karst. A thick residual soil, alluvium, or loess may mantle an active karst landscape, and the surface material will gradually settle or be carried into a buried doline. Frequently, small dolines (*swallow holes, ponors*) form in a stream bed, and although alluvium completely conceals the buried karst relief, the stream abruptly disappears and abandons the downstream segment of its former channel. These also have been called *alluvial streamsink dolines.*

Dolines in tropical regions develop rapidly and grow very large. Adjacent dolines intersect, fuse, or engulf smaller ones to form compound, reticulate, honeycomb, or even star-shaped depressions (Williams, 1972; Monroe, 1976). The floors are frequently swamps at the water table. The *cockpit country* of Jamaica is the classic locality for tropical dolines (Versey, 1972). One doline or cockpit in Puerto Rico is so nearly a perfect hemisphere in cross profile that it has been surfaced with metal mesh as a giant radio telescope (Fig. 7–12).

(a) Solution doline (funnel sink)

(b) Collapse doline

(c) Subsidence doline

(d) Subjacent karst collapse doline

(e) Cockpits (intersecting star-shaped dolines)

Figure 7-11.   Five major classes of dolines. (From Jennings, 1985, Figs. 37 and 43.)

Uvalas (Compound Sinkholes, Karst Windows, Blind Valleys). A genetic progression can be visualized in which dolines progressively abstract surface runoff into groundwater circulation and leave networks of dry valleys as relict surface forms. The Serbo-Croatian term **uvala** was given to a compound doline, or a chain of intersecting dolines (Cvijić, 1960). A **karst window** is similar, in that an unroofed segment of an underground stream channel becomes a surface valley in the window, and passes underground again downstream. If the exposed segment has normal stream valley form and an alluvial floor but disappears downstream into a swallow hole, the term **blind valley** is often used. A natural bridge (Fig. 7–12) may mark a segment of unroofed cave or may form by progressive underground abstraction of surface flow, either across the neck of an intrenched stream meander or from a high-level stream to a lower adjacent stream.

Figure 7-12. The radio telescope at the Arecibo Observatory, Puerto Rico. The reflector surface is a doline about 300 m in diameter and 70 m deep. The floor of the doline probably drains through a cave into the Tanama River, seen entering a natural bridge in the right foreground. The Arecibo Observatory is a part of the National Astronomy and Ionosphere Center, which is operated by Cornell University, under contract with the National Science Foundation. (Photo: R. C. Hamilton, Cornell University Office of Public Information.)

A **karst valley** is a special form of blind valley that occurs where a surface drainage pattern first developed on nonsoluble rock and later eroded into soluble rock, when subsurface drainage totally defeated the surface flow. It has all the properties of a stream valley except that it lacks a stream. Karst valleys are common in areas of nearly flat-lying, interbedded limestone, shale, and sandstone. Many good examples are shown on topographic maps of the Mammoth Cave area in Kentucky.

Poljes (Plans, Wangs, Hojos). The Dinaric karst region of Yugoslavia is in a belt of Alpine folding (Fig. 20–7). The karst topography is on carbonate rocks that are folded and faulted, often in close proximity to insoluble rocks. Structural control becomes a major factor of landscape evolution. Large blind valleys or karst valleys, enclosed by either soluble or insoluble rocks, usually elongated along tectonic axes, are called **poljes** there. Because of their orientation on tectonic trends, they have been interpreted as grabens or half grabens. Some are, but others are simply large-scale examples of structurally controlled karst processes. Their floors truncate folded limestone formations or deep sedimentary infills at a local base level determined by a narrow gorge through insoluble rock or a series of swallow holes (ponors) at the downstream end of the polje (White, 1988, p. 41). Poljes are large valleys; the largest in Yugoslavia are up to 60 km in length and 6 to 8 km wide. In general uvalas are measured in hundreds of meters; poljes in kilometers (Herak, 1972, p. 37).

Poljes have broad, flat floors that abut sharply against steep enclosing walls. On occasion, during floods, the swallow holes by which some valley floors are drained become choked with debris, and the entire polje floor becomes a shallow lake. Some Yugoslavian poljes flood annually when the water table is high and are fertile agricultural land during drier seasons. Their hydrology is the subject of sophisticated Yugoslavian research (Herak, 1972; White, 1988).

Residual Karst

There is a tendency among many geomorphologists to consider solution landforms as a genetic sequence: doline → uvala → polje. Dolines grow and merge into uvalas through time in a genetic sequence. However, poljes are more than large, complex uvalas. They have a structural control that does not lend itself to a genetic sequence following the much smaller, process-controlled forms.

The concept of a karst "cycle of erosion," beginning with the genetic sequence of doline and uvala and ending with a plain of solution dotted with residual limestone hills, can be traced to the historical impact of the visit with Albrecht Penck to the Dinaric karst by W. M. Davis in 1899 (Davis, 1901). At the time, the Davisian concept of a cycle of erosion was achieving great popularity. Cvijić (1893) had previously published a distinguished monograph on karst under the supervision of Penck. Penck and Davis inspired him to rework his observations into a cyclic concept, which he published in France while he taught there during World War I. The paper was not well known because of the wartime communication problems in Europe, but a review was published in English (Sanders, 1921) reproducing Cvijić's original block diagrams. A major manuscript by Cvijić (1960) was lost by his publisher and not rediscovered and published until 33 years after his death.

The deduced residual landforms of the karst cycle were not well illustrated by the Dinaric karst. Small limestone hills (hums) rise above the floors of Yugoslavian poljes, but the structural complexities of the region made it unsuitable as an area for a model of a karst erosion cycle. Later authors used the tropical karst of Jamaica and southeast Asia as models for the more advanced stages of a deduced karst cycle, neglecting the obvious differences in lithologic and climatic controls. As a result, the deduced karst "cycle of erosion" starts in a Mediterranean climate but ends in the tropics. Rather than persist in following the cyclic concept, it is more productive to review residual karst landforms simply as the products of extreme karst processes, especially common in tropical humid climates.

Lehmann (1936) introduced the term **kegelkarst** (cone karst) for a landscape in Java where the terrain is approximately equally shared by conical dolines and conical residual hills. In fact, both in Jamaica and in Java, the typical kegelkarst residual hills are more nearly hemispheric than conic, and the dolines are star-shaped or valleylike among the hills.

Tower karst (turmkarst, mogote, pepino hill) has been distinguished from kegelkarst in that the residual hills are steep-sided or vertical and are separated from each other by swamps or alluvial plains (Fig. 7–13). Marginal solution at the edge of the swampy plains may be largely responsible for the steep walls of tower karst. The residuals are riddled with caves; a swamp slot may deeply notch the base of the towers.

The term **tower karst** has become the general term for all residual karst hills, including those with conical or hemispheric shapes, that rise above adjacent plains (Williams, 1987; Ford and Williams, 1989, pp. 440–47). The plains are usually erosional surfaces that truncate the same soluble rocks that form the towers, but they can equally well be erosional plains across other rock types, or aggradational plains. In coastal regions and along the valleys of large rivers, late Quaternary sea level fluctuations have caused alternate cycles of aggradation and degradation, and the tower karst of these regions is clearly part of the *palimpsest* landscape (p. 16) (Williams, 1987).

Figure 7-13.   Tower karst in Guilin, Guangxi Province, China. Compare with Figure 7-14. (Photo: J. H. Ferger, M.D.)

The tower karst of northern Vietnam extends northwestward from near Haiphong harbor, where it is drowned by the postglacial rise of sea level, along the Red River to the Chinese frontier. It continues into southern China, especially in Guangxi Province. Many art historians believe that the strange, needlelike mountains of classical Chinese art originated in the subtropical karst landscape of the southern provinces of China (Fig. 7–14). The scenes seem exotic and dreamlike to Western eyes, but they are actually reasonable geomorphic sketches. It should be noted that later classical Chinese art idealized the forms of mountains so that all look equally exotic. The incident portrayed by Figure 7–14, "Emperor Ming-Huang's Journey to Shu," is known to have occurred in the karst region of southern China.

Tower karst requires thick formations of pure limestone uplifted rapidly in late Cenozoic time. Some of the towers in China are more than 500 m in height. Most tower karst is in the tropics, but a small example has been described in the cold, dry Mackenzie Mountains of northwestern Canada (Brook and Ford, 1978; Ford et al., 1988, p. 405).

The *mogotes* (haystack hills) of Sierra de los Organos, Cuba, are residual karst towers of classic beauty (Lehmann, 1960, Plate 1; Gèze and Mangin, 1980). The pepino hills of Puerto Rico are comparable (Meyerhoff, 1938; Monroe, 1976; Guisti, 1978), but many show distinct asymmetry, with a gentle slope toward the northeast from which most of the wind-driven rain strikes them. Thorp (1934) proposed that solution on the windward slopes lowered that side of the pepino hills more rapidly. Subsequently, Monroe (1966; 1976, p. 45) proposed that solution and redeposition of surficial secondary carbonate has "case hardened" the windward slopes, whereas the lee slopes are subject to more extensive basal solution and develop steeper cliffs.

Residual karst landforms have a curious relevance to the political and social histories of the regions in which they are found. Bandits, partisans, guerrilla troops, and fugitives who are native to a karst region are able to live safely and base their operations in the many caves that penetrate karst towers. The Viet Minh operated against French colonial troops from the karst of northern Vietnam for decades; Fidel Castro and his revolutionaries based their eventual control of Cuba on the karst of the Sierra Maestra; Yugoslav partisans prevented motorized German troops

Figure 7-14. A south China landscape as painted by an unknown artist of the T'ang dynasty (A.D. 618–907). Compare with Figure 7-13. (Collection of the National Palace Museum, Taipei, Taiwan, Republic of China.)

from controlling large parts of the Dinaric karst; the examples could be multiplied at length. On a less significant level, illegal distilleries are traditionally hidden in the Appalachian karst of the eastern United States, and Jesse James, a notorious bandit, hid his gang in the caves of Missouri. Whatever the merit of a cause, its supporters are likely to find a karst terrain that they know intimately an overwhelming advantage against numerically superior and better equipped, but foreign, opponents.

LIMESTONE CAVERNS

A limestone cavern (cave)[2] can arbitrarily be defined as a solution cavity large enough for people to enter. Access implies topologic continuity with the subaerial terrain, and therefore caverns are part of the karst landscape, even when they are hundreds of meters below the generally accepted "surface." Even the largest cavern is volumetrically insignificant as a landform (Thrailkill, 1968, p. 42), but the beauty of the deposits and the mysterious thrill of cave exploration give caverns a geomorphic significance out of proportion to their size.

Theories of Cave Development

This chapter has repeatedly stressed that karst is essentially a near-surface process. The maximum solution takes place immediately beneath or in the soil profile. How, then, can caves form hundreds or thousands of meters below the other landforms of a karst landscape?

Early theories of cave formation were vague about the chemistry of the process but implied that caves developed in the zone of aeration by downward-moving vadose water. Beginning with solution enlargement along bedding planes and joints, open conduits developed that were large enough to guide underground rivers and waterfalls. Great emphasis was placed on erosion by underground rivers. A surface river was necessary to accept the drainage from underground streams, and as the *vadose theory* developed, it became closely tied to theories of multiple erosion surfaces that had been rejuvenated by tectonic uplift. As the surface streams intrenched their valleys in each new cycle of erosion, the water table was lowered and caves were excavated in the newly drained vadose zone (Watson and White, 1985, p. 110).

The vadose theory of cave formation was accepted without serious question until 1930, when W. M. Davis wrote an extended evaluation of old theories and new observations and pointed out serious theoretical deficiencies. A fundamental problem was that many caverns contain extensive dripstone deposits, which are observed to grow. If vadose water is actively filling caverns with dripstone, how could it be the agent that formerly excavated them? Davis (1930, p. 549) revived interest in long-standing hydrogeologic theories of deep flow paths in the zone of saturation (Fig. 7–4) by proposing that because the phreatic water circulated to great depths, caverns could be excavated below the water table, perhaps far below it. Later, when the regional surface drainage network had eroded downward into the limestone terrane, the caves would be in the zone of aeration, rivers could flow through them, and dripstone could form. This became known as the *two-cycle theory* of cavern formation. Like much of his later work (he was 80 years old in 1930), Davis's paper was based on field work by others and his own deductions, largely colored by the success of his cyclical theory of landscape evolution 30 or more years earlier. His ideas of slow solution by stagnant water at great depth are contrary to all modern experiments and field evidence for the importance of CO_2 uptake, mixing effects, and flow rates (Watson and White, 1985, p. 112).

Two years after Davis published his study, A. C. Swinnerton (1932) proposed that the maximum solution takes place at or just beneath the water table, especially in the narrow zone within which the water table fluctuates. He called his variant the *water-table hypothesis.* Swinnerton postulated that water at the water table has a va-

---

[2] Caverns form in a variety of ways other than by solution. Lava tubes, tubes of headward "piping" by streams, and unusually wide overhanging ledges of any strong rock may be classed as caverns, or caves. Here the terms are meant to imply limestone solution features unless specifically stated otherwise.

Figure 7-15. Simplified cross section of Lehman Caves, Nevada. Horizontal passages intersect tilted strata. (Diagram courtesy of G. W. Moore.)

riety of paths along which it can flow toward a spring, but most of the volume moves along the shortest path, which is along the water table.

Swinnerton's water-table hypothesis explained the many observations that caverns are generally horizontal, even in deformed strata (Fig. 7–15). The water-table hypothesis permitted the observed development of three-dimensional networks of cavern passages, with blind passages that cannot be explained by through-flowing rivers or vadose percolation. It permitted, but did not require, second-cycle episodes of dripstone deposition and dissection by underground rivers as the regional water table and landscape was lowered. However, his hydrogeologic model was faulty (Hubbert, 1940). Speleologists continued to prefer theories of cavern excavation by flowing water in the vadose zone or at the water table. Few further theoretical studies were done until after World War II, except for an excellent extension of Davis's theory by Bretz (1942).

The present opinion is that caverns form in the zone of saturation but only a short distance below the water table (Moore, 1968; Thrailkill, 1968; Moore and Sullivan, 1977; Watson and White, 1985). The favored cause for renewed aggressiveness of groundwater is the mechanism of mixing (Fig. 7–3). Phreatic water is quite likely to have different chemical properties than those of vadose water, especially if the vadose water is flowing directly to the water table along a conduit from the floor of a doline above. With the addition of the temperature effect and the flow-rate effect, the mixing effect provides a theoretically sound and quantitatively reasonable solution process to excavate caverns (Thrailkill, 1968). However, by the mixing hypothesis, cavern excavation is only related to karst processes at the soil level insofar as dolines provide a mechanism for locally introducing vadose water into phreatic water at the water table. Diffuse seepage is less likely to produce contrasting water types at the water table. Those karst hydrogeologists who dislike the concept of a water table in soluble rocks, and point to evidence of multiple pipelike conduits with unlike water masses in them, flowing side by side or crossing each other, require an abrasion hypothesis for largely mechanical vadose enlargement of caverns. Their views are well summarized by Jennings (1985, pp. 66–67, 149–52).

Cave Deposits

Dripstone and **flowstone** are general popular terms for cave deposits from water; the equivalent term **speleothem** is gaining popularity. A variety of delicate and beautiful shapes are assumed by speleothems (Fig. 7–16). As limestone is dissolved by vadose water, the pH of the solution rises but in equilibrium with a relatively high CO_2 partial pressure. When the water enters an aerated cave, the excess dissolved CO_2 diffuses into the cave atmosphere, and calcite is deposited. The popular idea that cave deposits are the result of evaporation should be dispelled, because wet caves have essentially 100 percent relative humidity, and evaporation is negligible.

Figure 7-16. Speleothems in the Postojna Caves, northern Yugoslavia. This is one of the most popular tourist caves in the famous Dinaric karst region of Yugoslavia. (Photo: Yugoslav State Tourist Office.)

There is a possibility that biological processes may cause some speleothems. Fungal threads have been observed at the tips of growing stalactites in Lehman Caves, Nevada (Went, 1969).

The principal speleothems are *stalactites* and *helictites,* which grow downward from cave roofs, and *stalagmites,* which build upward from the floors. Helictites assume twisted or spiral forms that are related to strained calcite crystal structures. Curtains or ribbons of flowstone may coat cave walls or form a partition beneath a prominent joint-controlled seepage zone. In cave pools, a variety of delicate skeletal mineral "flowers" may form. Tufa rims and terraces line slow-moving, sediment-free underground streams.

In addition to speleothems, a variety of clastic deposits may accumulate in caves. Caves that were occupied by animals and prehistoric people may have artifact-bearing organic litter on their floors. Rock falls or weathering spalls from cave roofs and walls add sporadically to the thickness of cave-floor deposits. Caves that have flowing streams contain a variety of alluvial deposits. Many limestone caves have thick layers of red clay that was formerly thought to be the insoluble residue of subsurface solution. However, the volume of dissolved limestone has been judged insufficient to account for the amount of red clay, so it must also include washed-in surface sediment (Jennings, 1985, p. 166). Windblown silt (loess), volcanic ash, and organic detritus may accumulate in caves to produce stratigraphic records of great archeologic significance.

Speleothems and clastic deposits in caves offer excellent opportunities to date cave evolution. Radiocarbon dating of organic-rich biogenic and archeologic strata can date occupation history and intervals of climate changes back to about 50,000 years (Solecki, 1963, 1972). Oxygen-isotope variations in layered speleothems record climatic changes within the cave watershed, and Uranum-series dating of the same layers of calcite gives a chronology back to 250,000 years (Winograd et al., 1988). Similar dating of Bahama speleothems show that sea level was at least 45 m below present level during the interval of 160,000 to 139,000 years ago, during the Illinoian ice age (Gascoyne et al., 1979). Clastic sediments in several levels of Mammoth Cave show magnetic polarity reversals that document the downcutting of the cave for at least 1 million years and possibly 2 million years (Schmidt, 1982). Cave deposits are becoming of increased importance in documenting Pleistocene climatic changes on the continents, to correlate with the deep-sea record (Fig. 18–1).

REFERENCES

ADAMS, C. S., and SWINNERTON, A. C., 1937, Solubility of limestone: Am. Geophys. Un. Trans., pt. 2, pp. 504–8.

BÖGLI, A., 1980, Karst hydrology and physical speleology (transl. by J. C. Schmid): Springer-Verlag, Berlin, 284 pp.

BRAHANA, J. V., THRAILKILL, J., FREEMAN, T., and WARD, W. C., 1988, Carbonate rocks, *in* Back, W., Rosenshein, J.S., and Seaber, P. R., eds., Hydrogeology: The geology of North America, v. O–2, Geol. Soc. America, Boulder, Colo., pp. 333–52.

BRETZ, J. H., 1942, Vadose and phreatic features of limestone caverns: Jour. Geology, v. 50, pp. 675–811.

BROOK, G. A., FOLKOFF, M. E., and BOX, E. O., 1983, A world model of soil carbon dioxide: Earth Surface Processes and Landforms, v. 8, pp. 79–88.

———, and FORD, D. C., 1978, Origin of labyrinth and tower karst and the climatic conditions necessary for their development: Nature, v. 275, pp. 493–96.

CORBEL, J., 1959a, Vitesse de l'érosion: Zeitschr. für Geomorph., v. 3, pp. 1–28.

———, 1959b, Erosion en terrain calcaire: Annales de Geog., v. 68, pp. 97–120.

CVIJIĆ, J., 1893, Das Karstphänomen: Geographische Abhandlungen herausgegeben von A. Penck, v. 5, no. 3, pp. 218–329.

———, 1960, La géographie des terrains calcaires (French transl. by E. de Martonne): Serbe Acad. Sci. Arts Mon., v. 341, 212 pp.

DAVIES, W. E. and LeGRAND, H. E., 1972, Karst of the United States, *in* Herak, M., and

Stringfield, V. T., eds., Karst: Important karst regions of the northern hemisphere, Elsevier Publishing Co., Amsterdam, pp. 467–505.

DAVIS, W. M., 1901, An excursion in Bosnia, Hercegovina, and Dalmatia: Geog. Soc. Philadelphia Bull., v. 3, pp. 21–50.

———, 1930, Origin of limestone caverns: Geol. Soc. America Bull., v. 41, pp. 475–628.

DEIKE, R. G., 1969, Relations of jointing to orientation of solution cavities in limestones of central Pennsylvania: Am. Jour. Sci., v. 267, pp. 1230–48.

DEMEK, J., GAMS, I., and VAPTSAROV, I., 1984, Balkan Peninsula, in Embleton, C., ed., Geomorphology of Europe: John Wiley & Sons, Inc., New York, pp. 374–86.

DILL, R. F., 1977, The Blue Holes—geologically significant submerged sink holes and caves off British Honduras and Andros, Bahama Islands: Third International Coral Reef Symposium, Proceedings, v. 2, pp. 237–42.

FOLK, R. L., ROBERTS H. H., and MOORE, C. H., 1973, Black phytokarst from Hell, Cayman Islands (B.W.I.): Geol. Soc. America Bull., v. 84, pp. 2351–60.

FOOSE, R. M., 1953, Groundwater behavior in the Hershey Valley, Pennsylvania: Geol. Soc. America Bull., v. 64, pp. 623–46.

FORD, D. C., 1971, Characteristics of limestone solution in the southern Rocky Mountains and Selkirk Mountains, Alberta and British Columbia: Canadian Jour. Earth Sci., v. 8, pp. 585–609.

———, ed., 1983, Castleguard cave and karst, Columbia Icefields area, Rocky Mountains of Canada: A symposium: Arctic and Alpine Res., v. 15, pp. 425–554.

———, PALMER, A. N., and WHITE, W. B., 1988, Landform development; karst, in Back, W., Rosenshein, J. S., and Seaber, P. R., eds., Hydrogeology: The geology of North America, v. O-2, Geol. Soc. America, Boulder, Colo., pp. 401–12.

———, and WILLIAMS, P. W., 1989, Karst geomorphology and hydrology: Unwin Hyman Ltd., London, 601 pp.

FREEZE, R. A., and CHERRY, J. A., 1979, Groundwater: Prentice-Hall, Inc., Englewood Cliffs, N.J., 604 pp.

GARRELLS, R. M., and MACKENZIE, F. T., 1971, Evolution of sedimentary rocks: W. W. Norton & Company, Inc., New York, 397 pp.

GASCOYNE, M., BENJAMIN, G. J., SCHWARCZ, H. P., and FORD, D. C., 1979, Sea-level lowering during the Illinoian glaciation: Evidence from a Bahama "blue hole": Science, v. 205, pp. 806–8.

GÈZE, B., and MANGIN, A., 1980, Le karst de Cuba: Revue de Géologie Dynamique et de Géographie Physique, v. 22, pp. 157–66.

GUISTI, E. V., 1978, Hydrogeology of the karst of Puerto Rico: U.S. Geol. Survey Prof. Paper 1012, 68 pp.

HACK, J. T., and DURLOO, L. H., 1962, Geology of Luray Caverns, Virginia: Va. Div. Min. Resources Rept. Invest. 3, 43 pp.

HERAK, M., 1972, Karst of Yugoslavia, in Herak, M., and Stringfield, V. T., eds., Karst: Important karst regions of the northern hemisphere: Elsevier Publishing Co., Amsterdam, pp. 25–83.

HUBBERT, M. K., 1940, Theory of ground-water motion: Jour. Geology, v. 48, pp. 785–944.

JENNINGS, J. N., 1985, Karst geomorphology, 2nd ed.: Basil Blackwell, Inc., New York, 293 pp.

LEHMANN, H., 1936, Morphologische Studien auf Java: J. Engelhorn, Stuttgart, 114 pp.

———, 1960, International atlas of karst phenomena, Sheets 1, 1a, and 1b, "Sierra de los Organos, Cuba": Zeitschr. für Geomorph., Supp. no. 2, plates in pocket.

MEYERHOFF, H. A., 1938, Texture of karst topography in Cuba and Puerto Rico: Jour. Geomorphology, v. 1, pp. 279–95.

MONROE, W. H., 1966, Formation of tropical karst topography by limestone solution and reprecipitation: Caribbean Jour. Sci., v. 6, pp. 1–7.

————, 1970, Glossary of karst terminology: U.S. Geol. Survey Water-Supply Paper 1899-K, 26 pp.

————, 1976, Karst landforms of Puerto Rico: U.S. Geol. Survey Prof. Paper 899, 69 pp.

MOORE, G. W., 1968, Limestone caves, *in* Fairbridge, R. W., ed., Encyclopedia of geomorphology: Reinhold Book Corp., New York, pp. 652–53.

————, and SULLIVAN, G. N., 1977, Speleology: Zephyrus Press, Teaneck, N.J., 151 pp.

PENCK, A., 1894, Morphologie der Erdoberfläche: J. Engelhorn, Stuttgart, v. 1, 471 pp.; v. 2, 696 pp.

PLUHAR, A., and FORD, D. C., 1970, Dolomite karren of the Niagara escarpment, Ontario, Canada: Zeitschr. für Geomorph., v. 14, pp. 392–410.

POWER, F. D., 1925, Phosphate deposits of the Pacific: Econ. Geology, v. 20, pp. 266–81.

ROGLIC, J., 1972, Historical review of morphologic concepts, *in* Herak, M., and Stringfield, V. T., eds., Karst: Important karst regions of the northern hemisphere: Elsevier Publishing Co., Amsterdam, pp. 1–18.

SANDERS, E. M., 1921, Cycle of erosion in a karst region (after Cvijić): Geog. Rev., v. 11, pp. 593–604.

SAWKINS, J. G., 1869, Reports on the geology of Jamaica: Great Britain Geol. Survey Mem., Longmans, Green, and Co., London, 339 pp.

SCHMIDT, V. A., 1982, Magnetostratigraphy of sediments in Mammoth Cave, Kentucky: Science, v. 217, pp. 827–29.

SHACHAK, M., JONES, C. G., and GRANOT, Y., 1987, Herbivory in rocks and the weathering of a desert: Science, v. 236, pp. 1098–99.

SMITH, J. F., and ALBRITTON, C. C., 1941, Solution effects on limestone as a function of slope: Geol. Soc. America Bull., v. 52, pp. 61–78.

SOLECKI, R. S., 1963, Prehistory in Shanidar Valley, northern Iraq: Science, v. 139, pp. 179–93.

————, 1972, Shanidar: The humanity of Neanderthal man: Allen Lane, London, 322 pp.

SWEETING, M. M., 1973, Karst landforms: Columbia Univ. Press, New York, 362 pp.

SWINNERTON, A. C., 1932, Origin of limestone caverns: Geol. Soc. America Bull., v. 43, pp. 663–93.

THORP, J., 1934, Asymmetry of the "Pepino Hills" of Puerto Rico in relation to the trade winds: Jour. Geology, v. 42, pp. 537–45.

THRAILKILL, J., 1968, Chemical and hydrologic factors in the excavation of limestone caves: Geol. Soc. America Bull., v. 79, pp. 19–45.

————, 1972, Digital computer modeling of limestone groundwater systems: Univ. Kentucky Water Resources Inst. Res. Rept. 50, 70 pp.

TROESTER, J. W., and MOORE, J. E., 1989, Karst hydrology in the United States of America: Episodes, v. 12, pp. 172–78.

VERSEY, H. R., 1972, Karst of Jamaica, *in* Herak, M., and Stringfield, V. T., eds., Karst: Important karst regions of the northern hemisphere: Elsevier Publishing Co., Amsterdam, pp. 445–66.

VILES, H. A. 1987, Blue-green algae and terrestrial limestone weathering on Aldabra Atoll: An S.E.M. and light microscope study: Earth Surface Processes and Landforms, v. 12, pp. 319–30.

WATSON, R. A., and WHITE, W. B., 1985, History of American theories of cave origin, *in* Drake, E. T., and Jordan, W. M., eds., Geologists and ideas: A history of North American geology: Geol. Soc. America Centennial Special Volume 1, pp. 109–23.

WENT, F. W., 1969, Fungi associated with stalactite growth: Science, v. 166, pp. 385–86.

WHITE, W. B., 1988, Geomorphology and hydrology of karst terrains: Oxford Univ. Press, New York, 464 pp.

WILFORD, G. E., and WALL, J. R. D., 1965, Karst topography in Sarawak: Jour. Tropical Geog., v. 21, pp. 44–70.

WILLIAMS, P. W., 1972, Morphometric analysis of polygonal karst in New Guinea: Geol. Soc. America Bull., v. 83, pp. 761–96.

————, 1987, Geomorphic inheritance and the development of tower karst: Earth Surface Processes and Landforms, v. 12, pp. 453–65.

WINOGRAD, I. J., SZABO, B. J., COPLEN, T. B., and RIGGS, A. C., 1988, A 250,000-year climatic record from Great Basin vein calcite: Implications for Milankovitch theory: Science, v. 242, pp. 1275–80.

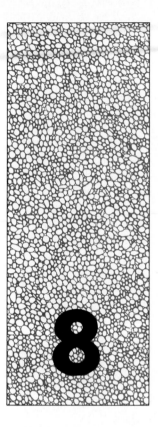

8

Mass Wasting and Hillslope Evolution

On the scale of individual landforms, where rheid flow is not a significant factor of deformation, unweathered rock resists the surface forces that act on it. Only when rock has been broken by mechanical stresses or has reacted with water and the atmosphere can the fragments by mobilized. Weathering, then, is a necessary precondition for the movement of rock fragments downslope and for the development of the slopes. A variety of surface forces can move loose rock particles. Whenever they move, ubiquitous gravity adds a downward component to the motions produced by other forces so that they move preferentially downhill.

MASS WASTING, GRAVITY, AND FRICTION

The collective term for all gravitational or downslope movements of weathered rock debris is **mass wasting.** The term implies that gravity is the sole important force and that no transporting medium such as wind, flowing water, ice, or molten lava is involved. Although flowing water is excluded from the process by definition, water nevertheless plays an important role in mass wasting by oversteepening slopes through surface erosion at their bases and by generating seepage pressures through groundwater flow. Sometimes, this second role of water is incorrectly referred to as "lubrication" of rock or weathered rock debris on a slope. In fact, water has a complex effect on the sliding friction between surfaces of rock-forming minerals, increasing the friction for quartz, feldspar and calcite, but decreasing it for the sheet-like mica minerals (Mitchell, 1976, pp. 310–13). It is the very important seepage pressures generated by groundwater flow, plus the additional weight of a soil mass when it is saturated with water, that have such great influence on slope instability and mass wasting.

Angle of Friction

A series of simple experiments illustrates the relative importance of gravity and pore-fluid pressure in determining downslope movement. A cone of noncohesive dry sand can be built by pouring sand onto a horizontal surface through a funnel. The cone will have a surface *angle of repose* of approximately 35°, depending on the size, shape, surface roughness, and other properties of the sand grains. The tangent of the angle of repose of dry granular materials is slightly greater than, but approximately equal to, the coefficient of sliding friction of the material, or its *mass friction* (φ) (Van Burkalow, 1945). Both are determined by the balance of forces between gravity and friction. The coefficient of sliding friction of a particle is equal to the ratio between the downslope component of weight and the component of weight acting perpendicular to the slope when the particle is moving.

If in a second experiment the same sand is deposited in a saturated condition under water, the angle of repose of the cone will be virtually identical to the angle of repose of the dry sand cone. In neither example will there be any fluid flow through the sand, and consequently neither cone will be subjected to seepage pressures tending to decrease its stability.

A third experiment demonstrates the influence of water flow. If water is poured gently onto the top of a cone of dry sand and allowed to seep downward and outward (the water must infiltrate and not erode the surface), failure occurs as soon as the water reaches the slope face. It can be demonstrated by experiment, or by adding a vector representing seepage pressure to a simple diagram (Fig. 8–1), that even the flow of rainwater runoff in the surface soil on a slope reduces the stable angle of repose to less than half of the dry value.

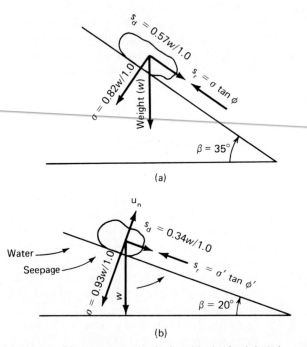

Figure 8-1. Resolution of forces on a cohesionless block of unit basal area resting on a slope. (a) σ (stress component perpendicular to surface) $= w \cos \beta$; s_d (stress component parallel to surface) $= w \sin \beta$. Shear strength (s_r) resists motion. When $\beta = \varphi$ (angle of repose), $s_d = s_r$, and the block is ready to move. (b) As above, but with u_n (component of seepage pressure normal to the surface) counteracting σ ($\sigma' = \sigma - u$) and reducing s_r. Particle will move when $\beta \ll \varphi$.

Cohesion

Fine-grained sediments, especially clay-size[1] particles, have strong *cohesive* surface forces between them. These forces are determined by electrostatic charges on grain surfaces that immobilize the water films between them. These cohesive forces act in addition to the mass friction to determine the angle at which slope failure occurs. Referring to the first experiment described previously, if the noncohesive dry sand is moistened but not saturated, it can be molded and shaped to a vertical slope, as any child in a sandbox will verify. Strength in that sense depends on the tension in the capillary water between sand grains.

The previous discussion was first summarized as an equation by C. A. Coulomb in 1776, in which

$$s = c + \sigma \tan \varphi$$

where s is the shear strength (in stress, or force per unit area, units), c is cohesion (also in stress units), σ is the stress component perpendicular to the surface, and φ is the soil mass angle of friction. This important equation, which relates the shear strength of a soil mass just at the threshold of failure to the stresses acting on it, is now called the *Mohr–Coulomb failure criterion* (Holtz and Kovacs, 1981, p. 453; Nash, 1987).

Engineers use an equivalent form of the Mohr–Coulomb equation to take into account the excess pressures in the pore fluids, including the seepage pressures. This concept is called the *principle of effective stress*. It is common engineering practice to add a superscript to the symbols in the Mohr–Coulomb equation when working with effective stresses

$$s = c' + \sigma' \tan \varphi'$$

where c' is the effective stress cohesion, $\sigma' = (\sigma - u)$, the effective stress component normal to the surface, as reduced by positive fluid pressure u, and φ' is the effective stress angle of friction. In noncohesive sand, gravel, or large talus blocks, $c' = 0$, and shear stress is proportional to the tangent of the angle of friction, although the relationship is complicated by shape factors such as the tendency of large angular blocks to stack and support steeper-than-predicted angles. Fine-grained silt and clay make cohesive soils for which c' can be significant. As cohesion increases, so does shear strength. Many moist clays hold vertical faces of modest height without failure, which noncohesive materials can never do.

Engineers also use an alternate form of the Mohr–Coulomb equation when studying short-term stability problems such as the initial cutting of a clay slope (Holtz and Kovacs, 1981, p. 556). In this case, complex fluid pressure distributions can be developed within the soil pores during loading or unloading. These fluid pressures can be difficult to calculate or predict, but they can be simulated fairly easily in the laboratory with *undrained strength* tests. The strength measured in these laboratory test simulations is the *undrained shear strength* (s_u). For this alternative form of the Mohr–Coulomb Equation, s_u replaces s and includes c', σ', φ', as well as the additional developed fluid pressures in the simulation. Although these matters are of importance to geomorphology, their technical details are the subject

---

[1] "Clay" is the name used for a group of residual minerals (Chap. 6) as well as for the fine-grained fraction of sediment regardless of its mineralogy. The distinction between "clay minerals" and "clay-size sediment" should always be made clear. The geologists' classification defines clay-size particles as less than 4 μm mean diameter; the engineers' classification defines clay particles as all those less than 2 μm in diameter.

of the related field of geotechnical engineering and are best studied in appropriate engineering texts (e.g., Mitchell, 1976; Holtz and Kovacs, 1981).

Slope angles on noncohesive weathered debris such as grus or loose rock rubble are usually between 25 and 40°. Natural slopes steeper than 40° are usually barren of rock debris, and in British terminology are classed as *cliffs.*

Rates of Downslope Movement

The geometry of Figure 8–1 demonstrates that the shear stress acting parallel to a slope, which causes a particle to move down slope, is proportional to the sine of the slope angle ($S_d = w \sin \beta$). Intuitively, we would expect that the rate of downslope movement of small, loose rock fragments would also be proportional to the sine of the slope angle (Strahler, 1956, p. 577). Most data support this inference (Fig. 8–2). Schumm (1967) measured the rate of migration of small plates of sandstone down a shale hillslope in semiarid western Colorado by marking 110 pieces with paint and measuring their distance from reference stakes driven into bedrock. Over a period of seven years, the rate of downslope movement ranged from 2 to 3 mm/yr to almost 70 mm/yr in proportion to the sine of the slope angle, which ranged from 3 to 40°. Rates of periglacial gelifluction (Chap. 15) over perennially frozen ground in eastern Greenland are much higher than the sliding rate of dry rock chips on a semiarid talus (Fig. 8–2), and the modes of movement are also very different, but nevertheless, the rates of surface movement on both rapidly flowing solifluction lobes and "undifferentiated wet slopes" in Greenland are proportional to the sine of the slope angles.

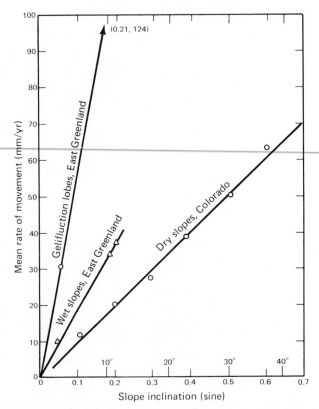

Figure 8-2. Rate of downslope movement is proportional to the sine of the slope inclination under a wide range of conditions. (Data from Schumm, 1967; Washburn, 1967, Fig. 43.)

It should be noted that for angles of less than 10°, the sine and tangent functions differ by only a few percent, and Figure 8–2 would not look much different if the movement rates were plotted against the tangent of slope angle as an approximation of the mass friction (tan φ). Whichever function is chosen, the conclusion is obvious that downslope movement is slowest on gentle slopes and increases rapidly with increase in slope. On slopes of 40 to 45°, few detached fragments can remain at rest. All roll or slide to the base of the slope. On vertical slopes, a loosened rock article drops under the free acceleration of gravity, its rate of fall limited only by air resistance.

DESCRIPTIVE CLASSIFICATION OF MASS WASTING

The explanation of mass wasting is so simple and obvious that a genetic, or explanatory-descriptive, classification of the processes and forms is not necessary. Instead, a simplified descriptive classification is generally adopted, which emphasizes certain features as diagnostic and recognizes other features as parts of continuous series. No combination of categories has been devised to classify totally the many landforms and processes of mass wasting, nor is such a classification likely or desirable. The essence of mass wasting is that it bridges the gap between *weathering,* which is defined as occurring in place, and *erosion,* which requires as one element transport by some agent or medium. Mass wasting combines elements of weathering and erosion, yet separates those two large categories of processes. In every sense of cause and effect, mass wasting is a category of transitional phenomena.

Two classifications of mass wasting are in common use in the United States and are comparable to schemes adopted in England, France, and Germany. There is broad international agreement that the definitive criteria of mass wasting should be (1) the type of material in motion, including its coherence and dimensions; and (2) the type and rate of movement, whether sliding, flowing, or falling.

The classification scheme most commonly seen in American geomorphology textbooks is from a valuable little book by Sharpe (1938) (Fig. 8–3). The factors by which Sharpe subdivided mass wasting were (1) the amount of included ice or water; (2) the nature of the movement, whether sliding or falling as a coherent mass, or flowing by internal deformation; and (3) the speed of the movement, ranging from the imperceptible to speeds of tens of meters per second under the full acceleration of gravity. The term **landslide** (or *landslip* in some British usage) was not given any specific definition by Sharpe (1938, p. 64) but remains a popular nontechnical term for all perceptible forms of relatively dry mass wasting.

Nature and Rate of Movement			With Increasing Ice Content	Rock or Soil ←→	With Increasing Water Content	
Flow	Imperceptible	Glacial Transport	Solifluction	Creep (rock creep, soil creep)	Solifluction	Fluvial Transport
	Slow to rapid		Debris avalanche		Earth flow Mud flow Debris avalanche	
Slide	Slow to rapid			Slump Debris slide Debris fall Rock slide Rock fall		

Figure 8-3. Classification of landslides and related phenomena. (Simplified from Sharpe, 1938, p. 96.)

TYPE OF MOVEMENT			TYPE OF MATERIAL		
			BEDROCK	ENGINEERING SOILS	
				Predominantly coarse	Predominantly fine
FALLS			Rock fall	Debris fall	Earth fall
TOPPLES			Rock topple	Debris topple	Earth topple
SLIDES	ROTATIONAL	FEW UNITS	Rock slump	Debris slump	Earth slump
	TRANSLATIONAL	MANY UNITS	Rock block slide Rock slide	Debris block slide Debris slide	Earth block slide Earth slide
LATERAL SPREADS			Rock spread	Debris spread	Earth spread
FLOWS			Rock flow (deep creep)	Debris flow (soil creep)	Earth flow
COMPLEX			Combination of two or more principal types of movement		

Figure 8-4.   Types of slope movements, abbreviated version. (Varnes, 1978, Fig. 2.2.)

A more detailed classification (Fig. 8–4) of slope movements (the favored engineering term for mass wasting) was published in the United States by a committee of engineering specialists (Schuster and Krizak, 1978). The committee's responsibility was to consider the relationship of slope movements to engineering practice, so they defined slope movement as a downward and outward movement of slope-forming bedrock, rock debris, and "earth" (fine-grained fragmental debris).

In the description that follows, Sharpe's classification (Fig. 8–3) is generally followed. Some of the excellent diagrams and examples from the committee report (Varnes, 1978) are used to illustrate certain terms. No uniform spelling or definitions of the various terms have been established. Some authors prefer single words such as "mudflow" or "earthflow," while others use two words, as in "debris flow". Others use various combinations of terms not listed in Figs. 8–3 and 8–4, such as "debris avalanche," "slump-earthflow," etc.

Creep

Creep is imperceptible and nonaccelerating downslope movement. It is most common, in fact ever-present, in weathered rock debris on slopes, but the term is also applied to the slow movement of otherwise unweathered joint blocks. **Soil creep** or **rock creep** are the respective subordinate terms. The material moving downslope is called **colluvium.**

Soil Creep.   Although soil creep is too slow to be observed, the cumulative results become obvious over a period of years (Fig. 8–5). Most stone walls and pavements on hillsides show downslope motion by tension cracks, downslope tilt, or visible displacement. Tree trunks are concave uphill, although the significance of their curvature to creep has been questioned (Phipps, 1974). All these features are due to the essentially laminar nature of soil creep. Each layer of soil is carried downhill by the motion of the layer beneath it, and the effect is cumulative, with the maximum rate at the surface exponentially decreasing to zero with depth (Kirby, 1967). As a result, soil creep does not shear across immobile rock or soil at depth and is not capable of abrading a buried surface. Strata in unweathered bedrock thin abruptly as they enter the surface layer that is creeping, and instead of breaking the surface as outcrops, they are drawn out in a downhill direction (Fig. 8–6). Creep is recognized as both a problem and an aid in prospecting. If fragments of the desired coal or ore mineral are found in the "float," or colluvium, the buried layer must be in the subsoil at some point farther upslope.

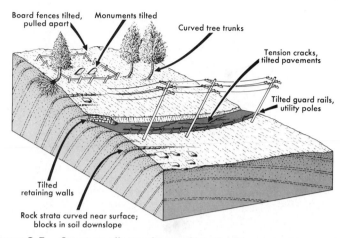

Figure 8-5. Common effects of creep. Not all will be present in one place.

Grass may be able to retain a continuous sod cover over an area of soil creep, because movement is typically only a few millimeters per year. The highest rates of creep are measured on steep slopes under tropical rainforests, where the entire surface layer of interlocking roots moves downhill at rates of up to 4 mm/yr. An anomalous deeper soil zone of high creep velocity was measured in a Puerto Rico rainforest, just below the root zone (Lewis, 1974).

Soil creep is aided by expansion and contraction of soil by heating and cooling, freezing and thawing, or wetting and drying. Volumetric expansion by any cause displaces particles toward the free face of the expanding mass, or perpendicular to the ground surface. On contraction, however, the particle is not pulled back

Figure 8-6. Rock creep in roadcut along Yarra Boulevard, Melbourne, Australia. (Photo: Renate R. Hodgson.)

1 to 3 =
net downslope transfer

Figure 8-7. Path of a surface particle during expansion and contraction of soil on a slope. See Figure 15-5 for more complex motions.

into its former position but settles with a gravitational component (Fig. 8–7). Only rarely are the cohesive forces of soil and water strong enough to draw particles back into the ground during contraction without some net downslope motion. Figure 8–7 illustrates that on most soil-covered slopes, which rarely have angles greater than 20°, the net downslope movement is only a small fraction of the displacement perpendicular to the surface. On steep slopes, fluvial erosion is likely to displace creep as the dominant process. Soil creep tends to maintain smooth, ungullied hillsides by filling the small channels eroded by seeping groundwater or surface rill wash (Chap. 9). It is strongly dependent on soil moisture conditions, and creep rates fluctuate seasonally, so creep is as much climate-controlled as it is gravity-controlled. Creep may be responsible for much of the net downslope movement even on terrains where slump and flow are more obvious, by providing the recharge supply for areas of repeated, but intermittent, slumps and flows (Sidle et al., 1985, p. 14).

Rock Creep. To geophysicists and engineers, rock creep connotes slow, permanent internal rock deformation (strain) under low stress. As a process of mass wasting, rock creep is less precisely used to describe slow movement of large coherent rock masses over a discrete basal surface. Most often, rock creep is observed on a cliff face, where a massive rock such as sandstone overlies shale. Pressure release normal to the cliff face (p. 120) produces large joint-bounded sheets or blocks, which slowly creep or lean outward until they become unstable and topple over or slide downhill (de Frietas and Watters, 1973).

The best-documented history of toppling failure that resulted from rock creep (Schumm and Chorley, 1964) is "the fall of Threatening Rock" in 1941 in Chaco Canyon National Monument, New Mexico. For five years prior to the fall, the gap between the cliff face and a free-standing monolith of sandstone 46 m long, 30 m high, and 9 m thick had been measured at monthly intervals. For three years the gap widened at approximately 25 mm/yr. During the next two years, the rate doubled, culminating in a violent rock fall that partly destroyed an ancient pueblo at the base of the cliff. The annual movement was concentrated during the wet winter season and was greater during years of greater precipitation. Movement had been in progress or at least recognized as possible for at least nine centuries, proved because the prehistoric occupants of Pueblo Bonito had built an earth and masonry terrace at the base of Threatening Rock and wedged logs beneath its base about A.D. 1000. By extrapolation, Schumm and Chorley (1964, p. 1052) estimated the time required for each foot (30.5 cm) of movement away from the cliff as follows:

Movement	Years Required
First foot	1600
Second foot	650
Third foot	200
Fourth foot	60
Fifth foot	8.0
Sixth foot	0.2

Rock creep illustrates the importance of pressure-release joints, or sheeting (p. 121), parallel to cliff faces or gorge walls in massive rocks. Rapid stream erosion can unbalance the stress distribution in rocks so that elastic expansion is outward toward the valley axis (Ferguson, 1967). When open joints have formed, it is only a matter of time until weathering destroys the support and a large mass of rock collapses from the cliff. Robinson (1970, p. 2802) demonstrated that the crushing strength of massive sandstone is sufficient to support free-standing joint columns several hundred meters high on cliff faces and that failure occurs primarily by loss of basal support. Even though these events seem random and catastrophic, and may occur only a few times in a thousand years, they are part of the "normal" process of valley-side evolution.

Flow

Incoherent rock debris may be mobilized sufficiently so that it flows like a viscous fluid. The criteria for defining flow are evidence of internal turbulence and either discrete boundaries or narrow marginal zones of shear. In both criteria, flow is distinct from creep. Subdivision of flow can be based either on the kind of material, the degree of saturation, or the speed of advance (Figs. 8–3 and 8–4). The factors of sediment concentration in water and the flow rate can be combined in a rheological classification, ranging from fluid flow with low sediment concentrations through plastic flow with a finite yield stress but essentially fluid behavior above that stress level, to granular flows that are dry or nearly dry. Flow rates for all rheological behaviors can range through many orders of magnitude (Pierson and Costa, 1987). Flows typically move as lobes or tongues that follow the pre-existing topography. Some are known to erode channels. The transitional nature of mass-wasting terminology is nowhere better illustrated than in attempts to distinguish between a water-saturated mud flow and a mud-laden stream. Thus far, attempts at rheological quantification use only approximate ranges of sediment concentration (Pierson and Costa, 1987, p. 6).

Solifluction. If soil or regolith is saturated with water, the soggy mass may move downhill a few millimeters or a few centimeters per day or per year. This type of movement is called **solifluction** (literally, "soil flow"). The original definition of solifluction excluded any connotation of climate but was included in a discussion of cold-climate landforms (Andersson, 1906; Washburn, 1967, pp. 10–14) and has frequently been misdefined as a cold-climate process. To clarify the confused terminology, Washburn (1967, p. 14) urged adoption of **gelifluction** for solifluction associated with frozen ground. That usage is adopted here and the subject is deferred to Chapter 15.

Solifluction is not restricted to frozen ground. It is a form of mass wasting common wherever water cannot escape from a saturated surface layer of regolith by percolation into deeper levels. A clay hardpan in a soil or an impermeable bedrock layer can promote solifluction as effectively as a frozen substratum. Under tropical rainforests, steep slopes in deeply weathered rock are scarred by large debris slides, and lower slopes are masses of solifluction lobes.

Debris Flow, Earth Flow, Mud Flow.   These three terms are applied to types of mass wasting very similar to solifluction. They are somewhat more rapid, and they commonly flow along valleys, whereas solifluction sheets or lobes cover an entire hillside with moving debris. The three terms form a series of progressively higher water content but are often used interchangeably. In the classification by Varnes (1978) (Fig. 8–4), debris flows have 20 to 80 percent particles coarser than sand sizes, whereas earth flows and mud flows are 80 percent or more mud and sand. Mud flow is the most liquid "end member" of the series.

Other authors (Johnson and Rodine, 1984, p. 257) lump all three terms under the general term *debris flow,* to stress the similarity of their mechanisms, despite differences in their flow rate or percentage of fine particles. The transition between a debris flow and a muddy river is illustrated by measurements of densities of 2.0 to 2.4 g/cm^3 in debris flows (dense enough to buoyantly support large blocks), densities of 1.3 to 1.8 g/cm^3 in more fluid "hyperconcentrated" flows, and densities of only slightly greater than 1.0 g/cm^3 in a stream with a heavy suspended load (Hooke, 1987, p. 508). The distinction is that in debris flows, earth flows, and mud flows, gravity acts directly on the particles that drag along interstitial fluid, rather than on a fluid that then entrains the particles (Hooke, 1987, p. 505).

Earth flows (the most general term based on heterogeneous grain size) almost always result from excessive rainfall, and oversaturation seems to be the major factor in their formation. They may form on planar hillsides or at the toes of large slumps, but most often they develop at the heads of small gullies, where groundwater seepage or surface runoff concentrates to initiate streams. Unusually deep weathering along "seepage lines" at the head of incipient drainage networks may be the reason for the high frequency of flow initiation there (Bunting, 1961).

Case histories of earth flows from Ireland (Prior and Stephens, 1972), New Zealand (Crozier, 1969), Japan (Machida, 1966), and comparable events called debris avalanches in the United States (p. 191) emphasize the combination of steep slopes (to 25°), weak or easily weathered rock or surficial cover, and abnormal precipitation in triggering earth flows. Artificial alteration, especially excavation at the foot of a slope, and vegetation clearing are very often additional factors. In seismic regions, earthquakes may trigger earth flows and avalanches (Sidle et al., 1985, pp. 66–72). Quite commonly, aerial photographs of earth flows show many scars of earlier flows of the same general form.

Earth flows are common along the valley of the St. Lawrence River and in British Columbia, Canada, and in Scandinavia, where thick deposits of late-glacial marine clay have been uplifted by postglacial isostatic movements and now form most of the lowlands. The clay-sized particles were deposited rapidly in saline water and were flocculated into large agglomerates that have chaotic internal structure. Instead of being dense and laminated, the late-glacial marine clays have unusually high porosity and are initially held together by electrostatic forces between clay-size particles and the saline pore water. After subaerial leaching for 10,000 years or more, the soluble salts are removed, and the clays may become "quick," or "sensitive" (Fig. 8–8). They retain their shear strength if they are not excessively or abruptly stressed, but with excessive vibration or shock, the pore water mobilizes, the weak framework of poorly arranged mineral grains collapses, and an apparently stiff plastic clay suddenly becomes a milky fluid. Along river valleys in Canada and Scandinavia, these clays in their cohesive unleached state form vertical river banks tens of meters high. Their strength degenerates, however, and sudden shocks such

Figure 8-8. Demonstration of "quick" clay. Undisturbed sample (left) supports 11 kg. Another sample of the same clay is poured from the beaker after being stirred. No water was added. (Photo: Div. Building Research, National Research Council of Canada.)

as earthquakes, thunder, highway traffic, blasting in excavations or log jams, or the spring breakup of ice on a river may cause an abrupt slope failure (Bentley and Smalley, 1984; Torrance, 1987).

The form of an earth flow in quick clay is a characteristic pear or bottleneck shape, with a relatively narrow neck at the point of initial failure. As each slump mass collapses, liquefies, and flows toward the adjacent lowland, new concentric slices collapse. The flow margin retrogresses to a bowl shape from the initial narrow flow, which becomes the gorge through which a slurry of liquefied mud flows. The pattern on aerial photographs (Fig. 8–9) has a distinctive ribbed pattern of slump blocks that retained their cohesive strength briefly and then collapsed. The earth-flow bowls may continue to grow for hours or weeks to a diameter of several kilometers and are as much as 100 m deep. Their floors are hummocky, including tilted slices and pinnacles of the former surface layer that failed to liquefy (Karrow, 1972). In 1971, a disastrous flow at Saint-Jean-Vianney, Quebec, claimed 30 lives in a series of expanding slumps that progressively engulfed houses in a small village (Tavenas et al., 1971). Personal narratives recount incidents of people outrunning the retrogressive expansion of earth-flow scarps in the Quebec lowland. It appears that the sod and soil profile stay coherent even after liquefaction has begun at depth, and people who have felt the road collapse beneath their automobile have yet had time to abandon the car and run uphill to safe ground.

A remarkable feature of Norwegian "quick" clay is the abruptness with which it loses plasticity if salt is restored and mixed into the flowing mass. Some earth flows across Norwegian roads can be stopped by throwing bags of sodium or magnesium chloride into the fluid mass and churning through it with a tracked vehicle. Almost instantly, the water–clay bonds are restored, the mass stabilizes, and it can be cleared with standard earth-moving equipment.

In semiarid and arid regions, torrential rains or *cloudbursts* may send flash floods roaring through mountain canyons. Dams of debris, including mud, rocks,

Figure 8-9.   Earth flow of May 1971 on the South Nation River, Ontario, Canada. View upstream. Head of the flow is 500 m from the river channel (Eden et al., 1971). (Photo: Div. Building Research, National Research Council of Canada.)

logs, and stumps, temporarily impede the flood water, then fail and surge downstream to strike the next obstacle with even greater force. Such processes, aided by landslides and bank collapse into the floodwater, quickly convert the mass into a high-density debris flow (Johnson and Rodine, 1984; Hooke, 1987). The deposits are poorly sorted and unstratified; they may choke valley floors or spread onto piedmont slopes as lobes. The role of flash floods and debris flows in shaping arid landscapes is reviewed at length in Chapter 13.

Lahar is an Indonesian word that describes a debris flow on the flank of a volcano (p. 77). Torrential rains, sometimes triggered by pyroclastic eruptions, saturate the unstable tephra on the slopes of composite cones. Other sources of mobilizing water are breached crater lakes, melted snow, and slide-dammed rivers. Lahars may be hot or cold. Their water content is comparable to that of wet concrete. One of the most serious lahars from Mount St. Helens in 1981 began more than four hours after the initial eruption, and was probably triggered when the large amount of ice in the initial debris avalanche had melted and a magnitude 5 earthquake within the volcano liquefied the mass (Fairchild, 1987, pp. 57–60). Numerous debris flows from destabilized watersheds continued for months following the 1981 eruption of Mount St. Helens (Chap. 4). Most followed former river channels, but rapidly spilled over adjacent floodplains. Their mode of flow typically involved a frontal lobe of boulders that rode the crest of each surge, carried forward on the surface of the flow like a conveyor belt, then overridden. The surface boulders moved forward faster than the internal mass, ensuring their concentration at the front of each surge (Pierson, 1986, p. 279).

Slide

Mass wasting wherein a mass of rock or weathered debris moves downhill along discrete shear surfaces is defined as a **slide.** The general word *landslide* is frequently extended to include all rapid forms of flow, slide, or fall, because most of the movement in all three categories is along surfaces of separation. Subcategories of slide in various classifications include *slump, rock slide, block glide, mud slide, debris slide,* and *debris avalanche* (Brunsden, 1984).

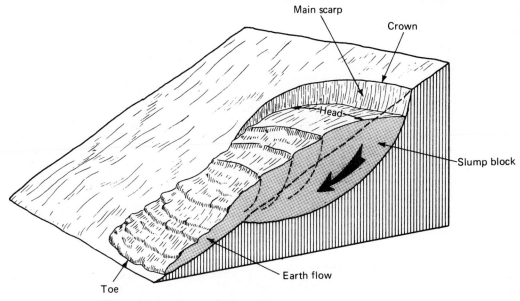

Figure 8-10. Slump-earth flow with principal parts labeled. Components can be recognized on Figures 7-10 and 8-9.

Slump. Figure 8–10 is a composite diagram of the nomenclature of the parts of a typical "landslide." The diagram portrays a **slump** (rotational slip along a con-cave-up surface of rupture) passing downhill into an earth flow. This is an ex-tremely common form of mass wasting (the *slump-earthflow* of Varnes, 1978), although some of the named components might be absent in any single example.

Slump is the form of slide most common in thick, homogeneous, cohesive materials such as clay. The surface of failure beneath a slump block is spoon-shaped, concave upward or outward (Fig. 8–10). The upper surface of a slump block commonly is tilted backward, because the entire mass rotates as the lower part moves outward and downhill. Vegetation or even houses may be carried intact on the surface of a large slump block. Slumps are tens or hundreds of meters wide and may be single blocks or consist of multiple slices. Ponds often form at the angle between the base of a slump scar and the head or top of the rotated slump block. This water, percolating down along the surface of rupture, may cause renewed in-stability on old slumps.

Slumps may be caused by currents or waves in water undercutting the foot of a slope. They are also a common result of faulty engineering design of cut embank-ments. They are sometimes controlled by loading the base of an unstable slope with a heavy layer of coarse rock rubble, which permits water to drain off the hill but off-sets the weight of unstable earth higher on the slope. An elaborate engineering tech-nology has been developed to predict the surface of rupture beneath a slump in order to drill into it and drain the water from the vicinity.

The most diagnostic feature of an ancient slide area is the hummocky or cha-otic landforms on it (Fig. 8–11). Unless the material moved as a single slump block or is thoroughly fluidized as a mud flow, the region below a major slide is covered with a mass of broken rock rubble in mounds and pits, sometimes in systematic transverse ridges or lobes, but often totally unsystematic. The debris is poorly sorted and can easily be mistaken for glacial till in a mountain valley (Porter, 1970, p. 1421; Porter and Orombelli, 1981). Ancient slump blocks along valley walls in the Wasatch Plateau, Utah, have been mistaken for lateral moraines (White, 1973).

Figure 8-11. Ancient landslide in Bailey Basin, Columbia River valley, northeastern Washington (Jones et al., 1961). (Photo: F. O. Jones, U.S. Geological Survey.)

Mud Slide. Depending on the slope gradient, degree of saturation, and grain size of sediment emerging from the toe of a slump, it can be called a mud flow or a *mud slide.* A mud slide moves over or between discrete shear surfaces in a lobate shape, somewhat slower than a more turbulent mud flow. In its midportion it follows a discrete track or chute between well defined walls. The flanks may be striated or slickensided by the sliding mass. At its terminus, it may form a bulbous lobe (Brunsden, 1984).

Rock Slide, Block Glide. The simplest form of slide is a rock slide or block glide. The movement is relatively rapid and most commonly occurs where steeply dipping bedded strata or sheeting nearly parallels the surface slope (Braun et al., 1989; Lee, 1989) (Fig. 8–12). Rock slides are generally shallow, because they occur along cohesionless planes, and because by the Mohr–Coulomb equation, the load perpendicular to the surface (σ') increases rapidly with depth. A heavy rain or freezing and thawing provides fluid pressure, or vibration breaks off obstructions and reduces the coefficient of friction on the glide plane, and a detached slab or block slides down. It may shatter at the base of the slope, or it may remain intact. Rock slides or block glides have no specified size, but their thickness is normally only about 10 percent of their downslope length.

Figure 8-12. Rock slide on dipping sandstone strata near Glenwood Springs, Colorado. Sheeting in massive rocks has the same effect as dipping strata (Fig. 6-3). (Photo: D. J. Varnes, U.S. Geological Survey.)

The dip of strata is an obvious factor in rock slides (Fig. 8–12). Other than the inherent strength of unstratified or massive rock, the controlling factor in such rock slides is the presence, spacing, and orientation of joints and other fractures. Sheeting joints in massive rocks roughly conform to the shape of the hill or mountain. If they dip less steeply than the hill slope, individual slabs are unsupported at their downshill edges and are more subject to sliding (Fig. 6–3). If they dip more steeply than the hillside, they may fail by buckling outward or toppling (Lee, 1989, Fig. 13). Rock slabs are noncohesive masses whose movement is controlled by the surface friction on joint faces (Moon, 1986; Selby, 1987).

Rock slides can be dramatic forms of sliding mass wasting if large masses of unweathered rock slide downhill along a sloping joint or bedding surface. Such a plane of weakness was probably involved in the Vaiont Reservoir disaster of October 1963 in northern Italy (Kiersch, 1964). On the night of October 9, a rock slide 2 km long, 1.6 km wide, and over 150 m thick moved suddenly down the south wall of the Vaiont Canyon and completely filled the 270-m-deep reservoir for 2 km upstream from the dam to heights of 180 m above the former water level. The movement took less than a minute, so rapid that the water in the reservoir was ejected 260 m up the north canyon wall and propelled in great waves both upstream in the reservoir and downstream over the dam. The resulting floods killed 3000 people, mostly around the town of Longarone, over 2.5 km downstream from the dam and across a broad valley from the mouth of the Vaiont Canyon. The dam itself survived, although damaged.

The geologic cross section of the reservoir is sketched in Figure 8–13. The most obvious feature of the geology is the bowl-shaped structure of the rocks, which dip inward toward the valley axis from both sides. The rocks are mainly limestone, with thin clay layers at intervals. The limestone is full of caves and smaller solution channels so that large amounts of rainwater can penetrate the rock and spread along the clay layers. A whole series of natural and man-made factors contributed to the disaster. The principal factors were: (1) The steeply dipping limestone beds and clay layers offered little frictional resistance to sliding; (2) all the rock types are inherently weak; (3) the river had eroded the steep inner canyon across the rock structure and removed lateral support long before the dam was built; (4) two weeks of heavy rainfall had raised the water level in the cavernous rocks and increased both fluid pressure and weight in the potential slide mass; and (5) the high water level in the reservoir had saturated the lower part of the slide, decreased frictional resistance, and increased buoyancy.

The Vaiont Reservoir disaster was a surprise only in its severity. The canyon is marked by the scars of ancient slides. In 1960 a smaller slide had occurred, and a pattern of cracks and slumps developed on the south valley wall that ultimately outlined the great slide of 1963. For six months prior to the slide, precise records of survey stations on the slide area showed that rock creep of 1 cm/week was in prog-

Figure 8-13.   Simplified geologic cross section of the Vaiont Reservoir, Italy. View is upstream. (Redrawn from Kiersch, 1964.)

ress. By three weeks prior to the disaster, the rate of creep had increased to 1 cm/day; during the last week of heavy rains prior to the slide, the rate of creep had increased from 20 to 40 cm/day. Wild animals that had grazed on the south wall of the valley sensed the danger and moved away about October 1. On the night of the disaster, 20 technicians were on duty in the control building on the south abutment of the dam, and 40 more were in the hotel and office building on the north abutment, but no one survived who witnessed the actual slide. Desperate measures were under way to lower the reservoir level, but the rock creep was apparently pinching the reservoir and actually raising the water level despite open outlet gates. No earthquake or other "trigger" for the Vaiont slide has been identified. The rock along the principal slide plane simply failed under excessive weight and excessive pore-water pressure.

Debris Slide and Avalanche. Debris avalanches in the temperate forested region of the eastern United States have claimed many lives and produced major geomorphic changes during torrential rainfalls (Williams and Guy, 1971, 1973; Clark, 1987; Kochel, 1987; Jacobson et al., 1989; Gryta and Bartholomew, 1989). Intense storms, usually in the summer and of local extent, are characteristic of the Appalachian region, and their results must be regarded as part of the normal land-forming processes of the region (Clark, 1987; Neary and Swift, 1987).

Debris avalanches of a 1969 torrential summer storm originated at the heads of first-order tributary streams on slopes of 16 to 39° (Fig. 8–14). The bedrock in the area is massive crystalline gneiss and granite, with foliation and sheeting joints as the major structures (Gryta and Bartholomew, 1989). The weathered layer is as much as 6 m thick. The debris avalanches stripped off the weathered layer but moved relatively little fresh rock. Bedrock, commonly a single joint sheet, is ex-

Figure 8-14. Debris-avalanche scars in the headwaters of Davis Creek, Nelson County, Virginia. Most of the slides occurred during a few hours near the end of a torrential rainstorm on August 19, 1969 (Williams and Guy, 1973, Fig. 11). (Photo: Edwin Roseberry.)

posed the entire length of many avalanche scars, which are as much as 250 m long downslope and 10 to 25 m wide. The avalanche debris was carried down the initial slope and onto the adjacent valley floor, where it became a debris fan or flow or was carried away later as alluvium (Fig. 8–14). The obvious general cause for temperate-climate debris avalanches is saturation by heavy rain, but the precise trigger is unknown. Summer storms are known to fluctuate manyfold in intensity, even over a time span of hours or less, and during the brief intervals of heaviest precipitation, water alone may be the ultimate cause of failure. Lightning striking trees, the vibration of thunder, and windthrow are also suggested trigger mechanisms.

Very large terrestrial debris slides and avalanches have received renewed interest in recent years because the shape and size of their deposits and their probable mode of origin are similar to those of debris lobes below crater walls and cliffs on Mars and the moon. Both the terrestrial and extraterrestrial large debris slides have "runout" distances—the horizontal distance traveled by the debris—measured in kilometers, very long relative to their vertical fall (Melosh, 1987, Fig. 2). For example, the deposit of the largest Quaternary debris avalanche known on earth extends 43 km northwest from the base of Mount Shasta in northern California (Crandell et al., 1984). Radiometric ages from volcanic rocks within the debris and from a basalt flow that overlies it bracket the age of the event between about 300,000 and 360,000 years ago. The surface morphology of hills, mounds, and ridges that covers 450 km² of valley floor had been interpreted previously as an area of glacial moraine (Chap. 17). All of the rock debris is volcanic and must have been derived from the ancestral Mount Shasta. The flat surface of lahar deposits that surround the mounds and hills has a slope of only about 5 m/km (0.3°).

The Saidmarreh slide in southwestern Iran occurred more than 10,000 years ago, but because of the aridity in the region, its surface morphology can still be seen (Fig. 8–15). It has been dissected by the gorge of the Saidmarreh River, which could have prepared conditions for the slide by undercutting the dip slope of a hogback (Fig. 11–9). The surface of the slide is a karst landscape formed by solution of gypsum from the bedrock beneath the slide debris (Watson and Wright, 1969).

The slide came from the northern flank of an anticlinal ridge with an average elevation of about 2000 m above sea level. On the northern flank of the anticline is a high hogback of limestone that dips toward the valley at about 20° and rests on thin-bedded limestone and marl. A segment of the hogback 15 km long, 5 km wide, and at least 300 m thick slid off the mountain into the valley below. Watson and Wright (1969, p. 129) proposed that an earthquake triggered the slide in late Pleistocene time, after the Saidmarreh River had undercut the base of the hogback ridge. The slide must have been extremely rapid. The vertical component of initial motion at the center of gravity of the slab was only about 900 m, yet the slide mass crossed a valley, rose 600 m across the nose of an adjacent plunging anticline, and continued northward for another 18 km. Shreve (1966, p. 1640) calculated that a minimum velocity of 300 km/h was required. The slide was not turbulent, for the debris is crudely stratified with fragments from the youngest strata nearest the top of the slide debris. Pore water under dynamic stress may have counterbalanced part of the weight of the slide, and possibly the gypsum over which it moved had a very low coefficient of sliding friction.

The well-documented Sherman slide (Fig. 8–16) was triggered by the March 27, 1964, Alaskan earthquake (Shreve, 1966; Marangunic and Bull, 1968). A sheet of debris averaging 1.3 m in thickness traveled 5 km from its origin across the gently sloping surface of the Sherman Glacier yet did not disturb the 2 m of fresh snow on the surface of the glacier. After the initial rock slide down a 40° cliff for a vertical height of 600 m, the slide crossed a bedrock ridge 150 m high and then spread out over 8.5 km² of glacier surface. The debris is thoroughly pulverized, probably by

Figure 8-15. Scar of the prehistoric Saidmarreh landslide, southwestern Iran. The slide moved toward the camera from the bedding plane scar in the background, and at the right margin of the photo it slid over the anticlinal ridge in the foreground. (Photo: Aerofilms Ltd.)

crossing the rock ridge if not in the initial slide. The ridge may have served as the "launching platform" to send the debris slide airborne over the glacier at a top speed of at least 185 km/h (Shreve, 1966, p. 1640).

In these and other large debris slides, the ratios of vertical drop to horizontal "runout" distance are small, equal to the tangents of angles ranging from 0.3° to 7°. These angles are typically one-tenth or less of the angle of sliding friction (p. 176) of the rock debris. So some kind of lubrication or additional transporting energy seems to be required for their motion. Furthermore, their debris slid like a flexible sheet over the landscape, rather than flowing turbulently. Shreve (1966, 1968) argued that these large debris slides moved on a layer of trapped and compressed air, but later experiments demonstrated that the debris sheet would be too permeable. Slides of similar dimensions on Mars and the moon were emplaced in a vacuum or near vacuum, so support on an air layer cannot explain them, either. An alternate hypothesis of elastic impact between particles, or *dispersive grain flow,* proposes that particles in the rapidly moving debris are separated by open space but support themselves by elastic collision in a manner analogous to the molecules in a gas (Hsü, 1975). A third hypothesis of *acoustic fluidization* is now favored (Melosh, 1983, 1987). In this hypothesis, the particles remain in contact with each other and transmit pressure (sound) waves by elastic compression and rarefaction. The energy of the initial fall is converted into ultrasonic vibrational energy that supports the mass, but does not turbulently mix it. It is called acoustic fluidization because, like the molecules in a fluid, the particles are in contact and can efficiently transmit sound energy, yet the mass can freely flow. Acoustic fluidization is only effective in very large debris flows, in which the ratio of volume to surface area is very large. Smaller slides, with relatively large surface areas, dissipate energy too fast (Melosh, 1983, p. 162). The implication is that the larger the flow, the more gentle can be its

Figure 8-16. Sherman landslide, Alaska. Debris fell from the high peak on the right, overtopped a spur in the right foreground, and spread more than 2.5 km across the Sherman Glacier during the earthquake of March 27, 1964. (Photo: Austin Post, U.S. Geological Survey.)

gradient. For a few dozen measured terrestrial, Martian, and lunar landslides, this relationship seems to be confirmed (Melosh, 1987, Fig. 2).

An **avalanche** usually begins with a free fall of a mass of rock or ice, which is pulverized on impact and flows at great speed. Two of the worst debris avalanches in history destroyed the region around Ranrahirca and Yungay, Peru, on January 10, 1962, and again on May 31, 1970 (McDowell and Fletcher, 1962; Cluff, 1971; Plafker et al., 1971; Plafker and Ericksen, 1978). Observers witnessed both catastrophes from the time huge ice cornices fell from the north peak of Huascaran (6700 m) until the debris came to rest against the opposite valley wall, 14.5 km away and 4 km lower in altitude. The initial rock and ice masses, of an estimated 3 million m³ in 1962 and 50 to 100 million m³ in 1970, tore loose other millions of tons of rock as they roared down the valley. The shock waves produced a noise like continuous growing thunder and stripped hillsides bare of vegetation. Rocks and ice were pulverized by the turbulent flow.

The 1962 avalanche required only 7 minutes to travel 20 km. It bounced from one side of a narrow gorge to the other at least five times before it emerged onto the fertile, heavily populated valley floor at the base of the mountain. As it spread to 1 km wide over the villages and fields of the valley, the mass slowed to an estimated

100 km/h and thinned to about 20 m. When the avalanche stopped, air and water spouted from the settling debris. Later, melting blocks of ice created pockets of soft mud in the flow that were added hazards to the nearly hopeless search operations. An estimated 3500 people were killed.

Eight years later, a major earthquake triggered another debris avalanche from the north face of Huascaran. The 1970 avalanche followed the same path as the 1962 avalanche for most of the 14 km down the gorge, but at the edge of the main valley a lobe of it jumped a bedrock ridge 200 to 300 m high. Within 3 minutes after the original rock fall, the debris avalanche had obliterated the town of Yungay, which had been protected from the earlier avalanche by the ridge. The average velocity must have been between 280 and 335 km/h. Possibly 40,000 people were killed in Yungay and neighboring towns by the 1970 avalanche, but because the total casualties of the earthquake in central Peru were estimated at 70,000 killed and 50,000 injured, the exact numbers are unknown.

Fall

The examples just cited prove that many avalanches in high mountains are initiated by a mass of rock or ice breaking off and free-falling from a glacier or cliff. Three years before the 1970 Yungay disaster, a French mountain-climbing team reported that Huascaran peak had well-developed vertical joints behind the nearly vertical north face. The earthquake may have done nothing more than topple or shake down some huge joint blocks, which in turn shattered the lower cliffs and started the avalanche. Fall is a distinct category of mass wasting, but it is rarely independent of subsequent events.

High, steep slopes are especially characteristic of mountains that have been glaciated, with their diagnostic sharp peaks and glaciated valley forms collectively termed *alpine topography* (Chap. 17). Their present slopes are oversteepened, relief joints have opened parallel to recently deepened valleys, and ice fields or glaciers cover summit areas. Whether in the Canadian Rocky Mountains, the Alps, the Andes, the Himalayas or any other high mountains, alpine topography promotes frequent and severe avalanches (Porter and Orombelli, 1981; Shroder, 1989). Fully 137 destructive avalanches have been documented during the recorded history of settlement in the European Alps alone (Eisbacher and Clague, 1984).

Submarine Mass Wasting

The largest mass movements on earth are probably under the sea. Enormous landslides on the growing Hawaiian volcanic domes have become a significant part of their mass (p. 85). Volumes of outer continental shelf sediment and rock hundreds of kilometers in length and width and hundreds of meters in thickness have moved rapidly downslope on gradients of less than 1° (Prior and Coleman, 1984, Table 10.1; Coleman and Prior, 1988). The scars and debris of these huge mass wasting events are seen in seismic reflection profiles and in side-scanning sonar images of the ocean floor (Fig. 8–17). The largest masses probably are slides, moving on planar surfaces at extremely low gradients. High pore pressures, caused by tectonic deformation, seismic shock waves, or rapid sediment loading and compaction, can equal the perpendicular load stress, so that σ' of the Mohr-Coulomb equation becomes zero and the shear strength of the mass also becomes zero. Although submarine sonar images show headwall scarps and other features typical of slumps (Fig. 8–10), Prior and Coleman (1984, p. 429) emphasized that the aspect ratio (height/length ratio) of large submarine slides is so small that "rotation" is an inappropriate term. Rather, a huge slab of sediment detaches, moves downslope, degenerates into smaller blocks, and in some instances becomes a turbulent submarine mud flow, or

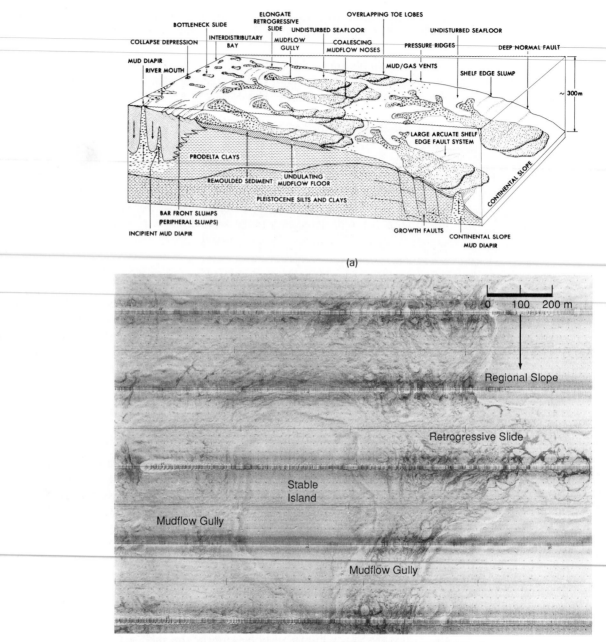

Figure 8-17.   (a) Types of subaqueous mass wasting in the offshore part of the Mississippi River delta. (Coleman, 1988, Fig. 13.) (b) Side-scan sonar image of delta-front ocean floor of about 1.5 km², showing retrogressive slides and mud flows moving down a very gentle gradient. (Photo: D.B. Prior.)

turbidity current. The density of the fluidized mass causes it to stay in contact with the ocean floor, following preexisting channels or eroding new ones, eventually spreading out as a thin layer of stratified sediment.

Submarine landslides down the continental slopes deliver massive quantities of shallow-water sediment to the continental rises and abyssal plains. Since the early days of submarine cable laying, cable breaks were believed to be associated with submarine slides. The classic example was the 1929 Grand Banks earthquake off Newfoundland, which broke 12 submarine cables in 28 places. About 15 of the breaks in six cables crossing the epicentral region are attributed to earthquake motion or immediate compaction of loose sediment. The explanation of the other breaks, which occurred progressively downslope for as long as 13 hours after the earthquake, has been the subject of controversy. Heezen and Ewing (1952) attributed the progressive breaks to a turbidity current that originated in a mud slide on the slope and flowed 800 km down the continental rise to the abyssal plain.

The inferred velocity of the turbidity current was 92 km/h, capable of powerful erosion (Fig. 9–10). Shepard (1973, p. 330) maintained that many of the cable breaks were caused by successive slumps of an expanding landslide and that the turbidity current did not exceed a velocity of 28 km/h. Emery et al. (1970) showed that all of the cables that broke within 59 minutes after the earthquake were within a zone of slumping and landslides on the lower continental slope. Cables that broke more than an hour after the earthquake were well out on the smooth continental rise and were probably broken by a turbidity current with about the velocity calculated by Shepard. The issue has emphasized the role of earthquakes in triggering submarine landslides, but simple sediment loading could just as easily result in similar slope failure.

Divers have observed sand flowing down the axes of submarine valleys off the California coast (Shepard and Dill, 1966). The heads of some California submarine canyons are at shallow depths within a few hundred meters of the shoreline and are major factors in intercepting sand that is drifting alongshore. The sand accumulates at the canyon heads, often temporarily stabilized by kelp beds. Then, during storms, earthquakes, or lesser triggering events of unknown origins, it cascades down the steep submarine valleys, producing visible scouring of the valley walls. Such movement is transitional from mass wasting to abrasive erosion.

HILLSLOPE DEVELOPMENT AND EVOLUTION

Most of a landscape consists of curved, sloping surfaces, largely shaped by mass wasting. How these slopes form, how they are maintained, and how they change with time are major topics of geomorphic research. Slopes are exceptionally difficult to study, because they are transitional both in processes and in form. Recall the arbitrary definitions of mass wasting: Add a little water, and soil creep becomes earth flow; add more water, and earth flow becomes a muddy stream.

It is surprisingly difficult even to describe the geometry of a natural slope. One is commonly provided with only a profile surveyed down the steepest part of a hillside as a description of the slope, but obviously a profile measured along the crest of a descending ridge will have a significance very different from that of a profile measured along the bed of an adjacent gully. Slopes are irregular surfaces that cannot be described by simple mathematical equations. The best topographic maps are only approximations of the infinite irregularities of hillsides, which perhaps will be best described by a Gaussian probability distribution function (Fig. 1–1) (Culling and Datko, 1987, pp. 371–72). We do not yet know what degree of irregularity is significant in the stability of slopes, so we are never sure that we are measuring the correct angles and distances.

Hillslope Analysis: Techniques and Theories

There have been two philosophies about studying slopes. An older school deduced the systematic changes of slope form that would accompany long-continued subaerial weathering and erosion. Because landscape evolution is too slow to be witnessed, deductions concerning the changes of slope form with time were based on assumptions that could not be tested until we had radioactive isotopic techniques for dating old land surfaces. It is no wonder that the deductive approach to slope analysis has enriched geologic literature with some remarkably opinionated and authoritarian writings. A majority of deductive geomorphologists have held that slopes, especially in humid regions, become lower and more broadly rounded with time. A vocal minority have insisted that slopes are stable forms with angles controlled by rock type and weathering processes, and when a stable slope has evolved, it persists through time, migrating backward parallel to itself unless it is eliminated by the intersection of other slopes. For an excellent review of the classic debate, see von Engeln (1942, pp. 256–67). The issues are evaluated further in Chapter 12.

Another group of geomorphologists have concerned themselves with the empirical description of slopes. With less regard for theoretical projection into the future, they have studied the processes of slope formation and the geometry of slopes. Innumerable slope profiles and descriptive texts have been published, but empirical study has suffered from the lack of a guiding theory.

Progress has been made in modeling the evolution of simple scarps a few meters high, such as those made by faulting or river erosion in unconsolidated sediment. Typically, a fresh scarp will be steeper than the angle of repose, but when it is abandoned, the scarp *ravels,* or collapses back, as masses of sediment from the upper part of the slope break loose and fall or slide to its base (Fig. 8–18). Within a few months or a few years, a straight slope is established at the angle of repose, which is approximately equal to the mass angle of friction and is usually about 25 to 35° (Colman and Watson, 1983; Nash, 1984; Pierce and Colman, 1986). Thereafter, the evolution of the slope through time by creep, raindrop impact, and slope wash can be described by the *diffusion equation*

$$\frac{\partial y}{\partial t} = c \frac{\partial^2 y}{\partial x^2}$$

At any point on the initial slope, the rate at which the elevation y changes with time t is the product of a diffusion constant c and the local slope curvature. Erosion occurs on the portion of the scarp that is convex upward and deposition on the portion that is concave upward. The rate of creep is proportional to the sine or tangent of the slope at any point (Fig. 8–2), so as the local curvature decreases, so does the further rate of change (Fig. 8–18). The equation is identical in form to the diffusion equation used to describe chemical diffusion, conductive heat flow, and fluid flow through a permeable medium. With time, the upper and lower part of the slope become less sharply curved and the midslope angle decreases. By the simplest assumptions of the diffusion equation, the midpoint of the slope does not change position and the initial scarp height does not decrease. If the diffusion constant for a particular sediment type in a particular climate can be established by measuring the profiles of scarps of various known ages and correcting for scarp height and exposure direction, the diffusion equation can be used to predict the age of an unknown scarp by its shape (Fig. 8–18). The method has had some success in predicting the ages of neotectonic fault scarps (p. 39).

Hillside slope profiles generally have an upper segment convex to the sky and a lower concave segment, some slope profiles show a straight segment between the upper and lower curves, and if a cliff interrupts the slope, an additional segment marked by free fall of weathered debris is introduced into the profile above the straight segment (Wood, 1942). A total of nine slope segments are recognized in one classification (Fig. 8–19).

The straight segment of a slope below a free face (Fig. 8–19, unit 5) is usually a **talus.** A talus is a slope landform, not a type of material. It is redundant to speak of a "talus slope," and the coarse rock debris on a talus should be called **sliderock** or *scree* (the common British usage). At first glance, a talus may appear to be a constructional landform, built of a great wedge of sliderock at the base of a steep cliff

Figure 8-18. Stages in the evolution of a terrace scarp. 1. Lateral undercutting of the terrace gravels by stream to form a scarp steeper than the angle of repose. 2. This oversteepened slope rapidly ravels to form a rectilinear scarp at the angle of repose, here inferred to be 33.5°, which is the starting form for diffusion-equation modeling. 3, 4, 5. Scarp profiles predicted by diffusion-equation model at stated time intervals (degradation rate coefficient, c, of 12×10^{-4} m^2/yr). (Pierce and Colman, 1986, Fig. 1.)

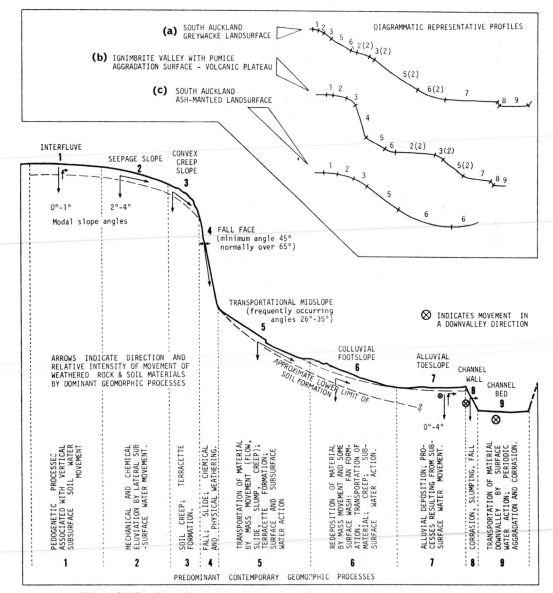

Figure 8-19. Diagrammatic representation of a hypothetical nine-unit land-surface model. (Dalrymple et al., 1968, Fig. 1.)

(Fig. 15–2). Rarely is such the case. Usually, a talus covers a convex-upward bedrock surface veneered with a layer of sliderock that is creeping. Debris that falls onto a talus from the cliff "armors" the talus and replaces fragments lost by weathering and creep. The angle of the talus is a function of fragment size and angularity, climate, vegetation, rate of cliff recession, and rate of sliderock supply and removal (Statham and Francis, 1986).

Gentler slopes with straight intermediate segments (not taluses at the bases of cliffs as in Figure 8–19) seem to form where erosion is unusually rapid. In the extreme examples of gullies cut on artificial embankments by single intense storms, most of the slopes are straight. Natural landscapes scored by closely spaced V-shaped gullies with straight sides that intersect as knife-edged ridges are called **badlands.** Strahler (1950, 1956) demonstrated that straight valley-side slopes are common in areas of recent or continuing tectonic uplift where downcutting by rivers is active.

The upper convex and straight segments of slope profiles (Fig. 8–18, units 1 to 3) are largely controlled by mass wasting, especially creep. Gilbert (1909) offered an explanation of summit convexity that is largely deductive but is still the best available explanation. He observed a reasonably uniform thickness of soil or regolith over the convex surface and assumed that in a given interval of time a uniform thickness of the weathered material is removed from the entire summit area. Under these conditions, larger quantities must move through cross sections progressively farther downhill. In other words, with the stated assumption, the amount of material that creeps past any point is proportional to the distance of the point from the summit. As creep is primarily a gravitational phenomenon, the slope angle must increase radially from the summit in order to move the progressively greater amount of debris. The summit curvature becomes convex to the sky. Carson and Kirkby (1972, pp. 108–9, 306–7) strongly supported Gilbert's conclusion that the hilltop convexity of humid temperate regions is primarily the result of soil creep, but they presented additional theoretical arguments that Gilbert's deduction is unnecessarily restrictive.

The mathematical treatment of creep as a simple diffusion phenomenon is consistent with Gilbert's qualitative explanation of hillslope profile convexity. As drawn in Figure 8–18, the diffusion curve is symmetrical around the center point on the profile and the lower half simply records the accumulation of an equivalent cross section of debris removed from the upper half. On slopes higher than a few meters, other processes of mass wasting and erosion remove some of the accumulated debris from the lower half of the profile, the midpoint of the curve migrates with the retreating slope, and the simple diffusion curve becomes asymmetric. Nevertheless, the fundamental relationship between the rate of creep and the convexity of the local upper slope is retained.

It has been argued persuasively that soil creep is rarely the predominant denudational process on slopes. An excessively large volume of sediment must move through successive cross-sectional areas of slope to lower the uphill surface appreciably. Apparently, the major function of creep is to maintain smooth upland slopes by removing soil from local mounds and filling the heads of incipient gullies and hollows (Young and Saunders, 1986, p. 8). The process is slow enough to span significant intervals of Quaternary climate change; colluvium-filled hollows in California have been accumulating creep debris for 9000 to 15,000 years, since the end of a late Pleistocene episode of gully erosion under different climatic conditions (Reneau et al., 1986).

On lower slopes (Fig. 8–19, units 6 and 7 and part of unit 5), transportation by flowing water assumes dominance over creep. Seepage pressure of groundwater emerging from lower slopes reduces the shear strength of the rock and sediment and encourages weathering to finer grain sizes, reducing the slope angle. Small rills also collect seepage water and rainwater and assume characteristic concave-up profiles. Two little rills flowing down a bare hillside during a rainstorm require a certain slope to keep flowing with their suspended-sediment load. When the two rills join, the resulting rivulet has a greater proportional increase of mass than its increase in wetted surface area. Friction is reduced in proportion to the discharge, and the larger trickle of water can transport the joint loads of the two lesser streams with no loss of velocity but on a more gentle slope. Thus slopes controlled by seepage, rain wash, sheet wash, or rill wash are generally concave skyward. At some position on a slope, rain wash becomes dominant over soil creep, and the slope profile inflects from convex near the top to concave near the base. (For an extended discussion of slope concavity, see Carson and Kirkby, 1972, pp. 310–15.)

Thus far we have considered only slope profiles. Even from profiles alone, one can see how soil creep near the top of a slope increases the gradient downhill until water begins to flow over the surface instead of penetrating and saturating the creeping soil. At that level on the hillside, sheet wash and slope concavity begin. To date, most of the research on hill slope development has been confined to analyses of profiles (Brunsden, 1971; Carson and Kirkby, 1972; Young, 1972; Nash, 1984; Pierce and Colman, 1986).

Slopes also curve in directions other than downhill, and these other curvatures also affect water movement. Where contour lines bulge convexly outward on a hillside around sloping spurs or **noses,** water is spread laterally as it flows downhill. Noses and ridge crests tend to be drier than adjacent **hollows** where the contour lines swing concavely into the hill. Concave contours tend to gather water from a large area higher on the slope, and the heads of streams are localized downhill from hollows (Hack and Goodlett, 1960, p. 6; Dietrich et al., 1986).

Profile curvature and contour curvature can be combined into a single diagrammatic classification of slopes (Fig. 8–20). The horizontal axis of the diagram

Figure 8-20. Classification of slope elements by profile curvature and contour curvature. (From Troeh, 1965.)

divides "water-gathering" slopes with concave contours (quadrants I and II) from "water-spreading" slopes with convex contours (quadrants III and IV). The vertical axis of the diagram separates slopes with convex profiles dominated by creep (quadrants II and III) from those with concave profiles dominated by wash (quadrants I and IV). Almost any segment of a land surface can be placed into one of the four quadrants of the diagrams (Troeh, 1965). The only exceptions are saddle-shaped surfaces, which require a higher order of mathematical analysis. Horizontal surfaces or planar slopes plot on the axes of the diagram.

Each of the slope elements of Fig. 8–20 can be represented mathematically by a quadratic equation. Each surface is generated by rotating a segment of a parabola around a vertical axis. In perfecting his classification, Troeh surveyed agricultural lands around Cornell University and calculated the best-fitting paraboloid of rotation for each small area of surveyed land. In areas where the actual land surface could be represented by a single quadratic equation with vertical deviations no greater than 10 to 15 cm, a single soil type was represented. Where the land surface departed from the calculated surface by more than 15 cm, a new soil type appeared. Usually, the difference in soil was controlled by slightly better or worse drainage (Troeh, 1964).

Mode and Rate of Slope Retreat

Most hillslopes are landforms that result when rivers cut valleys. Only tectonic scarps and sloping depositional surfaces (sand dunes, moraines, volcanoes, etc.) are not the secondary result of erosion. Ordinarily, a landscape is composed of small slope elements, each reacting in a particular way to the local effectiveness of weathering, mass wasting, and erosion. All elements are related, however, because an accidental disequilibrium in any part of a slope affects adjacent segments above and below the site of the accident. Suppose that an animal burrowing on a hillside raises a mound of loose soil on the downhill slope. The slope is locally oversteepened, and the debris is rapidly spread downhill. Simultaneously, rain wash from uphill is trapped by the burrow, and creep is accelerated by the undermining. A dynamically stable, or graded, slope is an example of an open physical system (Chap. 5) through which both energy and matter move, a system that tends by self-regulating processes to maintain itself in the most efficient possible configuration. Slopes constantly change but always tend toward some central graded state appropriate to the environment of the moment.

Questions that have stimulated generations of geomorphologists concern the retreat of slopes as river valleys deepen and widen. Are slope profiles stable, migrating backward parallel to themselves, or do slope angles decrease through time as summits are lowered? Are there climatic variations in the way slopes evolve? Do slopes persist through epochs of geologic time, or are present slopes the result of present-day processes? Can slope retreat be measured, or must it be deduced from indirect evidence? These questions are so deeply embedded in the fabric of geomorphic thought that they must be repeatedly reexamined in subsequent chapters. Only a brief preview is appropriate here, while the processes that shape slopes are fresh in the reader's mind.

Some generalizations about slopes seem reasonably established. Straight valley-side slopes are associated with rapidly deepening valleys in a variety of tectonic and climatic settings and on a wide scale of sizes. In humid regions, slope profiles are commonly sigmoid, convex skyward near summits and concave on lower slopes. Creep is the dominant process on the upper convexity of the profile, and sheet wash or rill wash dominates on the concave lower elements. In plan view, well-drained noses alternate with water-gathering hollows in which tributary streams originate. Weathering is intensified along the axes of hollows. Hollows enlarge by erosion at the expense of intervening noses, but are also filled with colluvium over intervals of 1000 to 10,000 years.

Creep and slope wash, which determine the relative proportion of convex and concave profile elements, are in part climate-dependent. In some climates, convexities extend across low interfluves from one valley floor to the next (Fig. 15–3). In other climates, concave slopes sweep upward from valley floors to knife-edge divides (Fig. 12–1). Unless the climate changes, the regional dominance of one slope component or another will persist, although declivities might change with time.

Cliffs and taluses are usually structurally determined (Chap. 11). Vertical cliffs can be cut by valley glaciers, waves, or rivers, but they are quickly buried under their own taluses unless some erosional agent removes the debris from their bases. If this does not happen, the talus grows in height until the entire source of sliderock at the top of the cliff is buried. Thenceforth, the slope becomes vegetated as a fossil or relict talus and gradually diffuses into a gently curved segment of a normal slope profile. If the sliderock is removed from the base of the talus as fast as it is supplied from the cliff at the top, the cliff and talus landforms migrate backward as long as the structural factors that define the cliff persist.

Colluvium of coarse rock debris on a slope may weather and change its angle of repose as it creeps downslope. The persistence of slope angles as valley sides retreat is thus related to the weathering character of the rock ("structure") and climate ("process"). If debris becomes finer as it moves downslope, eventually it is entrained in slope wash and transported over a concave profile to a stream. If, as is often the case with granitic rocks on desert slopes, large joint blocks "armor" slopes until they break down to coarse sandy grus, a sharp reentrant angle marks the foot of the hillslope and the head of an alluvial fan or pediment (Fig. 13–4).

Much has been said here of the climatic and structural controls of hillslope development and retreat. Slope evolution through time is deferred until Chapter 12, and additional aspects of climatic morphogenesis are reviewed in Chapters 13 to 18. In concluding this chapter, a few rates of slope retreat are quoted, only to suggest that such measurements can be made or deduced. Each of the five examples is from a region of dramatic slope development. Many more such examples have been compiled by Young (1974) and Saunders and Young (1983).

1. Cliffs of massive red or brown sandstone formations, generally of Mesozoic age, form much of the striking scenery of the Colorado Plateau. Schumm and Chorley (1964; 1966, p. 30) estimated that one cliff had retreated 30 feet in the last 2500 years (3 to 4 mm/yr) by rock fall. However, the cliff was exposed to weathering for an indeterminate length of time before the measured blocks began to move. The Colorado River crosses the Colorado Plateau, and its canyon is now an average of 12 km wide at the rim. By assuming that the Grand Canyon began to erode about 20 million years ago, these authors estimated the long-term widening rate of the canyon at about 0.6 mm/yr, or a rate of cliff retreat of 0.3 mm/yr on each canyon wall.

2. LaMarche (1968) measured denudation rates in the White Mountains of California by the root exposure of bristlecone pines as much as 4000 years old (Fig. 12–3). Steep hillslopes there are being lowered (parallel to the surface) at about 0.2 mm/yr; if the slope angle were 30°, horizontal recession at a rate of 0.4 mm/yr would result. These slopes are on heavily jointed dolomite in a montane climate of intense frost action, cloudbursts, and mud flows.

3. From a comprehensive study of slope development in a subarctic moun-

tain valley in northern Sweden, Rapp (1960, p. 186) concluded that rock fall alone causes steep slopes there to retreat at about 0.04 to 0.15 mm/yr, or about 1 m in the 10,000 years of postglacial time. In the cold maritime climate, frost cracking and hydrofracturing are the most active processes. The bedrock is mica schist and limestone.

4.   Mountain slopes under rainforest in Papua New Guinea are intensely scarred by active landslides, which on some slopes cover as much as 40 percent of the total area (Simonett, 1967, p. 73). Judging by the rates of postwar vegetation recovery, slide areas are not revegetated for 40 to 60 years. Therefore, an estimate of regional denudation rates can be made from the area and volume of landslide scars measured on aerial photographs. A combination of intense weathering, frequent and intense earthquakes, and steep slopes results in an estimated regional denudation rate of 1.0 to 1.4 mm/yr by landslides alone, on granitic terrane (Simonett, 1967, p. 83; Pain and Bowler, 1973). Assuming average forested slopes of clifflike 45° declivity, slope retreat here would approach 1 mm/yr. Similar rapid slope retreat and valley enlargement have been reported from Japan (Machida, 1966) and Panama (Garwood et al., 1979).

5.   As a final example, the extreme deductions of King (1962, pp. 154, 158) concerning the South African landscape can be cited. King asserted that tectonic movements initiate cycles of erosion by sending scarps retreating wavelike across the continent, each leaving a lower-level pediment (Chap. 13) behind while an older, higher land surface is consumed by the scarp. He believed that the Drakensberg scarp in Natal, 1200 m high, is one such scarp that was initiated at the coast in late Cretaceous time and has since migrated 250 km inland at a rate of about 3 mm/yr. King's hypothesis represents an extreme view of the importance of parallel scarp retreat and regional pedimentation, but the rates he cited are not excessive as compared to the previously cited four examples.

One can conclude that steep valley walls and cliffs retreat by landsliding at rates of 0.5 to 5.0 mm/yr in a variety of climates on diverse lithologies. Other processes of slope retreat are slower and more variable; even on steep slopes and cliffs, recession rates range over two orders of magnitude from 0.01 to 1.0 mm/yr (Saunders and Young, 1983, p. 487).

REFERENCES

ANDERSSON, J. G., 1906, Solifluction, a component of subaerial denudation: Jour. Geology, v. 14, pp. 91–112.

BENTLEY, S. P., and SMALLEY, I. J., 1984, Landslips in sensitive clays, *in* Brunsden, D., and Prior, D. B., eds., Slope Instability: John Wiley & Sons Ltd., Chichester, England, pp. 457–90.

BRAUN, D. D., GILLMEISTER, N. M., and INNERS, J. D., 1989, Postglacial to historic dip-slope rock block slides in the Valley and Ridge province of northeastern Pennsylvania, *in* Schultz, A. P., and Jibson, R. W., eds., Landslide processes of the eastern United States and Puerto Rico: Geol. Soc. America Spec. Paper 236, pp. 75–87.

BRUNSDEN, D., 1971, Slopes—form and process: Inst. Brit. Geographers Spec. Pub. no. 3, 178 pp.

———, 1984, Mudslides, *in* Brunsden, D., and Prior, D. B., eds., Slope Instability: John Wiley & Sons Ltd., Chichester, England, pp. 363–418.

BUNTING, B. T., 1961, The role of seepage moisture in soil formation, slope development, and stream initiation: Am. Jour. Sci., v. 259, pp. 503–18.

CARSON, M. A., and KIRKBY, M. J., 1972, Hillslope form and process: Cambridge Univ. Press, Cambridge, England, 475 pp.

CLARK, G. M., 1987, Debris slide and debris flow historical events in the Appalachians south of the glacial border, *in* Costa, J. E., and Wieczorek, G. F., eds., Debris flows/avalanches: Process, recognition, and mitigation: Geol. Soc. America Reviews in Engineering Geology, v. 7, pp. 125–38.

CLUFF, L. S., 1971, Peru earthquake of May 31, 1970; engineering geology observations: Seismol. Soc. America Bull., v. 61, pp. 511–21.

COLEMAN, J. M., 1988, Dynamic changes and processes in the Mississippi River delta: Geol. Soc. America Bull., v. 100, pp. 999–1015.

———, and PRIOR, D. B., 1988, Mass wasting on continental margins: Ann. Rev. Earth Planet. Sci., v. 16, pp. 101–19.

COLMAN, S. M., and WATSON, K., 1983, Age estimated from a diffusion equation model for scarp degradation: Science, v. 221, pp. 263–65.

CRANDELL, D. R., MILLER, C. D., GLICKEN, H. X., CHRISTIANSEN, R. L., and NEWHALL, C. G., 1984, Catastrophic debris avalanche from ancestral Mount Shasta volcano, California: Geology, v. 12, pp. 143–46.

CROZIER, M. J., 1969, Earthflow occurrence during high intensity rainfall in eastern Otago (New Zealand): Eng. Geology, v. 3, pp. 325–34.

CULLING, W. E. H., AND DATKO, M., 1987, The fractal geometry of the soil-covered landscape: Earth Surface Processes and Landforms, v. 12, pp. 369–85.

DALRYMPLE, J. B., BLONG, R. J., and CONACHER, A. J., 1968, An hypothetical nine-unit landsurface model: Zeitschr. für Geomorph., v. 12, pp. 60–76.

DIETRICH, W. E., WILSON, C. J., and RENEAU, S. L., 1986, Hollows, colluvium, and landslides in soil-mantled landscapes, *in* Abrahams, A. D., ed., Hillslope processes: Allen & Unwin, Inc., Boston, pp. 361–88.

EDEN, W. J., FLETCHER, E. B., and MITCHELL, R. J., 1971, South Nation River landslide, 16 May 1971: Canadian Geotech. Jour., v. 8, pp. 446–51.

EISBACHER, G. H., and CLAGUE, J. J., 1984, Destructive mass movements in high mountains: Hazard and management: Geol. Survey Canada, Paper 84-16, 230 pp.

EMERY, K. O., UCHUPI, E., PHILLIPS, J. D., BROWN, C. O., BUNCE, E. T., and KNOTT, S. T., 1970, Continental rise off eastern North America: Am. Assoc. Petroleum Geologists Bull., v. 54, pp. 44–108.

VON ENGELN, O. D., 1942, Geomorphology: The Macmillan Company, New York, 655 pp.

FAIRCHILD, L. H., 1987, The importance of lahar initiation processes, *in* Costa, J. E., and Wieczorek, G. F., eds., Debris flows/avalanches: Process, recognition, and mitigation: Geol. Soc. America Reviews in Engineering Geology, v. 7, pp. 51–61.

FERGUSON, H. F., 1967, Valley stress release in the Allegheny Plateau: Eng. Geology, v. 4, no. 1, pp. 63–71.

DE FRIETAS, M. H., and WATTERS, R. J., 1973, Some field examples of toppling failure: Géotechnique, v. 23, no. 4, pp. 495–513.

GARWOOD, N. C., JANOS, D. P., and BROKAW, N., 1979, Earthquake-caused landslides: A major disturbance to tropical forests: Science, v. 205, pp. 997–99.

GILBERT, G. K., 1909, The convexity of hilltops: Jour. Geology, v. 17, pp. 344–50.

GRYTA, J. J., and BARTHOLOMEW, M. J., 1989, Factors influencing the distribution of debris avalanches associated with the 1969 Hurricane Camille in Nelson county, Virginia, *in* Schultz, A. P., and Jibson, R. W., eds., Landslide processes of the eastern United States and Puerto Rico: Geol. Soc. America Spec. Paper 236, pp. 15–28.

HACK, J. T., and GOODLETT, J. C., 1960, Geomorphology and forest ecology of a mountain region in the central Appalachians: U.S. Geol. Survey Prof. Paper 347, 66 pp.

HEEZEN, B. C., and EWING, M., 1952, Turbidity currents and submarine slumps, and the 1929 Grand Banks earthquake: Am. Jour. Sci., v. 250, pp. 849–73.

HOLTZ, R. D., and KOVACS, W. D., 1981, An introduction to geotechnical engineering: Prentice-Hall, Inc., Englewood Cliffs, New Jersey, 733 pp.

HOOKE, R. L., 1987, Mass movement in semi-arid environments and the morphology of alluvial fans, *in* Anderson, M. G., and Richards, K. S., eds., Slope stability: Geotechnical engineering and geomorphology: John Wiley & Sons Ltd., Chichester, England, pp. 505–29.

HSÜ, K. J., 1975, Catastrophic debris streams (sturzstroms) generated by rock falls: Geol. Soc. America Bull., v. 86, pp. 129–40.

JACOBSON, R. B., CRON, E. D., and MCGEEHIN, J. P., 1989, Slope movements triggered by heavy rainfall, November 3–5, 1985, *in* Schultz, A. P., and Jibson, R. W., eds., Landslide processes of the eastern United States and Puerto Rico: Geol. Soc. America Spec. Paper 236, pp. 1–13.

JOHNSON, A. M., and RODINE, J. R., 1984, Debris flow, *in* Brunsden, D., and Prior, D. B., eds., Slope instability: John Wiley & Sons Ltd., Chichester, England, pp. 257–361.

JONES, F. O., EMBODY, D. R., and PETERSON, W. L., 1961, Landslides along the Columbia River valley, northeastern Washington: U.S. Geol. Survey Prof. Paper 367, 98 pp.

KARROW, P. F., 1972, Earthflows in the Grondines and Trois Rivières areas, Quebec: Canadian Jour. Earth Sci., v. 9, pp. 561–73.

KIERSCH, G. A., 1964, Vaiont reservoir disaster: Civil Engineering, v. 34, pp. 32–39.

KING, L. C., 1962, Morphology of the earth: Hafner Publishing Company, New York, 699 pp.

KIRKBY, M. J., 1967, Measurement and theory of soil creep: Jour. Geology, v. 75, pp. 360–74.

KOCHEL, R. C., 1987, Holocene debris flows in central Virginia, *in* Costa, J. E., and Wieczorek, G. F., eds. Debris flows/avalanches: Process, recognition, and mitigation: Geol. Soc. America Reviews in Engineering Geology, v. 7, pp. 139–55.

LAMARCHE, V. C., 1968, Rates of slope degradation as determined from botanic evidence, White Mountains, California: U.S. Geol. Survey Prof. Paper 352–I, pp. 341–77.

LEE, F. T., 1989, Slope movements in the Cheshire Quartzite, southwest Vermont, *in* Schultz, A. P., and Jibson, R. W., eds., Landslide processes of the eastern United States and Puerto Rico: Geol. Soc. America Spec. Paper 236, pp. 89–102.

LEWIS, L. A., 1974, Slow movement of earth under tropical rain forest conditions: Geology, v. 2, pp. 9–10.

MACHIDA, H., 1966, Rapid erosional development of mountain slopes and valleys caused by large landslides in Japan: Tokyo Metropolitan Univ. Geog. Repts., no. 1, pp. 55–78.

MARANGUNIC, C., and BULL, C., 1968, The landslide on the Sherman Glacier, *in* The great Alaskan earthquake of 1964, v. 3, "Hydrology": Natl. Acad. Sci. Pub. 1603, Washington, D.C., pp. 383–94.

MCDOWELL, B., and FLETCHER, J. E., 1962, Avalanche! 3,500 Peruvians perish in seven minutes: Natl. Geographic, v. 121, pp. 855–80.

MELOSH, H. J., 1983, Acoustic fluidization: American Scientist, v. 71, pp. 158–65.

———, 1987, The mechanics of large rock avalanches, *in* Costa, J. E., and Wieczorek, G. F., eds., Debris flows/avalanches: Process, recognition, and mitigation: Geol. Soc. America Reviews in Engineering Geology, v. 7, pp. 41–49.

MITCHELL, J. K., 1976, Fundamentals of soil behavior: John Wiley & Sons, Inc., New York, 422 pp.

MOON, B. P., 1986, Controls on the form and development of rock slopes in fold terrane, *in* Abrahams, A. D., ed. Hillslope processes: Allen & Unwin, Inc., Boston, pp. 225–43.

NASH, D. B., 1984, Morphologic dating of fluvial terrace scarps and fault scarps near West Yellowstone, Montana: Geol. Soc. America Bull., v. 95, pp. 1413–24.

NASH, D. F. T., 1987, Comparative review of limit equilibrium methods of stability analysis, *in* Anderson, M. G., and Richards, K. S., eds., Slope stability: Geotechnical engineering and geomorphology: John Wiley & Sons Ltd., Chichester, England, pp. 11–75.

NEARY, D. G., and SWIFT, L. W., JR., 1987, Rainfall thresholds for triggering a debris avalanching event in the southern Appalachian Mountains, *in* Costa, J. E., and Wieczorek, G. F., eds., Debris flows/avalanches: Process, recognition, and mitigation: Geol. Soc. America Reviews in Engineering Geology, v. 7, pp. 81–92.

PAIN, C. F. and BOWLER, J. M., 1973, Denudation following the November 1970 earthquake at Madang, Papua New Guinea: Zeitschr. für Geomorph., Supp. no. 18, pp. 92–104.

PHIPPS, R. L., 1974, The soil creep-curved tree fallacy: U.S. Geol. Survey Jour. Res., v. 2, pp. 371–77.

PIERCE, K. L., and COLMAN, S. M., 1986, Effect of height and orientation (microclimate) on geomorphic degradation rates and processes, late-glacial terrace scarps in central Idaho: Geol. Soc. America Bull., v. 97, pp. 869–85.

PIERSON, T. C., 1986, Flow behavior of channelized debris flows, Mount St. Helens, Washington, *in* Abrahams, A. D., ed., Hillslope processes: Allen & Unwin, Inc., Boston, pp. 269–96.

———, and COSTA, J. E., 1987, A rheologic classification of subaerial sediment-water flows, *in* Costa, J. E., and Wieczorek, G. F., eds., Debris flows/avalanches: Process, recognition, and mitigation: Geol. Soc. America Reviews in Engineering Geology, v. 7, pp. 1–12.

PLAFKER, G., AND ERICKSEN, G. E., 1978, Nevados Huascarán avalanches, Peru, *in* Voight, B., ed., Rockslides and avalanches, I: Natural phenomena: Developments in Geotechnical Engineering 14A, Elsevier Scientific Publishing Company, Amsterdam, pp. 277–314.

———, and CONCHA, J. F., 1971, Geological aspects of the May 31, 1970, Peru earthquake: Seismol. Soc. America Bull., v. 61, pp. 543–78.

PORTER, S. C., 1970, Quaternary glacial record in Swat Kohistan, West Pakistan: Geol. Soc. America Bull., v. 81, pp. 1421–46.

———, and OROMBELLI, G., 1981, Alpine rockfall hazards: American Scientist, v. 69, pp. 67–75.

PRIOR, D. B., and STEPHENS, N., 1972, Some movement patterns of temperate mudflows: Examples from northeastern Ireland: Geol. Soc. America Bull., v. 83, pp. 2533–43.

PRIOR, D. B., and COLEMAN, J. M., 1984, Submarine slope instability, *in* Brunsden, D., and Prior, D. B., eds., Slope instability: John Wiley & Sons Ltd., Chichester, England, pp. 419–55.

RAPP, A., 1960, Recent development of mountain slopes in Kärkevagge and surroundings, northern Scandinavia: Geograf. Annaler, v. 42, pp. 65–200.

RENEAU, S. L., DIETRICH, W. E., DORN, R. I., BERGER, C. R., and RUBIN, M., 1986, Geomorphic and paleoclimatic implications of latest Pleistocene radiocarbon dates from colluvium-mantled hollows, California: Geology, v. 14, pp. 655–58.

ROBINSON, E. S., 1970, Mechanical disintegration of the Navajo Sandstone in Zion Canyon, Utah: Geol. Soc. America Bull., v. 81, pp. 2799–805.

SAUNDERS I., and YOUNG, A., 1983, Rates of surface processes on slopes, slope retreat and denudation: Earth Surface Processes and Landforms, v. 8, pp. 473–501.

SCHUMM, S. A., 1967, Rates of surficial rock creep on hillslopes in western Colorado: Science, v. 155, pp. 560–61.

———, and CHORLEY, R. J., 1964, The fall of Threatening Rock: Am. Jour. Sci., v. 262, pp. 1041–54.

———, and CHORLEY, R. J., 1966, Talus weathering and scarp recession in the Colorado Plateaus: Zeitschr. für Geomorph., v. 10, pp. 11–36.

SCHUSTER, R. L., and KRIZAK, R. L., 1978, Landslides: Analysis and control: Transportation Research Board, Natl. Acad. Sci., Natl. Res. Council Spec. Rept. 176, 234 pp.

SELBY, M. J., 1987, Rock slopes, *in* Anderson, M. G., and Richards, K. S., eds., Slope stability: Geotechnical engineering and geomorphology: John Wiley & Sons Ltd., Chichester, England, pp. 475–504.

SHARPE, C. F. S., 1938, Landslides and related phenomena: Pageant Books, Inc., Paterson, N.J., 137 pp. (reprint 1960).

SHEPARD, F. P., 1973, Submarine geology, 3rd ed: Harper & Row, Publishers, New York, 517 pp.

———, and DILL, R. F., 1966, Submarine canyons and other sea valleys: Rand McNally & Company, Chicago, 367 pp.

SHREVE, R. L., 1966, Sherman landslide, Alaska: Science, v. 154, pp. 1639–43.

———, 1968, The Blackhawk landslide: Geol. Soc. America Spec. Paper 108, 47 pp.

SHRODER, J. F., JR., 1989, Hazards of the Himalaya: American Scientist, v. 77, pp. 564–73.

SIDLE, R. C., PEARCE, A. J., and O'LOUGHIN, C. L., 1985, Hillslope stability and land use: American Geophysical Union Monograph Series, v. 11, 140 pp.

SIMONETT, D. S., 1967, Landslide distribution and earthquakes in the Bewani and Torricelli Mountains, New Guinea, *in* Jennings, J. N., and Mabbutt, J. A., eds., Landform studies from Australia and New Guinea: Australian Natl. Univ. Press, Canberra, pp. 64–84.

STATHAM, I., and FRANCIS, S. C., 1986, Influence of scree accumulation and weathering on the development of steep mountain slopes, *in* Abrahams, A. D., ed., Hillslope processes: Allen & Unwin, Inc., Boston, pp. 245–67.

STRAHLER, A. N., 1950, Equilibrium theory of erosional slopes approached by frequency distribution analysis: Am. Jour. Sci., v. 248, pp. 673–96, 800–14.

———, 1956, Quantitative slope analysis: Geol. Soc. America Bull., v. 67, pp. 571–96.

TAVENAS, F., CHAGNON, J. Y., and LAROCHELLE, P., 1971, The Saint-Jean–Vianney landslide: Observations and eye-witness accounts: Canadian Geotech. Jour., v. 8, pp. 463–78.

TORRANCE, J. K., 1987, Slope stability, *in* Anderson, M. G., and Richards, K. S., eds., Slope stability: Geotechnical engineering and geomorphology: John Wiley & Sons Ltd., Chichester, England, pp. 447–73.

TROEH, F. R., 1964, Landform parameters correlated to soil drainage: Soil Sci. Soc. America Proc., v. 28, pp. 808–12.

———, 1965, Landform equations fitted to contour maps: Am. Jour. Sci., v. 263, pp. 616–27.

VAN BURKALOW, A., 1945, Angle of repose and angle of sliding friction: An experimental study: Geol. Soc. America Bull., v. 56, pp. 669–707.

VARNES, D. J., 1978, Slope movement types and processes, *in* Schuster, R. L., and Krizak, R. J., eds., Landslides: Analysis and control: Transportation Research Board, Natl. Acad. Sci., Natl. Res. Council Spec. Rept. 176, pp. 11–33.

WASHBURN, A. L., 1967, Instrumental observations of mass-wasting in the Mesters Vig District, northeast Greenland: Meddelelser om Grønland, v. 166, no. 4, 296 pp.

WATSON, R. A., and WRIGHT, H. E., JR., 1969, The Saidmarreh landslide, Iran, *in* Schumm, S. A., and Bradley, W. C., eds., U.S. contributions to Quaternary research: Geol. Soc. America Spec. Paper 123, pp. 115–39.

WHITE, S. E., July 14, 1973, Personal communication.

WILLIAMS, G. P., and GUY, H. P., 1971, Debris avalanches—a geomorphic hazard, *in* Coates, D. R., ed., Environmental geomorphology: State Univ. of New York Publications in Geomorphology, Binghamton, N.Y., pp. 25–46.

———, 1973, Erosional and depositional aspects of Hurricane Camille in Virginia, 1969: U.S. Geol. Survey Prof. Paper 804, 80 pp.

WOOD, A., 1942, Development of hillside slopes: Geologists' Assoc. Proc., v. 53, pp. 128–40.

YOUNG, A., 1972, Slopes: Oliver & Boyd Ltd., Edinburgh, 288 pp.

———, 1974, The rate of slope retreat, *in* Brown, E. H., and Waters, R. S., eds., Progress in geomorphology: Inst. British Geographers Special Pub. no. 7, pp. 65–77.

———, and SAUNDERS, I., 1986, Rates of surface processes and denudation, *in* Abrahams, A. D., ed., Hillslope processes: Allen & Unwin, Inc., Boston, pp. 3–27.

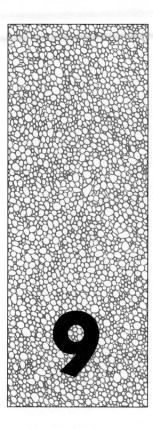

Fluvial Erosion and Transport: River Channels

THE FLUVIAL GEOMORPHIC SYSTEM

Water, flowing down to the sea over and immediately beneath the land surface, is the dominant agent of landscape alteration. Near-surface weathering and groundwater solution provide a load for flowing streams, and mass wasting may dump great quantities of rock debris at the foot of slopes, but eventually rivers must carry all but a small fraction of the total rock waste from the lands to the sea. Estimates of the relative importance of various agents of continental denudation can hardly be better than orders of magnitude, but one careful evaluation (Garrels and Mackenzie, 1971, p. 114) credited rivers with 85 to 90 percent of the total present sediment transport to the sea, glaciers with about 7 percent, groundwater and waves with about 1 to 2 percent, and wind and volcanoes with less than 1 percent each. Therefore, to understand how landscapes evolve, one must understand how rivers do their work. The useful adjective **fluvial** (from L. *fluvius:* "river") pertains to the work of rivers, but in the context of landscape development, also includes the water-dominated preconditioning of rock debris by weathering and mass wasting before reaching a river channel (Leopold et al., 1964). Schumm (1977, p. vii) included in the **fluvial system** all of the area and processes extending from drainage divides in the source areas of water and sediment, through the channels and valleys of the drainage basin, to depositional areas in valleys, at the base of mountains, or at the coast.

Rivers are the return to the ocean of the excess water that falls as rain or snow on the landscape and does not evaporate during the hydrologic cycle (Fig. 5–5). Hence the fluvial system is powered by the conversion of the potential energy of solar distillation and gravity to the kinetic energy of motion and heat. Most of the energy is lost to friction of internal turbulence in flowing water, but perhaps 2 to 4 percent (Rubey, 1938, p. 138) of the total potential energy of water flowing downhill over an erodible bed is converted to the mechanical work of erosion and transportation.

Fluvial processes obviously vary in intensity among climatic regions and along gradients of temperature, precipitation, altitude, and seasonality (Chap. 5). The annual flow of the Amazon River alone is estimated to be almost 15 percent of the total annual runoff from all the land (Oltman et al., 1964; Gibbs, 1967, p. 1206; Oltman, 1968). Oddly, on this hydrous planet, perhaps one-third of the land surface has no runoff to the ocean (Chap. 13). However, even arid regions with drainage into closed intermontane basins have landscapes of branching stream valleys. Infrequent or brief seasonal stream flow can shape otherwise dry landscapes. Furthermore, on the time scale in which landscapes evolve, climatic regions have shifted and changed in intensity, so that many regions now intensely arid show evidence of previous fluvial erosion (Chap. 18).

The hydrologic cycle (Fig. 5–5) offers abundant evidence that more water annually falls on the land than evaporates from it and that most landscapes evolve under conditions of excess water runoff. The fluvial system is so widespread and effective that it has been commonly regarded as the "normal" process of landscape evolution on earth (Cotton, 1948). W. M. Davis (1905a, reprinted 1954, p. 288; 1905b, reprinted 1954, p. 296) based his classic geomorphic cycle on the certainty that "the greater part of the land surface has been carved by . . . the familiar processes of rain and rivers, of weather and water." He defined a "normal" climatic region as "not so dry but that all parts of the surface have continuous drainage to the sea, nor so cold but that the snow of winter all disappears in summer."

It is regrettable that Davis's definition betrays a certain provinciality that was common in eastern North America and western Europe at the time he wrote. Nevertheless, the subsequent recognition that vast areas of terrestrial landscape are shaped by processes not dominant in the cool, humid climates of Cambridge (Massachusetts), London, and Paris has not diminished the importance to geomorphology of fluvial, or "normal," processes. Use of the word "normal" to describe fluvial processes should cause no more inconvenience to the geomorphologist than it does when it is used to describe a "normal" fault.

INFILTRATION, SHEET FLOW, THROUGHFLOW, RILL FLOW, AND SOIL EROSION

The proportion of surface and subsurface water that feeds a stream varies greatly with climate, soil type, bedrock, slope, vegetation, and many other factors. One estimate is that one-eighth of the annual runoff of the hydrologic cycle goes directly overland to the sea, and seven-eighths of the water goes underground at least briefly. The rapidity with which infiltrating water reacts with minerals (Chap. 6) explains why "pure" spring water closely reflects the chemistry of the local rocks. Even as river water begins its downhill flow, its chemical energy has been largely expended.

During a gentle rain on bare soil, all the water may infiltrate into the ground. The *infiltration capacity* of soil is controlled by duration and intensity of precipitation, prior wetted condition of the soil, vegetation, soil mineralogy and texture, slope, and other factors (Horton, 1945, p. 307). When the surface layer becomes saturated, overland flow begins, and soil particles loosened by raindrop impact or turbulence are entrained in the flow (Smith and Wischmeier, 1962; Young and Wiersma, 1973). Direct runoff from well-vegetated slopes rarely exceeds a few percent of the precipitation, but on sun-dried, clay-rich surface crusts, runoff may begin as soon as a thin surface layer is wetted by capillary absorption.

In laboratory experiments of real soils or artificial sediment mixtures compacted into tilted boxes under simulated rainfall of varying intensity, the erosion and transport of soil particles can be studied. The impact energy of raindrops easily disturbs and entrains sand and small aggregates of sand, silt, and clay, and if the ground is saturated, water and sediment begin to move downslope. In experiments, a wire screen placed a few centimeters above the surface to break the raindrop impact without decreasing the precipitation intensity decreased soil losses by 90 percent or more (Young and Wiersma, 1973). On bare gentle slopes, the scattering effect of impact suppresses channel formation, and the muddy water flows as a thin, slow-moving surface layer called **sheet flow.** As slope increases, helical threads of secondary currents develop in the layer of water, which are capable of additional sediment scour (Moss et al., 1982). At some threshold of erosional and transportational energy that combines the factors of rainfall intensity, grain size, cohesion, infiltration capacity, slope, and surface microrelief, the threads of higher velocity and more turbulent current erode small channels, or **rills** (Dunne and Aubrey, 1986, p. 35).

Adding to raindrop impact as a source of mobilized sediment grains is the seepage pressure of infiltrated groundwater moving downslope within the soil profile. Variously defined as **throughflow** or **interflow** (McCaig, 1985, p. 123), this water moves relatively rapidly through interconnected cracks, animal burrows, decayed root channels, and other soil voids at shallow depth, whether the deeper soil and parent material are saturated or not. Whereas groundwater flow is theoretically laminar, as defined by Darcy's law (p. 156), throughflow is within relatively large spaces in the soil (macropores) and flows with turbulence that can erode and further enlarge the soil voids (Bevin and Germann, 1982). Where this water moves downslope but outward toward the soil surface, the seepage pressure (Chap. 8) reduces the effective particle weight and initiates sediment motion on a saturated surface. The process quickly focuses at points of more rapid seepage discharge, undermining a small pit or rill head. The headward propogation of the rill is then limited only by the ability of conventional fluvial transport processes to remove the material provided by the seepage pressure. The process, call *piping,* is a common method of rill initiation (Howard and McLane, 1988), especially on vegetated slopes where very little direct overland runoff occurs. Piping in soil is similar to the larger-scale process of *sapping* that occurs where the land surface intersects the water table and springs or seepage lines emerge. There, erosion by groundwater seepage and flow undermines the overlying sediment and rock and cuts a steep-headed rill or channel that can expand headward and branch into a drainage network (p. 213) even without overland flow. Sapping has been invoked to explain drainage networks on Mars as well as those on earth (Higgins, 1984; Howard and McLane, 1988).

When cut to a size large enough to survive the interval between rains (Fig. 9–1) a rill collects surface runoff at its steep headward end, and the resulting waterfall migrates upstream, extending the rill into previously undissected upland and capturing other rills. On barren slopes, rills can be observed growing headward at a rate of centimeters per minute or faster. They tend to parallel each other, rather than to form dendritic networks, because each is growing headward on the steepest local gradient. By intrenching the surface a few centimeters deeper than the adjacent interfluve areas, each rill side and head becomes a local discharge area for the shallow throughflow. Water flow and piping continue at the rill heads even after the surface soil has dried out, encouraging maintenance and growth of the evolving rill network. Although raindrop impact, creep, and many other processes can obliterate rills up to several centimeters in width and depth, on convex hillslopes a few hundred meters downslope from a ridge crest rill development becomes dominant over creep and sheet flow (Dunne and Aubrey, 1986) and slope concavity begins (Fig. 8–18).

Fluvial soil erosion is the collective result of creep, raindrop impact, sheet flow, throughflow, piping, rill wash, and erosion by larger channels, plus losses by

Figure 9-1.   Rill erosion from less than 4000 m² of drainage on late-seeded winter wheat after summer fallow, near Mica, Washington. Rill in foreground is 45 cm deep. (Photo: U.S. Dept. Agriculture—Soil Conservation Service.)

Figure 9-2.   Gully formed in 26 years, Madison County, Georgia. (Photo U.S. Dept. Agriculture—Soil Conservation Service.)

groundwater solution. All five factors of soil formation (p. 136) contribute to the rate at which mass is removed from the landscape. Catastrophic events such as floods, droughts, fire, and land clearance for agriculture and construction will increase erosion rate by many orders of magnitude. A continuing problem for geomorphologists are the observations that soil losses measured from fields or undissected landscapes are always much higher than the sediment loads of the rivers that drain the sites would indicate (p. 305). The scaling problems of space and time (p. 6) are certainly part of the problem of the "missing" sediment in river systems; soil-loss measurements are restricted to relatively small plots, measured at most for a few years or decades. Another problem may be the thousand-year storage time of colluvium in hillside hollows (p. 200). Even though careful inventories are maintained of the dissolved and solid components leaving carefully measured test sites, the resulting soil losses are always much larger than the regional losses.

Only some vague notion of size and permanency distinguishes a **gully** from a rill (Fig. 9–2). A gully is a steep- or vertical-sided ephemeral stream valley with a steep head that is actively eroding headward, usually into a water-gathering wash slope (Fig. 8–20). Gullies are known to develop rapidly as a result of poor agricultural practices. One in Georgia, the size of which impressed the great British geologist Sir Charles Lyell in 1846, subsequently grew to a length of 230 m, a width of 150 m, and a maximum depth of 20 m, in little more than 100 years. It is eroded in the deeply weathered crystalline rocks of the Georgia piedmont and began to form after the area was deforested (Ireland, 1939). Gullies grow headward rapidly enough to capture adjacent rills and commonly develop a branching network of tributaries. They are the first step in the fluvial dissection of landscapes.

DRAINAGE BASINS AND NETWORKS

Drainage Basins

In nature and in laboratory experiments, rain falls over an entire surface, then infiltrates or flows overland under the influence of gravity until it collects in surface rills, gullies, and channel systems (Fig. 9–3). All of the water and the sediment that it has entrained eventually leaves a unit of surface area through a single channel. The **drainage basin,** or watershed, of a river system is the surface upstream and uphill from a channel that sheds water and sediment into that channel. A drainage basin is an open system into which and from which energy and matter flow, and its boundaries are normally well defined. Because so much of the terrestrial landscape is drained by rivers, drainage basins are a fundamental unit of geomorphic analysis. The network of branching stream channels includes only a small percentage of the total basin area, but their role in eroding and transporting sediment is critical to the basin's behavior and evolution. Drainage basins and their drainage networks are studied by a combination of the techniques used in measuring mass wasting (Chap. 8) and those used in hydrogeology.

Quantitative Analysis and Probability

Every river appears to consist of a main trunk, fed from a variety of branches, each running in a valley proportioned to its size, and all of them together forming a system of valleys, communicating with one another, and having such a nice adjustment of their declivities, that none of them join the principal valley, either on too high or too low a level; a circumstance which would be infinitely improbable, if each of these valleys were not the work of the stream that flows in it (Playfair, 1802, p. 102).

(a)

(b)

Figure 9-3. Rainfall-erosion facility (REF), Colorado State University. (a) Start of first experiment. (b) Drainage network development. (Schumm et al., 1987, Fig. 2-9.)

Playfair's law, quoted above, in addition to being written in a style only rarely achieved in scientific prose, illustrates the mathematician's appreciation of probability theory. In discussing the quotation, Playfair noted (1802, pp. 353, 355):

> The truth of the proposition . . . is demonstrated on a principle which has a close affinity to that on which chances are usually calculated . . . we must conclude, that the probability of such a constitution having arisen from another cause, is, to the probability of its having arisen from the running of water, in such a proportion as unity bears to a number infinitely great.

For nearly 150 years, Playfair's law was quoted as a defensible and acceptable logical proposition, without any attempt to quantitatively test it. In 1945, Horton (1945, p. 280) demonstrated that enough hydrologic measurements were available to quantify the description and theories of developing drainage basins and river networks. "Horton's laws," a series of exponential or geometric equations, were expanded by Strahler (1952a, b, 1954) and others, and a new subject of *quantitative*

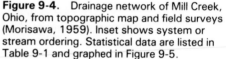

- – – 1st order
- – – – 2nd order
- – – – – 3d order
- – · – 3d order
——— 4th order (trunk stream)

0   300   600

Meters

N

Figure 9-4. Drainage network of Mill Creek, Ohio, from topographic map and field surveys (Morisawa, 1959). Inset shows system or stream ordering. Statistical data are listed in Table 9-1 and graphed in Figure 9-5.

fluvial geomorphology was established (Salisbury, 1971; Gregory and Walling, 1973). The goal was to establish quantitative, rather than qualitative, relationships between geomorphic processes and landforms. The technique is called **morphometry,** the measurement and mathematical analysis of various landform parameters.

One of the early successes of fluvial morphometry was the analysis of branching drainage networks (Fig. 9–4). An ingenious numbering technique was devised whereby the fingertip tributaries of streams, those that originate from overland or groundwater flow and flow to a junction without receiving any permanent tributaries themselves, are called **first-order** streams. The drainage basins of first-order streams can cover 50 percent of the area of higher-order basins (Marcus, 1980), and as illustrated by the location of groundwater seepage and catastrophic mud slides (Fig. 8–14), are the region in which mass wasting merges into fluvial processes. A **second-order** stream begins at the junction of two first-order streams; it may receive additional first-order tributaries, but if it joins another second-order tributary, a **third-order** stream is formed, and so forth. In subsequent jargon (Shreve, 1966, 1967; James and Krumbein, 1969), first-order streams were called **external links** in topological networks, and higher-order segments were called **internal links.**

The statistical analysis of parameters such as link lengths, bifurcation (branching) frequency, and bifurcation angles can become a highly technical and practical subject. Similar botanical morphometric analysis has demonstrated that the growth forms of all species of trees can be classified into only 23 architectural modes or categories. Computer simulations proved that the pattern of branching that most closely approximates a real tree is one that maximizes the interception of light by the leaves and makes possible the most efficient photosynthesis (Honda and Fisher, 1978; Tomlinson, 1983). To illustrate a practical problem, suppose that a cluster of houses is to be provided with water by a network of underground pipes. Each size of pipe has a stated cost per unit length, the cost of laying each size of pipe is specified, and each coupling costs a certain additional amount. What is the distribution array from a trunkline to individual houses that can be laid for the least cost? If we rephrase such a question in terms of natural drainage networks and make the important initial *assumption* that nature is inherently conservative in expending energy, we ask: What is the most efficient array of branching channels that can carry water and sediment from an area downhill to the sea?

Early in the studies of drainage networks it was discovered that on homogeneous terranes they are highly ordered systems in which parameters such as mean link length, bifurcation ratio, downlink slope, drainage density (total length of links per unit area of drainage basin), stream azimuth, and many other dimensions and dimensionless ratios (Table 9–1; Fig. 9–5) could be predicted from information about some segment of the system (Strahler, 1958). Hydraulics engineers and geomorphologists have a common interest in being able to predict such practical facts as the annual and peak discharges of a river, flood frequency, and duration of flood crest by combining a minimum number of observation stations within a river system with morphometric analysis.

TABLE 9–1

Analysis of Some Characteristics of the Mill Creek Drainage Network[a]

Dimension:	(a)	(b)	(c)	(d)	(e)
		Average	Average	Stream	Average Channel
Stream Order	Number of Streams	Length (ft)	Basin Area (10^5 ft²)	Density (mi/mi²)	Slope (tan $\angle \times 10^3$)
1	104	364	6.97	5.45	396
2	22	993	33.73	7.02	123
3	5	3432	161.97	6.06	39
4	1	6283	747.14	5.66	10

Source: Morisawa (1962), Table 1.

[a] See Figures 9–4 and 9–5.

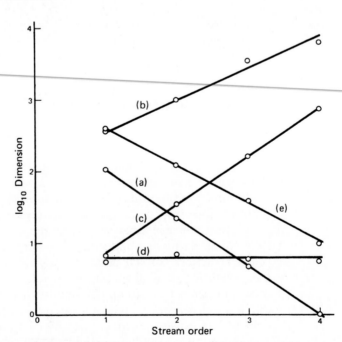

Figure 9-5. Graph of stream order in the Mill Creek drainage network compared to the channel and basin dimensions in Table 9-1.

Random Walks and Infinite Topologically Random Networks

Quantitative analysis of fluvial systems took a turn from the empirical (and tedious) analysis of real drainage basins to theoretical and statistical considerations of two-dimensional networks or systems in general. Leopold and Langbein (1962, p. 14) used a "random walk" technique to generate drainage networks that had dimensions and ratios similar to the natural networks analyzed by Horton, Strahler, and others. It is an instructive exercise to generate a river system by a "random walk" game. Start a number of markers at equal spaces along one edge of a piece of graph paper. Move the markers either ahead, left, or right, one space at a time, by the cast of a die. One restraint to random movement is that a marker cannot move backward. This is the game equivalent of gravity. If a marker intersects the path of another marker, it must thereafter follow that path. This rule of the game represents the surface tension and viscosity of water. The paths of the markers trace drainage nets in which first-order tributaries join to form second-order streams, and so on, until either all paths have merged into a single master stream or have diverged beyond any probable junction. Because the possible moves of all the markers are equally probable, if one part of a large net is known, other parts can be predicted.

The significance of "random walk" models is that they demonstrate the highly probable organization of drainage networks. High probability, in the concept of general systems theory (Chap. 5), represents the minimum energy expenditure to the system (maximum entropy) and the optimal distribution of energy expenditure within the system. The highly probable nature of drainage networks was repeatedly and emphatically demonstrated by Shreve (1966, 1967, 1969, 1975). He showed that all of the empirical relationships of links and junctions that had been determined by earlier workers analyzing real drainage networks could also be demonstrated in "infinite topologically random channel networks." Although random processes do not perfectly generate all the details of branching drainage networks (Fig. 9–6), models based on randomness are an excellent standard against which natural networks can be compared (Abrahams, 1984, p. 185). If some property of a natural network deviates significantly from randomness, a search should be made for structural controls, or inheritance from a previous condition.

Orderliness and randomness are not mutually exclusive (Mann, 1970). Although it is not possible to predict the exact position or origin of a component in a random system, it is quite possible to state precisely the general properties and subdivisions of the system and the number of components of a certain rank that are present, especially if some information is available about a small part of the system. For example, it is surprisingly easy to estimate accurately the number of leaves on a large tree by sampling the average number of leaves on a twig, the average number of twigs on a branch, and the average number of branches on the trunk. The serious deficiency of fluvial morphometry for explanatory description is that even though real drainage networks on homogeneous or structureless terranes can be demonstrated to be topologically random and are highly probabalistic systems, there is no way to prove or disprove that real networks form in the same way that random network models are generated (Leopold et al., 1964, p. 421). The techniques of quantitative fluvial geomorphology give excellent *descriptions* and *predictions* of drainage networks but no *explanation* for the manner in which they evolved.

The deficiency was illustrated by a morphometric study (Howard, 1971a) in which stream networks were simulated by headward growth and branching. The technique is essentially the reverse of a random walk in which paths merge when they touch; at intervals specified by the rules of headward growth, each headward-expanding channel branches randomly. Howard's models, which are the equivalent

Figure 9-6. Ephemeral branching networks of intertidal creeks draining the mud flats at the mouth of the Colorado River near San Felipe, Mexico. Tide range is 8 to 10 m. (Photo: U.S. Navy.)

of gully systems expanding into undissected uplands, satisfactorily simulate natural systems. His results were even better with models in which growth took place on a matrix that corresponded to an area of previously dissected upland. Howard (1971a, pp. 48–49) concluded that it is dangerous to generalize about genetic processes from computer simulations. Many different simulation procedures generate similar topologically random networks. It is not possible to determine whether a given network was formed by (1) headward-branching growth of a gully system, or (2) progressive intersection and consolidation of small rills and channels as water flowed downhill, with or without preexisting irregularities (noses and hollows) that guided the evolution of the network. Leopold et al. (1964, p. 421) agreed that "a river or a drainage basin might best be considered to have a heritage, rather than an origin."

The Principle of Least Work

Some important generalizations have emerged from the evaluation of drainage networks as random, highly ordered systems (Abrahams, 1984). For example, Howard (1971b) provided a new explanation for the old observation (Playfair, 1802, pp. 113–14) that tributary streams normally enter the trunk stream at an acute angle that points downstream. The only previous explanation was that the angle of junction was determined at the time of origin of the network by the ratio between the hillslope gradient and the gradient of the trunk stream (Horton, 1945, pp. 349–50). However, junction angles have been observed to change during dissection of rapidly eroding terrains and thus cannot be totally inherited, and the predicted angles were not supported by field measurements on finely dissected shale badlands in Utah (Howard, 1971b, p. 865). Instead, a relationship was found between the junction angles and discharge of the tributary and other tributaries adjacent to it, so that the angles minimize the rate of work (power) expended by gravity at the junction. Erosion and sedimentation change the junction angle with time, so that the tributary neither enters at an angle so steep that excessive velocity and bank erosion occur nor so gentle that loss of velocity and deposition at the junction occur. This principle of *least work* is reconsidered frequently in the remainder of this chapter.

Another important generalization to emerge from computer models of drainage networks concerns the balance of energy expenditure between overland flow and channel flow (Woldenberg, 1969). Suppose that two watersheds of equal area but different shape are each crossed by a single stream channel of equal size. If the watershed is long and narrow parallel to the channel, overland flow is minimized, but length of channel flow is maximized. In an equant or circular watershed, the channel length from the center to the margin is decreased, but overland flow is greater. Because the energy consumption per unit area of overland flow is not in the same ratio as the consumption per unit length of channel flow, the "least work" shape of watersheds can be deduced. Woldenberg (1969, p. 100) argued that circular watersheds, compressed by the requirement of continuity into honeycomblike hexagons, offer the advantages of least work. He further deduced that if first-order streams and their watersheds can be approximated by hexagons, higher-order drainage networks are also least-work assemblages of close-packed, or nested, hexagons.

To Woldenberg and others, drainage networks, like branching trees, are not random but are determined by physical laws of conservation of energy and energy distribution in open systems. The issue of randomness versus determinism in natural systems is a fundamental topic in the philosophy of science, and many interesting discussions have used fluvial drainage networks as examples (Mackin, 1963; Leopold et al., 1964, pp. 420–28; Mann, 1970; Smart, 1979; Mandelbrot, 1983).

RIVERS AND CHANNEL GEOMETRY

Whether formed by chance or necessity, by headward erosion or downslope convergence, whether inherited or newly formed, systems of branching river channels dissect most of the subaerial landscape, each "in a valley proportioned to its size." Some channels are occupied by *permanent streams* that flow throughout the year, other have *intermittent streams,* and still others are used only briefly during and immediately after precipitation by *ephemeral streams.* Channel size is obviously related to river discharge, but it is common for rivers to overflow their channels at *flood stage.*

Fluvial Classification Systems

Rivers and their channels can be descriptively categorized in several ways. Some river segments have *bedrock channels,* eroded into the regional rock with little or no sedimentary veneer. Others have *alluvial channels,* cut in river-transported rock debris, or **alluvium.** Bedrock channels are likely to be much more irregular than alluvial channels.

Whether in bedrock or alluvium, the **thalweg,** or line connecting points of maximum water depth in a general downstream direction along the channel, is rarely straight. Structural control usually determines the details of bedrock channels. Plucking action by flowing water can break away exposed layers of bedrock, or mechanical abrasion from the solid load can grind the channel into bedrock. An abrasive bed load may be caught in eddies and abrade circular pits in the channel floor. This process is called *pothole drilling.* If the bedrock is soluble, it may be chemically dissolved by flowing water as well as mechanically abraded. The net effect is a series of pools alternating with ledges or depositional bars on alternating sides of the channel.

In humid regions, the mean annual discharge of rivers increases downstream unless structural controls such as karst divert some of the water. Not all the annual runoff from the land is poured into rivers at the heads of first-order tributaries, equally spaced from river mouths like the markers at the start of a game. Water can be added anywhere along a river from surface runoff and from groundwater flow. Rivers in humid regions are called **effluent,** because they receive contributions of groundwater. Rivers in arid regions generally lose water to the ground in addition to losing it by evaporation, and often they dry up entirely without reaching the sea. These are called **influent** streams; their distinctive channel characteristics are described in Chapter 13.

The shape or pattern of river and channel segments, or *reaches,* offers another set of descriptive terms. Some segments may be *straight* in plan view. *Crooked* or *zigzag* channels are known, but where found they are structurally controlled, and the shape is externally determined rather than being a description of the channel character. *Sinuous* channels develop naturally in bedrock or alluvium, either by rounding the corners of crooked channels or alternately eroding and depositing sediment as *pools* and *riffles* along former straight reaches. In alluvial reaches, smooth free-swinging sinuosities call **meanders** (Fig. 10–4) may form. Rivers that carry large amounts of coarse sediment (sand and gravel) build midstream bars at frequent intervals and divide into numerous separate but intersecting and shifting minor channels. Such a channel is anastomosing or **braided** (Figs. 15–10 and 17–12). The meandering and braiding habits are important aspects of channel geometry, interrelated with channel shape, river discharge, slope, and other hydraulic properties. These two basic habits may alternate in space or time but do not intergrade. They seem to be separate and distinct modes of fluvial activity.

Longitudinal Profile

One of the interesting and significant aspects of stream flow is that as the quantity of water in a stream increases, the downstream slope of the water surface decreases. As an empirical rule, *slope is an inverse function of discharge* (Richards, 1982, p. 225). Water flows more efficiently in larger channels and therefore requires less slope to maintain its velocity. Langbein and Leopold (1964, p. 786) expressed the relationship between longitudinal slope (s) and mean annual discharge (Q_m) as lying within the range of two power functions: $s \propto Q_m^{-0.5}$, and $s \propto Q_m^{-1.0}$ (Fig. 9–14). That is, slope varies in a range between the reciprocal of discharge and the reciprocal of the square root of discharge. For effluent rivers, in which discharge increases downstream, the long profile should be concave to the sky, as a parabolic curve. For the Red River in Arkansas and Louisiana, a representative alluvial channel that is well adjusted to its discharge and sediment load, the relationship between slope (s) and mean annual discharge (Q_m) is $s \propto Q_m^{-0.55}$ (Fig. 9-7). Carlston (1969b) measured the long profiles of numerous rivers in the Mississippi and Atlantic coast watershed in the eastern United States and confirmed that their longitudinal profiles are overall concave upward unless dominated by structural or tectonic controls (Fig. 9-8). The average relationship of a group of midwestern and eastern U.S. rivers is $s \propto Q_m^{-0.65}$. However, he noted that many rivers have straight segments or even convex-skyward segments as part of the overall concavity. For instance, the Missouri River has a nearly constant slope despite increasing discharge ($s \propto Q_m^{-0.0}$) from the base of the Rocky Mountains nearly to the Mississippi River (Carlston, 1968, 1969b). Downstream from the confluence of the Missouri and the Mississippi Rivers, the relatively coarse sediment load of the Missouri steepens the slope of the Mississippi (Fig. 9-8).

The inverse relationship between longitudinal river slope and average discharge can be illustrated by a practical application, if not explained by a general theory. In irrigation systems, each ditch must slope steeply enough to keep water

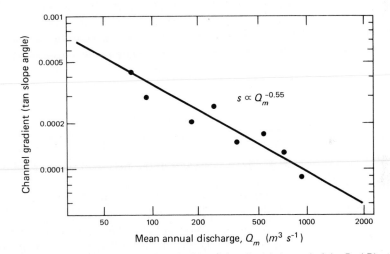

Figure 9-7. Slope–discharge relationship of the alluvial channel of the Red River in Arkansas and Louisiana. (From Carlston, 1968.)

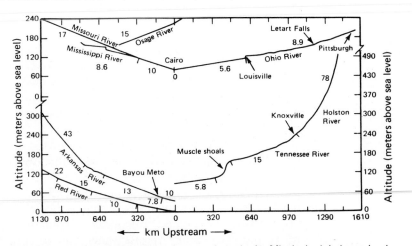

Figure 9-8. Longitudinal profiles of seven rivers in the Mississippi drainage basin. Gradient units are slope tangents $\times 10^5$ (meters of drop per 100 km of stream length). (From Carlston, 1969b, Fig. 1.)

moving and keep mud from settling in the channel but not so steeply as to cause the water to erode the banks of the ditch. A delicate energy balance must be maintained. Through centuries of trial and error and experiment, farmers learned that successively smaller distributary irrigation ditches must be given successively steeper gradients to keep the water moving at the proper speed. The mathematical concept of a **regime canal,** which neither scours nor fills its channel, was first worked out for canals in India (Leopold and Maddock, 1953, pp. 43–48; Hey, 1988, p. 101), but the principles must date from the beginning of human history on the irrigated alluvial plains of the Tigris–Euphrates, Nile, Indus, and Ganges Rivers. In an irrigation system, the carefully controlled discharge decreases downstream, as the water from a single large feeder canal is divided and subdivided, but the inverse rule between slope and discharge holds, even in this reversal of the pattern of a normal river system. The added complexity of highly variable discharge makes river systems much harder to model than canals, however (Yu and Wolman, 1987).

Base-Level Control

Where rivers enter the sea, the gravity potential of the falling water reaches zero. No further conversion of potential energy to stream work is possible, so sea level, and its projection under the land, is called the **ultimate base level** of stream erosion. Actually, most streams enter the sea with a considerable velocity and therefore have kinetic energy available to maintain their channel floors well below sea level, but this observation does not invalidate the use of sea level as the reference level for the limit of potential energy conversion. Within river systems there are also **local or temporary base levels** that delay stream erosion but do not halt it (p. 239).

Because rivers have their greatest discharge and therefore their most gentle gradients near their mouths, they enter the sea at smooth, almost undetectable tangents. The Mississippi River at New Orleans is at sea level 170 km upstream from its multiple mouths in the Gulf of Mexico. The mighty Amazon River is only about 65 m above tidewater at the Peruvian border, 3000 km from its mouths (Oltman et al., 1964, p. 4).

Repeated continental glaciations during the Quaternary Period introduced serious complications into the concept of base-level control. All large rivers now enter the sea either in *estuaries,* which are obviously the drowned segments of their former lower valleys, or across *deltas,* where borings demonstrate that 100 m or more of aggradation to a rising base level has filled former valleys (Fig. 10–2). Sea level has fluctuated through a range of at least 100 m in harmony with glaciation, introducing cyclic perturbations in the hydrologic cycle on a time scale shorter than the age of most river valleys (Fig. 17–16). Thus, when glaciers expanded, sea level fell and rivers eroded their valleys deeper. During the retreat of glaciers, sea level rose, and rivers with minor solid loads were drowned. Those with large loads, especially if glacier meltwater contributed to the discharge, built deltas. It is not possible to demonstrate ultimate base-level control by modern sea level in any river system, yet the theoretical significance of base-level control is an important concept of geomorphology.

Dams on rivers impose an unusual form of local or temporary base-level control. The *trap efficiency* of a dam and a large reservoir can exceed 99 percent of the sediment load (Williams and Wolman, 1984, p. 9). The adjustment of a river's long profile to a dam is generally a flattening of the profile by deltaic deposition at the head of the reservoir and a steepening of the former profile by erosion immediately downstream of the dam (Williams and Wolman, 1984, p. 26). However, dams have many purposes. Some release water almost continuously for power generation. Others store water for most of the year and release it only during one season for irrigation. Still others are open most of the time and are closed only when a flood is anticipated. No single set of predicted responses is appropriate for all dams. In general, dams tend to reduce peak discharges and augment low flows, so that the river discharge becomes more regular. Landforms and channel shapes based on the episodicity of natural fluctuating discharge are most likely to be modified by a dam.

Channel Width, Depth, and Current Velocity

For many years, agencies of governments have maintained *gaging stations* along rivers all over the world. At these stations, the water-surface level, channel shape, stream velocity, amount of dissolved and suspended mineral matter, and other variables are periodically or continuously recorded. The discharge of a stream is measured indirectly by multiplying the cross-sectional area of the channel at the gaging station by the average velocity of the current. Discharge is expressed in ft^3/s (cross-sectional area in square feet $\times$ velocity in feet per second), or m^3/s[1].

The records of stream gaging stations are essential for the prediction of potential flood damage, stream pollution, and other disasters (Fig. 10–1). These records also provide a voluminous history of stream flow. In 1953, L. B. Leopold and Thomas Maddock published an analysis of thousands of measurements from stream gaging stations all over the United States. They called their analysis of the relationship among stream discharge, channel shape, sediment load, and slope the **hydraulic geometry of stream channels.**

The first step in analyzing the hydraulic geometry of stream channels was to study the changes in channel width and depth, stream velocity, and suspended load at selected gaging stations during conditions ranging from low flow to bankfull discharge and flood (Leopold and Maddock, 1953, pp. 4–9). Over a wide range of conditions, it was found that for width, depth, velocity, and suspended load, each increases as some small, positive power function of discharge. Some of the gratifyingly simple equations are

$$w = aQ^b, \qquad d = cQ^f, \qquad v = kQ^m$$

where Q = water discharge, w = water-surface width, d = mean water depth, and v = mean current velocity. The changes of load with discharge are described later in this chapter.

The numerical values of the arithmetic constants a, c, and k are not very significant for the hydraulic geometry of streams. They are simply constants of proportionality that convert a given value of discharge into the equivalent values of width, depth, and velocity. The numerical values of the exponents b, f, and m, which describe the changes of width, depth, and velocity for a specified change of discharge, are very important. On log-log graphs, b, f, and m are the slopes of the linear regression lines of width, depth, and velocity plotted against discharge. Leopold and Maddock (1953, p. 9) found that the average of 20 representative gaging stations in central and southwestern United States gave values of the exponents b, f, and m as follows:

$$b = 0.26, \qquad f = 0.40, \qquad m = 0.34$$

These values signify that as the discharge of water past a gaging station increases, as after a heavy rain, on the average, the width of the channel (in feet) increases approximately as the fourth root of discharge (in ft^3/sec) ($w = aQ^{0.26}$), the mean depth increases approximately as the square root of discharge ($d = cQ^{0.40}$), and the velocity

---

[1] In this section, all equations are expressed as in the original publications, most of which used the nonmetric U.S. units of measurement.

increases approximately as the cube root of discharge ($v = kQ^{0.34}$). Channel width, depth, and current velocity all increase at gaging stations during rising water. This conclusion is no surprise to anyone who has seen a river in flood, but the regularity of the average changes is significant. A larger sample of 315 at-a-station values of b, f, and m showed that the range of measured values is quite large, and that certain types of rivers have diagnostic values of b, f, and m (Fig. 9–11; pp. 230–32). Their interpretation is deferred until the role of sediment transport by rivers has been reviewed.

More surprising are the results of comparing the changes in channel shape and stream velocity in a downstream direction (Leopold and Maddock, 1953, pp. 9–16). River discharge in humid areas increases downstream. When the mean annual discharge of many rivers past successive gaging stations was compared to the width, depth, and velocity at each station, the same equations were found to apply that had been derived for changes in flow past a single point. *As mean discharge of a river increases downstream, channel width, channel depth, and mean current velocity all increase.*

Everyone knows that effluent rivers get both wider and deeper as they grow larger downstream, but until Leopold and Maddock published their work, few had guessed that average current velocity also increases downstream. The conclusion violates our poetic impressions about wild, rapidly flowing mountain streams and deep, wide, placid rivers such as the Mississippi. We do not immediately realize that much of the high-velocity current in a mountain torrent flows in circular eddies, with backward as well as forward motion. The mean current velocity that is calculated for a succession of gaging stations down a river is the average velocity throughout each channel cross section at the time when the discharge at each of the stations is equal to the calculated mean annual discharge for that station. Both the mean velocity and the mean annual discharge are mathematical abstractions. Figure 9–9 reproduces an example of the evidence that mean current velocity increases downstream with mean annual discharge (Leopold, 1953). Both velocity and discharge are plotted on logarithmic scales to show as a straight line the power function relationship between the variables.

The numerical values of the three exponents b, f, and m that describe the variations in width, depth, and velocity with variable discharge at a gaging station (p. 223) are not the same as those that describe the downstream increases in width, depth, and mean velocity with progressively increasing mean annual discharge. In

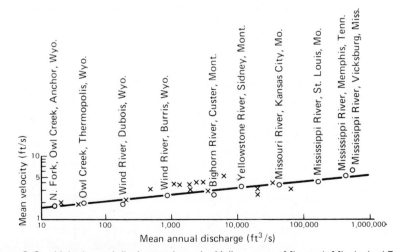

Figure 9-9. Velocity and discharge along the Yellowstone–Missouri–Mississippi River system, demonstrating that mean velocity increases downstream with increasing mean discharge. Additional gaging stations, unnamed, are shown by X. (Data from Leopold and Maddock, 1953, App. A.)

the downstream direction, at mean annual discharge, the average values for the exponents were found by Leopold and Maddock (1953, p. 16) to be

$$b = 0.5, \quad f = 0.4, \quad m = 0.1$$

These values demonstrate that in the downstream direction, channel width increases most rapidly with mean annual discharge (approximately as the square root of discharge), depth next most rapidly, and mean velocity increases only slightly, although it definitely increases. Leopold and Maddock (1953, p. 14) proposed that the increasing depth downstream permits more efficient flow in a river and overcompensates for the decreasing slope, thus providing a slight net increase in velocity at mean annual discharge.

A mathematical test of the hydraulic geometry equations suggests useful applications of the principles. Discharge is defined as cross-sectional area times mean velocity, or $Q = wdv$. If, as previously determined,

$$w = aQ^b, \quad d = cQ^f, \quad v = kQ^m$$

then by substitution,

$$Q = (aQ^b)(cQ^f)(kQ^m)$$

or

$$Q = ackQ^{b+f+m}$$

It follows that

$$a \times c \times k = 1.0$$

and

$$b + f + m = 1.0$$

The arithmetic constants a, c, and k satisfy this test but are not of concern here. It is more interesting to verify that in the examples given for both single gaging stations and a downstream succession of gages, $b + f + m = 1.0$.

Carlston (1969a, p. 500) reviewed the statistical problems of evaluating these hydraulic geometry equations and confirmed their validity. He recalculated least-square solutions for the same data from 10 river basins that were used to determine downstream changes by Leopold and Maddock. His more precise values of the exponents are

$$b = 0.461, \quad f = 0.383, \quad m = 0.155$$

For a larger set of 91 rivers, the downstream changes in width, depth, and velocity described by the values of the exponents b, f, and m are similar to those defined by Leopold and Maddock, and Carlston (Rhodes, 1987). The scatter of values in the larger sample shows a normal distribution around mean values that are similar to those calculated by Carlston. About 75 percent of the 91 rivers widen downstream faster than they deepen, as predicted by the mean values of b and f. Most of the 91 rivers also increase slightly in mean velocity downstream, although the few that decrease cause a statistical skewing. Significantly, the most common values of b, f, and m are close to the values predicted by a theory of minimum variance. That is, they are close to the predicted optimal conditions of stream-power expenditure per unit of stream length, volume, bed area, and time (Rhodes, 1987, pp. 154–59). The significance of this additional example of the tendency toward least work will be reviewed later in this chapter.

Leopold (1953) and many subsequent analysts have noted that at flood-stage discharge, the exponent $m \simeq 0$ in the downstream direction. That is, flood velocity is essentially constant downstream along the length of a river. The downstream increase in flood discharge is accommodated by increasing depth and especially width, so that when $m = 0$, $b + f = 1$. Mackin (1963, pp. 143–45) and Carlston (1969a) argued that because the significant work of rivers is accomplished during flood, when downstream velocity is constant, the increase in mean velocity downstream at mean annual discharge is of statistical, but not practical, significance. Nevertheless, to the average person, the fact that many rivers accelerate as they flow downstream is a striking and sometimes surprising revelation.

SEDIMENT TRANSPORT AND EROSION

Solid and Dissolved Loads

Weathered rock is carried by rivers in three forms. The compounds in solution or colloidal mixtures are the **dissolved load.** The solid matter is either fine-grained particles in suspension **(suspended load)** or coarse-grained particles that slide, roll, or bounce along the stream bed **(bed load,** or traction load). The division of the load varies greatly, controlled by climatic and structural factors. Table 12–1 gives some estimates for major rivers of the world. Rivers that are subject to large fluctuations in flow, or that drain poorly vegetated regions, or that have generally large loads, tend to have high percentages of solid load. Rivers that drain heavily forested regions, or overflow from lakes that act as settling basins, or drain karst regions tend to have mostly dissolved loads. Their total loads are generally small.

Both the dissolved and suspended loads in river water are routinely measured at gaging stations, and excellent data are available. Bed load defies attempts to measure it accurately. Any device lowered to the bed of a stream to measure or collect the sediment in motion also deflects the boundary-flow conditions and distorts the measurement. Bed loads as great as 50 to 55 percent of the total solid load have been estimated for alluvial rivers on the Great Plains (Leopold and Maddock, 1953, pp. 29–30), but for most purposes bed load is simply assumed to be 10 percent of the total solid load, or about an additional 11 percent of the measured suspended load.

Dissolved load has no detectable effect on stream flow. The concentrations average only 130 ppm (Livingstone, 1963), so the solutions are too dilute to affect viscosity, turbulence, or the density of river water. Therefore, dissolved load, which can represent over half of the total work of fluvial denudation (Table 12–1), gets a "free ride" to the sea by rivers. No kinetic energy is required to move it.

Competence of Rivers

Competence is the measure of a stream's ability to transport a certain maximum grain size of sediment. Competence depends primarily on velocity, although channel shape, shape and degree of sorting of the sediment particles, amount of suspended load, and water temperature can also affect competence. Figure 9–10a shows graphs of the relationship between mean velocity in a river, and the size of particle that can be eroded or transported by it (Hjulström, 1935, 1939; Nevin,

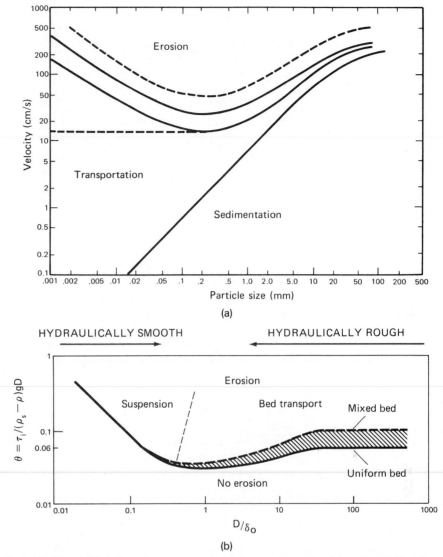

(a)

(b)

Figure 9-10. (a) Relation of mean current velocity in water at least 1 m deep to the size of mineral grains that can be eroded from a bed of similar-sized grains and transported by the current. Dashed lines above and below upper curve are revisions proposed by Sundborg (1956). (b) The Shields curve, relating the Shields parameter Θ, or threshold of particle motion (erosion), to the grain Reynolds number D/δ_0, a measure of bed roughness.

1946; Sundborg, 1956, pp. 177–80). The assumptions in the figure caption should be noted, because the average velocity in a stream is not the velocity at the bank or bed of the stream, where much of the erosion and transport occur. Furthermore, the curves were experimentally derived from flume experiments on well-sorted sediments. If coarse and fine particles are mixed, the fine particles fill spaces between the coarse ones and are thereby protected, but they also help prevent the coarse grains from moving (Blatt et al., 1972, Chap 4).

Figure 9–10a demonstrates that the grain size of sediment that will stay in suspended transport is a power function of mean velocity (approximately a straight line on the log-log graph). A point on the lower graphed line, between the transport and sedimentation fields, represents the mean velocity in a stream at least 1 m deep that will provide sufficient turbulence to keep a particle of a given size from settling out of suspension. The upper curve ("Hjulström's curve") is actually a field or zone of values for the velocity necessary to initially dislodge or erode particles. For pebbles, cobbles, and boulders, the erosion threshold is only slightly greater than the transportation velocity, because they barely rise from the bottom as they bounce or roll as bed load. Actually, large stones on the bed of a stream are overturned or rolled by mean velocities somewhat less than those indicated by Hjulström's curve. The exact relationship between eroding velocity and large particle sizes has yet to be worked out (Novak, 1973). Probably 95 percent of the total work done by stream transport is attributable to the few percent of the load that is in the coarsest grain sizes (Bagnold, 1968, p. 53). The minimum threshold eroding velocity of 15 cm/s is for fine sand, but sand can be kept in transport by a mean velocity an order of magnitude less if some other process stirs it from the bottom. Silt and clay sizes (collectively called **mud** when wet and **dust** when dry) are kept in suspension by very slight currents, but the velocities necessary to erode a consolidated clay bank are comparable to those that move pebbles and cobbles, because of the strong cohesiveness of clay-size particles and their relative smoothness.

Most formulas for the flow of water and sediment were experimentally determined in closed pipes or open flumes with fixed walls. The geomorphic equivalent would be a bedrock channel with fixed sides and floor of specified hydraulic resistance, or *roughness.* Alluvial channels have mobile beds, so that width, depth, velocity, roughness, and other channel parameters constantly adjust to discharge and sediment load. Furthermore, whereas the concept of the regime canal (p. 222) applies to flows with almost constant discharge, rivers have extremely variable discharge, fluctuating through several orders of magnitude at a gaging station. The dynamic behavior of rivers and their alluvial channels may result in channel dimensions that are not appropriate for the discharge and sediment load at the time of measurement, the channel retaining a kind of "memory" of the most recent effective channel-shaping discharge (Yu and Wolman, 1987).

Despite the complexity of natural rivers and their channels, geomorphologists continue to rely heavily on hydraulic engineering formulas to explain how fluvial systems work. For example, the *Reynolds number (Re)* is a dimensionless number that distinguishes laminar from turbulent flow by the ratio of inertial and viscous forces. It is defined as

$$Re = \frac{\rho v R}{\mu}$$

where ρ is the fluid density, v is the mean velocity, R is the *hydraulic radius* of the channel (cross-sectional area divided by length of the wetted perimeter; approximated by stream depth), and μ is the fluid viscosity. When Re is <500 in water, viscous forces dominate and flow is *laminar,* with each element of fluid moving in parallel paths with uniform velocity. When Re is >2500 in water, inertial forces of mass and velocity dominate and flow is *turbulent.* Complex transitional flow occurs when $500<Re<2500$ (Richards, 1982, p. 59; Knighton, 1984, p. 48). Water flow is turbulent in almost all natural channels.

The *Froude number (Fr)* is another dimensionless number that defines the type or degree of turbulence. It distinguishes *subcritical* or *tranquil* flow ($Fr <1$) from *supercritical* flow ($Fr >1$) with breaking surface waves and increased resistance to flow. Supercritical flow is uncommon in rivers with mobile-bed channels (Richards, 1982, p. 61). The formula for the Froude number is

$$Fr = \frac{v}{\sqrt{gd}}$$

where v is the mean velocity, g is the gravitational constant, and d is the depth. Both the Reynolds number and Froude number are velocity- and depth-dependent.

Several other hydraulic formulas relate the velocity of nonaccelerating channel flow to the total resistance of the bed, so that the gravitational acceleration of water in an inclined channel is balanced by the surface area and roughness of the bed. One of them, the *Manning equation,* is

$$v = \frac{kR^{2/3}s^{1/2}}{n}$$

where v = mean velocity, k is a constant appropriate to SI (metric) or foot-pound units, R is the hydraulic radius (defined as the cross-sectional area divided by the wetted perimeter, but commonly approximated by mean channel depth), s is the longitude slope, and n is the *Manning roughness coefficient,* another dimensionless number that defines the flow resistance of a unit of bed surface. This resistance includes not only particle size, but also the shape of the bed and constructional bed forms such as ripples. As in the formula for the Reynolds number and Froude number, the Manning equation relates mean velocity to depth (as an approximation of R; but the approximation may be a poor one: Tinkler, 1981), as well as to surface slope and bed roughness (Richards, 1982, p. 63). It is apparent that channel depth is an especially critical dimension that affects velocity, turbulence, and bed shear stress.

For coarse sediment moving on the bed of a channel, the boundary shear stress may be a better measure of competence than the average velocity (Baker and Ritter, 1975). The boundary shear stress τ_o exerted on its bed by a stream is approximated by

$$\tau_o = \gamma R s$$

where γ is the specific weight of water (ρg), R is the hydraulic radius, and s is the slope. When $\tau_o = \tau_i$, the threshold shear stress, particle motion on the bed will begin. The *Shields curve* (Fig. 9–10b) relates two dimensionless variables, the *Shield's parameter,* or threshold of particle motion,

$$\theta = \frac{\tau_i}{(\rho_s - \rho)\, gD}$$

to the *grain Reynolds number,*

$$D/\delta_o$$

where

τ_i = threshold bed shear stress
ρ_s = grain density
ρ = fluid density
g = acceleration of gravity
D = grain diameter
δ_o = thickness of the laminar-flow boundary layer (a function of the threshold shear stress and the fluid viscosity).

The Shields curve relates θ, the threshold of sediment motion, which includes particle and fluid density, particle grain size, grain shape, and packing, to bed roughness, expressed by the ratio D/δ_o, which measures the extent that individual particles project above the laminar-flow boundary layer into the turbulent flow. While difficult to visualize because it plots two dimensionless variables against each other, the Shields curve is similar to Hjulström's curve in that both define a threshold of sediment erosion. In the Shields curve the threshold is at a minimum value of the Shields parameter θ corresponding to a certain bed roughness (Knighton, 1984, pp. 56–60; Komar, 1988). Hjulström's curve assumes clear water, but the Shields parameter includes the density difference between sediment grains and the fluid. A heavy suspended load can greatly increase the fluid density and thereby increase the Shields parameter, but it can also increase δ_o, the thickness of the laminar-flow boundary layer. For the rough beds typical of most streams (large values of D/δ_o), the Shields parameter becomes nearly constant and the threshold shear stress is proportional to the sediment/fluid density ratio and grain size (Fig. 9–10b). At grain sizes finer than about 0.7 mm (medium sand) the Shields curve rises, illustrating that fine sand and mud beds, in which the bed roughness is less than the thickness of the laminar-flow boundary layer, are more difficult to erode. This effect reinforces the effect of cohesiveness in mud in resisting erosion as shown by Hjulström's curve (Fig. 9–10a), and also helps explain the decrease in competence of streams in the downstream direction, as grain size decreases.

Recalling that the exponents $b + f + m = 1$ in the hydraulic geometry equations that describe width, depth, and velocity variations as power functions of discharge (p. 225), Rhodes (1977) used ratios derived from various hydraulic formulas to define various responses of channel cross sections to changing discharges (Fig. 9–11, Table 9–2). On a triangular diagram, b, f, and m values for 315 sets of single-station measurements were plotted so that at a labeled apex of the triangle, that exponent equals 1.0, and on the opposite side, the same exponent equals zero (Fig. 9–11a). Rhodes divided the triangular diagram by five lines, defined as follows:

1. $b = f$. If $b = f$, with variable discharge the width/depth ratio stays constant. For all sets of values that plot in the left side of the diagram ($b > f$) channel width increases faster than depth with increasing discharge. Such streams typically carry mostly bed load and commonly have a braided pattern. Streams on the right half of the diagram ($b < f$) deepen faster than they widen, presumably because they have resistant banks and mobile beds.

2. $m = f$. Changing competence with changing discharge is related to the increase in the velocity/depth ratio (m/f). For points that plot above the line $m = f$ ($m > f$), velocity increases faster than depth and competence should increase with increased discharge (at constant width). This inference has not been proved by hydraulic theory, but is a reasonable hypothesis.

3. $m = f/2$. The significance of this line is derived from the formula for the Froude number. With increasing discharge, if velocity increases faster than the square root of depth (that is, if m, the exponent that describes the change in velocity, is greater than $f/2$, the exponent that describes the change in the square root of depth, or $m > f/2$) the Froude number increases. Supercritical flow may result. If the Froude number decreases with increased discharge ($m < f/2$), supercritical flow will not be attained, and transport of particles larger than pebbles will be unlikely.

4. $m = b + f$. From the requirement that $b + f + m = 1$, for points plotting above this line, $m > 0.5$, and with increasing discharge the increase in velocity past a point is greater than the sum of the increase in width and depth ($b + f$). Such streams experience rapid increase in velocity relative to cross-sectional area, and probably flow in stable bedrock channels that

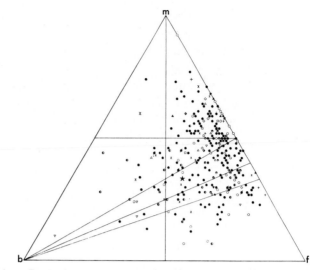

Figure 9-11a. The b–f–m diagram showing plotting position of 315 sets of at-a-station hydraulic geometry exponents. Some points represent more than one station. (Rhodes, 1977, Fig. 1.)

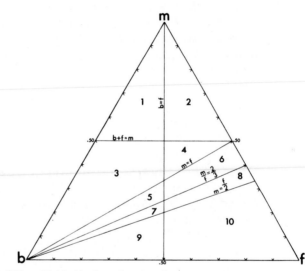

Figure 9-11b. Divided b–f–m diagram showing numbered areas in Table 9-2 that delineate channel types. (Rhodes, 1977, Fig. 2.)

TABLE 9–2
Expected Direction of Change of Morphologic and Hydrodynamic Parameters with Increasing Discharge for Different Channel Types

Channel Type	Width/Depth Ratio (w/d)	Competence	Froude Number	Velocity/ Area Ratio (v/A)	Slope/ Roughness Ratio ($s^{1/2}$/n)
1	Increases	Increases	Increases	Increases	Increases
2	Decreases	Increases	Increases	Increases	Increases
3	Increases	Increases	Increases	Decreases	Increases
4	Decreases	Increases	Increases	Decreases	Increases
5	Increases		Increases	Decreases	Increases
6	Decreases		Increases	Decreases	Increases
7	Increases		Increases	Decreases	Decreases
8	Decreases		Increases	Decreases	Decreases
9	Increases		Decreases	Decreases	Decreases
10	Decreases		Decreases	Decreases	Decreases

Source: Rhodes (1977, Table 2).

exhibit decreased resistance with increasing discharge (Baker and Costa, 1987, p. 7).

5. $m/f = \frac{2}{3}$. This division of the triangle diagram is derived from an approximation of the Manning equation in which depth is substituted for hydraulic radius. In the Manning equation, the power-function relationships of velocity with depth, slope, and the Manning roughness coefficient are such that if $m > \frac{2}{3}f$, the ratio of slope to roughness ($s^{1/2}/n$) increases. Observations of slope demonstrate that it changes very little with increasing discharge, so for points that plot above this line, bed roughness probably decreases with increased discharge. Possible causes are greater depth, which reduces bottom shear stress, increased suspended load that reduces turbulence, or a more direct flow path in a larger channel.

The 10 channel types marked out on the triangular diagram by the lines defined above are numbered on Figure 9–11b and the predicted effect of increasing discharge for each of these is summarized in Table 9–2. Rhodes' compilation demonstrates the wide variety of morphologic and hydrodynamic changes that can be predicted in a single cross section of a channel because of the relative changes in only three variables: width, depth, and velocity (slope changes being negligible). Competence, bed roughness, channel shape, and habit (straight, meandering, or braided) all interact complexly. The average values of b, f, and m calculated by Leopold and Maddock in 1953 ($b = 0.23$, $f = 0.40$, $m = 0.34$) happen to plot in Rhodes' channel type 6 (large star, Fig. 9–11a) but they are in no way representative of the 315 widely scattered sets of data compiled by Rhodes. Obviously, even at different gaging stations along a single river, the possible responses are enormously variable.

Capacity of Rivers; Variations of Suspended Load

Capacity is the theoretical maximum amount or mass of sediment load that a stream can transport. The grain size of the detritus may partly determine how much can be carried, but capacity is primarily a measure of the maximum amount, not grain size, of the load. It must be considered carefully, because of frequent confusion of the idea of capacity with that of competence (Nevin, 1946). The maximum possible suspended load cannot be specified. At some arbitrary concentration of mud in water, a muddy river is simply called a mudflow instead (p. 183). One of the greatest reported suspended loads is that of a tributary to the Hwang He (Yellow River) of northern China. It dissects a great region of windblown silt and is reported to carry a suspended load of 48 percent silt by weight during high discharge (Eliassen, *in* Lane, 1937, p. 154). Capacity might have significance for evaluating bed loads, but it can also be argued that if sufficient sediment of the proper size is available to be moved as bed load, it will be moved. Any alluvial stream is likely to be moving its capacity bed load at all times.

Because dissolved load has no effect on the hydraulic geometry of rivers, and bed load defies accurate measurement, the load of a river that is usually measured is the amount of suspended load. The unit of measurement is the dry weight of sediment per volume of water or more commonly units such as tons per day of sediment for a specified basin area. Some annual river loads are listed in Table 12–1.

The hydraulic geometry of streams involves, in addition to discharge, slope, width, depth and velocity, the amount of solid load as well as bed roughness and grain size of the load. The power-function equations relating these additional variables to discharge are similar in form to the equations for slope, width, depth, and velocity. For example, the power-function equation for the sediment load is $L = pQ^j$, where L is suspended-sediment load, Q is discharge, and p and j are numerical constants (Leopold and Maddock, 1953, p. 21). Values for the exponent j range from 2.0 to 3.0. These large exponential values mean that as discharge at a station increases tenfold, the suspended load may increase a hundredfold to a thousandfold! Figure 9–12 is a typical graph of suspended-sediment load compared with discharge. The suspended-sediment load at a station increases much more rapidly with discharge than either channel width or depth; therefore, the enlargement of the channel by erosion cannot account for all of the increased load. Most of the suspended sediment comes from the watershed upstream from the gaging station. The sediment is newly delivered to the stream by mass wasting and rill wash during the same rains or snowmelts that swell the discharge of the river.

Measurements of channel shape and suspended-sediment load confirm that streams move most of their loads during times of higher than average discharge. As the water rises during a flood, alluvial channels initially aggrade their bed (Fig. 9–13). Presumably, the aggradation results from the sudden influx of new sediment from the uplands and the decreased roughness effect of the more dense water–sediment mixture. Aggradation increases the bed slope in order to move that sediment downstream. During the peak flood discharge (Fig. 9–13, Oct. 14 cross section) the channel may be scoured even deeper than during the preflood level, and then aggrade again during the declining discharge late in the flood. Thus competence is not to be simply correlated with velocity. At a given velocity during increas-

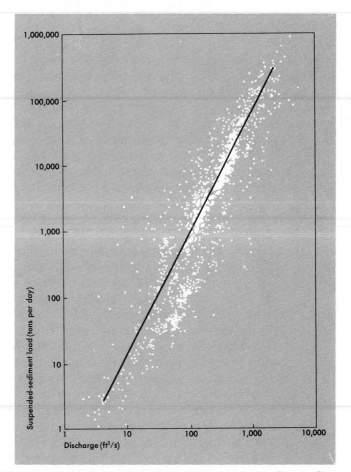

Figure 9-12.   Relation of suspended sediment load to discharge, Powder River at Arvada, Wyoming. (Leopold and Maddock, 1953, Fig. 13.)

Figure 9-13. Channel cross sections during progress of a flood, September–October 1941, San Juan River near Bluff, Utah. Horizontal line is the water surface on the date indicated. (Leopold and Maddock, 1953, Fig. 22.)

ing discharge, bed aggradation occurs, but when that same velocity is reached during decreasing discharge after the flood crest has passed, the bed may be eroded (Leopold and Maddock, 1953, pp. 30–35).

In some rivers the net channel erosion of the spring runoff is gradually restored by deposition of alluvium during the low-water summer season. If channel shape is measured at the same location at the same season in successive years, one might believe that no permanent change has taken place in the river channel during a period of high discharge, unless one realizes that the mud, sand, and gravel that previously formed the stream bed at the location have been moved downstream toward the sea and have been replaced by new sediment from upstream. The geomorphology machine has surged ahead slightly.

The change of suspended-sediment load downstream with increasing discharge has not been measured directly, but it can be estimated indirectly from other parameters of downstream hydraulic geometry (Leopold and Maddock, 1953, pp. 21–26; Rhodes, 1987, p. 152). In the equation $L = PQ^j$, the value for exponent j in a downstream direction is 0.8, which implies that the total suspended-sediment load increases downstream slightly less rapidly than discharge, and the concentration of suspended sediment becomes more dilute toward the river mouth. Two factors are probably involved. First, few first-order tributaries and negligible overland flow enters a trunk stream directly, and these are the original source of most of the sediment load. Second, effluent groundwater enters the trunk stream to increase discharge but carries no suspended load, thus diluting the concentration.

In contrast to the decreasing sediment concentration downstream in throughflowing rivers of humid regions, the concentration of suspended load increases downstream in intermittent and ephemeral streams, and also in rivers of dry regions which decrease in discharge downstream as they lose water by infiltration and evaporation (Rhodes, 1987, p. 157). As emphasized in Chapter 13, the hydraulic geometry of fluvial systems in dry regions is very different from that of rivers in humid regions.

Channel Shape, Habit, and Solid Load

The bed of an alluvial channel is similar in grain size to the solid load of the river at that point, for if the stream is competent to move a fragment on the stream bed, it will do so, and that fragment becomes part of the bed load or suspended load. If the stream lacks competence to move fragments of a given size, they settle and become part of the alluvial bed. A constant exchange between bed and load takes place. Even though the channel is shaped primarily during high discharge (Fig. 9–13), for every lesser discharge there is a combination of depth, velocity, and other variables that has a certain competence to move grains selectively up to a certain size limit (Fig. 9–10). To the extent that channel roughness is determined by the grain size or bed forms of the alluvial bed, and roughness affects turbulence and competence, a complex readjustment between discharge, grain size of the load (and the bed), channel shape, velocity, and the amount of load is constantly in progress.

Rivers that carry primarily sand and gravel as bed load develop wide, shallow alluvial channels, often braided (Figs. 15–10 and 17–12). It could be visualized that the river optimizes bed-surface area as a device to move the load efficiently by shear. However, a more accurate statement is that because bed load is carried near the bottom of the stream, the velocity gradient increases rapidly upward in such a way that the relatively clear, faster-moving upper layer of water can erode banks made of grain sizes similar to those already in transport as bed load and thereby widen the channel. A related factor is that sand and gravel are noncohesive, and banks of coarse alluvium collapse readily, to be spread out on the flat, shallow bed.

Rivers that carry mostly mud (silt and clay) as suspended load develop deeper, narrower channels with a catenary or trapezoid cross section. The wetted perimeter approaches the minimum length, that of a semicircle. A channel shape of this sort minimizes surface area and friction and therefore provides for the maximum transport of suspended load, which is carried by internal turbulence, not bed shear. A deep, narrow channel has steep banks, which can be undercut, but the cohesive strength and smoothness of consolidated mud (Fig. 9–10) resists bank erosion.

Schumm (1960a, b; 1977, p. 109) found that at a given discharge, the width-to-depth ratio of rivers on the Great Plains is inversely proportional to the percentage of fine-grained alluvium in the banks, and therefore also in the load. The relationship between channel shape and sediment grain size is given by the equation

$$F = 255M^{-1.08}$$

where F is the width/depth ratio, and M is the weighted mean percent of silt and clay in the sediment. In many Great Plains rivers, the bed load may exceed half of the total load, so the value for M is low and, inversely, F is high. The braided habit of these rivers is partly an adjustment among interdependent variables of width, depth, and sediment grain size and partly a response to the external variables of discharge and load.

The meandering habit of many rivers, especially those that flow on fine-grained alluvium in humid regions, is also related to the width/depth ratio of the channel and sediment grain size. As the suspended-sediment load (mud) increases in proportion to bed load, F decreases and the channel narrows and deepens. By these interrelated adjustments, more of the energy of the stream is expended against the banks and less against the deep bottom. The sinuosity of the channel increases, and meanders form.

Alluvial meanders are remarkably regular, and their dimensions are proportional to channel width. The radius of curvature of meanders is usually between two and three times the channel width. The wavelength of most meanders varies between 7 and 10 times the channel width. Meandering involves inherent properties of flowing water, as well as size and shape of the channel, erodibility of the stream banks, proportion of suspended and bed load, and probably other factors. In turn, meandering increases the channel length between two points and thus decreases the slope of the stream. Slope influences velocity and sediment-transporting capacity, so meanders not only are affected by other variables of stream flow but in turn affect those same variables. An analogous argument can be made for the role of the braided habit as one of the interdependent variables of hydraulic geometry.

In flume experiments, Schumm and Khan (1972) attempted to produce meandering channels by varying slope, discharge, and sediment loads. They were able to produce channels with alternating bars and pools so that the thalweg meandered, but the channel banks remained essentially straight until the stream abruptly developed midchannel bars and became braided. The sediment in the flume was poorly sorted sand. A truly meandering channel could not be formed until 3 percent by weight of kaolinite clay was mixed into the water. The clay coated and stabilized the banks and bars and allowed the thalweg to deepen and expose the stabilized sandbars on alternating sides of the channel. True meanders formed. The experiment illustrates well the action of alluvial grain size as a factor in the hydraulic geometry of rivers.

It has been noted frequently that transitional forms are rare between wide, shallow channels with a braided habit and narrower, deeper channels that meander (Ikeda and Parker, 1989). Some reaches of alluvial rivers have changed historically from braiding to meandering or back (Schumm, 1969), but the changes are abrupt. Perhaps the lack of intermediate forms is related to the fact that the minimum threshold of sediment erosion on both Hjulström's curve and the Shields curve (Fig. 9–10) is in the sand size. If the alluvium is noncohesive sand or coarser sediment, rising velocity drags it as bed load, and a wide, shallow braided channel develops. If the alluvium is cohesive and smooth, an equivalent velocity increase may erode it, but it is transported as suspended load in a deep, meandering channel.

THE CONCEPT OF GRADE IN FLUVIAL SYSTEMS

The concept of an open system that maintains itself in a steady state of most efficient configuration by internal self-regulation among variables was introduced in Chapter 5, by analogy to a rotary cement kiln. A river is an excellent example of an open system through which matter and energy flow, but within which are inherent tendencies toward self-regulation. Numerous examples of the interaction between discharge, load, channel shape, and other variables of hydraulic geometry have been cited. It is appropriate now to review the interaction of all the known variables of fluvial systems in the tendency to achieve **grade**, or long-term self-regulation, in a river channel. Although reduction of a regional landscape to low relief near sea level might take millions of years, fluvial processes might achieve mutual adjustment in much shorter time, perhaps in centuries or millenniums (Schumm, 1977, p. 12).

Stream Power

In Chapter 5, the potential power (rate of doing work) of terrestrial runoff was calculated to be $9 \times 10^9 \, kW$, equivalent to the continuous expenditure of 1 horsepower on each 3 acres of land. Water entering a river has potential energy

$$PE = mgh$$

where m is the mass of water, g is the acceleration of gravity and h is the height above sea level. As the water moves downstream, the potential energy is progressively converted to kinetic energy

$$KE = \frac{mv^2}{2}$$

The principle of conservation of energy requires that the loss of potential energy between two points on a river must be balanced by an equivalent increase in kinetic energy. However, all but a few percent of the converted energy is loss to heat by turbulent flow. The rate of kinetic energy conversion (kinetic power, Ω) per unit length of channel is

$$\Omega = \gamma Q s$$

where $\gamma = \rho g$, the specific weight of water, Q is discharge and s is the slope (Bagnold, 1977; Knighton, 1984, p. 55). The power supply/unit bed area (ω) is then

$$\omega = \gamma Q s/w = \gamma dvs = \tau_o v$$

because of the relationship $Q = wdv$, where v is the mean velocity and depth is the approximate equivalent of hydraulic radius in the formula for boundary shear stress τ_o (p. 229).

These equations relate stream power/bed area to the shear stress at the bed and determine whether the river will erode, transport its existing load, or deposit sediment on that portion of the bed. A **critical power** can be defined as that rate of useful energy conversion necessary to transport the existing load (Bull, 1979). If more power is being expended on a portion of river bed, erosion or *degradation* will result. If too much sediment is supplied, the river will *aggrade*. Continuous fluctuations by the variables of hydraulic geometry insure that no simple conversion of stream power to conditions of erosion or deposition is possible, but over an appropriate interval of time, grade may be achieved or maintained. Bull (1979) emphasized that sections of rivers that are at or near the threshold of critical power are especially sensitive to changes of climate, base level, or human interference. This point will be repeatedly illustrated in subsequent chapters.

Variables of the Graded River

At least 10 variables are involved in the tendency for a river to achieve or maintain a graded state. Not all are of equal significance, and some are not within the self-regulatory ability of the river. Leopold and Maddock (1953) divided the variables of hydraulic geometry into three classes: *independent, semindependent,* and *dependent.*

Discharge, sediment load, and ultimate base level are the three *independent* variables in the graded state. The stream has little control over these factors; rather, it must adjust to them. Discharge is determined by precipitation and evaporation in the drainage basin, the permeability of the soil, the amount and type of vegetation, and the area of the drainage basin. Only the area of the drainage basin is affected by other changes in the river system. Headward erosion of first-order tributaries can enlarge the drainage basin and thereby increase discharge, but even this process is limited, because adjacent drainage nets are probably enlarging, too. Loss of part of a network through capture by a competing system can cause an abrupt decrease in discharge, but quite early in the erosional development of a landscape, the boundaries of each river's drainage basin become defined.

Sediment load is also nearly independent of the other variables of stream flow. Many of the same climatic, soil, and biologic factors that determine discharge also determine the amount of sediment that slopes deliver to streams. The type of bedrock is an additional powerful control for sediment load. Some rocks weather quickly to sand-size particles; others produce only silt and clay. Limestone weathering produces mostly a dissolved load, with little solid detritus. Streams erode their channels and thereby have some self-regulation of their load, but we have seen that most of the load reaches the river "ready-made" by weathering and mass wasting on the hillslopes of the drainage basins and in hollows at the heads of first-order tributaries. A few terranes yield sediment so slowly that sediment load is said to be *supply limited,* or *source limited.* In most alluvial rivers, much more sediment is potentially available than can be moved, and the sediment load is then said to be *transport-limited.* When the river is competent to move the sediment, the sediment is there to be moved.

The ultimate base level of erosion is the third independent variable of stream flow. When a stream reaches the sea, it loses its identity. The potential energy of the stream is set by the altitude above sea level at which precipitation falls. Regardless of the discharge, load, or any other variable, a river that rises on a coastal plain only a few hundred feet above sea level will never be a mountain torrent.

The *semidependent* variables that interact to achieve the graded state include channel width, channel depth, bed roughness, grain size of the sediment load, velocity, and channel habit (the tendency for a stream to either meander or braid). These are semidependent inasmuch as they are partly determined by the three independent variables, but they are also partly capable of mutual self-regulation in a river. Width, depth, and mean velocity have been shown to be power functions of discharge. Competence, or grain size of the load, is a function of depth and velocity, but the mixture of grain sizes, especially the "mud ratio" M, has been shown to determine the width/depth ratio. Bed roughness is determined by the alluvial grain size, bed forms, and channel shape. The alluvial bed is built of grain sizes determined by the stream competence, yet that competence is in turn determined by such factors as velocity, channel shape and roughness, and the amount of load and discharge. In turn, bed roughness creates turbulence that affects the competence of the stream to entrain and move the load. Channel habits also interact with the other semidependent variables, as discussed in previous paragraphs.

Only one variable in the hydraulic geometry of a fluvial system, the downstream slope of the water surface, is regarded as being *dependent* on all other variables. Slope can be changed only by building up one part of the channel, cutting down another, or changing channel length as by meandering or delta building. All these changes take time, so slope is usually the final adjustment the stream makes in becoming graded. If slope could change suddenly, it would be mutually semidependent with the variables previously listed. Because it ordinarily cannot do so, it is subject to the influences of all the other known variables, and evolves on a time scale at least an order of magnitude longer than the mutual adjustments of the others.

Progressive Development of Grade

We can visualize the achievement of grade, or the graded state, in a river by first imagining an ungraded river that flows over a tectonic landscape, newly raised from the floor of the sea. Rainfall, and therefore discharge, varies over the new landscape. Water flows downhill along chance hollows and other water-collecting slopes. Lakes form in closed depressions. Drainage networks are eroded, and valleys are widened by mass wasting on slopes along the entrenching channels. Variables of the stream system are wildly out of equilibrium. Half of the potential energy of such a river might be expended in one great vertical waterfall. The load of the stream is determined by chance rock slides or mud flows anywhere along the banks. Channel width and depth are restricted by the erodibility of the rocks that are crossed by the channel. The slope of the river approximates the initial slopes of the landscape.

Rather quickly, the semidependent variables of stream flow interact to form a channel system appropriate to the work at hand. Many of the hydraulic geometry equations for stream flow apply to both graded and ungraded streams. The independent variables of discharge, load, and base level may be the same for a graded and an ungraded river. The critical factor in the achievement of the graded condition is that the stream must flow on "adjustable" materials, so that changes of one variable can produce the appropriate changes in others. Alluvium is an excellent buffer for variations in stream energy, because every fragment on the floodplain has some critical energy threshold of transportation. Having been once transported by the river, it will move again when the energy level is high enough. The establishment of an alluvium-lined channel on a continuous floodplain signals the achievement of grade in that part of a river.

Alluvium not only permits stream channels to adjust toward equilibrium; it is also a powerful absorbent for peak energy inputs into the river system. Discharge may increase by many orders of magnitude during a flood (Fig. 9–12). If the excess energy of the flood is not fully absorbed by the increased load carried into the river from headwater slope, alluvium is eroded until the ability of the river to do work is balanced by the work it is doing (Fig. 9–13). Alluvium is analogous to a chemical buffering compound that is added to a solution in great excess to ensure that some property of the solution remains constant during a reaction.

Grade, as defined by an alluvial bed on an eroding valley flat (suballuvial rock floor), is established first in the downstream segments of eroding rivers and is gradually extended upstream (Davis, 1902). Larger rivers attain grade earlier than smaller streams. A trunk stream may be at grade when its first-order tributaries are still gnawing headward into undissected slopes. A *graded reach,* or graded segment of a river, may form on easily eroded material upstream from a gorge across a resistant rock barrier. The resistant rock then forms the local or temporary base level for the graded reach. As the barrier is slowly eroded, the graded reach remains at grade, because it can adjust progressively during the slow lowering of the temporary base level.

Grade as a Thermodynamic Equilibrium

Another approach to the concept of grade is to regard the graded condition of a river from the viewpoint of theoretical thermodynamics. In any steady-state physical system through which matter and energy move, we find a tendency for the least possible work to be done and also a tendency toward a uniform distribution of work. Nature is inherently conservative in these matters. In a river system, which derives its energy from water flowing downhill, the tendency for minimum work opposes the tendency for a uniform distribution of work in several different aspects of hydraulic geometry.

The Long Profile as a Graded Form.   If all the water of a river were added at the head of a single tributary, the least-work profile would be a waterfall straight down to sea level. In humid regions, where rivers gain water downstream, the least-work profile is a curve in which the greatest loss of altitude takes place where the discharge is least, near the head of the river. The profile is very steep near the head and almost horizontal near the mouth (Fig. 9–14). Typically, in such a river, slope is almost inversely proportional to mean annual discharge ($s \propto Q_m^{-1}$).

In contrast to the sharply concave theoretical profile of least work, the theoretical profile for uniformly distributed work has a more nearly constant slope (Fig. 9–14). The profile is only slightly concave skyward, because as the rate of doing work increases downstream with discharge, the surface area of the stream bed also increases. The power expended per unit area of stream bed is constant on a river that widens downstream but decreases its slope only gradually. Typically, $s \propto Q_m^{-0.5}$.

If two tendencies toward equilibrium oppose each other, the resulting equilibrium is most likely to be a predictable intermediate condition. Langbein and Leopold (1964) concluded from their studies of river dynamics that both the downstream profiles (Fig. 9–14) and channel cross sections of rivers approach the equilibrium form predicted from the principles of least work and uniform distribution of work, *provided that the channels are in material that is adjustable.* The presence of an alluvial floodplain is again implied as a condition of grade. The long profile of a river flowing in an alluvial channel is a smooth concave-skyward profile that allows water to flow downhill to the sea with slight or negligible acceleration yet to expend only a few percent of the potential energy of the system on channel erosion or sediment transport (Bagnold, 1968).

Meanders in Graded Rivers.   The free, regular meanders of alluvial rivers have aroused curiosity and artistic senses for a long time. It seems a waste of potential energy for a river to avoid the most direct path downhill, which in plan view would be a straight channel. However, meanders, like the long profile of alluvial rivers, have been demonstrated to be the result of balance between least work and uniform distribution of work (Langbein and Leopold, 1966; Leopold and Langbein, 1966). Meanders are *sine-generated curves* that have the statistical property of minimum variance. That is, the work done on each segment of stream bank is uniform, and a minimum amount of total work is done. Furthermore, the sine-generated curve is the shape of the average, or most probable, random walk between

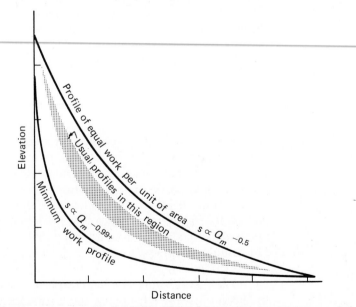

Figure 9-14.   Schematic longitudinal profiles of rivers showing tendency to balance minimum total work with maximum distribution of work. (From Langbein and Leopold, 1964, Fig. 1.)

two points. As in the random walk game described earlier in this chapter, randomness is not most likely to take the straightest path. Thus meanders represent highly probable paths of fluid flow. Meltwater streams on glaciers meander, although they may be sediment-free between ice walls. The Gulf Stream meanders up the east coast of the United States and eastward into the Atlantic, and high-altitude jet streams meander in the atmosphere.

Accordant Junctions and Least-Work Drainage Networks. Two other examples of the balance between least work and uniform distribution of work were described in an earlier section of this chapter on drainage networks. The acute angles at stream junctions were shown to be most probable, least-work configurations that minimize erosion and deposition at the junctions (Howard, 1971b). Overland flow and the ratio of basin area to first-order stream length was also shown to be a highly ordered, most-probable condition (Woldenberg, 1969). In these examples, alluvium is not required to achieve the graded condition, but as the drainage systems evolve, the junction angles and basin area change as appropriate to maintain the self-regulation.

The application of thermodynamic principles to the description of graded streams is not simple, and rivers are sufficiently complex that rigorous "laws of stream flow" may never be written, but it is satisfying to discover that graded rivers with alluvial floodplains apparently meet certain broadly defined conditions of equilibrium that are typical of the steady-state open systems maintained during carefully controlled laboratory experiments.

Summary of Grade

The most concise definition of a graded stream was carefully phrased by Mackin (1948, pp. 471, 484). The hydraulic geometry of Leopold and Maddock required only a slight transposition of phrases in Mackin's definition. It is quoted here, as modified by Leopold and Maddock (1953, p. 51), as a summary of the concept of the graded river:

> A graded river is one in which, over a period of years, slope and channel characteristics are delicately adjusted to provide, with available discharge, just the velocity required for the transportation of the load supplied from the drainage basin. The graded stream is a system in equilibrium; its diagnostic characteristic is that any change in any of the controlling factors will cause a displacement of the equilibrium in a direction that will tend to absorb the effect of the change.

A few additional comments must be made about the concept of grade. Grade is a condition, not an altitude or a certain slope angle. It develops first near the mouths of rivers and gradually extends headward. Erosional lowering of the land continues for a long time after grade is achieved, because as long as rivers carry sediment to the sea, they continue to lower the landscape over which they flow. A graded river is in a steady state only with regard to a certain time scale ("graded time" of Schumm and Lichty, 1965). Over a time scale of millions of years, typical of the time intervals in which landscapes evolve ("cycle time" of Schumm and Lichty, 1965), the potential energy of an undisturbed river system gradually approaches zero, and the rate of change of the system also decreases. The river remains at grade, but the characteristics of the graded condition change with time. The significance of the concept of grade to the life history of regional landscapes is emphasized in Chapter 12.

REFERENCES

ABRAHAMS, A. D., 1984, Channel networks: A geomorphological perspective: Water Resources Res., v. 20, pp. 161–68.

BAGNOLD, R. A., 1968, Deposition in the process of hydraulic transport: Sedimentology, v. 10, pp. 45–56.

———, 1977, Bed load transport by natural rivers: Water Resources Res., v. 13, pp. 303–12.

BAKER, V. R., and COSTA, J. E., 1987, Flood power, *in* Mayer, L., and Nash, D., eds., Catastrophic flooding: Allen & Unwin, Inc., Boston, pp. 1–21.

BAKER, V.R., and RITTER, D. F., 1975, Competence of rivers to transport coarse bedload material: Geol. Soc. America Bull., v. 86, pp. 975–78.

BEVEN, K., and GERMANN, P., 1982, Macropores and water flow in soils: Water Resources Res., v. 18, pp. 1311–25.

BLATT, H., MIDDLETON, G. V., and MURRAY, R. C., 1972, Origin of sedimentary rocks: Prentice-Hall, Inc., Englewood Cliffs, N.J., 634 pp.

BULL, W. B., 1979, Threshold of critical power in streams: Geol. Soc. America Bull., v. 90, pp. 453–64.

CARLSTON, C. W., 1968, Slope–discharge relations for eight rivers in the United States: U.S. Geol. Survey Prof. Paper 600-D, pp. D45–D47.

———, 1969a, Downstream variations in the hydraulic geometry of streams: Special emphasis on mean velocity: Am. Jour. Sci., v. 267, pp. 499–509.

———, 1969b, Longitudinal slope characteristics of rivers of the midcontinent and the Atlantic East Gulf slopes: Internat. Assoc. Sci. Hydrology Bull., v. 14, no. 4, pp. 21–31.

COTTON, C. A., 1948, Landscape, as developed by the processes of normal erosion, 2nd ed: Whitcombe and Tombs Ltd., Wellington, N.Z., 509 pp.

DAVIS, W. M., 1902, Base-level, grade and peneplain: Jour. Geology, v. 10, pp. 77–111 (reprinted 1954 *in* Geographical essays: Dover Publications, Inc., New York, pp. 249–78).

———, 1905a, Complications of the geographical cycle: Internat. Geog. Cong., 8th, Washington 1904, Repts., pp. 150–63 (reprinted 1954 *in* Geographical essays: Dover Publications, Inc., New York, pp. 279–95).

———, 1905b, Geographical cycle in an arid climate: Jour. Geology, v. 13, pp. 381–407 (reprinted 1954 *in* Geographical essays: Dover Publications, Inc., New York, pp. 296–322).

DUNNE, T., and AUBRY, B. F., 1986, Evaluation of Horton's theory of sheet wash and rill erosion on the basis of field experiments, *in* Abrahams, A. D., ed., Hillslope processes: Allen & Unwin, Inc., Boston, pp. 31–35.

GARRELS, R. M., and MACKENZIE, F. T., 1971, Evolution of sedimentary rocks: W. W. Norton & Co., Inc., New York, 397 pp.

GIBBS, R. J., 1967, Geochemistry of the Amazon River system: Geol. Soc. America Bull., v. 78, pp. 1203–32.

GREGORY, K. J., and WALLING, D. E., 1973, Drainage basin form and process: A geomorphological approach: John Wiley & Sons, Inc., New York, 456 pp.

HEY, R. D., 1988, Mathematical models of channel morphology, *in* Anderson, M. G., ed., Modelling geomorphological systems: John Wiley & Sons Ltd., Chichester, England, pp. 99–125.

HIGGINS, C. G., 1984, Piping and sapping: Development of landforms by groundwater outflow, *in* LaFleur, R. G., ed., Groundwater as a geomorphic agent: Allen & Unwin, Inc., Boston, pp. 18–58.

HJULSTRÖM, F., 1935, Studies of the morphological activity of rivers as illustrated by the River Fyris: Geol. Inst. Univ. Uppsala Bull., v. 25, pp. 221–527.

———, 1939, Transportation of detritus by moving water, *in* Trask, P. D., ed., Recent marine sediments: Am. Assoc. Petroleum Geologists, Tulsa, pp. 5–31.

HONDA, H., and FISHER, J. B., 1989, Tree branching angle: Maximizing effective leaf area: Science, v. 199, pp. 888–90.

HORTON R. E., 1945, Erosional development of streams and their drainage basins: Geol. Soc. America Bull., v. 56, pp. 275–370.

HOWARD, A. D., 1971a, Simulation of stream networks by headward growth and branching: Geog. Analysis, v. 3, pp. 29–50.

———, 1971b, Optimal angles of stream junction: Geometric, stability to capture, and minimum power criteria: Water Resources Res., v. 7, pp. 863–73.

———, and MCLANE, C. F., III, 1988, Erosion of cohesionless sediment by groundwater seepage: Water Resources Res., v. 24, pp. 1659–74.

IKEDA, S. and PARKER, G., eds., 1989, River meandering: American Geophys. Union, Water Resources Mono. 12, 485 pp.

IRELAND, H. A., 1939, "Lyell" gully, a record of a century of erosion: Jour. Geology, v. 47, pp. 47–63.

JAMES, W. R., and KRUMBEIN, W. C., 1969, Frequency distributions of stream link lengths: Jour. Geology, v. 77, pp. 544–65.

KNIGHTON, D., 1984, Fluvial forms and processes: Edward Arnold (Publishers) Ltd, London, 218 pp.

KOMAR, P. D., 1988, Sediment transport by floods, *in* Baker, V. R., Kochel, R. C., and Patton, P. C., eds., Flood geomorphology: John Wiley & Sons, Inc., New York, pp. 97–111.

LANE, E. W., 1937, Stable channels in erodible materials: Am. Soc. Civil Engineering Transactions, v. 102, pp. 123–94 (including discussion).

LANGBEIN, W. B., and LEOPOLD, L. B., 1964, Quasi-equilibrium states in channel morphology: Am. Jour. Sci., v. 262, pp. 782–94.

———, 1966, River meanders—theory of minimum variance: U.S. Geol. Survey Prof. Paper 422-H, 15 pp.

LEOPOLD, L. B., 1953, Downstream change of velocity in rivers: Am. Jour Sci., v. 251, pp. 606–24.

———, and LANGBEIN, W. B., 1962, Concept of entropy in landscape evolution: U.S. Geol. Survey Prof. Paper 500-A, 20 pp.

———, and LANGBEIN, W. B., 1966, River meanders: Sci. American, v. 214, no. 6, pp. 60–70.

———, and MADDOCK, T., 1953, Hydraulic geometry of stream channels and some physiographic implications: U.S. Geol. Survey Prof. Paper 252, 57 pp.

———, WOLMAN, M. G., and MILLER, J. P., 1964, Fluvial processes in geomorphology: W.H. Freeman and Company, San Francisco, 522 pp.

LIVINGSTONE, D. A., 1963, Chemical composition of rivers and lakes: U.S. Geol. Survey Prof. Paper 440-G, 64 pp.

MACKIN, J. H., 1948, Concept of the graded river: Geol. Soc. America Bull., v. 59, pp. 463–512.

———, 1963, Rational and empirical methods of investigation in geology, *in* Albritton, C. C., ed., The fabric of geology: Freeman, Cooper and Company, Stanford, Calif., pp. 135–63.

MANDELBROT, B. B., 1983, The fractal geometry of nature: W.H. Freeman and Company, New York, 468 pp.

MANN, C. J., 1970, Randomness in nature: Geol. Soc. America Bull., v. 81, pp. 95–104 (discussions, pp. 3185–96).

MARCUS, A., 1980, First order drainage basin morphology—definition and distribution: Earth Surface Processes, v. 5, pp. 389–98.

MCCAIG, M., 1985, Soil properties and subsurface hydrology, *in* Richards, K. S., Arnett, R. R., and Ellis, S., eds., Geomorphology and soils: George Allen & Unwin (Publishers) Ltd., London, pp. 121–40.

MORISAWA, M. E., 1959, Relation of quantitative geomorphology to stream flow in representative watersheds of the Appalachian Plateau province: Office of Naval Research Project NR 389-042, Tech. Rept. 20, 94 pp.

————, 1962, Quantitative geomorphology of some watersheds in the Appalachian Plateau: Geol. Soc. American Bull., v. 73, pp. 1025–46.

MOSS, A. J., GREEN, P., and HUTKA, J., 1982, Small channels: Their experimental formation, nature, and significance: Earth Surface Processes and Landforms, v. 7, pp. 401–15.

NEVIN, C., 1946, Competency of moving water to transport debris: Geol. Soc. America Bull., v. 57, pp. 651–74.

NOVAK, I. D., 1973, Predicting coarse sediment transport: The Hjulström curve revisited, *in* Morisawa, M., ed., Fluvial geomorphology: State Univ. of New York Publications in Geomorphology, Binghamton, N.Y., pp. 13–25.

OLTMAN, R. E., 1968, Reconnaissance investigations of the discharge and water quality of the Amazon River: U.S. Geol. Survey Circ. 552, 16 pp.

————, STERNBERG, H. O'R., AMES, F. C., and DAVIS, L. C., 1964, Amazon River investigations reconnaissance measurements of July, 1963: U.S. Geol. Survey Circ. 486, 15 pp.

PLAYFAIR, J., 1802, Illustrations of the Huttonian theory of the earth: Dover Publications, Inc., New York, 528 pp. (facsimile reprint, 1964).

RHODES, D. D., 1977, The *b–f–m* diagram: Graphical representation and interpretation of at-a-station hydraulic geometry: Am. Jour. Sci., v. 277, pp. 73–96.

————, 1987, The *b–f–m* diagram for downstream hydraulic geometry: Geograph. Annaler, v. 69A, pp. 147–61.

RICHARDS, K. S., 1982, Rivers: Form and process in alluvial channels: Methuen & Co. Ltd., London, 358 pp.

RUBEY, W. W., 1938, The force required to move particles on a stream bed: U.S. Geol. Survey Prof. Paper 189-E, pp. 120–41.

SALISBURY, N. E., 1971, Threads of inquiry in quantitative geomorphology, *in* Morisawa, M., ed., Quantitative geomorphology: Some aspects and applications: State Univ. of New York Publications in Geomorphology, Binghamton, N.Y., pp. 9–60.

SCHUMM, S. A., 1960a, The effect of sediment type on the shape and stratification of some modern fluvial deposits: Am. Jour. Sci., v. 258, pp. 117–84.

————, 1960b, Shape of alluvial channels in relation to sediment type: U.S. Geol. Survey Prof. Paper 352-B, pp. 17–30.

————, 1969, River metamorphosis: Hydraulics Div. Jour., Am. Soc. Civil Engineers Proc., Paper 6352, v. 95, no. HY 1, pp. 255–73.

————, 1977, The fluvial system: John Wiley & Sons, Inc., New York, 338 pp.

————, and KHAN, H. R., 1972, Experimental study of channel patterns: Geol. Soc. America Bull., v. 83, pp. 1755–70.

————, and LICHTY, R. W., 1965, Time, space, and causality in geomorphology: Am. Jour. Sci., v. 263, pp. 110–19.

————, MOSLEY, M. P., and WEAVER, W. E., 1987, Experimental fluvial geomorphology, John Wiley & Sons, Inc., New York, 413 pp.

SHREVE, R. L., 1966, Statistical law of stream numbers: Jour. Geology, v. 74, pp. 17–37.

————, 1967, Infinite topologically random channel networks: Jour. Geology, v. 75, pp. 178–86.

————, 1969, Stream lengths and basin areas in topologically random channel networks: Jour. Geology, v. 77, pp. 397–414.

————, 1975, Probabilistic-topologic approach to drainage-basin geomorphology: Geology, v. 3, pp. 527–29.

SMART, J. S., 1979, Determinism and randomness in fluvial geomorphology: EOS, Trans. American Geophys. Union, v. 60, pp. 651–55.

SMITH, D. D., and WISCHMEIER, W. H., 1962, Rainfall erosion: Advances in Agronomy, v. 14, pp. 109–48.

STRAHLER, A. N., 1952a, Dynamic basis of geomorphology: Geol. Soc. America Bull., v. 63, pp. 923–38.

————, 1952b, Hypsometric (area-altitude) analysis of erosional topography: Geol. Soc. America Bull., v. 63, pp. 1117–42.

————, 1954, Quantitative geomorphology of erosional landscapes: Internat. Geol. Cong., 19th, Algiers 1952, Comptes Rendus, sect. 13, fasc. 15, pp. 341–54.

————, 1958, Dimensional analysis applied to fluvially eroded landforms: Geol. Soc. America Bull., v. 69, pp. 279–300.

SUNDBORG, Å., 1956, The River Klarälven, a study of fluvial processes: Geograf. Annaler, v. 38, pp. 125–316.

TINKLER, K. J., 1981, Avoiding error when using the Manning equation: Jour. Geology, v. 90, pp. 326–28.

TOMLINSON, P. B., 1983, Tree architecture: American Scientist, v. 17, pp. 141–49.

WILLIAMS, G. P., and WOLMAN, M. G., 1984, Downstream effects of dams on alluvial rivers: U.S. Geol. Survey Prof. Paper 1286, 83 pp.

WOLDENBERG, M. J., 1969, Spatial order in fluvial systems—Horton's laws derived from mixed hexagonal hierarchies of drainage basin areas: Geol. Soc. America Bull., v. 80, pp. 97–112.

YOUNG, R. A., and WIERSMA, J. L., 1973, The role of rainfall impact in soil detachment and transport: Water Resources Res., v. 9, pp. 1629–36.

YU, B., and WOLMAN, M. G., 1987, Some dynamic aspects of river geometry: Water Resources Res., v. 23, pp. 501–9.

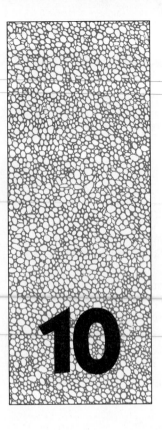

Fluvial Deposition: Processes and Landforms

10

An important group of fluvial landforms are created not by erosion but by alluvial deposition. Although constructional rather than destructional in origin (Chap. 2), alluvial landforms are so closely related to fluvial erosion that this chapter is logically part of the larger review of fluvial processes in general (Price, 1947; Russell, 1958). Schumm (1977) divided the entire fluvial system into three zones, zone 1 being the drainage basin of water and sediment production, zone 2 a zone of river transport, where the input of sediment can balance the output, and zone 3 regions of deposition on an alluvial plain, delta, or in deeper ocean water. Zones 1 and 2 were reviewed in Chapters 8 and 9. In this chapter, the processes and landforms of zone 3 are described.

For many reasons we consider alluvium to be in transit toward the sea, and therefore the landforms on alluvium must be ephemeral as compared to the erosional forms. Yet in terms of human lifetimes, or even the length of human history, landforms built of alluvium appear permanent. Written human history began more than 6000 years ago on the alluvial plains and deltas of the Middle East. Even today, an overwhelming majority of the human population lives on alluvium.

FLOODS

The water surface of a river rises during times of high discharge. At some particular discharge, depending on the channel geometry at the point of observation, a river will exceed the height of the channel banks, and overbank flow, or flooding, will occur. A **flood** is synonymous with discharge greater than bankfull capacity, although for general descriptive purposes it is common to refer to any high discharge as a flood.

Most floods have meteorological causes occasioned by anomalies of atmosphere circulation that can vary greatly in temporal and spatial scales. An hour or less of intense thunderstorm precipitation can cause flooding in small stream valleys, or a decade can separate seasons of heavy precipitation and floods such as the El Niño floods in the Peruvian coastal desert. The spatial extent of flood precipitation can range from 1 km² for a local storm to 10^6 km² for a major tropical cyclone, or hurricane (Hirschboeck, 1987, 1988).

Flood Frequency

The magnitude and frequency of meteorological floods can be estimated. River basin planning studies and the design of bridges and road networks rely upon flood frequency analyses (U.S. Water Resources Council, 1981; Chow et al., 1988). Fundamentally, the goal of flood frequency analysis is to determine, or at least estimate, the probability that various discharge levels will be exceeded in the future. Hydrologic designs commonly employ estimates of the discharge exceeded with a 1 percent probability. The flood discharge exceeded in any year with a probability of 1 percent is called the 100-year flood. In general, if a flood discharge is exceeded with a probability p, then on average that discharge level will be exceeded once every $T = 1/p$ yr. T is called the **recurrence interval** or **return period** of the flood. However, this term has the unfortunate connotation that only one T-year flood will occur in a T-year period, or that these floods occur regularly every T years. Neither connotation is correct; exceedances will actually occur at irregular randomly spaced intervals.

To establish a relationship between the frequency and magnitude of floods it is desirable to have a record at least as long as the return period of the event of interest. With such a record, analytical and graphical methods can be used to estimate the frequency with which various discharge levels are exceeded. This is often done using an **annual flood series,** a list of the largest flood that occurred in each water year within the available record. A water year is a conveniently chosen 12-month period which need not begin on January 1. In the United States, water years generally start on October 1. The annual flood series is often sorted by magnitude so that one can refer to the ith largest. It is easy to show that, on average, the probability p_i that the ith largest flood will be exceeded in the future is (Loucks et al., 1981, pp. 108–9)

$$p_i = \frac{i}{n+1}$$

Thus the ith largest flood is often assigned a return period of

$$r_i = 1/p_i = (n+1)/i$$

For example, the largest flood in a 25-year annual series would be assigned a return period of $(25 + 1)/1 = 26$ years, and the fifth greatest annual discharge would be assigned a return period of $(25 + 1)/5 = 5.2$ years.

The magnitude and return period (or alternatively, the exceedance probability) of floods in an annual series can be plotted on probability paper that can be constructed for particular families of probability distributions, such as the normal, log-normal, or Gumbel EV-1 distributions (Gumbel, 1958). Each distribution describes a different relationship between the frequency and magnitude of flood events (Chow et al., 1988; Chap. 12). From such a graph (Fig. 10–1), the exceedance probability of a flood of any intermediate magnitude can be estimated by interpolation. The relative magnitudes of floods in an annual series should plot approximately as a straight line if the type of paper corresponds to the distribution from

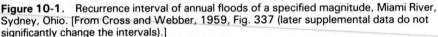

Figure 10-1. Recurrence interval of annual floods of a specified magnitude, Miami River, Sydney, Ohio. [From Cross and Webber, 1959, Fig. 337 (later supplemental data do not significantly change the intervals).]

which the flood series was drawn (Loucks et al., 1981). However, sampling fluctuations often result in larger deviations at the low and high ends of the graph. A graphic solution is usually considered reliable for events with recurrence intervals of up to one-half the length of the annual series.

The *median flood* has an exceedance probability of 50 percent, corresponding to a return period of 2 yrs. Thus the sample median magnitude for the 44-year flood series in Fig. 10–1 would be the average of the floods ranked 22nd and 23rd in the series. For the Gumbel distribution, the *most probable annual flood* has an exceedance probability of 63 percent, corresponding to a return period of 1.58 years; for the normal distribution, the most probable flood is the same as the median flood.

An alternative analytical procedure ranks all discharges above a selected magnitude or threshold. This approach can include more than one flood peak in some water years, and none in others. A minimum time duration for each event is commonly specified. Such a record is called a **base-stage series** or **partial-duration series.** When modeling flood frequency employing partial duration series, one can model both the frequency with which the specified threshold is exceeded and the likely magnitude of such exceedances. Alternatively, one can directly estimate the arrival rate (or frequency in events/yr) of all floods exceeding specified thresholds.

If the arrival rate of floods exceeding a threshold discharge is υ events per year, then for statistically independent events the annual probability p with which that discharge is exceeded is $p = 1 - \exp[-\upsilon]$. For υ on the order of 0.05 events/yr or less, the annual exceedance probability p essentially equals the exceedance frequency υ. The two approaches are often considered to be equivalent (Langbein, 1949).

Dury hypothesized, and demonstrated for several examples, that in graded rivers the most probable annual flood, with a return period of 1.58 years, is similar in magnitude to bankfull discharge (Dury et al., 1963). This would imply that bankfull discharge corresponds to a flood level equaled or exceeded once per year in a partial duration series ($\upsilon = 1$ event/yr). Dury (1967) concluded that bankfull discharge is the dominant channel-shaping flow and is likely to be reached or exceeded once per year on average by streams in alluvial channels. Overall the adjustments of natural channels to both large floods and smaller discharges is a complex compromise between the frequency of different flow rates, their duration, the river's competence, the bed and bank material, and many other factors (Richards, 1982, pp. 122–144; Brakenridge, 1988).

Catastrophic Floods and Paleofloods

Graphs of flood-frequency relationships may include one or more major floods of far greater discharge than would be predicted from the magnitude of lesser discharges. The 1913 flood in Fig. 10–1 is such an example. Even larger catastrophic events can be due either to truly unusual and anomalous meteorological events within the present climatic setting, to relict former climatic conditions such as glaciation, or to "accidents" such as the failure of a landslide or glacier dam that temporarily impounded great volumes of water (Schuster, 1986; Schuster and Costa, 1986; Mayer and Nash, 1987). Such an event, even if it occurred thousands of years earlier, may have been the principle event that shaped the present-day river channel and valley (p. 228). Many, perhaps most, rivers in glaciated regions have observable channel and valley forms that were shaped by catastrophic paleofloods from lakes that had been temporarily impounded by glaciers (Kehew and Lord, 1986; Lord and Kehew, 1987; Teller, 1987).

Any lake must be regarded as a temporary feature in the fluvial system. It is a highly efficient sediment trap and thus tends to fill with deltaic alluvium. At the same time, the local and temporary base level that causes the lake is certain to be eroded, so the long profile of the river channel tends to become smoothly concave and eliminate the lake basin by erosion.

Lakes may be caused by tectonic movements, volcanism, glacial erosion and deposition, landslides, aggradation by a trunk stream, or a host of other causes. Rivers that are temporarily dammed by landslides create large lakes that break out with catastrophic severity usually within a few days or a few months (Schuster, 1986; Costa and Schuster, 1988).

The greatest floods known to have occurred on earth are the Pleistocene Lake Missoula paleofloods that repeatedly broke through a glacial lobe in eastern Washington and gouged thousands of square kilometers of anastomosing river channels across the Columbia Plateau to the Columbia River (Fig. 11–1) (Bretz, 1969; Baker, 1973; Waitt, 1985). Another great flood resulted from the overflow of pluvial (p. 453) Lake Bonneville into the Snake River (Malde, 1968; Jarrett and Malde, 1987). These great floods occurred more than 12,000 years ago, but their eroded channels and enormous alluvial deposits are the dominant features of the present landscape on the Columbia Plateau and the Snake River Plain (Baker et al., 1987). The maximum discharge of the Lake Missoula floods (Baker et al., 1987, p. 426) was at least $17.5 \pm 2.5 \times 10^6$ m^3/s ($= 618 \times 10^6$ ft^3/s; compare with Fig. 9–9). An alluvial fan built by the Lake Bonneville flood includes well-rounded basalt boulders 3 m in diameter (Malde, 1968, p. 13). A considerable extrapolation of Hjulström's curve (Fig. 9–10a) is required to estimate the flood velocity that could move such boulders. One estimate is that the peak discharge through a constricted reach of the Snake River canyon was about 1 million m^3/s, for a duration of 8 weeks (Jarrett and Malde, 1987). Midchannel alluvial bars as much as 100 m thick, 2 km long, and 1 km wide were built of boulder gravel. Postflood weathering and erosion have been insignificant in modifying the shape of the flood deposits. Accounts of these great natural catastrophes should be read by all geomorphologists as a basis for speculation and contemplation on the role of infrequent but spectacular events in the shaping of "normal" landscapes.

FLOODPLAINS AND ALLUVIUM

Hydrologic Significance of Floodplains

Several of the semidependent variables of hydraulic geometry (Chap. 9) change abruptly when flood stage is reached. Most obvious are the changes of width and depth with discharge. Prior to overbank flooding, with increasing discharge at a station, depth typically increases more rapidly than width (p. 223). However, when overbank flooding begins, width abruptly increases manyfold, so that depth and velocity need not increase at all in order that $Q = wdv$. In the power functions for width, depth, and velocity during flood

$$w = aQ^b, \; d = cQ^f, \; v = kQ^m,$$

the exponents $b \simeq 1.0$, $f \simeq 0.0$, and $m \simeq 0.0$. With those values of b, f, and m, the width of the flooded plain increases in linear proportion to discharge, and if the floodplain is very wide relative to the normal channel, depth and velocity need not increase at all. Actually, depth continues to increase slightly, but the mean velocity of a river in flood decreases to less than the mean velocity at bankfull stage, because the broad, shallow sheet of water over the floodplain has a very low velocity that reduces the average of the entire cross section.

The slope along a meandering channel is always less than the down-valley slope of the floodplain because of the greater length of the meandering channel over the straight-line distance along the floodplain. Thus, rivers in flood have a down-valley surface gradient that is several times as steep as when the river is confined in a meandering channel. This increased slope enables the floodwaters to move down the valley efficiently, even though the mean flood velocity need not be as high as in channeled flow—thus the peculiarity that rivers in flood flow on an increased average down-valley slope but at decreased mean velocity. The greater width of the floodplain more than compensates for the loss of velocity. Mean current velocity in the flood stage is approximately constant along the entire length of rivers (p. 226).

The sediment-transporting competence and capacity of a river also change abruptly in flood. Although an enormously greater sediment load is in transit (Fig. 9–12) along the axis of the permanent channel, the shallow, low-velocity sheet of floodwater over the adjacent floodplain has little competence. As velocity is checked, sand, silt, and clay may settle out of the floodwater, and the shallowness over the floodplain ensures that the fine sediment will reach the bottom of the water column before turbulence sweeps it downstream. People who live on flood plains know that a single flood may deposit a meter or more of muddy alluvium on their fields or in their houses. It may seem anomalous at first reading that a river in flood is flowing down a steeper gradient but may be less competent to transport sand and mud. Yet these interactions of the several semidependent variables of hydraulic geometry of rivers in flood illustrate well the fundamental role of channels and floodplains: to provide a low-gradient, efficient, sinuous or braided channel for average discharges and loads, and a steep, shallow, broad, constant-mean-velocity channel for overbank flood discharges. Truly, floodplains are for rivers. People occupy floodplains only at great risk, but judging from historical patterns, it is a risk that many find tolerable or inevitable.

Downstream Changes in Alluvium Grain Size

Coarse alluvium decreases exponentially in mean grain size downstream by sorting and abrasion. The complex interplay among bed roughness, depth, velocity, slope, and grain size was discussed in Chapter 9. Size sorting by progressive loss of competence downstream is the most important of several causes for decreasing mean grain size in alluvium. A given grain size simply cannot be transported downstream beyond the point where the stream is competent to move it.

Abrasion also reduces the grain size of bed-load alluvium in a downstream direction, although abrasion is not a factor in reducing the grain size of the suspended load. In 260 km of transport down the Colorado River of Texas below Austin, granitic pebbles and cobbles from a known source area diminish in mean diameter by about 50 percent, quartz by about 30 percent, and chert by about 20 percent (Bradley, 1970, p. 68). The glacial outwash (p. 426) of the Knik River in Alaska shows a more dramatic decrease in grain size downstream (Bradley et al., 1972). Cobbles and boulders that are readily moved by periodic floods from an ice-dammed lake are reduced in size by 87 percent (to 13 percent of initial mean diameter) in only 26 km of river transport. This rapid reduction is consistent with many other observations of the rapid downstream change in grain size of glacial outwash.

An attempt was made to verify both sets of field measurements just cited by abrading gravel in laboratory conditions. Selected stones were dragged over the roughened bed of a toroidal abrasion tank by a current of water until they had traveled a distance equivalent to the downstream movement in the Colorado and Knik Rivers. Fresh granitic pebbles from the Colorado were reduced by only 10 percent of mean grain size in the tank in contrast to their 50 percent reduction in an equivalent field transport distance. Knik River gravel was reduced by only 8 percent in 26 km of laboratory transport in contrast to the 87 percent reduction in the field.

The authors of these studies concluded that sorting rather than abrasion is the major cause of the rapid downstream decrease in grain size in an aggrading glacial outwash stream bed. Sorting was also judged to be important to the downstream change of grain size in the Colorado River alluvium, but even more important in that river is the abrasion of gravel that weathers slightly during temporary storage in the floodplain. Although fresh pebbles and cobbles were not notably abraded in the tank experiments, stones with a surficial weathered layer were abraded much more rapidly. A third factor in the downstream reduction of bed-load alluvium is the chipping and abrasion of cobbles as they shift and vibrate under current velocities that are not strong enough to transport them. Conceivably, cobbles could be abraded and rounded by vibration "in place" until they become small enough to be transported. The greater size reduction observed in actual rivers as opposed to flume experiments may be due in part to this kind of abrasion without net transport (Schumm and Stevens, 1972).

It can be concluded that sorting, abrasion, and weathering during temporary storage all contribute to the observed decrease in mean grain size of bed-load alluvium in a downstream direction. The abraded and weathered detritus is also transported but as dissolved and suspended load. As a very crude guess, the average alluvial fragment might take 1000 years to move downstream to the sea, most of that time being at rest (and weathering) in some part of the floodplain (Leopold et al., 1964, p. 328).

Active and Relict Alluvium

In the geologic record, alluvial conglomerates and sandstones are well known, but they comprise only a small fraction of the total volume of sedimentary rocks. Alluvial deposits, ancient and modern, are studied in greater proportion than their abundance would indicate, because they may contain valuable placers of heavy ores, precious metals, or gems. Furthermore, terrestrial vertebrate fossils and the artifacts of ancient people are most commonly found in late Cenozoic alluvium. Quaternary tectonism and climatic fluctuations have made alluvium a complex material to interpret. The distinction between alluvium actively "in transit" and "relict" is especially critical in areas of former glaciation, where glacial outwash gravel or sand underlies much finer recent alluvium. If the modern stream is not competent to move all grain sizes in the outwash, then properly speaking, the outwash is not the alluvium of the present stream but rather the substrate on which the modern alluvium is deposited. The Mississippi River illustrates that interpretive problem well (Fig. 10–2 and p. 451). In full-glacial time, the lower river entrenched a deep valley to the Gulf of Mexico. As the ice margin retreated from the upper midwestern states, sand and gravel alluvium of a braided Mississippi River filled the lower trench. Later, when the ice margin was fringed by proglacial lakes in the present Great Lakes region, the coarse alluvium dropped out far upstream, and only mud and fine sand went down the Mississippi. By that stage, rising sea level was causing swampy deltaic deposition in Louisiana. In the last few thousand years, with sea level rising only slowly and with the modern delta prograding across the continental shelf, mud and organic deposition predominate on the floodplain. How much of the cross section in Figure 10–2 should be called alluvium, or what qualifications to the term are required to explain the entire cross section?

As a working definition, we can restrict the definition of "active" alluvium to that sediment capable of being transported by the stream that now flows on it. Such a definition relates to both the depth of the alluvium and its grain size. The thickness of the modern alluvium is then defined by the flood-scour depth of the modern channel, and its maximum grain size by the maximum competence of the modern stream. Anything deeper or coarser is "relict" alluvium. The depth of modern channel scour should not be underestimated, however. Todd (1900) cited soundings of the bed of the Missouri River during high discharge that proved the channel to be 15 to 30 m deep in flood where it is only 3 to 15 m deep at low water (compare Fig. 9–13). The flood-eroded channel was floored by bedrock, thus proving that the entire alluvial thickness is potentially in transit. The great vertical exaggeration of Figure 10–2 shows that the modern Mississippi channel is as deep as the uppermost stratigraphic unit of braided-stream and backswamp deposits, which thereby become classed as the modern alluvium.

Floodplain Morphology

Floodplains and their minor morphologic subdivisions are primarily depositional landforms. As such, their forms are genetically related to specific depositional processes (Fig. 10–3; Table 10–1). Not all deposits create landforms, however. The lag deposits shown beneath the channel in Figure 10–3 might well be relict alluvium, for instance.

Lateral Accretion.   Floodplains are built in two fundamental ways, by *lateral accretion* and by *vertical accretion.* **Point bars** (Fig. 10–3) are the most important component of lateral accretion. As a meandering channel migrates across the floodplain, the **cut bank** on the convex side is undercut and eroded. Piping by groundwater flow out of the alluvium during falling stages of river discharge is an important factor in bank collapse and mobilization of alluvium. As banks collapse, the derived bed load is carried a short distance downstream and deposited as a submerged bar, usually on the same side of the stream. The result is a cross-stratified deposit, with a subdued relief of low ridges and intervening swales, that may record many episodes of meandering channel migration (Figs. 10–4 and 10–5). The crests of point bars approach the level of the former floodplain on the cut-bank side unless the stream is actively entrenching or aggrading, so the level of the floodplain need not change significantly by lateral accretion. In streams that carry relatively

Figure 10-2. Stages in the late-glacial and postglacial history of the Mississippi River valley near Natchez, Mississippi. (Fisk, 1947, Plate 5.)

1 Late Glacial Entrenched Stage - Sea Level 400 ft. lower than present. Valley slope 0.83 ft. per mile. Master stream overloaded, braided, carrying gravels to Gulf of Mexico.

2 Valley Aggradation Stage 1 - Sea Level 100 ft. lower than present. Valley slope approx. 0.75 ft. per mile. Master stream overloaded, braided, carrying sands and fine gravels as far downstream as present Gulf shoreline.

3 Valley Aggradation Stage 2 - Sea Level 20 ft. lower than present. Valley slope 0.68 ft. per mile. Final period of overloaded braided stream. Mississippi River carrying silts and sands. Thick deposits of backswamp silts and clays building up in marginal lowlands.

4 Valley Aggradation Stage 3 - Sea Level at present elevation. Construction of Cocodrie Meander Belt. Valley slope approx. 0.60 ft. per mile. Meandering deep-channel stream with great stage variation gradually replaced braided shallow-channel constantly flooding stream.

5 Valley Aggradation Stage 4 - Mississippi River occupying Walnut Bayou Meander Belt. Diversion from Cocodrie Meander Belt brought river into floodbasin west of Cocodrie Meander Belt Ridge. Backswamp deposits destroyed within zone of river migration.

6 Present Stage-Mississippi, diverted from Walnut Bayou Meander Belt west of Vicksburg, established present course in lowland east of the Cocodrie Meander Belt Ridge.

253

Figure 10-3. Typical associations of valley sediments (S. C. Happ, *in* Vanoni, 1971, Fig. 2-Q-1). A, alluvial fan; B, backland; BS, backswamp; C, colluvium; F, channel fill; L, lag deposit; LA, lateral accretion; N, natural levee; P, point bar; S, splay; T, transitory bar; VA, vertical accretion. See Table 10-1 for sediment characteristics.

coarse alluvium and lack natural levees, lateral accretion may account for 80 to 90 percent of the floodplain deposition (Wolman and Leopold, 1957, p. 96).

The age of point-bar deposits along the Little Missouri River was determined by Everitt (1968) by tree-ring chronology of cottonwood trees growing on various parts of the floodplain. An age map of the valley floor (Fig. 10–5) shows that only relatively small areas of floodplain are older than 200 years. Everitt concluded that the reach of the river he studied redistributes every century a volume of sediment equal to the total present volume of alluvium along that reach, yet the floodplain has probably been within 1 m of its present elevation for at least 150 years. Even

TABLE 10–1

Genetic Classifications of Valley Sediments[a]

Place of Deposition	Name	Characteristics
	Transitory channel deposits	Primarily bed load temporarily at rest; part may be preserved in more durable channel fills or later accretions.
Channel	Lag deposits	Segregations of larger or heavier particles, more persistent than transitory channel deposits, and including heavy mineral placers.
	Channel fills	Accumulations in abandoned or aggrading channel segments; ranging from relatively coarse bed load to fine-grained oxbow lake deposits.
Channel margin	Lateral accretion deposits	Point and marginal bars that may be preserved by channel shifting and added to overbank floodplain by vertical accretion deposits at top.
Overbank flood plain	Vertical accretion deposits	Fine-grained sediment deposited from suspended load of overbank flood-water; including natural levee and backland (backswamp) deposits.
	Splays	Local accumulations of bed-load materials, spread from channels onto adjacent floodplains.
Valley margin	Colluvium	Deposits derived chiefly from unconcentrated slope wash and soil creep on adjacent valley sides.
	Mass movement deposits	Earth flow, debris avalanche, and landslide deposits commonly intermix with marginal colluvium; mud flows usually follow channels but also spill overbank.

Source: S. C. Happ, *in* Vanoni, 1971.
[a] Compare with Figure 10–3.

(a) (b)

(c)

Figure 10-4. Floodplain and channel of the Little Missouri River near Watford City, North Dakota. For orientation and scale, see Figure 10-5. (a) 1939 vertical aerial photograph; (b) 1949 vertical aerial photograph; (c) 1958 vertical aerial photograph of the northernmost meander. (Photos: U.S. Dept. Agriculture, from Everitt, 1968.)

though a floodplain has retained a stable form for centuries, the alluvium that composes it is intermittently moving through the system, weathering while it is in storage and being abraded as it moves in the river. Sigafoos (1964) gave an excellent summary of the technique of using floodplain trees to document flooding and floodplain deposition on a time scale of about a century.

Vertical Accretion. Large rivers with gentle gradients and a predominantly suspended load typically have meandering channels on broad floodplains. When

Figure 10-5. Age map of the floodplain of the Little Missouri River near Watford City, North Dakota, based on ages of cottonwood trees. (Everitt, 1968.)

they flood, the velocity in the overbank water may be very low, and the floodwaters may be weeks in receding back into the channel. Under these conditions, flood-plain deposition is largely by *vertical accretion.* The deposits are usually sand or mud, although splays from breached channel banks may bring coarse sediment. From April to June 1973 the Mississippi River flooded large areas of its lower floodplain to depths of up to 4 m. During the two-month interval, vertical accretion averaged 53 cm along natural levees but decreased rapidly to less than 2 cm beyond 400 m from the natural levees and averaged only slightly more than 1 cm in the backswamps (Kesel et al., 1974). The abrupt loss of velocity at the edge of the flooded channel and the abundant sediment supply there commonly cause more and coarser deposition on the natural levees, which grade laterally into finer back-swamp deposits. A result of vertically accreting natural levees may be that the river surface within the levees is many meters above the level of adjacent backswamps. Under these conditions, a flood level high enough to overtop a segment of natural levee has a pronounced gradient into the backswamp, and floods are violent and severe. The current at a point of overflow erodes a gap or **crevasse** in the levee, and a **splay** spreads far across the backswamp (Fig. 10–3). Sometimes the erosion is sufficient to begin an entirely new master channel, and a segment of the former channel is abandoned, to fill slowly with channel deposits and peat along a curved **oxbow lake.** If a meander recurves until it intersects an upstream portion of the channel, a **neck cutoff** occurs, wherein the abrupt shortening of channel length locally increases the gradient and velocity in the cutoff, and eventually seals the truncated ends of the abandoned meander with bank deposits or **plugs.**

If a master stream flows between high natural levees, few tributaries can join it but must flow parallel to the trunk stream along the lowest axis of the backswamp. In older American terminology, such streams were called **Yazoo tributaries,** from the Yazoo River in Mississippi. A Yazoo tributary may be forced to extend downstream by the migrating meander belt of the trunk stream, but eventually it will be intersected by a meander and will be shortened again.

Vertical accretion cannot continue indefinitely unless the base level is rising. The postglacial rise of sea level has accentuated vertical accretion and the formation of natural levees and backswamps during the later history of the lower Mississippi River (Fig. 10–2). The striking regular frequency of the most probable annual flood on most upland rivers suggests that vertical accretion is not a major contributor to floodplain development; otherwise, rivers with larger floodplains should flood progressively less frequently. We may suspect that thick alluvium deposited by vertical accretion, such as the relict alluvium found beneath large rivers in their lower valleys (Fig. 10–2), is the result of postglacial rise of sea level and would not be typical of hypothetical rivers graded to a stable sea level. Thin vertical-accretion alluvium may accumulate slowly on a floodplain if the river is constrained from migrating laterally (Ritter et al., 1973).

DELTAS AND ALLUVIAL FANS

Two of the most common depositional landforms of alluvium result from the same cause: an abrupt loss of competence in a stream. *Alluvial fans* are subaerial forms, deposited at an abrupt decrease in velocity where a stream channel loses water by infiltration or widens abruptly. *Deltas* are mostly subaqueous forms, although their upper surfaces are usually slightly above water level. They result when a river enters a lake or the ocean and its velocity is abruptly reduced. The two forms may merge; many deltas are capped by low-gradient alluvial fans. The "delta" of the Colorado River into the Gulf of California has separated the Salton Basin from the upper end of the Gulf, and its north side is the apex of a sea-level alluvial fan 70 m above the floor of the closed, below-sea-level Salton Basin.

Deltas

A level surface of water, whether a lake or the sea, has a powerful trapping effect on the suspended and bed loads of a river. A river enters standing water as a jet that maintains its central velocity for a distance equal to about four times the width of the river mouth. Then the flow begins to disperse by marginal turbulence and the mixing effects of wind, waves, and currents. Because of density differences produced by salinity, temperature, and sediment load, the river water and sediment may (a) sink to the bottom and flow downslope as a dense *turbidity current,* (b) flow directly forward into equally dense still water and mix turbulently within a distance equal to a few times the channel width, or (c) spread forward over the surface of the standing water (Bates, 1953). The third option is the most common where rivers enter the ocean because the suspended sediment load can only rarely increase the fluid density to equal the density of seawater. Eventually, velocity is sharply reduced, and competence drops. All of the bed load, and most of the suspended load, is dropped close to the point where a stream enters a lake or the sea. The bed load is dropped quickly as velocity is checked and usually forms a shallow arcuate **lunate bar** across the river mouth. To maintain cross-sectional area, the channel commonly breaks into multiple **distributary channels** across the accumulating alluvium and distributes new sediment over a broad arc concentric to the original river mouth. The resulting landform is a **delta.**

Deltas may be the subject of the oldest geomorphic description. Herodotus (fifth century B.C.) observed the growth of the Nile Delta by the annual floods and correctly inferred that all of lower Egypt was the delta of the Nile. The name "delta" comes from the Greek capital letter delta, Δ, which is the shape of the Nile delta (Fig. 10–6).

Not all deltas have the same shape (Fig. 10–7). The shape is influenced by river discharge and load, grain size, flood frequency and intensity, climate, vegetation, wave energy, tidal range, salinity, water temperature, and tectonics (Morgan, 1970). Some deltas have shifting sublobes that mark separate periods of growth (Fig. 10–8). Others have been fairly stable in postglacial time, gradually prograding

Figure 10-6.   The Nile Delta, Suez Canal, and the Sinai Peninsula, viewed toward the southeast. Rosetta and Damietta distributary mouths are prominent. (Photo: NASA Technology Application Center, Gemini IV S-65-32752.)

(a)

(b)

(c)

Figure 10-7.   Some common delta forms. (a) Digitate (''birdfoot'') delta of the Mississippi River. Another common form, the bay-head delta, is shown in Figure 3-15. (b) Cuspate (concave seaward) Tiber Delta. See also the Rosetta and Damietta mouths of the Nile River (Fig. 10-6). (c) Lobate (convex seaward) Niger Delta.

Figure 10-8. Chronology of the deltas that compose the Mississippi delta plain (from Coleman, 1988, Fig. 2). The modern or Balize digitate delta is shown in Figure 10-7(a).

1. Maringouin/Sale Cypremort 7500–5000 yrs. B.P.
2. Teche 5500–3800 yrs. B.P.
3. St. Bernard 4000–2000 yrs. B.P.
4. Lafourche 2500–800 yrs. B.P.
5. Balize 1000–present yrs. B.P.
6. Atchafalaya 50–present yrs. B.P.

(building forward) along their entire seaward perimeter through a maze of distributaries (Wright,1985). Deltas are almost entirely Holocene landforms.

The surface shape of deltas can be broadly classified into (a) fluvial- or river-dominated, (b) wave-dominated, and (c) tidal-dominated (Galloway and Hobday, 1983, pp. 90–113; Coleman et al., 1986). Very large rivers extending into sheltered or shallow gulfs or lakes build *digitate* or *bird-foot* deltas, with long, projecting natural levees little modified by waves or currents. The Mississippi delta building across a very shallow continental shelf into the nearly tideless Gulf of Mexico is an exceptional example of a river-dominated delta (Fig. 10–7a) (Coleman, 1988). If waves and currents are strong enough to erode the river-mouth bars and levees, the sand is carried laterally to form barrier beaches (Chap. 19) and the delta becomes cuspate or broadly rounded and smooth in a map view (Fig. 10–7b). The Nile delta is wave-dominated, with barrier beaches enclosing a series of lagoons along its margin, and small cuspate deltas at the active Damietta and Rosetta distributary mouths (Fig. 10–6). The Niger delta is also wave-dominated, although tidal currents are significant. It has an arcuate shoreline of barrier beaches enclosing a maze of mangrove-covered distributary channels, none of which projects seaward (Fig. 10–7c). In seas with high tidal range, the ebb and flow of the tides may distribute sediment rapidly offshore, and the distributary mouths become trumpet shaped, opening broadly to the sea between marsh islands or sand bars. Most of the nation of Bangladesh occupies the vast tidal-dominated merged deltas of the Ganges and Brahmaputra Rivers (Coleman et al., 1986, pp. 326–29).

The cross section of a simple or small delta into a lake or reservoir of water with density equal to the density of the river discharge shows three basic depositional units: *bottomset beds, foreset beds,* and *topset beds* (Gilbert, 1890). Only the third is subaerial. A surface gradient must be maintained to carry alluvium to the perimeter, and aggradation may extend upstream as a delta grows forward and reduces the river gradient. The *Gilbert-type* delta is relatively rare, restricted to freshwater lakes and reservoirs, although widely described in introductory texts as the most common type of delta.

The Mississippi River has one of the largest deltas and certainly the most studied one (Morgan, 1970; Coleman, 1988; Penland et al., 1988). It has built forward to the edge of the continental shelf after having filled a deeply entrenched glacial-age valley (Fig. 10–2). The internal structure consists of linear, branching channel sands flanked by paired natural levees of poorly sorted sand and mud, which in turn grade laterally into highly organic backswamp mud. The entire delta complex, called a **delta plain** (Fig. 10–8), resembles a pile of maple leaves, with the veins represented by channel and levee deposits and the intervein areas represented by backswamps. When a set of channels is abandoned, compaction and subsidence lower the area until only the drowned levee crests are exposed. Simultaneously, waves erode the delta front and concentrate the sand as barrier spits and islands. These reworked shallow marine sediments of the erosion phases that follow the abandonment of a distributary system contribute up to 50 percent of the total sediment thickness in the Mississippi River delta plain. The geologic record of such shallow-water deltas could be largely of marine, rather than fluvial, sediments (Penland et al., 1988). Some later generations of distributary channels reoccupy segments of earlier drowned levee systems, in a few instances flowing in the opposite direction. Gradients are negligible throughout the modern delta.

Location of Major Deltas

On a relief map of the world, many of the largest rivers are seen to rise within orogenic belts but then flow across continents to the trailing margins where they build their deltas (Inman and Nordstrom, 1971; Audley–Charles et al., 1977; Potter, 1978). The Chang Jiang (Yangtze), Amazon, Mississippi, and Rhine Rivers are representative of this pattern. Others, like the Ganges, Brahmaputra, Yukon, and Mackenzie Rivers, flow parallel to the orogenic grain, either within the orogen or along one of its borders. A few rivers rise near a rifted continental margin, drain toward the interior of the continent, and then turn and find their way through the highlands of the rifted margin to the sea, often along ancient transverse fault valleys, as in the case of the Niger. It is uncommon to find large river deltas on a converging continental margin, for the local orogenic relief is too young and active for a large river system to have evolved. Deltas are usually part of the coastal plain and continental shelf environments on the trailing borders of continents. Because deltaic sediments and sedimentary rocks are important sources of petroleum, modern deltas are studied carefully to understand the depositional environments of ancient sediments (Galloway and Hobday, 1983; Whateley and Pickering, 1989).

Alluvial Fans

In its simplest classic form, an **alluvial fan** is a segment of a low cone with its apex at the mouth of a gorge or canyon (Fig. 10–9). Contours are arcs of circles concentric on the head of the fan. In longitudinal or radial profile, fans are slightly concave skyward. They are excellent examples of water-spreading wash slopes (Fig. 8-20).

An alluvial fan usually forms where a high-gradient stream leaves a confining gorge and discharges directly onto the floodplain of a trunk stream or the flat floor of an intermontane basin. A simple abrupt decrease in downstream gradient is not an adequate explanation for alluvial fans. More important is the abrupt widening

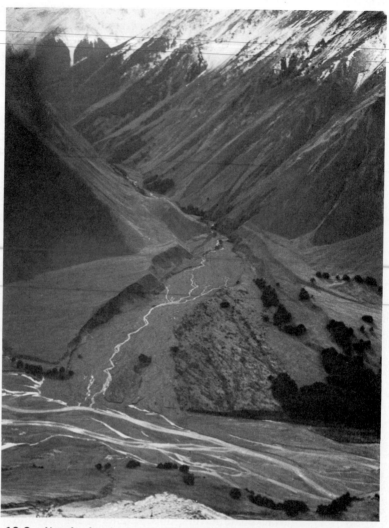

Figure 10-9.   New fan forming on a braided flood plain in front of an erosionally truncated earlier fan at the mouth of Centre Creek, Avoka River valley, Canterbury, New Zealand. (Photo: C. O'Loughlin, Ministery of Forestry, New Zealand.)

and shoaling of the channel cross section. Deposition results from the abrupt loss of competence at the gorge mouth through great widening, infiltration, loss of depth, and in deserts, evaporation.

Fans are normally built of bed-load alluvium, especially gravel. Stratification may be sheetlike parallel to the fan surface or may contain many cut-and-fill structures of buried channels. Their surface morphology is characterized by radiating, branching distributary channels, most often braided rather than meandering. The permeable character of fan alluvium allows considerable infiltration, so that fans may contain mudflow deposits recording the progressive loss of water after the suspended load became too thick to flow as a muddy stream.

The slope of an alluvial fan increases with increased grain size and concentration of the sediment load, and decreases with increased discharge (Hooke and Rohrer, 1979). Therefore, fans in arid regions, where the sediment load is coarse and concentrated, tend to be steeper than fans in humid regions, where water discharge is larger and more continuous and sediment loads are constrained by vegetation. If the sediment load is less than the capacity and competence of a fan stream, the stream will entrench the upper fan and transport the sediment out to the margin of the fan, thereby reducing the stream gradient. Stable channels are possible if the capacity and competence of the fluvial system are balanced by the sediment load. On actively growing fans, the sediment is dropped at the apex of the fan, filling preexisting channels and perhaps spreading over the fan surface until the steeper gradient permits transport. Fan-building streams typically have extreme variations in both discharge and load, so fans commonly show complex surface morphology (French, 1987, pp. 170–72).

Fans are *aggradational,* or built-up, forms. One radial channel may carry most of the discharge for a time until its gradient becomes oversteep by deposition, when the stream abruptly shifts to a lower-gradient radius. The process ensures the uniform slope development along all radii and a resulting smooth conical form. The Kosi River of the Ganges Plain in northern India has an exceptionally low-gradient alluvial fan at the base of the Himalaya Mountains. It is 154 km long and 143 km wide (Fig. 10–10). Repeated surveys in the past 228 years record a continuous westward shifting of the Kosi River for 113 km across its fan, in a series of episodic diversions of up to 19 km/yr (Wells and Dorr, 1987). Major westward diversions do not correlate with major earthquakes in the region, but seem to record a persistent tendency of the river when it abandons its channel to migrate to the previously unoccupied fan surface rather than shift eastward to a channel recently abandoned.

Many underfit trunk streams of glaciated valleys have low gradients that cannot transport the coarse alluvium delivered by valley-side tributaries, and the mas-

Figure 10-10.   Map of Kosi drainage system, to resolution limit of about 80 m, constructed from March 1977 dry-season false-color satellite imagery. Dates for occupation of channels 1 to 12 are 1731, 1770–1775, 1807–1839, 1840–1873, 1873–1893, 1893–1921, 1921–1926, 1926–1930, 1930–1936, 1936–1942, 1942–1948?, and 1977 respectively. (For sources, see Wells and Dorr, 1987, Fig. 1.)

ter stream is pushed laterally from one valley wall to the other by fans. Tributary fans may block the trunk stream and cause a lake. Very steep alluvial fans are called **alluvial cones** and are comparable in form, if not in internal structure, to talus cones. Alluvial fans are especially abundant and significant land forms in arid intermontane regions (Figs. 13-2 and 13–3), but are common in humid regions as well.

Piedmont Alluvial Plains

Large regional landforms develop from the lateral merging of adjacent low-gradient alluvial fans at the base of dissected mountains. Most of the Great Plains province of the central United States is veneered by an apron of late Cenozoic sand and gravel named the Ogallala Formation (Frye, 1971; Seni, 1980; Gustavson and Winkler, 1988), a region 1300 km long southward from South Dakota to Texas and 500 km wide (Fig. 10–11). Most of the alluvium is sandy, but gravel channel deposits and fine-grained vertical-accretion deposits are also known. The Ogallala Formation is as much as 150 m thick along the axes of buried valleys. Some combination of tectonic upwarp of the Rocky Mountains, and perhaps climatic change as well, caused the rivers of the Great Plains to aggrade until they entirely filled their valleys and coalesced across former interfluves. Eolian sand and loess (Chap. 14) blew off the braided floodplains and blanketed adjacent uplands, indicating that the climate was semiarid, at least in the southern part of the Ogallala depositional re-

Figure 10-11.   Generalized distribution of the Ogallala Formation in the Great Plains province of the United States. (Frye, 1971, Fig. 1.)

gion (Gustavson and Winkler, 1988). The modern Kosi fan of India has been suggested as a model for the environment of the Ogallala Formation deposition (Fig. 10–10). Pleistocene climatic changes and renewed tectonism probably ended the aggradation phase (Stanley and Wayne, 1972). The present relief of the unglaciated Great Plains is essentially the depositional surface of the Ogallala Formation and similar, but slightly younger, Pleistocene alluvium. The rivers that now flow east to the Mississippi are at grade on this ancient alluvial surface. The regional slope is only about 0.1 percent.

Similar piedmont alluvial plains form the Argentine Patagonian plains on the eastern side of the Andes Mountains, the Indo-Gangetic Plain on the southern foreland of the Himalaya Mountains, and the Canterbury Plains of New Zealand. Late Cenozoic tectonism has played an important role in initiating these and other great alluvial plains. Climatic changes during the Quaternary Period, including glaciation in the mountains, repeatedly caused the rivers to change their hydraulic geometry, however, and much of the alluvium is probably relict.

AGGRADATION IN THE FLUVIAL CYCLE

Aggradation as a Normal Phase of Valley Evolution

The primary function of river systems is to carry away weathered debris fed into them in great quantity by mass wasting and overland flow. If any potential energy is available beyond that required to move the load, the river will erode its channel. The deduced normal cross section of a river valley thus should have an erosional rock surface, the **valley flat,** beneath alluvium in transit or temporarily stored in the floodplain. Examples of relict alluvium have previously been described in this chapter, where aggradation has been caused by rising base level, climatic change, or tectonism. All three of these factors have been so important in shaping the present landscape that we are hard pressed to find a river that has a "normal" floodplain underlain by alluvium in transit to the sea and that is slowly eroding a valley flat beneath the alluvium. For example, compare the discussion of alluvium and grade (p. 239) with Figure 10–2.

In view of the abundant accidental events that can cause aggradation, it is appropriate to consider whether an aggradational phase can occur during the normal evolution of a fluvial landscape. A trunk stream might become graded early in the dissection history, while headward erosion is still in progress by first-order streams. Then, as the drainage network expands, greater discharge and load might require the longitudinal profile to be steepened by aggradation in the middle portion of the river system. When the uplands are totally dissected by competing drainage networks, no additional expansion can occur except by capture, and presumably the trunk stream will resume its slow downcutting (Davis, 1899, reprinted 1954, p. 260). Alluvial terraces might be left along the valley walls to record an aggradational phase that was a normal part of the continuous regional erosional evolution.

Cotton (1948, pp. 80–82) considered other possible causes of aggradation as a temporary phase of normal fluvial landscape development. Channel lengthening by both meandering and delta progradation causes a decrease of gradient, but the principles of hydraulic geometry suggest that aggradation need not result. Neither does it seem likely that the increased channel size of a graded river or subsurface flow through alluvium could cause a loss of competence and result in aggradation (Cotton, 1948, p. 81). The interaction of the many semidependent variables of the fluvial system is much too well balanced to permit such a disequilibrium to develop. Thus the only likely cause for an aggradational phase during fluvial dissection of a landscape, barring climatic and tectonic "accidents," seems to be the development of a graded trunk stream while drainage networks are still expanding. It should be recalled that most rivers are transport-limited, rather than supply-limited (p. 238).

Climatic Change and Aggradation

Rivers in alluvial channels on the Great Plains have been observed to change from a braiding to a meandering habit in response to decade-length climatic changes (p. 235). A generalization about the response of fluvial systems in semiarid or sub-humid regions to slight shifts in climate was set forth by Huntington (1907, p. 358) and is now called *Huntington's principle* (Fairbridge, 1968, p. 1125). Huntington postulated that increased aridity in areas of marginal rainfall leads to a loss of vege-tative cover and increased mass wasting (Fig. 12–7). Stream channels become choked with the excessive loads supplied to them concurrent with a reduction in discharge. Aggradation results as rivers build up their alluvial beds until gradients are steep enough to move the greater load of often larger grain size. Huntington's enunciation of this principle of fluvial response to increased aridity was condi-tioned by his observations in the interior of Asia, where decreased total rainfall often coincides with increased seasonality of precipitation. This is often true, but seasonality need not be incorporated into the principle as a requirement. Clearly, climates with an intense dry season followed by concentrated precipitation are con-ducive to a maximum sediment load in streams (Chap. 13).

Huntington's principle further postulates that during times of greater precipi-tation, vegetation will stabilize landscapes, inhibit mass wasting, promote infiltra-tion, and result in more sustained flow of rivers. Peak flood discharges will be minimized, and rivers will erode channels of more gentle longitudinal gradients. Most geomorphologists would mentally associate a braided habit with the aggrada-tional phase and meandering with the more humid valley-deepening phase. Hun-tington suggested that the great gravel terraces that line the valley of central Asia represent multiple cycles of arid or semiarid aggradation alternating with entrench-ment during more humid intervals. The implied general correlation of humid and arid intervals with Quaternary glacial and interglacial episodes is obvious, but the current evidence suggests that in some climatic zones the wetter intervals were asso-ciated with high sea levels of interglacial ages, and the arid aggradational phases are recorded by coarse alluvium graded to low sea levels that are correlative with glacial ages (pp. 457–59). In other regions, especially the presently arid southwestern United States, glacial ages were the more humid, or pluvial, ages as inferred by Huntington. Huntington's principle, with allowance for important regional varia-tion, is an important key to relating Quaternary climatic change to landscape evo-lution (Chap. 18).

Fluvial Terraces

Most floodplains have a microrelief of a few meters or less, including low *flood-plain terraces* eroded during floods of varying severity. The conditions of fluvial deposition during variable discharge, by either meandering or braided rivers, ensure such minor floodplain terraces. However, minor low terraces are not likely to be preserved for more than a few centuries before channel migration removes them (Fig. 10–5).

If a river is slowly downcutting as its channel meanders laterally from one side of the valley to the other, portions of older floodplain or valley flat may be pre-served as **unpaired terraces** (Fig. 10–12), especially if they are *rock-defended* by a

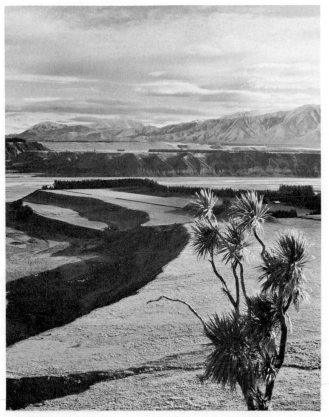

Figure 10-12.   Terraces of the lower Rakaia River valley, Canterbury, New Zealand. Two high terraces in the distance are aggradational glacial outwash surfaces. Terraces in the foreground are unpaired erosional terraces, cut in glacial deposits. (Photo: National Publicity Studios, New Zealand.)

buttress of the valley wall. Unpaired terraces need not represent a change in the rate or nature of fluvial processes. They are little more than accidentally preserved fragments of former floodplains. Long ago, Playfair (1802, p. 103) cited such terraces as strong proof that valleys are indeed the work of the streams that flow in them.

Paired terraces, which occupy the same level on opposite valley walls, are significant evidence that a river has cut downward in an intermittent fashion. Like unpaired terraces, paired terraces may be remnants of alluvium or of the former valley flat cut into rock. The distinction in the nature of terrace material is useful for descriptive purposes but does not have much genetic significance, because most rock-cut terraces retain at least a veneer of alluvium. Paired alluvial terraces may represent climate-controlled aggradation alternating with downcutting, as inferred by Huntington, or simply pulses of entrenchment alternating with times when the valley flat was veneered with alluvium in transit. Even though a former floodplain was built by aggradation, it does not become a terrace except by subsequent erosion. It is important to remember that *fluvial terraces are cut, not built, by rivers* (Cotton, 1948, p. 187).

Flights of terraces on a weak-rock lowland may end downstream at a gorge through resistant rock that is the local and temporary base level for the upstream graded reach (p. 239). Because of slow downcutting at the resistant structure, the river upstream has time to open out a broad terraced valley. Such terraces are a special class of rock-defended terrace, where a single resistant obstruction controls and preserves multiple terraces (Cotton, 1948, p. 197). They are normally unpaired.

Thus far, only the cross-valley profile of fluvial terraces has been considered. The longitudinal profiles of terraces should also be studied for clues to river history. For instance, terraces may converge or diverge downstream. Convergence suggests progressive rejuvenation of the headwaters, perhaps by continuing tectonic uplift (Fig. 3–23). Divergence suggests a progressive lowering of base level, either temporary or ultimate, more rapidly than the regional erosion rate. Changes in the amount and grain size of the alluvium carried by a river when it flowed at each successive terrace level may also explain differences in longitudinal slope of terraces.

Fluvial terraces on a single river system may be misinterpreted if the regional setting is ignored. Ritter (1967, 1972) described examples of rivers at the foot of the Beartooth Mountains in Montana. Rivers that rise in the mountain range have relatively steep gradients to carry coarse alluvium, whereas rivers that rise on the plains carry finer alluvium and have lower gradients. Therefore, even though two adjacent rivers drain to the same base level, near the mountain front the "mountain bred" stream is on an alluvial fan at a higher level than, and is likely to be captured by, the adjacent inter-fan stream that heads at the mountain front. However, the low-gradient inter-fan river cannot carry the coarse alluvium, and when capture occurs, an active aggradational phase begins, while the beheaded stream on the abandoned fan adjacent to it begins downcutting and forming terraces (see also p. 340).

Valleys that now head in the mountains and also those that now originate at the mountain front have prominent terraces of coarse alluvium derived from the Beartooth Range, proving that capture has caused alternating episodes of aggradation and entrenchment, with terrace formation in each valley. Although a single valley would seem to record a complex history that might be given a tectonic or climatic explanation, the larger view demonstrates normal or "single-cycle" regional geomorphic development. The regional history of the Beartooth Range terraces is complicated further by outwash terraces from former valley glaciation.

REFERENCES

AUDLEY-CHARLES, M. G., CURRAY, J.R., and EVANS, G., 1977, Location of major deltas: Geology, v. 5, pp. 341–44.

BAKER, V. R., 1973, Paleohydrology and sedimentology of Lake Missoula flooding in eastern Washington: Geol. Soc. America Spec. Paper 144, 79 pp.

————, GREELEY, R., KOMAR, P. D., SWANSON, D. A., and WAITT, R. B., JR., 1987, Columbia and Snake River plains, in Graf, W. L., ed., Geomorphic systems of North America: Geol. Soc. America, Centennial Special Vol. 2, pp. 403–68.

BATES, C. C., 1953, Rational theory of delta formation: Am. Assoc. Petroleum Geologists Bull., v. 37, pp. 2119–62.

BRADLEY, W. C., 1970, Effect of weathering on abrasion of granitic gravel, Colorado River (Texas): Geol. Soc. America Bull., v. 81, pp. 61–80.

————, FAHNESTOCK, R. K., and ROWEKAMP, E. T., 1972, Coarse sediment transport by flood flows on Knik River, Alaska: Geol. Soc. America Bull., v. 83, pp. 1261–84.

BRAKENRIDGE, G. R., 1988, River flood regime and floodplain stratigraphy, in Baker, V. R., Kochel, R. C., and Patton, P. C., eds., Flood geomorphology: John Wiley & Sons, Inc., pp. 139–56.

BRETZ, J. H., 1969, The Lake Missoula floods and the channeled scablands: Jour. Geology, v. 77, pp. 503–43.

CHOW, V. T., MAIDMENT, D. R., and MAYS, L. W., 1988, Applied hydrology: McGraw-Hill, Inc., New York, 572 pp.

COLEMAN, J. M., 1988, Dynamic changes and processes in the Mississippi delta: Geol. Soc. America Bull., v. 100, pp. 999–1015.

————, ROBERTS, H. H., and HUH, O.K., 1986, Deltaic landforms, *in* Short, N. M., and Blair, R. W., Jr., eds., Geomorphology from space: A global overview of regional landforms: NASA publication SP-486, pp. 317–52.

COSTA, J. E., and SCHUSTER, R. L., 1988, The formation and failure of natural dams: Geol. Soc. America Bull., v. 100, pp. 1054–68.

COTTON, C. A., 1948, Landscape as developed by the processes of normal erosion: Whitcombe and Tombs Ltd., Wellington, New Zealand, 509 pp.

CROSS, W. P., and WEBBER, E. E., 1959, Floods in Ohio, magnitude and frequency: Ohio Department of Natural Resources, Division of Water, Bull. 32, 325 pp.

DAVIS, W. M., 1899, The geographical cycle: Geog. Jour., v. 14, pp. 481–504 (reprinted 1954 *in* Geographical essays: Dover Publications, Inc., New York, pp. 249–78).

DURY, G. H., 1967, Bankfull discharge and the magnitude-frequency series: Australian Jour. Sci., v. 30, p. 371.

————, HAILS, J. R., and ROBBIE, M. B., 1963, Bankfull discharge and the magnitude-frequency series: Australian Jour. Sci., v. 26, pp. 123–24.

EVERITT, B. L., 1968, Use of the cottonwood in an investigation of the recent history of a flood plain: Am. Jour. Sci., v. 266, pp. 417–39.

FAIRBRIDGE, R. W., 1968, Terraces, fluvial-environmental controls, *in* Fairbridge, R. W., ed., Encyclopedia of geomorphology: Reinhold Book Corp., New York, pp. 1124–38.

FISK, H. N., 1947, Fine-grained alluvial deposits and their effects on Mississippi River activity: U.S. Army Corps of Engineers, Mississippi River Commission, Vicksburg, Miss., v. 1, 82 pp., v. 2, 74 plates.

FRENCH, R. H., 1987, Hydraulic processes on alluvial fans: Developments in water science, 31, Elsevier Science Publishers B. V., Amsterdam, 244 pp.

FRYE, J. C., 1971, The Ogallala Formation—a review, *in* Ogallala Aquifer Symposium: Texas Tech. Univ. Internat. Center for Arid and Semi-arid Land Studies Spec. Rept. no. 39, pp. 5–14a.

GALLOWAY, W. E., and HOBDAY, D. K., 1983, Terrigenous clastic depositional systems: Springer-Verlag New York, Inc., New York, 423 pp.

GILBERT, G. K., 1890, Lake Bonneville: U.S. Geol. Survey Mon. 1, 438 pp.

GUMBEL, E. J., 1958, Statistical theory of floods and droughts: Inst. Water Engineers Jour., v. 12, pp. 157–84.

GUSTAVSON, T. C., and WINKLER, D. A.,1988, Depositional facies of the Miocene–Pliocene Ogallala Formation, northwestern Texas and eastern New Mexico: Geology, v. 16, pp. 203–6.

HIRSCHBOECK, K. K., 1987, Catastrophic flooding and atmospheric circulation anomalies, *in* Mayer, L., and Nash, D., Catastrophic flooding: Allen & Unwin, Inc., Boston, pp. 25–56.

————. 1988, Flood hydroclimatology, *in* Baker, V. R., Kochel, R. C., and Patton, P. C., 1988, Flood geomorphology: John Wiley & Sons, Inc., New York, pp. 27–49.

HOOKE, R. L., and ROHRER, W. L., 1979, Geometry of alluvial fans: Effect of discharge and sediment size: Earth Surface Processes, v. 4, pp. 147–66.

HUNTINGTON, E., 1907, Some characteristics of the glacial period in non-glaciated regions: Geol. Soc. America Bull., v. 18, pp. 351–88.

INMAN, D. L., and NORDSTROM, C. E., 1971, On the tectonic and morphologic classification of coasts: Jour. Geology, v., 79, pp. 1–21.

JARRETT, R. D., and MALDE, H. E., 1987, Paleodischarge of the late Pleistocene Bonneville Flood, Snake River, Idaho, computed from new evidence: Geol. Soc. America Bull., v. 99, pp. 127–34.

KEHEW, A. E., and LORD, M. L., 1986, Origin and large-scale erosional features of glacial-lake spillways in the northern Great Plains: Geol. Soc. America Bull., v. 97, pp. 162–77.

KESEL, R. H., DUNNE, K. C., McDONALD, R. C., ALLISON, K. R., and SPICER, B. E., 1974, Lateral erosion and overbank deposition on the Mississippi River in Louisiana caused by 1973 flooding: Geology, v. 2, pp. 461–64.

LANGBEIN, W. B., 1949, Annual floods and the partial-duration series: Am. Geophys. Un., Trans., v. 30, pp. 879–81.

LEOPOLD, L. B., WOLMAN, M. G., and MILLER, J. P., 1964, Fluvial processes in geomorphology: W.H. Freeman and Company, San Francisco, 522 pp.

LORD, M. L., and KEHEW, A. E., 1987, Sedimentology and paleohydrology of glacial-lake outburst deposits in southeastern Saskatchewan and northwestern North Dakota: Geol. Soc. America Bull., v. 99, pp. 663–73.

LOUCKS, D. P., STEDINGER, J. R., and HAITH, D. A., 1981, Water resources systems planning and analysis: Prentice-Hall, Inc., Englewood Cliffs, New Jersey, 559 pp.

MALDE, H. E., 1968, The catastrophic late Pleistocene Bonneville flood in the Snake River Plain, Idaho: U.S. Geol. Survey Prof. Paper 596, 52 pp.

MAYER, L., and NASH, D., 1987, Catastrophic flooding: Allen & Unwin, Inc., Boston, 410 pp.

MORGAN, J. P., 1970, Depositional processes and products in the deltaic environment, *in* Morgan, J. P., ed., Deltaic sedimentation, modern and ancient: Soc. Econ. Paleontologists and Mineralogists Spec. Pub. 15, pp. 31–47.

PENLAND, S., BOYD, R., and SUTER, J. R., 1988, Transgressive depositional systems of the Mississippi delta plain: A model for barrier shoreline and shelf sand development: Jour. Sed. Petrol., v. 58, pp. 932–49.

PLAYFAIR, J., 1802, Illustrations of the Huttonian theory of the earth: Dover Publications, Inc., New York, 528 pp. (facsimile reprint, 1964).

POTTER, P. E., 1978, Significance and origin of big rivers: Jour. Geology, v. 86, pp. 13–33.

PRICE, W. A., 1947, Geomorphology of depositional surfaces: Am. Assoc. Petroleum Geologists Bull., v. 31, pp. 1784–1800.

RICHARDS, K., 1982, Rivers: Form and process in alluvial channels: Methuen & Co. Ltd., London, 358 pp.

RITTER, D. F., 1967, Terrace development along the front of the Beartooth Mountains, southern Montana: Geol. Soc. America Bull., v. 78, pp. 467–83.

————, 1972, Significance of stream capture in the evolution of a piedmont region, southern Montana: Zeitschr. für Geomorph., v. 16, pp. 83–92.

————, KINSEY, W. F., and KAUFFMAN, M. E., 1973, Overbank sedimentation in the Delaware River Valley during the last 6000 years: Science, v. 179, pp. 374–75.

RUSSELL, R. J., 1958, Geological geomorphology: Geol. Soc. America Bull, v. 69, pp. 1–21.

SCHUMM, S. A., 1977, The fluvial system: John Wiley & Sons, Inc., New York, 338 pp.

————, and STEVENS, M. A., 1972, Abrasion in place: A mechanism for rounding and size reduction of coarse sediment in rivers: Geology, v. 1, pp. 37–40.

SCHUSTER, R. L., 1986, Landslide dams: Processes, risk, and mitigation: Am. Soc. Civil Engineers Geotechnical Spec. Pub. no. 3, 164 pp.

————, and COSTA, J. E., 1986, A perspective on landslide dams, *in* Schuster, R. L., ed., Landslide dams: Processes, risk, and mitigation: Amer. Soc. Civil Engineers Geotechnical Spec. Pub. no. 3, pp. 1–20.

SENI, S. J., 1980, Sand-body geometry and depositional systems, Ogallala Formation, Texas: Texas Bureau of Economic Geology, Rpt. of Invest. no. 105, 36 pp.

SIGAFOOS, R. S., 1964, Botanical evidence of floods and floodplain deposition: U.S. Geol. Survey Prof. Paper 485-A, 35 pp.

STANLEY, K. G., and WAYNE, W. J., 1972, Epeirogenic and climatic controls of early Pleistocene fluvial sediment dispersal in Nebraska: Geol. Soc. America Bull., v. 83, pp. 3675–90.

TELLER, J. T., 1987, Proglacial lakes and the southern margin of the Laurentide ice sheet, *in* Ruddiman, W. F., and Wright, H. E., Jr., eds., North America and adjacent oceans during the last deglaciation: Geol. Soc. America, The geology of North America, v. K-3, pp. 39–69.

TODD, J. E., 1900, River action phenomena: Geol. Soc. America Bull., v. 12, pp. 486–90.

U. S. WATER RESOURCES COUNCIL (now called Interagency Advisory Committee on Water Data), 1981, Guidelines for determining flood frequency: U. S. Geol. Survey Office of Water Data Coordination Bull. 17B, 28 pp.

VANONI, V. A., 1971, Sediment transportation mechanics: Q. Genetic classification of valley sediment deposits: Hydraulics Div. Jour., Am. Soc. Civil Engineers Proc., Paper 7815, no. HYI, pp. 43–53.

WAITT, R. B., JR., 1985, Case for periodic, colossal jokulhlaups from Pleistocene glacial Lake Missoula: Geol. Soc. America Bull., v. 96, pp. 1271–86.

WELLS, N. A., and DORR, J. A., JR., 1987, Shifting of the Kosi River, northern India: Geology, v. 15, pp. 204–207.

WHATELEY, M. K. G., and PICKERING, K. T., 1989, Deltas: Sites and traps for fossil fuels: Geological Society (London) Spec. Pub. no. 41, 360 pp.

WOLMAN, M. G., and LEOPOLD, L. B., 1957, River flood plains: Some observations on their formation: U.S. Geol. Surv. Prof. Paper 282-C, pp. 86–109.

WRIGHT, L. D., 1985, River deltas, in Davis, R. A., Jr., Coastal sedimentary environments, second revised, expanded edition: Springer-Verlag New York Inc., New York, pp. 1–76.

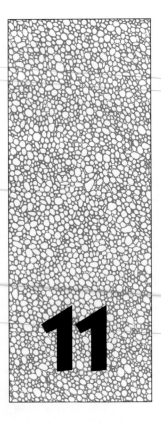

11

Structural Control
of Fluvial Erosion

INTRODUCTION

No geomorphologist doubts the importance of rock structure to the shape of erosional landforms. *Structure,* along with *process* and *time,* is a cornerstone of explanatory description (Chap. 1). Structural factors of rock weathering were discussed in Chapters 6 and 7. This chapter is concerned with the effects on fluvial erosion of macroscopic or large-scale rock structure, including stratification and other primary layering, foliation and similar metamorphic fabrics, joints, faults, and folds.

The preceding two chapters have emphasized processes and depositional forms, so it has not been necessary to consider structural control of fluvial processes except incidentally. Drainage networks were assumed to develop over homogeneous terranes in order to simplify the study of fluvial processes. However, on the real earth, rivers cross belts or zones of contrasting structures, and their hydraulic geometries change in response to the erodibility of each structure and the type of load it delivers to the river.

The present subaerial landscape is nothing more than an instantaneous and arbitrary interface between atmosphere and lithosphere, marking a transitory phase of the erosion process. Erosion has probably removed more than 10 km from many continental landscapes and as much as 30 km from some mountain ranges (p. 120). During such tremendous amounts of erosion, whole structural terranes can be removed. For instance, coastal plain sediments can be stripped off as streams entrench into the older rocks beneath. Folded and faulted sedimentary structures can be stripped from basement complexes of igneous and metamorphic rocks. The relief that has been buried under unconformities and thrust sheets, perhaps part of ancient landscapes, can be exhumed.

During the progress of erosion, it is inevitable that some structures will be more resistant to destruction than others. Relative erodibility, or rock resistance, is a subjective property, not simply defined. For instance, gravel alluvium may survive on very old river terraces, not because it became strongly cemented or lithified but because it is so permeable that rainfall percolates rapidly through it and no surface streams can form to erode it (Rich, 1911). Emerged coral reefs on tropical coasts (Fig. 2–3) may have their relief preserved or even enhanced, because the permeable reef rock is lowered only slightly while solution lowers the adjacent less permeable lagoon-floor carbonate sand and mud. In both examples, passive resistance is more effective than strength.

A loose distinction can be made between the sedimentary *cover strata* of continents and the igneous and metamorphic *basement* rocks that underlie them. Some cover strata, for example thick-bedded quartz-rich sandstones, are significantly less erodible than many igneous and metamorphic rocks. More commonly, the cover strata are more erodible and are easily stripped off the basement rocks by fluvial, glacial, and eolian processes. In the Pampean Ranges of Argentina (p. 47), crystalline basement rocks have been truncated by a very flat surface during several intervals of erosion since late Paleozoic time, when South America was part of the Gondwana supercontinent. Subsequently, the erosion surface was buried by no more than 2.8 to 4.2 km of poorly consolidated continental clastic sedimentary strata ranging in age from late Paleozoic to late Cenozoic. In the last 10 million years, thrust faulting associated with the Andean orogeny has raised fault-block mountains of basement rocks to heights of 1 to 6 km above sea level (Jordan and Allmendinger, 1986; Jordan et al., 1989). The easily eroded cover strata have been stripped off, yet the erosion surface truncating the basement rocks has barely been modified in the process (Fig. 3–13). Its exhumation with negligible dissection is a result of the extreme erodibility contrast between the cover strata and basement rocks. A less extreme erodibility contrast is to be seen in the eastern United States along the **Fall Zone,** a seaward-dipping exhumed erosion surface between crystalline rocks of the Piedmont geomorphic province and the overlying coastal plain sediments of Cretaceous and younger ages.

FRACTURES AS STRUCTURAL CONTROLS

Joints

Systematic or nonsystematic joint sets cut most indurated rocks. Joints are probably the most common rock structures, breaking igneous, metamorphic, and sedimentary rocks in all kinds of orogenic and epeirogenic settings (Pollard and Aydin, 1988), They control weathering, mass wasting, and erosion on a wide range of scales, from microcracks (Chap. 6) to regional lineations that help characterize entire geomorphic provinces (Fig. 11–6). Some are the result of contraction on cooling (Fig. 11–1). Others are sheeting joints, the result of pressure release (Fig. 6–3). Still others are due to regional tectonic stresses (Fig. 11–2). Whatever their origin, joints provide openings for groundwater movement, and weathering and mass wasting are intensified. Joints that open parallel to gorges and steep-sided valleys probably aid in enlarging and perpetuating the valleys.

During weathering, joint-bounded blocks become progressively more rounded. **Corestones** of fresh rock, initially surrounded by weathered spheroidal exfoliation shells (Fig. 11–3), may actually weather free to become loose boulders on the surface or in regolith. It has been proposed that the large boulders of crystalline rocks that are found in glacial drift are not necessarily quarried free and rounded by glacial erosion but might be corestones from a preglacial or interglacial weathering profile (Feininger, 1971).

Masses of jointed bare rock, stacked like well-worn childrens' blocks, are called **tors** (Fig. 11–4). They are not transported but result from differential weathering and erosion along joints, at least one set of which is subhorizontal. Tors may

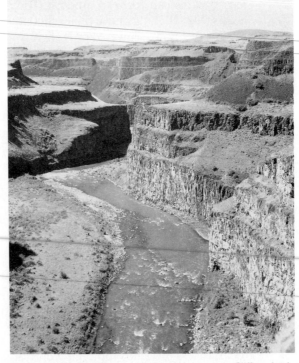

Figure 11-1. Downstream from Palouse Falls, Washington. Cliff-and-talus canyon walls are controlled primarily by vertical columnar joints in multiple lava flows. (Photo: F. O. Jones, U.S. Geological Survey.)

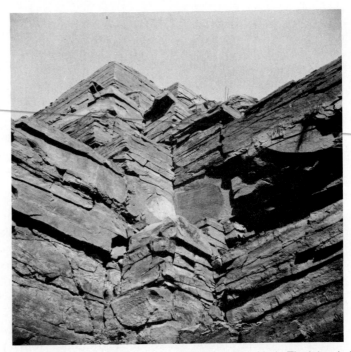

Figure 11-2. Jointed sandstone and shale near Ithaca, New York. The joints belong to two vertical sets, one striking east-northeast and the other north-northwest. The joint sets are systematically related to fold axes in the Appalachian Mountains, which in this region strike east-northeast (Engelder and Geiser, 1980).

Figure 11-3. Corestones of fresh quartz diorite surrounded by concentric shells of partly decomposed rock. Sheeting joints exert some control on weathering patterns. Roadcut 35 km northeast of Medellin, Colombia. (Photo: Tomas Feininger.)

form at depth by joint-controlled weathering within the regolith, or they may form subaerially by differential weathering and erosion of structures (Wahrhaftig, 1965; Eggler et al., 1969). Both the tropical savanna (p. 343) and periglacial (p. 370) climates have been claimed to be especially favorable for tors (Thomas, 1968). A regional landscape of schist tors in New Zealand has been described and illustrated by

Figure 11-4. Tors in a tropical savanna landscape, Townsville, northern Queensland, Australia. See also Figures 13-8 and 13-9.

Figure 11-5. Two classes of fault-line scarp. In both examples, the stipulation is made that no tectonic movement has occurred in the present erosion cycle. (a) Resequent fault-line scarp. The former extent of a post-faulting, but dissected, lava flow has been added to suggest that evidence can sometimes be found to prove that such a landform is a fault-line scarp rather than a fault scarp. (b) Obsequent fault-line scarp. At a still later stage of erosion, the scarp might become resequent if bed 1 resists erosion on the front half of the block and beds 2 and 3 are eroded from the rear half.

Ward (1952), Turner (1952), and McCraw (1965). Tors are of interest to geomorphologists because they indicate that structurally controlled subsurface weathering may precondition the shape of a landscape long before the subaerial surface is lowered by erosion to expose the differentially weathered terrane (Fig. 13–9).

Fault-Line Scarps

The distinction between *fault scarps,* of tectonic origin, and **fault-line scarps,** formed by structurally controlled erosion at a fault contact, has been called "the indispensable prerequisite to a correct interpretation of geological history" (p. 39). The distinction is not based on the observable morphology of a scarp but on full understanding of the regional geology. Any scarp, whether formed by recent tectonic movement or differential erosion of unlike rock types, is likely to be a facet or surface at an appropriate angle of repose.

Two genetic classes of fault-line scarps are distinguished (Fig. 11–5). **Obsequent** ("opposite to the consequent") scarps are unambiguous, because the geologic relationships prove that movement on the fault would have produced a scarp facing in the opposite direction. **Resequent** ("renewed consequent") fault-line scarps are difficult to distinguish from tectonic fault scarps, because in both, the upthrown fault block is also the high side of the scarp. Geomorphic criteria alone cannot distinguish the two.

Differential erosion along faults does not necessarily produce scarps. In a faulted terrane of massive igneous or metamorphic rock, only the *crushed zone* along the faults may be subject to more rapid weathering and erosion. On aerial photographs, crystalline terranes show a characteristic network of differentially eroded zones on fault lines and joints (Figs. 3–13 and 11–6).

A confused or chaotic landscape is likely to result from structurally controlled erosion of tectonic **mélanges.** In many regions of the Mediterranean and circum-Pacific tectonic belts, great masses or blocks of rigid igneous, metamorphic, and

Figure 11-6. Satellite image of central New York. Water is white on this image. Lake Oneida, left center. Mohawk River drains east (right) across the center. Metamorphic terrane of the southwest Adirondacks, upper right, shows strong fault-line and joint control of weathering and erosion. (NASA ERTS E-1080-15180.)

sedimentary rocks, some several kilometers in length, are embedded in a more plastic matrix of shale or low-grade, fine-grained metamorphic rock (Hsü, 1968). The tectonic interpretation of mélanges is a matter of debate, but they probably are formed by thrust faulting and submarine slumping in an orogenic marine sedimentational environment (Raymond, 1984; Cowan, 1985). Little has been written about the geomorphology of mélange terranes. Hsu (1971) described a mélange landscape in Cyprus and compared it with parts of eastern Taiwan. In both regions, resistant blocks of graywacke, chert, ultrabasic igneous rock, or limestone form hills or boulder-strewn ridges surrounded by claystone lowlands. Maxwell (1974, p. 1201) described the mélange topography of northern California as "accidental, characterized by scattered blocks jutting out of a rounded or land-slid surface." The mélange landscape in the Napa Valley, California, is "hummocky topography, with resistant blocks protruding above the surface throughout the terrane. Where streams have cut deeply into it, blocks in profusion crown knobs high above the stream valleys, jut out from valley walls, and stick up from stream bottoms, commonly deflecting the streams around them. Blocks range from pebble-sized or even smaller to mountain-sized" (Phipps, 1984, p. 107). The mélange structure may extend to a depth of 10 km. A unique geomorphic aspect of mélange terrane is that the hills of resistant rock have no "roots," and no prediction of sequential landscape evolution can be made. Although not strictly an example of fault-line erosional scarps, the abrupt structural boundaries of mélange hills do not seem to fit into any other available descriptive category.

DIFFERENTIAL FLUVIAL EROSION ON LAYERED ROCKS

Most sedimentary rocks, and many igneous and metamorphic rocks, have layers of contrasting erodibility. The layering may be the result of sedimentary stratification, igneous intrusion, metamorphic differentiation, or a host of other causes. Weathering emphasizes even slight contrasts in lithology, but on the larger scale, belts of distinctive erosional topography form on the exposed edges of rock layers. Where layers are horizontal, their exposed edges, or *outcrops,* are parallel to contours and form altitudinally zoned structural landforms. Where layers are tilted, much more complex geometry develops. Much of the geomorphic diversity of landscapes is due to the structural control of layered rocks, and an extensive descriptive and explanatory terminology has evolved. Much of the terminology is shared with the field of structural geology, because geomorphic form is one important way of recognizing structures in the field.

Layered Igneous and Metamorphic Rocks

Volcanic plateaus are built of piles of lava and tephra sheets and intrusive sills crossed by numerous dikes. Monogenetic or rapidly sequential eruptions may cover very large areas with fresh flows of lava or ignimbrite, which then cool as a unit. The interior of the cooling unit looses heat both upward and downward and as it cools, it contracts and fractures into a crudely hexagonal pattern of columnar joints (p. 74). However, the upper and lower marginal zones usually cool more rapidly and shatter into unsystematic blocks of glassy or vesicular rock. Columnar jointing provides high-angle sets of closely spaced joints that promote rock falls, cliffs, and taluses. The glassy or shattered marginal zones commonly form taluses separating joint-controlled cliffs (Fig. 11–1). The spectacular effects of the Missoula and Bonneville floods (p. 250) were in large part structurally controlled by the plateau basalts of the region (Baker, 1973; Baker et al., 1987).

Metamorphic terranes commonly have zones or layers of more resistant rock alternating with more erodible belts. A typical example would be quartz-rich gneiss alternating with biotite schist. Weathering and erosion more rapidly lower the schistose belts, and the resistant layers are etched out. The scale may be measured in meters, or as in the coastal region of Maine, in strike ridges of resistant gneiss 100 m or more in height and as much as 50 km in length (Fig. 11–7). Metamorphic terranes have a dominant regional structural grain. Hills have their long axes on strike. Streams follow erodible belts or cross ridges at fracture zones. Entire regional landscapes such as the New England geomorphic province of the northeastern United States become a series of structurally controlled subparallel ridges and valleys (Flint, 1963). The Precambrian Grenville metamorphic terrane of eastern North America provides some especially impressive examples of differential erosion. Marble belts between schists and gneisses have formed weak-rock lowlands that curl with the metamorphic grain (Fig. 11–8).

Folded Sedimentary Rocks

Most of the present-day subaerial landscape is eroded on stratified sedimentary rocks. In the United States, for instance, only 8.4 percent of the area is of Precambrian terrane, but strata of the Carboniferous, Cretaceous, and Quaternary Systems each cover about 14 percent of the area (Gilluly, 1949, p. 576). Sedimentary rocks cover 66 percent of the total continental area (excluding ice-covered Antarctica), although the coverage varies greatly from 87 percent of Europe to only 52 percent of

Figure 11-7. Satellite image of Casco Bay, Maine. North is to the top; water is black on this image. Gneiss ridges form long peninsulas and islands projecting into the Gulf of Maine. (NASA ERTS E-1472-14530.)

North America (Blatt and Jones, 1975). The sedimentary cover strata range from easily eroded, unconsolidated Cenozoic glacial drift or coastal-plain sedimentary rocks to stronger, better lithified, and possibly folded and faulted Mesozoic and Paleozoic rocks. Thickness ranges from a few meters to more than 10 km. Usually, the thickest sections of cover strata are now contorted in orogenic belts. Epeirogeny has, however, uplifted large continental areas with only slight warping of the cover strata, which are only 1 to 2 km thick (p. 284). Thus epeirogeny and orogeny have created regionally extensive subaerial landscapes that are largely eroded from sedimentary structures. The differential erodibility of shale, limestone, and sandstone is the dominant structural control of landscapes over two-thirds of the continental area.

Flat-lying sedimentary rocks of contrasting erodibility form *structural benches* on valley walls. Many plateau surfaces are structurally controlled by resistant *caprock* that persists at a high elevation, while the terrain around it is being lowered. The caprock is usually a sedimentary rock, but it can also be a lava flow or a concretionary weathered layer such as calcrete (p. 333) or laterite (p. 143).

Tilted or folded strata provide more complex structural controls to erosion. Generally, the outcropping edge of a resistant stratum will form an escarpment or cliff, often joint controlled. The *dip slope,* away from the crest of the escarpment, is parallel to the stratification and tends to resist dissection as weaker overlying strata are eroded. A single thin layer of sandstone may cap an extensive shale dip slope, for instance.

Popular terminology has evolved to differentiate structurally controlled ridges on tilted resistant layers (Fig. 11–9). If the more resistant unit is horizontal and caps a broad flat-topped hill, the hill is called a **mesa.** If the diameter of the cap rock is less than the height of the hill above the surrounding terrain, the term **butte** is more often applied. If the resistant bed dips gently, an asymmetric **cuesta** results, with a steep escarpment and a gentle dip slope. When the dip of the layer is comparable to the angle of repose of the sliderock on the escarpment face, the ridge becomes ap-

Figure 11-8. Contorted drainage pattern following marble and calcsilicate lowlands among gneiss hills, northwestern Adirondack Mountains, New York. (Part of Hammond, N.Y. 15′ topographic map quadrangle, 1935 edition.)

proximately symmetric and is known as a **hogback.** "Flatirons" are interfluve remnants of dissected hogbacks.

The structural control of tilted strata imposes a powerful asymmetry on drainage networks. Escarpment streams are steep, short, and have high gradients. Dip-slope streams are likely to have more gentle gradients, large watersheds, more tributaries, and more sustained flow (Fig. 11–10). As erosion progresses, a trunk stream flowing along a *weak-rock lowland* between two cuestas will migrate laterally in the down-dip direction, because the escarpment face can deliver waste to the

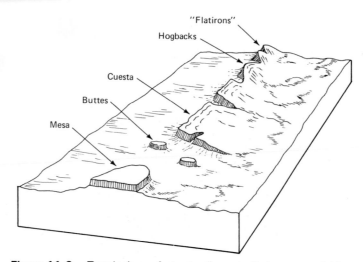

Figure 11-9. Terminology of structurally controlled scarps and ridges.

trunk river faster than the lower gradient dip-slope streams can. The entire ridge and valley system migrates laterally as well as downward with time in a process termed **homoclinal shifting.** The significance of the process was first recognized more than a century ago by Gilbert (1877, pp. 135–40) in his report on the Henry Mountains (p. 57), although his original term was *monoclinal shifting.* The terminology is not important, except that the lateral migration will stop when a monoclinal flexure flattens with depth, whereas it will continue indefinitely as long as strata have a uniform (homoclinal) dip. By this process, structurally controlled ridges and valleys must migrate laterally many times their own width as a region of dipping strata is eroded.

Beyond homoclinal dips, the next degree of structural complexity in stratified rocks is a series of parallel anticlines and synclines with horizontal fold axes (Fig. 11–11). Tectonic hills (Chap. 3) are normally anticlines or domes, but in structurally controlled eroded landscapes, the hills are just as commonly on synclinal structures. An early observation of this fact led to the concept of **topographic inversion,** whereby structural depressions such as synclinal axes become ridges or hills (Fig. 11–11). An impressive amount of erosion is implied, but as demonstrated earlier (p. 120), the amount of rock eroded from the continental masses can be meas-

Figure 11-10. Asymmetric drainage networks on homoclinal ridges. Ridge crests and valleys will migrate laterally downdip as the landscape is lowered by erosion.

Figure 11-11. Block diagram of dominantly synclinal ridges in a terrane of folded sedimentary rocks. (From von Engeln, 1942, Fig. 183.) It is unlikely that basement rocks would be folded, as suggested in this figure. More likely a *décollement,* or detachment zone, would separate basement blocks from the folded cover strata.

ured in tens of kilometers, and multiple topographic inversions must have resulted during the dissection of fold mountains such as the central Appalachians. Therefore, it is doubtful whether synclinal mountains are any more a measure of the depth of erosion than anticlinal mountains.

One may safely infer that erosion is in progress while an orogeny is in progress. The rising anticlines of tectonic landscapes (Fig. 3-23) are cut by gullies as soon as runoff begins to drain from them. When a resistant stratum becomes exposed in a rising anticline, it is still deeply buried in the axes of adjacent synclines. By the time the regional landscape has been eroded sufficiently to expose the resistant stratum at the level of the synclinal axes, the anticlines have been breached, and the synclinal floors become flat or gently basined summits, flanked by escarpments that face outward. If only a single resistant unit is embedded in a thick sequence of erodible rocks such as shale, the resistant stratum is very likely to cap synclinal ridges as erosion progresses. If resistant and erodible strata are interlayered at close intervals, the landscape will have numerous synclinal and anticlinal ridges formed on the various resistant units, as is shown by the Zagros Mountains of Iran (Figs. 8–15 and 11–14). Where folding becomes extremely complex and metamorphism has reduced the contrast in erodibility, dissection favors neither synclines nor anticlines (Fig. 11–12). The abundance of synclinal mountains has led some authors to suggest that unusually intense strain during folding encourages the subsequent erosional breaching of anticlines. However, breached anticlines are not the result of some peculiar fracture pattern at the crests of folds but are simply a consequence of Powell's dictum (p. 98) that early-formed topographic prominences, most commonly anticlines, are the first structures to be attacked by erosional agents.

Plunging structures introduce a still higher level of complexity to structurally controlled landforms. Cuesta- and hogback-forming resistant strata not only dip laterally from anticlinal toward adjacent synclinal axes, but the structures plunge along strike. The result is subparallel belts of *zigzag ridges* such as characterize the folded central Appalachians (Fig. 11–13). The same resistant rock, such as the Tuscarora quartzite of the Appalachians, may form ridge after ridge as it is repeatedly intersected by the land surface. Faults that cause strata to be offset laterally also displace ridge crests controlled by the resistant layers.

Structurally controlled *domes* and *basins* are the result of erosion on double-

Figure 11-12. Tight folding in Cambrian limestone and shale, Canadian Rockies, Kootenay District, British Columbia. (Photo: Geological Survey of Canada.)

Figure 11-13. Zigzag ridges in the Valley and Ridge province of the Appalachian Mountains in Pennsylvania. North is toward the top of the image; Harrisburg, Pennsylvania, is on the Susquehanna River just south of the major fold belt. Low-angle lighting from the southeast may cause the ridges to appear inverted. If so, invert the page. Most of the ridges in the north-central part of the image are breached anticlines. (NASA ERTS E-1513-15220.)

Figure 11-14. Doubly plunging anticlines and domes of limestone east of Shiraz, Iran, in the Zagros Range. Some domes are breached, others show only consequent valleys down their flanks. (NASA ERTS E-1221-06293.)

plunging folds. Although common in fold belts of orogenic regions (Fig. 11–14), broad domes and basins are important features of epeirogeny as well. In the Central Lowland of the United States, the Michigan basin has a central structural depression of more than 4000 m. To the south, the Nashville dome is 80 km wide and 200 km long but has only a gentle central uplift. Resistant strata form cuestas encircling the crests of many epeirogenic structures, facing inward onto an eroded lowland on the domes and facing outward from the basins. The Paris basin of northern France is ringed by outward-facing escarpments, especially in the northeast quadrant. The low wooded ridges and intercuesta lowlands between Paris and the Rhine graben were the scene of bitter fighting during World Wars I and II. Control of the cuesta ridges meant tactical advantage to one of the opposing armies. The English Channel nearly transects a broad structural dome that is ringed by inward-facing cuestas of chalk, which are called **downs** in England. The core of the dome is deeply eroded in Jurassic claystones that underlie the chalk, and Cenozoic sediments were subsequently deposited on the eroded basin floor.

In most of the examples just cited, topographic inversion has produced a topographic basin on erodible shale in the center of an epeirogenic dome. However, in the Black Hills of South Dakota, the Henry Mountains of Utah, and elsewhere, sharp domal uplift over a core of strong crystalline rock has created a central mountain range with ringing hogbacks and cuestas. Similar rings of upturned strata are common along the fronts of many mountain ranges (Fig. 11–15).

The most extreme orogenic deformation of layered rocks consists of great overturned, recumbent, often thrust-faulted folds known as **nappes** from their type locality in the Alps. Alpine mountain sides may show strata more than 1000 m in thickness, all overturned. Detached pieces of nappes or thrust plates may cap ridges

Figure 11-15. Sheep Mountain, Wyoming, looking southeast. A plunging anticlinal ridge with surrounding hogbacks. (Photo: J. S. Shelton.)

underlain by much younger rocks. Nappes, like mélanges, produce complex structurally controlled landforms, usually well exposed because they are within great mountain ranges. As the structures are eroded to progressively deeper levels, the increasing metamorphic alteration, joints, and igneous intrusions assume the dominant structural control, and stratification is of reduced importance to the landforms.

STRUCTURALLY CONTROLLED DRAINAGE PATTERNS

In regions of folded sedimentary rocks, structurally controlled strike ridges form most of the higher terrain. As with all upland landforms, these are best regarded as residuals of less-rapid fluvial erosion, with the intervening valleys the more rapidly lowered parts of the landscape. Systems of valleys eroded on contrasting structures conform to those structures, giving to both valley profiles and drainage networks distinctive, structurally controlled geometry. Both empirical and genetic classifications of structurally controlled drainage networks have evolved and are among the more useful and important tools of geomorphic analysis.

Empirical Classification by Form

Arthur D. Howard (1967) compiled and classified drainage patterns into categories of basic patterns, modifications to the basic patterns, and varieties of the modified patterns. His very thorough classification is summarized in Table 11–1 and Figs.

11–16, 11–17, and 11–18. The classification refers to the regional patterns of an aggregate of valleys and gullies, not to the patterns of single stream channels, which are described by such terms as crooked, straight, meandering, and braided.

Basic drainage patterns (Fig. 11–16) follow regional structural trends and cover a wide range of scales. They are subjectively defined and transitional from one to another. Nevertheless, the basic and modified terms in Table 11–1 are part of the working vocabulary of geomorphologists and aerial photographic interpreters. They can best be studied by comparing Figures 11–16 through 11–18 with a selection of topographic maps of various scales.

TABLE 11–1

Significance of Basic and Modified Basic Drainage Patterns

Basic	Significance	Modified Basic	Added Significance or Locale
Dendritic	Horizontal sediments or beveled, uniformly resistant, crystalline rocks. Gentle regional slope at present or at time of drainage inception. Type pattern resembles spreading oak or chestnut tree.	Subdendritic	Minor secondary control, generally structural.
		Pinnate	Fine-textured, easily erodible materials.
		Anastomotic	Floodplains, deltas, and tidal marshes.
		Distributary (dichotomic)	Alluvial fans and deltas.
Parallel	Generally indicates moderate to steep slopes but also found in areas of parallel, elongate landforms. All transitions possible between this pattern and type dendritic and trellis.	Subparallel	Intermediate slopes or control by subparallel landforms.
		Colinear	Between linear loess and sand ridges.
Trellis	Dipping or folded sedimentary, volcanic, or low-grade metasedimentary rocks; areas of parallel fractures; exposed lake or seafloors ribbed by beach ridges. All transitions to parallel pattern. Type pattern is regarded here as one in which small tributaries are essentially the same size on opposite sides of long parallel subsequent streams.	Subtrellis	Parallel elongate landforms.
		Directional trellis	Gentle homoclines. Gentle slopes with beach ridges.
		Recurved trellis	Plunging folds.
		Fault trellis	Branching, converging, diverging, roughly parallel faults.
		Joint trellis	Straight parallel faults and/or joints.
Rectangular	Joints and/or faults at right angles. Lacks orderly repetitive quality of trellis pattern; streams and divides lack regional continuity.	Angulate	Joints and/or faults at other than right angles. A compound rectangular-angulate pattern is common.
Radial	Volcanoes, domes, and erosion residuals. A complex of radial patterns in a volcanic field might be called multiradial.	Centripetal	Craters, calderas, and other depressions. A complex of centripetal patterns in area of multiple depressions might be called multicentripetal.
Annular	Structural domes and basins, diatremes, and possibly stocks.		Longer tributaries to annular subsequent streams generally indicate direction of dip and permit distinction between dome and basin.

TABLE 11-1 (cont.)

Basic	Significance	Modified Basic	Added Significance or Locale
Multibasinal	Hummocky surficial deposits; differentially scoured or deflated bedrock; areas of recent volcanism, limestone solution, and permafrost. This descriptive term is suggested for all multiple-depression patterns whose exact origins are unknown.	Glacially disturbed Karst Thermokarst Elongate bay	Glacial erosion and/or deposition. Limestone. Permafrost. Coastal plains and deltas.
Contorted	Contorted, coarsely layered metamorphic rocks. Dikes, veins, and migmatized bands provide the resistant layers in some areas. Pattern differs from recurved trellis (Fig. 11–17, H) in lack of regional orderliness, discontinuity of ridges and valleys, and generally smaller scale.		The longer tributaries to curved subsequent streams generally indicate dip of metamorphic layers and permit distinction between plunging anticlines and synclines.

Source: Howard (1967).

Genetic Classification of Drainage Patterns

Consequent and Subsequent Rivers and Their Valleys. River systems that develop as a direct result of, and in harmony with, a preexisting land surface are said to be **consequent.** One may consider that the flow of water was a direct consequence of whatever slopes were initially present. (A less common term for small valleys or gullies is **insequent,** shortened from "initial consequent.") Even newly emerged seafloors or fresh basalt plateaus are not likely to be devoid of initial relief (Fig. 12–8), and runoff will fill initial hollows to the lowest point on the rim and then drain down to the next hollow. Only in arid regions will the depressions on an undissected landscape remain free of lakes. Each basin joins the next lower one until an entire drainage network is established.

Initially, all runoff is consequent. As valleys are eroded, however, structural control may be exerted. Rivers obliquely crossing belts of rocks with contrasting erodibility tend to follow the zone of easy erosion before crossing the next resistant structure (Fig. 11–19). Tributaries that expand into initial uplands underlain by belts of erodible rocks expand their drainage basins more rapidly than those that erode resistant rocks. Thus structurally controlled **subsequent** valleys evolve, which develop independent of and subsequent to the primary, or consequent, drainage (Fig. 11–20). Of the eight basic drainage patterns in Table 11–1, only dendritic, parallel, and radial are likely to be consequent. The others are all linked closely to control by structures. The contorted pattern of Figure 11–8 is a good example of subsequent drainage on a complex terrane, although deranged somewhat by glacial erosion and deposition.

Two special categories of consequent streams are often confused. **Resequent** (a shortened form of the phrase "renewed consequent") and **obsequent** (a shortened form of the phrase "opposite to the consequent") streams are low-order tributaries that develop on new structurally controlled hillslopes as the result of expanding and deepening fluvial erosion. If, as a result of subsequent erosion along a weak-rock belt, new escarpments and dip slopes are formed, the gullies on these slopes are

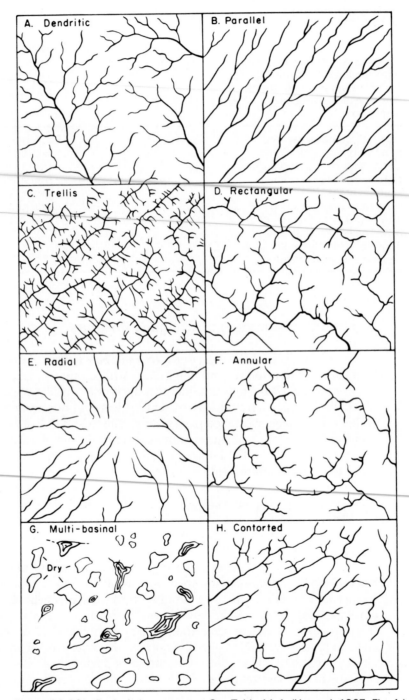

Figure 11-16.   Basic drainage pattern. See Table 11-1. (Howard, 1967, Fig. 1.)

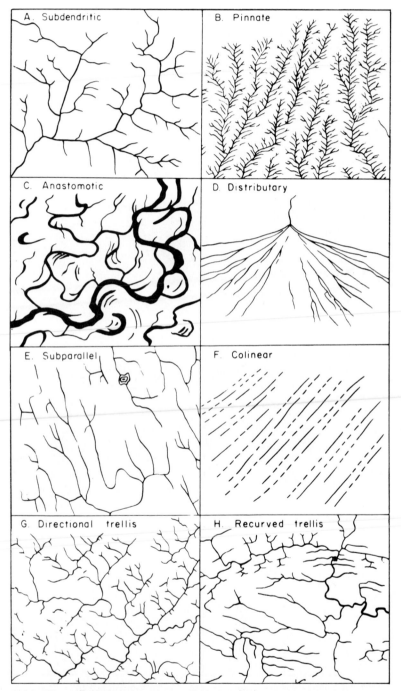

Figure 11-17. Modified basic patterns. Each pattern occurs on a wide range of scales. See Table 11-1. (Howard, 1967, Fig. 2.)

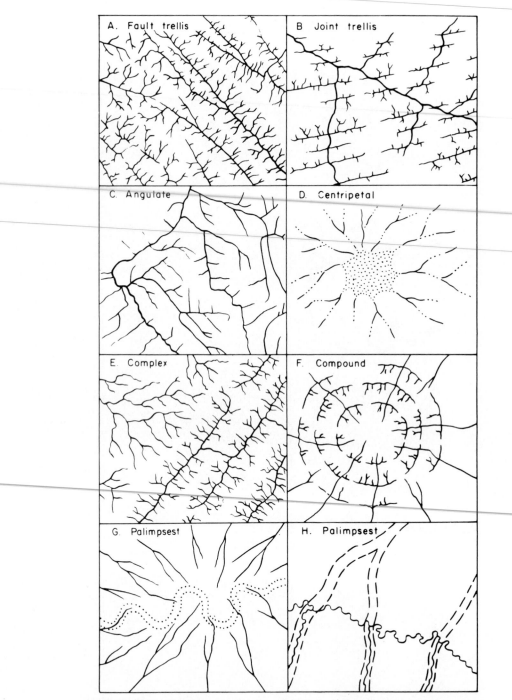

Figure 11-18.   Other modified basic patterns (A–D). See Table 111-1. Complex, compound, palimpsest patterns (E–H) are combinations or superpositions of basic or modified basic patterns. (Howard, 1967, Fig. 3.)

Figure 11-19. Structural control of the Pilcomayo River on the Argentina–Bolivia frontier across the Subandean fold and thrust belt of the eastern Andes. The general southeasterly consequent course of the river has evolved into a series of short reaches parallel to resistant sandstone hogbacks, separated by shorter, high-gradient reaches across the resistant strata. View approximately 25 km wide. Landsat Thematic Mapper image no. 50213-1342.)

called either resequent or obsequent by their orientation with respect to the original consequent trunk stream. A resequent stream flows approximately parallel to and in the same direction as the trunk stream, and an obsequent stream flows in the opposite direction. The terms should not be applied with reference to the dip of the controlling strata, as is frequently and erroneously done. Figure 11–20 illustrates that resequent or obsequent streams can be either on an escarpment or on a dip slope.

The terms "resequent" and "obsequent" are troublesome, because their use requires prior knowledge or assumption of the direction of the original consequent drainage. It is possible that capture or some other diversion has reversed the direction of the trunk stream. The entire geomorphic history of a region must be studied carefully before the relatively trivial designations of resequent and obsequent can be applied to drainage patterns. The words are also used with reference to fault-line

Figure 11-20. Expansion of a consequent valley system over an initial landscape, with a later state of erosion when subsequent tributaries have become dominant. Letters C, S, R, and O signify consequent, subsequent, resequent, and obsequent rivers and valleys. Arrows project downward from valley segments of the consequent system that were defeated by resistant layers and stream capture to become wind gaps.

scarps (Fig. 11–5) with similar connotation, although the direction of fault motion is relatively easy to establish.

Antecedent, Superposed, and Captured Drainage. In orogenic regions, the valleys that are first eroded over a landscape may later have new tectonic ridges rise across their trends. The rivers may be *defeated* (forced to abandon their valleys), or they may be able to maintain gorges across the landmass rising in their paths. Streams that have maintained their valleys across tectonic ridges are said to be **antecedent,** because the river is older than or antecedent to the deformation (p. 27).

Yet another class of rivers and valleys are called **superposed.** The word is a contraction of "superimposed" and implies that a river system has been let down or laid over a landscape. Suppose that a land newly emerged from the sea has a thin cover of sediments that bury an older terrane of complexly folded rocks. On emergence, a consequent drainage system, perhaps with a dendritic pattern, might form. As the valleys deepen, they are eventually eroded into the old land under the covering strata. The dendritic pattern becomes superposed on the older rocks without regard to structural control. When the covering beds have been entirely removed by erosion, the only clue to superposition is the anomalous positions of the river valleys.

Superposition is extremely common in glaciated regions, where interglacial and preglacial drainage networks are filled with glacial drift, and the present streams almost randomly occupy alternating segments of old reexcavated valleys and new, postglacial valleys. Short gorgelike segments superposed on bedrock form scenic attractions in the Finger Lakes region of central New York and in the "Dells" region of southern Wisconsin, among other places.

A common problem in geomorphology is the distinguishing between antecedent and superposed rivers. In the Middle Rocky Mountains, for instance, major rivers cross mountain ranges in narrow gorges, although easy alternative routes are available around the ends of the ranges a few miles away. Early geologists thought the rivers were antecedent, implying that the mountain blocks had risen very recently, after the rivers had become established. Later work proved that a mid-Cenozoic interval of deep lacustrine and alluvial filling buried large parts of the mountains. Rivers such as the Yellowstone, Bighorn, and Laramie formerly flowed over alluvium and were superposed on the resistant rocks of the buried ranges as the area was epeirogenically uplifted and dissected. The evidence for superposition is (1) remnants of former alluvium now on the mountainsides at suitable altitudes, and (2) the similarity of the alluvium on opposite sides of the ranges, both of which suggest a former continuous cover.

The eastern Subandean Belt of the Andes Mountains is a thin-skinned fore-land fold and thrust belt (p. 31), where Paleozoic to Cenozoic strata are being folded and thrust eastward onto the Amazon shield by active orogeny that began in the Miocene Epoch (Jordan et al., 1983, pp. 346–47). The major rivers draining east-ward from the higher interior of the Andes are clearly antecedent to the folds of the Subandean Belt, although subsequent tributaries are rapidly excavating strike val-leys along erodible shale beds (Figs. 11–19 and 11–21). The folds of the Subandean Belt are propagating eastward, and a difficult semantic problem arises when a live anticline has risen across the valley of a river that was formerly flowing on a thin al-luvial cover. Is the river then superposed, or antecedent? The usually obvious dis-

Figure 11-21. Antecedent Pilcomayo River flowing eastward across active folds of the Subandean Belt in southern Bolivia. The youngest folds are to the east, and some that are known at depth from petroleum exploration have not yet propagated through surficial alluvium. View approximately 25 km wide. (Landsat Thematic Mapper image no. 50213-1342.)

tinction between previously folded rocks at depth and overlying undeformed cover strata, diagnostic of superposition, is blurred if the cover strata is also involved in the deformation, in which case the river is clearly antecedent.

Stream capture (or "piracy") has been a popular concept of geomorphologists for generations. Either by chance or aided by a structural advantage, one stream may erode into the drainage basin of another and either progressively or abruptly divert, or capture, new tributaries. The trunk stream that is thereby deprived of a former watershed is said to have been **beheaded.**

Quite commonly, subsequent drainage systems capture segments of earlier consequent drainage. Figure 11–20 suggests how the process could occur, especially if the original drainage on cover strata had been superposed on a folded belt. As subsequent valleys expand along weak-rock belts, they progressively intersect the earlier consequent drainage, which by that stage would be a series of graded reaches alternating with gorges or **water gaps** across the resistant strata. After capture, the former course across the resistant ridges might be marked by **wind gaps,** notches that mark the former drainage line. Figure 11–20 shows only a single trunk stream that continues to drain the same region during the change from consequent to subsequent pattern. It is easy to visualize, however, that in the competition among regional trunk streams, substantial areas of watershed might be permanently gained or lost by capture.

The effects of a capture on the hydraulic geometry of both the capturing and the beheaded system can be dramatic. Most commonly, the capturing stream has a relatively steep longitudinal gradient in a low-order tributary at the point of capture, whereas the defeated stream is likely to have a lower gradient. The sudden diversion of a greatly increased discharge through the capturing system causes rapid entrenchment, perhaps with a waterfall or rapids that migrate rapidly up the captured segment from the point of initial capture. Any sharp change in direction of a stream, downstream from an anomalously steep section, suggests an **elbow of capture.**

Another common mechanism of capture is provided by the unequal gradients of two parallel streams emerging from a mountain front. The one carrying the greater bed load may build an alluvial fan until its channel has been aggraded above the level of the adjacent stream into which it then easily diverts. The terraces on rivers at the foot of the Beartooth Mountains in Montana, described in the preceding chapter (p. 268), exemplify this process.

Cenozoic orogeny has created many mountain belts that generally parallel the converging margins of continents (Fig. 11–21). During deformation, antecedent streams may maintain their valleys across foreland fold belts (p. 31), or cross-axial drainage may be superposed on deeper structures from thin cover strata. In either case, drainage networks are likely to develop with their long axes roughly transverse across orogenic belts (Oberlander, 1965, 1985). Later, as subsequent valleys deepen and lengthen, regional drainage networks are likely to develop trellis patterns, parallel to the orogen. Their great length makes the long subsequent tributaries of a trellis system susceptible to capture by a short, high-gradient stream valley expanding into the orogenic belt from the adjacent coast or foreland basin, if it happens to find a weak or narrow zone in the resistant ridge that separates it from the subsequent valley inland behind the ridge. If a capture occurs, the length of flow of the captured tributaries is shortened, and as the steep-gradient segment migrates upstream from the elbow of capture and smooths out, stream terraces may be left behind, graded toward the beheaded trunk stream rather than toward the capturing river.

Gilbert (1877) proposed a "law of equal declivities," which stated that if two streams of unequal longitudinal gradient drain the opposite sides of a ridge, the divide between them will shift laterally, or *migrate,* until their gradients are equal. If one of the streams is deepening its valley more rapidly than the other, by virtue of a structural advantage, the divide may shift laterally until the expanding stream entirely captures the drainage of the lesser competing stream. Possibly, the migrating divide may intersect a trunk stream, suddenly diverting a larger headwater region into the capturing stream. At that instant the divide "leaps" abruptly to a new regional position. Slowly or abruptly migrating divides are part of the regional adjustments among drainage systems that result from progressive adjustments to structure as a landscape evolves by fluvial erosion.

REFERENCES

BAKER, V. R., 1973, Erosional forms and processes for catastrophic Pleistocene Missoula floods in eastern Washington, *in* Morisawa, M., ed., Fluvial geomorphology: State Univ. of New York Publications in Geomorphology, Binghamton, N.Y., pp. 123–48.

———, GREELEY, R., KOMAR, P. D., SWANSON, D. A., and WAIT, R. B., JR., 1987, Columbia and Snake River plains, *in* Graf, W. L., ed. Geomorphic systems of North America: Geol. Soc. America Centennial Special Vol. 2, pp. 403–68.

BLATT, H., and JONES, R. L., 1975, Proportions of exposed igneous, metamorphic, and sedimentary rocks: Geol. Soc. America Bull., v. 86, pp. 1085–88.

COWAN, D. S., 1985, Structural styles in Mesozoic and Cenozoic mélanges in the western cordillera of North America: Geol. Soc. America Bull., v. 96, pp. 451–62.

EGGLER, D. H., LARSON, E. E., and BRADLEY, W. C., 1969, Granites, grusses, and the Sherman erosion surface, southern Laramie Range, Wyoming: Am. Jour. Sci., v. 267, pp. 510–22.

ENGELDER, T., and GEISER, P., 1980, On the use of regional joint sets as trajectories of paleostress fields during the development of the Appalachian Plateau, New York: Jour. Geophys. Res., v. 85, pp. 6319–41.

VON ENGELN, G. D., 1942, Geomorphology: The Macmillan Company, New York, 655 pp.

FEININGER, T., 1971, Chemical weathering and glacial erosion of crystalline rocks and the origin of till: U.S. Geol. Survey Prof. Paper 750-C, pp. 65–81.

FLINT, R. F., 1963, Altitude, lithology, and the Fall Zone in Connecticut: Jour. Geology, v. 71, pp. 683–97.

GILBERT, G. K., 1877, Report on the geology of the Henry Mountains (Utah): U.S. Geog. Geol. Survey Rocky Mtn. Region (Powell), 160 pp.

GILLULY, J., 1949, Distribution of mountain building in geologic time: Geol. Soc. America Bull., v. 60, pp. 561–90.

HOWARD, A. D., 1967, Drainage analysis in geologic interpretation: A summation: Am. Assoc. Petroleum Geologists Bull., v. 51, pp. 2246–59.

HSÜ, K. J., 1968, Principles of mélanges and their bearing on the Franciscan–Knoxville paradox: Geol. Soc. America Bull., v. 79, pp. 1063–74.

HSU, T. L., 1971, Mélange occurrence in Cyprus: Geol. Soc. China Proc. for 1970, no. 14, pp. 155–64.

JORDAN, T. E., and ALLMENDINGER, R. W., 1986, The Sierras Pampeanas of Argentina: A modern analogue of Rocky Mountain foreland deformation: Am. Jour. Sci., v. 286, pp. 737–64.

JORDAN, T.E., ISACKS, B. L., ALLMENDINGER, R. W., BREWER, J. A., RAMOS, V. A., and ANDO, C. J., 1983, Andean tectonics related to geometry of subducted Nazca plate: Geol. Soc. America Bull., v. 94, pp. 341–61.

JORDAN, T.E., ZEITLER, P., RAMOS, V., and GLEADOW, A. J. W., 1989, Thermochronometric data on the development of the basement peneplain in the Sierras Pampeanas, Argentina: Jour. South American Earth Sci., v. 2, pp. 207–22.

MAXWELL, J. C., 1974, Anatomy of an orogen: Geol. Soc. America Bull., v. 85, pp. 1195–1204.

McCRAW, J. D., 1965, Landscapes of central Otago, *in* Lister, R. G., and Hargreaves, R. P., eds., Central Otago: New Zealand Geog. Soc. Misc. Ser., no. 5, pp. 30–45.

OBERLANDER, T. M., 1965, The Zagros streams: A new interpretation of transverse drainage in an orogenic zone: Syracuse Geog. Ser., no. 1, Syracuse Univ. Press, Syracuse, N.Y., 168 pp.

———, 1985, Origin of drainage transverse to structures in orogens, *in* Morisawa, M., and Hack, J. T., eds., Tectonic geomorphology: Allen & Unwin, Inc., Boston, pp. 155–82.

PHIPPS, S. P., 1984, Ophiolitic olistostromes in the basal Great Valley sequence, Napa County, northern California Coast Ranges, *in* Raymond, L. A., ed., Mélanges: Their nature, origin and significance: Geol. Soc. America Spec. Paper 198, pp. 103–25.

POLLARD, D. D., and AYDIN, A., 1988, Progress in understanding jointing over the past century: Geol. Soc. America Bull., v. 100, pp. 1181–1204.

RAYMOND, L. A., ed., 1984, Mélanges: Their nature, origin and significance: Geol. Soc. America Spec. Paper 198, 170 pp.

RICH, J. L., 1911, Gravel as a resistant rock: Jour. Geology, v. 19, pp. 492–506.

THOMAS, M. F., 1968, Tor, *in* Fairbridge, R. W., ed., Encyclopedia of geomorphology: Reinhold Book Corp., New York, pp. 1157–59.

TURNER, F. J., 1952, "Gefugerelief" illustrated by schist tor topography, central Otago, New Zealand: Am. Jour. Sci., v. 250, pp. 802–07.

WAHRHAFTIG, C., 1965, Stepped topography of the southern Sierra Nevada, California: Geol. Soc. America Bull., v. 76, pp. 1165–89.

WARD, W. T., 1952, The tors of central Otago: New Zealand Jour. Sci. and Technology, ser. B., v. 33, pp. 191–200.

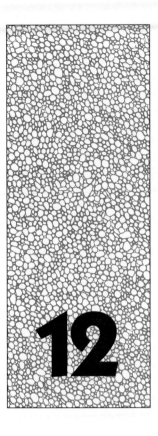

12

Landscape Evolution by Fluvial Processes

DEDUCTIVE GEOMORPHOLOGY

Chapters 5 through 11 have been concerned with the destructive processes that change landscapes, the rates at which the processes work, and the observational evidence that landscapes differ from place to place because of geologic structure and the nature, intensity, and duration of the processes of change. The reasoning has been largely from specific examples or experiments toward statements of general principles. This is the *inductive* approach, and it is a vital part of the system of logical thinking we call science.

The other side of the logical coin of science is the *deductive* approach. Here we reason from general principles toward analysis of specific problems. Chapters 5 through 11 have also provided numerous examples of this approach. For example, the Manning equation, Hjulström's curve, and the Shields curve (Chap. 9), while compiled from extensive laboratory and field measurements, are well established and now serve as generalizations for deducing future responses and events at a particular place, given a particular set of conditions or changes. Geomorphologists, as do all scientists, constantly use both inductive and deductive reasoning in solving problems, usually without consciously considering the distinction. The deductive approach to geomorphology primarily concerns the changes of landscapes through time. We cannot watch a landscape evolve, even though we have abundant reasons to think that it does. We wish to make a motion picture of a landscape's life history, but our source material is a series of still photographs of many different landscapes. If we had a large enough collection of such photographs so that we could say with confidence that a certain percentage of all landscapes are in a certain state at one moment of time, we could apply the **ergodic theorem** from statistical mechanics and deduce that any single landscape is statistically likely to be in that same state for the same percentage of its life history (Paine, 1985). The ergodic theorem asserts that sampling an assemblage of systems at one time is equivalent to sampling one system at a succession of times. Unfortunately, rather than having a statistically valid sample of all landscapes, geomorphologists must attempt to deduce a life history for a landscape from only a limited number of case studies. Ergodic reasoning is a powerful tool for geomorphic deduction, even though the open systems of the land surface and the processes of change are too complex for rigorous statistical prediction.

Biologists or philosophers might argue that a landscape is nonliving and therefore has no "life history." But in the same sense that one commonly speaks of the life of an automobile, an electric light bulb, or a book, life history is used here for the events that occur during the recognizable duration of a landscape. The initial form and time of origin of a landscape, as well as the conditions of its ultimate destruction, must be specified. Between these limits, it has a life history or sequential evolution to be deduced.

Deductions about landscape evolution have been severely criticized for having exceeded the restraints of observational evidence. It is true that in the first decades of this century the deductive system of "explanatory description" was pushed well beyond what had been proved experimentally. A great danger of deductive reasoning is that if the general principles are wrong, inadequate, or incorrectly applied, even the most careful logical procedures will inevitably lead to erroneous final deductions. Suppose, for instance, that a general principle had been established through repeated observations or application of the ergodic theorem in humid regions (it has not) to the effect that hillside slopes decrease in steepness through time. If this principle were incorrectly used to predict the future shape of a desert slope, the deduced shape would have little relation to real landforms, because the various processes of mass-wasting operate at different relative rates in arid and in humid climates.

In recent decades, much new evidence has been accumulated that the repeated glaciations of lands in the middle and high latitudes were but one aspect of repeated general changes in climate during the late Cenozoic Era (Chap. 18). We do not know the causes of these climatic changes, but we recognize their effects on landscapes. Other evidence supports the theory that some continents have been carried across substantial arcs of latitude and into different climatic zones by the motions of lithospheric plates, even while the subaerial portions of the continents were being reduced by erosion.

Some geomorphologists feel that we should not attempt to deduce the advanced stage of landscape evolution under a certain climatic condition, because long before that advanced stage is reached, new climatic conditions will be imposed that will alter the course of landscape development. They forget that we recognize the effects of climatic change on landscapes only because the landscapes are not what we predict should have formed under the conditions we now observe. Figure 5–5 gives adequate evidence that more water annually falls on the land than evaporates from it and that "normal" or "typical" landscapes evolve under conditions of excess water runoff. In this chapter the sequential evolution of landscapes under conditions of humid climate and abundant flowing rivers is given first importance. Subsequently, Chapters 13 through 18 evaluate the climatic controls of landscape evolution in detail, especially the impressive and widespread landscapes dominated by processes that are emphasized by aridity, perennially frozen ground, and glaciers. As in a musical composition, the complexities of the variations cannot be appreciated until the theme is known.

PROOFS THAT LANDSCAPES EVOLVE SEQUENTIALLY

No one who has seen a muddy river or a gullied hillside doubts that erosional processes consume the subaerial landscape. Nevertheless, the burden of proof rests on geomorphologists to demonstrate that landforms are produced, or that landscapes evolve, in a systematic fashion through time. As was noted in Chapter 1 (p. 15), one of the fundamental justifications for geomorphology is that changes can be predicted. There are at least six experimental or observational proofs that landscapes evolve in predictable sequences of forms. Some are more rigorous than others, but all are useful.

Small-Scale Experiments

Small-scale experiments on landscape evolution can be conducted on sand tables or other models. A groove molded in a sloping sand surface conducts water and, as it does so, changes its shape to accommodate the discharge. A mound of fine sediment under a spray mist develops drainage networks down its slopes. By careful choice of materials and experimental conditions, many landforms can be reproduced in miniature in a laboratory.

The difficulty with all scale models is that changes of dimensions in length, mass, and time do not alter the intrinsic properties of matter. Also, physical constants such as gravitation cannot be scaled down. For example, water in a model channel a few inches deep has the same density and viscosity as water in a real river; therefore, turbulence in the water due to these properties is grossly out of scale in the model. Dimensionless parameters such as D/δ_0, the ratio of bed grain size to the thickness of the surface laminar-flow layer (Fig. 9–10b), cannot be scaled. Furthermore, as the linear dimension of an object is decreased, its volume decreases as the cube of the length, but its surface area decreases only as the square of the length. Thus small particles have much larger surface areas in proportion to their masses than large particles of the same shape. Surface effects cause very fine-grained wetted particles in a model landscape to cling strongly together in a fashion totally inappropriate to the sand or gravel of the real landscape that they are meant to represent in scale. An experimental river channel on a sand table proves how river channels form on sand tables but proves very little more.

One constructive approach to scaling problems is to build large enough models so that they approximate small samples of real systems (Fig. 9–3) (Schumm et al., 1987, p. 5). Another approach is to use mathematical rather than "hardware" models such as sand tables. Geometric, dynamic, and kinematic variables can be programmed in accordance with known or assumed laws in deterministic models; probabilities can be stated and evaluated in probabilistic models (as in the case of random walk drainage networks, p. 217); or optimization criteria can be stipulated (such as the principles of least work or uniform distribution of work in river systems, p. 239) (Woldenberg, 1985, p. xv). Predicting the progressive degradation of initially steep scarps by the diffusion equation (Fig. 8–18) is a good example of simple mathematical modeling in two dimensions. More elaborate two- and three-dimensional computer models of slope evolution include numerous variables of weathering, mass wasting, and erosion, and feedback loops among them (Ahnert, 1988, p. 379). Modeling aided by computers is a growing subject of geomorphology as well as of related engineering fields of hydrology, rock mechanics, and soil mechanics (Anderson, 1988).

Real Landscapes Evolving Under Accelerated
Conditions

When a former equilibrium is disturbed by natural or man-made disaster, new landforms may develop quickly. By observing the changes, we can infer the sequences to be expected when changes are much slower. Complex drainage networks may evolve on tidal mud flats during the few hours of each low tide (Fig. 9–6). After volcanic tephra falls, new drainage systems develop within a period of months or years. Everyone has seen gullies erode the artificial fill at construction sites. Larger gullies, up to 10 m deep and several hundred meters in length, have been known to develop in less than a century after land was cleared for agriculture (Fig. 9–2).

Accelerated erosion due to natural or man-made catastrophe is good proof that landscapes evolve, because the changes are full-scale except in time. Unfortunately, the "accident" that induces accelerated change commonly affects only one or a few processes of change. Accelerated erosion due to forest clearing and farming is caused by more rapid and concentrated runoff, but soil formation is not speeded up correspondingly, and mass wasting may change from soil creep to slumping and earth flow. Thus the resulting landforms are not those that would have formed if all processes of change had been accelerated proportionally. Accelerated erosion cannot be directly extrapolated to interpret landscape evolution on the scale of geologic time.

Playfair's Law of Accordant Junctions, and Other Probabilistic Deductions

A third defense for the proposition that fluvial landscapes evolve sequentially is embodied in Playfair's law (p. 214) and more modern statistical studies of drainage networks (Chap. 9). Playfair's statement is not a rigorous proof of a natural law but only an observation about a highly probable condition. If valleys did not evolve by the work of the streams that flow in them, the "nice adjustment of their declivities" would be a highly improbable state. Having considered the hydraulic geometry of streams and the concept of graded rivers, the average reader of this book is better equipped to appreciate the significance of Playfair's words than the greatest scholars of 1802.

However, the inherent problems of interpreting topologically random networks should be recalled (p. 218). An integrated system of river valleys can be explained by either headward growth and branching of a stream network or by progressive integration of small gullies and rills down an initial hillslope. Many computer simulations generate branching systems that look like drainage networks, yet the programs do not tell us how real networks have evolved or how they might change in the future.

Progressive Loss of Potential Energy in Fluvial Systems

As long as a river is carrying sediment to the sea, it is lowering the landscape that provides the gravity potential for flow. There can be no long-term steady state in a physical system of declining energy supply (the "cyclic time" of Schumm and Lichty, 1965). We observe sediment in transport to the sea by fluvial systems and can prove that the sediment is derived from erosion of the landscape. These facts are powerful support for the assertion that landscapes evolve with time and, furthermore, that the rate of evolutionary change is likely to decrease with time as the energy of the system declines.

Just as the sediment in transport by rivers is a powerful proof that landscapes evolve, so is the alluvium that is temporarily left behind as river terraces during valley deepening (Chap. 10). Alluvial terraces have surface features and internal structure that prove they were once part of a river's floodplain. If they are not now reached by floods, either the valley has been deepened since they formed or there has been some diversion of water from the river. This proof of valley erosion is another of Playfair's contributions (1802, p. 103).

Climatic morphogenesis introduces a serious complication into deductions of sequential landscape evolution. Climatic change can cause a river to aggrade when it formerly was eroding (p. 266). River terraces that mark a climate-controlled aggradational phase might be misinterpreted as part of a continuous valley evolution. It has even been proposed (Cotton, 1963; Büdel, 1982, p. 49) that the "normal" system of landscape evolution, deduced for mid-latitude cool-humid climates (p. 372), is largely relict from a time of much colder climate with perennially frozen ground and intense frost action. We must learn to recognize the influence of climatic change on landforms, without losing sight of the fundamental observation that rivers carry by far the bulk of the eroded debris from the lands to the sea except under the most extreme climatic disruptions.

Landscapes Can Be Arranged in Series

The fifth method of proving that landscapes evolve is not based on rigorous logic but on pragmatic observation. It is still the best practical demonstration of landscape evolution. At the beginning of this chapter it was noted that the conceptual goal is a motion picture of a single landscape evolving, but the source material is a series of still photographs of many landscapes. The fact that such a motion picture can be made, even though it cannot be made to run at a uniform rate, is the most convincing evidence of its validity. We can assemble any group of photographs of eroding landscapes into progressive or transitional sequences, making due allowance for regional differences of climates and rock types, and always being alert for relict or exhumed forms. However, the ability to arrange landscapes in some sequential order does not tell us in which direction the order proceeds. Even if we make our motion picture, we do not know which way to run the film. For the establishment of a unidirectional series, we return to the preceding proof of landscape evolution, that rivers carry sediment. If landscapes form a developmental series, it must be in the direction of larger valleys and smaller residual hills.

The practical fact that landscapes can be sequentially arranged has a theoretical basis in thermodynamic principles. As noted in Chapter 9, the graded state results from a tendency toward equilibrium between opposing principles of least work and uniform distribution of work. As a landscape erodes, progressively better adjustment between opposing tendencies must result. In their discussion of river-channel morphology, Langbein and Leopold (1964, p. 793) made the significant inference that as rivers tend toward equilibrium between the two opposing tendencies, "deviations in time will approximate those which can be observed in an ensemble of places." Some conditions of landscapes are more probable than others, and nature favors tendencies toward the probable. The two most probable changes in landscapes are, first, that the rivers will become graded, and second, that their potential energy will decline through time. Both changes permit predictions to be made about the changes in appearance of individual landscapes with the passage of time. A difficulty remains in specifying whether the changes of the landscape lie within "graded time" or "cyclic time" (p. 241). On the shorter time scale of "graded time," fluvial systems evolve from the ungraded to the graded condition, then remain in the graded condition thereafter. The ergodic theorem cannot be applied to graded time, because the condition of grade is essentially timeless. On the longer time scale of cyclic time, possibly involving many millions of years, landscape change is progressive, and the concepts of the ergodic theorem can be applied although not with the necessary statistical rigor.

It may seem a waste of time and intelligence to outline these first five proofs of landscape evolution, because they are all quite apparent. Yet only 200 years ago, the concept of slow, orderly development of landscapes under the same conditions that operate today was a challenge to the established religious and philosophical order. The acceptance of Hutton's concept of uniformity and continuity of process, and the corollary concept of the enormity of geologic time, set the intellectual stage for the theory of organic evolution in the mid-nineteenth century.

The Sixth Proof: Radiometrically Dated Landscapes

A few landscapes, such as those eroded from volcanic rocks or emerged coral reefs, can have their structure dated by radiometric means. Knowing the length of time that has elapsed since inception, a rate of denudation can be determined and used to estimate the survival time of the subaerial landscape. Knowing that a landscape has a finite life does not aid in deducing its appearance at various stages of the aging process, but by dating the rocks in landscapes of similar structure but of different ages, a precise chronology of the erosional stages can be made.

For example, some of the oldest volcanic rocks on the island of Hawaii have been dated by the potassium–argon method at about 0.7 million years old (Dalrymple, 1971). Most of the island has been built since then. The island is a type example of a basalt shield volcano, and the initial constructional shape is obvious. By contrast, the original shield volcano of northeastern Oahu, from which the Koolau Range is carved (Fig. 12–1), is now about one-half gone, perhaps in part by caldera collapse. The basalts of the Koolau Range are 2.5 million years old, but the younger Castle Volcanics, with an age of 0.8 million years, rest on the eroded surface near present sea level (Gramlich et al., 1971). Thus the windward side of Oahu was eroded to low relief, and a substantial mass of the original subaerial shield was removed, in little over 1.5 million years. One could confidently predict that when the eruptive phase on the island of Hawaii has ended, the view after a million years or so will be similar to that of Figure 12–1.

Other datable erosional landscapes, such as dissected till plains of the last glaciation, surfaces covered by tephra of a known age, rock varnish (p. 333), or coral reefs elevated above sea level (Chappell, 1974), are available to measure the actual progress of erosional dissection. Radiometric dating methods now being developed will give the ages of rocks and soils by measuring their accumulated content of a va-

Figure 12-1.   View northwest from the Nuuanu Pali, Honolulu, Hawaii. Erosion on the northeastern side of Oahu is rapid because of high rainfall brought by the trade winds.

riety of cosmic-ray-produced nuclides. When perfected, these techniques should be able to define the time interval during which surface rocks and soil, perhaps to a depth of about 1 m, have been exposed to cosmic radiation. These methods give a proof of landscape evolution that was not available to the great geomorphologists of earlier generations. Numerous examples are described in subsequent paragraphs.

RATES OF LANDSCAPE EVOLUTION

The progress of sequential evolution of landscapes traditionally has been described in relative, not absolute, terms. The words "youth," "maturity," "old age," and "senility" were applied to landforms by analogy to organisms, noting that the absolute interval of time required to attain maturity may vary by orders of magnitude among the various forms of life. The stages were defined only by observed criteria, not by time in years. There are a number of ways, however, by which **denudation rates,** or the volume of rock removed in a specified time interval, can be measured. By applying the appropriate rate to the total volume of a landscape, quantified estimates of the durability of subaerial landforms can be made.

Methods of Measuring Denudation Rates

Degradation and Exposure of Historical Monuments. The stone monuments and walls of civilizations more than a few centuries old commonly show loss of surface detail and mass through weathering and erosion (Fig. 6–4). Marble tombstones in the humid northeastern United States become illegible after 150 to 175 years of exposure (Judson, 1968b, p. 357). Red sandstone monuments in cemeteries near Middletown, Connecticut, are weathering at a rate of about 6 mm per century (Matthias, 1967; Rahn, 1971), almost the same rate as the present average rate of landscape lowering for the entire United States. Carved and polished tombstones may not weather in the same way as naturally exposed rock, but they are exposed to equivalent weathering conditions and provide a good index of relative weathering by various lithologies.

The U.S. Department of Energy is studying the design of an underground repository for civilian nuclear waste. The design life for the repository is to be 10,000 years, and consideration must be given to designing a surface monument that will warn future generations of the dangers of site disturbance. The oldest surviving human structures are the pyramids of Egypt, which are less than half as old as the proposed life of the warning monument at the nuclear waste repository. They and other ancient structures have been seriously degraded, but the primary cause for their damage has been human, not natural. Ancient monuments have been quarried for reused building stone, used as military practice targets, and otherwise despoiled. Even without human defacing, an inscription carved in stone at the nuclear waste site would have to be protected from weathering to survive for 10,000 years. No precedent exists.

In addition to losses of their own surface detail, ancient monuments and walls may mark the former ground level, so that regional denudation can be measured. Neolithic earthworks on the chalk hills of southern England have preserved the chalk from solution, while around the structure the land surface has been lowered at a rate of 10 to 12.5 cm/1000 yr (Pitty, 1971, p. 191). Soil erosion has exposed the foundations of Roman ruins and lowered the land surface adjacent to paved Roman roads at a rate of 20 to 50 cm/1000 yr in about the last 2000 years (Judson, 1968a). The post-Roman rates are high, probably as the direct consequence of human interference.

Sediment Loads of Rivers.   The most obvious way to measure denudation rates by fluvial processes is to monitor the amount of load a river carries from its watershed. Although simple in principle, the actual measurements are difficult. Only the suspended load of rivers is routinely measured. Dissolved load is less frequently recorded, and bed load has thus far defied direct measurement (p. 226). Corrections for volumetric changes during weathering must always be kept in mind, because as noted in Chapter 6, chemical weathering produces mostly dissolved bicarbonate compounds, but the bicarbonate ions are derived from the atmosphere, not from the weathered minerals.

Careful estimates of river loads and regional denudation for the entire United States were first attempted by Dole and Stabler (1909). They concluded that the United States was being lowered at a rate of 1 inch in 760 years or 3.3 cm/1000 yr. Although their sampling was limited, their analysis was very thorough, and their study is still widely quoted. A later study (Judson and Ritter, 1964) had much more data available, especially for dissolved loads. The solid load of Table 12–1 includes an assumed bed load equal to 10 percent of the suspended load. The regional denudation rates in Table 12–1 vary from 17 cm/1000 yr lowering of the Colorado River Basin to 4 or 5 cm/1000 yr for Atlantic and Gulf Coast drainage basins. Rates of denudation by limestone solution are comparable (Fig. 7–6). Judson (1968b) estimated that owing to erosion induced by agriculture the present average denudation rate for the United States of 6 cm/1000 yr is approximately double the precolonial rate.

Total sediment loads for most of the world's rivers are poorly known, but some major rivers at least have had their average annual load of suspended sediment estimated (Milliman and Meade, 1983). The variation of annual suspended load with discharge for major rivers in various geographic regions shows certain regional trends (Fig. 12–2). The tonnages of suspended load were converted to cm/1000 yr of denudation over the drainage area on the assumption that the density of the denuded rock is 2.65 g/cm^3, a typical value for silicate and carbonate rocks. Non-Arctic Asian rivers and rivers draining glaciated mountains have the highest denudation rates. Africa, Arctic drainage, and Australia have generally low rates. The research on which Figure 12–2 was based should be carefully read as an illustration of the assumptions and extrapolations that are required for such summaries. For example, the Colorado River basin was estimated in 1964 to have a denudation

TABLE 12–1
Rates of Regional Denudation in the United States[a]

Drainage Region	Drainage Area (10^3 km^2)	Runoff (m^3/s)	Annual Load				Total (tonnes/ km^2)	Denudation Rate (cm/1000 yr)
			Dissolved (tonnes/ km^2)	(%)	Solid[b] (tonnes/ km^2)	(%)		
Colorado	629	0.6	23	5.2	417	94.8	440	17
Pacific Slopes, California	303	2.3	36	14.7	209	85.3	245	9
Western Gulf	829	1.6	41	28.9	101	71.1	142	5
Mississippi	3238	17.5	39	29.2	94	70.8	133	5
S. Atlantic and E. Gulf	736	9.2	61	55.5	48	44.5	109	4
N. Atlantic	383	5.9	57	45.2	69	54.8	126	5
Columbia	679	9.8	57	56.4	44	43.6	101	4
Totals	6797	46.9	43	26.5	119	73.5	162	6

Sources: Judson and Ritter (1964) and Judson (1968b).
[a] Great Basin, St. Lawrence, and Hudson Bay drainage are not included.
[b] Solid load includes assumed bed load equal to 10 percent of suspended load.

rate of 17 cm/1000 yr (Table 12–1). In 1968, another authority estimated the Colorado River denudation rate at 14 cm/1000 yr (Holeman, 1968). By 1983, the series of dams on the Colorado River had reduced the load that it carries to the sea to less than 0.1 percent of the previous measurements (Milliman and Meade, 1983, p. 5). Since the construction of the Aswan dam on the Nile River, it too carries essentially no suspended load to the Mediterranean Sea, and the delta coast is actively eroding. Some of the gaging stations used for compiling Figure 12–2 are upstream of river mouths, so sediment load may be added or removed farther downstream. The best documented example is the Huang He (Yellow River), where 33 percent of the suspended load that leaves the loess-covered hills of northern China is deposited on the lower alluvial plain and another 43 percent is dropped in the delta region. Only 24 percent of the upstream river load actually reaches the ocean (Milliman and Meade, 1983, p. 3). With all of their cautious assumptions, Milliman and Meade (1983, p. 18) concluded that the average annual sediment yield from the continents is about 180 t/km², which is equivalent to a bedrock denudation rate of 6.8 cm/ 1000 yr.

Abundant evidence has been presented in previous chapters that most of the sediment transported by rivers is derived from the slopes adjacent to first-order tributaries. The smaller and more headward the drainage basin, the larger the proportion of total stream load it supplies. Whereas soil erosion in the conterminous United States is estimated at more than 5×10^9 t/yr, the suspended sediment load of rivers draining the same area is less than 0.4×10^9 t/yr, only 8 percent of the reliably estimated soil loss (Milliman and Meade, 1983, p. 18). In fact, the sediment load of a trunk stream never is as great as the sum of the loads from its tributaries. This anomaly may be explained as a phenomenon of either climatic change or the explo-

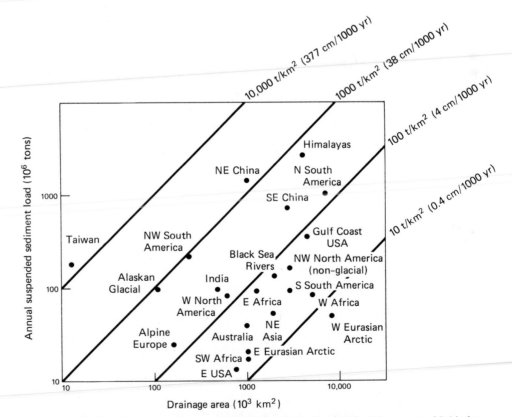

Figure 12-2. Variation of annual suspended sediment load with drainage area. Yields for Asian and glacial rivers are large, but desert rivers (Australia, southwestern Africa) have small yields. (Milliman and Meade, 1983, Fig. 3.)

sive impact of human beings on landscapes during the last few centuries and millenniums (Trimble, 1975, 1977, 1988). Alternatively, floodplains and hollows on hillslopes (p. 200) may naturally store alluvium much longer than has been suspected.

The rates of slope retreat by mass wasting (p. 202), when converted to approximate denudation rates, imply regional lowering at a rate an order of magnitude more rapid than the estimates based on stream suspended-sediment loads. The contradiction could be explained by hypothesizing that only about 10 percent of the area of a fluvial eroding landscape actually furnishes the stream load, and the other 90 percent is nearly immobile under colluvial or alluvial cover. Some evidence supports that hypothesis. In the Amazon Basin, for instance, 84 percent of the total stream load is furnished by the 12 percent of the drainage area that is high in the Andean headwaters (Gibbs, 1967). River-water chemistry and submarine cores off the mouths of the Amazon provide no hint that the largest river on earth traverses an enormous rainforest lowland for most of its length.

The contradictions between present-day upland erosion rates and river sediment loads remain to be resolved. New hypotheses could lead to drastic revisions of deductions about the sequential evolution of fluvial landscapes.

Submarine Sedimentation Rates. Very little fluvial sediment ever reaches the abyssal plains of the ocean floor. Nearly all of it is deposited on the continental margins. Therefore, geophysical studies of the sedimentary thickness on the continental margins can be used to derive longer-term rates of denudation. More assumptions are involved, including major uncertainties about the exact age of the sediments in the wedge under the continental margins, the area of continent that furnished the sediment, the contribution of marine biogenic sediments, volume and density changes during diagenesis, loss by uplift and recycling of previously deposited sediments, and many other factors. Representative papers on the subject are those by Menard (1961), Gilluly (1964), Curray and Moore (1971), Mathews (1975), and Bell and Laine (1985). This approach is promising, because as seismic reflection profiles and offshore exploratory drilling records become more available, the record of continental denudation should become better known. Unconformities and abrupt changes in grain size in offshore sediments should be capable of being correlated with significant changes in the geomorphic evolution of the adjacent landmass. Although the record of previous erosional episodes may be lost when a landscape is uplifted and vigorously eroded, the sediments deposited around the landmass should preserve that record.

Depth of Erosion Over a Known Interval of Time. In the arid to semiarid White Mountains of California, basalt lava flows with an age of 10.8 million years were erupted along valleys in a dissected landscape of low relief. Later, the area was block faulted and eroded. The lava flows now form sinuous ridges that are 50 m or more above most of the surrounding landscape. By drawing topographic profiles of the present landscape and projecting the base of the former lava flows above the profiles, cross-sectional areas can be obtained that when divided by profile lengths, give the mean lowering of the landscape since the flow was removed. The average thickness of the basalt, 37 m, must be added to the average amount of lowering to give the total denudation since the flows stopped and dissection began. By this technique, a denudation rate of 1 to 3 cm/1000 yr was derived for the region (Marchand, 1971). The rates were probably greater on granitic igneous rocks than on basalt and dolomite in this cold, dry upland climate. Another study of the same region (LaMarche, 1968) concluded that the denudation rate during the last few thousand years has been at least an order of magnitude more rapid, because the roots of bristlecone pine trees as old as 4000 years have been progressively exposed by mass wasting and erosion at rates of 15 to 20 cm/1000 yr (Fig. 12–3). Holocene

Figure 12-3. Bristlecone pines, White Mountains, Inyo County, California. Centers of roots extend about 60 cm above ground level, a measure of degradation while the tree has been growing. (Photo: V. C. LaMarche, Jr., U.S. Geological Survey.)

climatic change may have intensified the denudation rates, however. Possibly the rapid recent rates need correction for their duration interval, as noted on page 6.

The foregoing studies, from the same region but varying greatly in the scales of time that were studied and the evidence that was used, have a common basis in that a reference horizon, either the base of a lava flow or the former ground level on roots of long-lived trees, was used to measure regional denudation over an extended time interval. In Chapter 6 several other studies of reconstructed fold geometry, metamorphic facies, and geothermal heat flow were cited as proof that many landscapes are now developed on rocks that were formerly buried to depths of 20 km or more (p. 120). Denudation rates derived from those studies are: 2 cm/1000 yr for 1 billion years of erosion in the Adirondack Mountains; 4 cm/1000 yr for 200 million years of erosion in the folded Appalachians, 10 cm/1000 yr for post-Cretaceous erosion of the Sierra Nevada, and 100 cm/1000 yr for 30 million years of Alpine erosion. The rates, although they are gross approximations that do not take into account any changes in rate of erosion over great intervals of geologic time, are rather similar to those derived by the methods described previously.

When etched by hydrofluoric acid, mineral grains such as apatite and zircon commonly show small etch pits or *fission tracks,* the result of damage to the crystal lattice by radioactive decay. Fission tracks are commonly annealed and destroyed at temperatures of 105 to 300°C, so their density in a prepared mineral grain is a function of the time that has elapsed since the grain cooled below the specified closure temperature along the geothermal gradient as a result of uplift and erosion. In the Nanga Parbat massif of the western Himalaya Mountains in Pakistan, fission-track ages require erosion and uplift rates as high as 5 mm/yr (5 m/1000 yr) between 2.0 and 0.5 million years ago, and as high as 9 mm/yr (9 m/1000 yr) during the last 0.5

million years (Zeitler et al., 1982). The erosion rate is derived from a cooling model that requires rapid erosion to strip off surface rocks to steepen the geothermal gradient; uplift rates are assumed to be equal to the denudation rates on the assumption that such extreme erosion rates could not persist without continuing orogeny. The inferred denudation rates for the Nanga Parbat region are 2 to 10 times faster than the denudation rates derived from sediment loads of Himalayan rivers (Fig. 12–2) and are among the most rapid inferred by any technique. Their significance to continental erosion is discussed in a later section of this chapter.

Papuan Erosion: A Mathematical Exercise

Only one study has been made of an eroding mountain mass where denudation rates were established for various altitudinal zones (Ruxton and McDougall, 1967). The study and some extended conclusions from it are reviewed here as an example of the usefulness of known denudation rates to geomorphic research.

The Hydrographers Range is a dissected andesitic composite cone in northeastern Papua New Guinea. Annual rainfall in the region probably ranges from 2250 mm at sea level to more than 3000 mm above an altitude of 1000 m. Most of the mountains are covered by tropical rainforest. Potassium–argon dates establish that the volcano ceased eruption about 650,000 years ago, when it began its erosional history. Most of the central summit portion is now dissected, and its initial shape is not known, but it probably was at an altitude of about 2000 m and had a considerable amount of constructional relief around multiple eruptive centers. Remnants of the youngest lava flows form the interfluves below an altitude of 975 m, and the predissection shape of the volcano can be reconstructed with confidence below that altitude. By measuring the areas of numerous valley cross sections at selected altitudes, the volume of rock eroded in 650,000 years was calculated. Using assumptions similar to those used in the previous example of the White Mountains in California, denudation rates were tabulated for each of a succession of reconstructed contour lines between 975 and 60 m above sea level (Fig. 12–4). Above 1000 m, in the summit area, only rough estimates of the amount of erosion could

Figure 12-4.   Denudation rate as a function of altitude, eastern Papua New Guinea. (From Ruxton and McDougall, 1967.)

be made. The reported denudation rates range from 52 cm/1000 yr above 900 m altitude to only 8 cm/1000 yr at 60 m above sea level (Ruxton and McDougall, 1967). The best-fitting equation for the relationship between denudation rate and initial altitude is shown in Figure 12–4.

Because denudation rate is so strongly correlated with altitude in the range of 60 to 1000 m, it is possible to extrapolate the equation given in Figure 12–4 to 2000 m, the probable initial constructional height of the Hydrographers Range. By such extrapolation, the denudation rate near the original summit must have been about 86 cm/1000 yr, sufficient to lower the central region about 560 m since it was built. This extrapolation is close to the 490 m of summit-area lowering that can be roughly estimated from reconstructing the original cone-shaped profile. Both estimates suggest that the central part of the range has been lowered by fully one-fourth of its initial altitude in only 650,000 years.

The equation (Fig. 12–4) that links denudation rate to altitude can be integrated between any successive altitudes to give the amount of time necessary to lower the range the specified amount. Figure 12–5 graphs the denudation history of the region but assumes an initial altitude of only 1000 m to avoid the uncertainties of the unknown configuration of the actual summit region. At the denudation rates reported by Ruxton and McDougall (1967), the mountain range could be lowered from 1000 m to 500 m in 1.1 million years, to 100 m in 2.8 million years, to 10 m in 4.0 million years, and to sea level in 4.9 million years. Although the denudation rate approaches zero as altitude approaches zero, the rate remains finite, and the time for total erosion of the volcanic terrane to sea level is also finite.

The denudation rates for the Hydrographers Range are not excessive as compared to other regions (compare the rates shown in Figure 12–4 with Tables 12–1 and 12–2 and Fig. 12–2), but the integrated time for denuding the entire region is unusually short. Judson and Ritter (1964) estimated that the entire United States could be removed to sea level in 11 to 12 million years if the present rates (Table 12–1) are maintained and if no isostatic or other tectonic uplift occurs. Schumm (1963) estimated that with corrections for isostatic recovery and decreased rates as denudation proceeds, regional denudation to near sea level might require a time interval of 15 to 110 million years. The apparent rapid time for total removal of the Hydrographers Range might be caused by the initial assumption that denudation rate is only a function of altitude and not of some other variable such as area of wa-

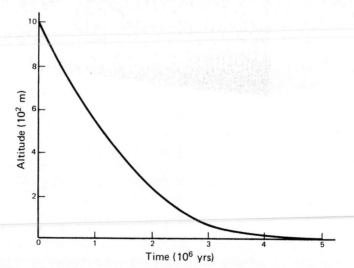

Figure 12-5.   Time required to lower the summit of the Hydrographers Range, eastern Papua New Guinea, from an initial assumed altitude of 1000 m. Rates are shown on Figure 12-4.

tershed above a certain altitude. As a mountainous area is reduced, the total orographic rainfall might decrease, even at low elevations.

Figure 12–6 gives additional evidence for inferring the sequential evolution of a mountainous terrain during denudation. In the Hydrographers Range, local relief (vertical difference between ridge crests and valley floors), slope gradients, and average slope length all decrease with decreasing denudation rates, and therefore also with altitude. Extrapolating in time, as the Hydrographers Range is lowered, its valleys will become broader and shallower and will develop decreased side-slope gradients. Highest areas are eroding most rapidly, so the original conic form should become lower and more dissected, and the interstream ridges should become lower and less steep-sided. After only a few million years, this constructional volcanic cone will become a low range of radiating ridges, separated by broad, shallow valleys. No mention is made in the original description of structural controls being ex-

Figure 12-6. Relation between denudation rate and (a) local relief, (b) average maximum slope, and (c) average slope length in the Hydrographers Range. Data from more dissected sectors are shown as closed circles; data from less dissected sectors are shown as open circles. (From Ruxton and McDougall, 1967, Fig. 5.)

erted on the denudation rate by massive lava flows, dikes, or sills, but such controls could convert the initial radial pattern to an annular or centripetal pattern (Table 11–1).

Climatic Factors in Denudation Rates

Table 12–2 gives some estimates of regional denudation rates characteristic of various climates in areas of steep and "normal" lowland relief. As expected from the preceding data compilations, mountainous regions erode at rates of at least 10 to 100 cm/1000 yr, an order of magnitude more rapid, regardless of climate, than lowlands with typical rates of 1 to 10 cm/1000 yr. The figures offer impressive support for Powell's dictum (p. 98) that mountains are ephemeral landforms. Mountains with valley glaciers seem to erode most rapidly of all, probably because freezing water and glacier erosion add to the erosional work of rivers. Studies of denudation rates in cold climates are complicated by the fact that many of the regions are only recently deglaciated, and landforms shaped primarily by glaciers are now out of equilibrium with fluvial processes. Under such conditions, rapid geomorphic change is inevitable (Mills, 1976). Temperate and tropical humid climates have comparable ranges of rates, although some of the most extreme denudation rates are recorded in southeast Asia and the tropical islands of Taiwan, Philippines, and Indonesia, where a combination of small drainage basins with high relief, seismicity, intense weathering, and human interference of the natural ecosystems all combine (Fig. 12–2).

Another way of comparing the climatic control of fluvial denudation rates is to compare the sediment yield per unit area of drainage basins of comparable size but located in a variety of climates (Langbein and Schumm, 1958; Douglas, 1967; Wilson, 1973). Figure 12–7 (upper curve) shows the sediment yields of 94 small watersheds in the United States, grouped into six categories of similar runoff (Langbein and Schumm, 1958). The lower curve of Figure 12–7 is based on a compilation of data from a wide range of published sources, calculated on the same basis as the upper curve (Douglas, 1967). Both curves show that the sediment yield per unit area, or denudation rate, has a maximum value at a mean annual runoff of about 30 mm, which, at a mean annual temperature of 10° C, requires an *effective precipitation* (the amount of precipitation required to produce the observed runoff) of

TABLE 12–2

Denudation Rates for Various Climate and Relief Conditions

Climate	Relief	Typical Range for Rate of Denudation (cm/1000 yr)
Glacial	Normal (= ice sheets)	5–20
	Steep (= valley glaciers)	100–500
Polar/montane	Mostly steep	1–100
Temperate maritime	Mostly normal	0.5–10
Temperate continental	Normal	1–10
	Steep	10–20+
Mediterranean	—	1–?
Semiarid	Normal	10–100
Arid	—	1–?
Subtropical	—	1?–100?
Savanna	—	10–50
Rainforest	Normal	1–10
	Steep	10–100
Any climate	Badlands	100–100,000

Source: Saunders and Young (1983), p. 497.

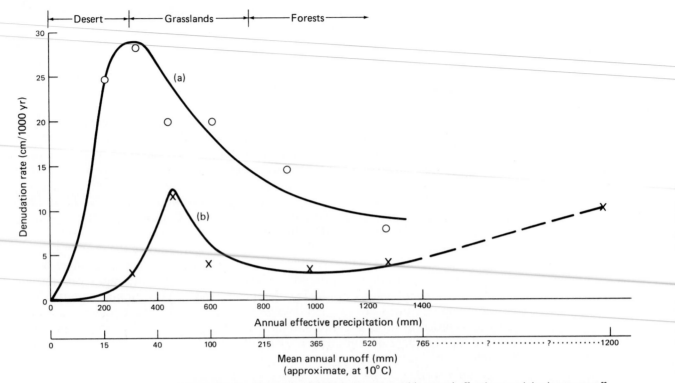

Figure 12-7. Variation of denudation rates with annual effective precipitation or runoff. Each symbol represents the arithmetic average of a group of similar stations. Effective precipitation is the precipitation necessary to produce a given amount of runoff at a mean annual temperature of 10°C. The vegetational boundaries apply primarily to temperature conditions in the United States. The conversion graph does not extend beyond 1400 mm of effective precipitation. (a) 94 stations in the United States in 6 groups. (After Langbein and Schumm, 1958.) (b) 90 stations in Australia and elsewhere (derived from suspended sediment yield) in 6 groups. (After Douglas, 1967.)

about 250 to 450 mm/yr. The lower graph shows low sediment yields for runoff of 200 to 600 mm/yr, or an effective precipitation of 760 to 1270 mm/yr. At still higher mean annual runoff, the sediment yield rises again (Douglas, 1967, Fig. 2).

The disparity in absolute values of the sediment yields shown by the two graphs has been attributed to the fact that most of the 94 United States sites used to compile the upper curve were in agricultural areas, whereas the lower curve was compiled from drainage basins in Australia and elsewhere that were in more nearly natural vegetation. Agricultural activity is known to at least double the denudation rate, so the two curves may be in reasonable agreement. For temperatures typical of the United States, the peak sediment yield corresponds to the vegetational boundary between desert shrubs and grasslands.

Several other researchers have drawn similar conclusions (summarized in Yair and Enzel, 1987, Fig. 1). In general, the denudation rate increases with precipitation and decreases with vegetation cover. As mean annual runoff increases from zero in deserts to about 30 mm in semiarid regions, fluvial denudation increases, because although there is more runoff, the precipitation is inadequate to maintain continuous vegetation cover. As precipitation further increases, grass and forest becomes more effective in stabilizing the soil and the denudation rate decreases. At precipitation levels in excess of the amount required to maintain a closed forest cover, the denudation rate begins to rise again, although not to the levels measured in semiarid to subhumid climates (Douglas, 1967; Kirby, 1980, pp. 3–5). The amount of denudation by chemical solution would also increase in very wet regions, although this was not considered in these denudation studies. These inferences are crudely supported by the denudation rates listed in Table 12–2, with lowlands in semiarid, subtropical, and savanna climates yielding a range of denudation rates approximately 10 times greater than lowlands in either arid or temperate and tropical humid climates. Clearly, type and extent of vegetation cover is at least as important as precipitation in determining denudation rates, and temperature is a minor factor. The importance of vegetation cover is emphasized by the contrasting denudation rates of the two curves in Figure 12–7, with agricultural disturbance clearly being a major factor in the denudation rates in the United States (Larson et al., 1983).

Although opinions may conflict concerning denudation rates under moderate to high precipitation, all authors agree that as precipitation approaches zero, the fluvial denudation rate must also approach zero. With no net runoff from the land, no net lowering of the landscape can occur unless by wind. The lowering of mountains by denudation is neutralized or even exceeded by aggradation of desert basins with sediments that have bulk density lower than that of the eroded rock. Arid mountains are literally buried in their own detritus. Erosion to a rising base level of aggradation is the most important factor in the deduced erosional history of desert landscapes (Chap. 13).

Conclusions from Figure 12–7 invite comparison with Huntington's principle (p. 266). In semiarid shrub or grassland environments with high denudation rates, an increase in precipitation should increase the vegetation cover and reduce erosion, as deduced by Huntington. In an extremely arid region, however, increased precipitation might actually increase the denudation rate, if it does not exceed the amount required to stabilize the landscape with grass.

RATES OF DENUDATION AND TECTONISM COMPARED

In Chapters 2 and 3, numerous examples were cited of orogenic and epeirogenic uplift increasing the height of landscapes by as much as a few millimeters per year, rarely even 1 cm/yr (10 m/1000 yr). The denudation rates reported in the preceding sections of this chapter should be compared with constructional rates in order to determine whether newly rising landscapes can outpace denudation and increase regional relief or whether landscapes are doomed to be kept low by erosion despite tectonic uplift. Most denudation rates for mountainous regions in various climates range from 10 to 100 cm/1000 yr, roughly 10 percent of the reported orogenic uplift rates. Denudation rates for large watersheds such as the Amazon or Mississippi are generally between 1 and 10 cm/1000 yr, again roughly an order of magnitude less than long-term epeirogenic uplift rates. It would appear that only extremely high mountain ranges might suffer denudation rates so great that their absolute elevation could not increase despite strong uplift. For example, an intriguing but unwarranted extrapolation of the best-fitted equation for Figure 12–4 would give a denudation rate of 233 cm/1000 yr for a hypothetical 10,000-m mountain range in Papua New Guinea. The rate is consistent with the denudation rate of more than 100 cm/1000 yr for the Kosi River, which drains the highest parts of the Himalaya Mountains, and the remarkable denudation rates of 500 to 900 cm/1000 yr for the Nanga Parbat Range. Because of the power function increase of denudation rate with relief and altitude (Fig. 12–4), perhaps the Himalayas are roughly as high as any mountain range on earth can be, given the present effectiveness of destructive geomorphic processes. (The rheidity of rocks may also set a comparable limit to terrestrial relief; see page 13.)

All of these estimates and extrapolations lead to the conclusion that if tectonic movements raise new land above the sea, the land is likely to increase in area and altitude much faster than erosion can destroy it. However, denudation rates set an upper limit to the height of subaerial relief on this planet of about 10 km.

At least two end members of the contest between tectonic uplift and erosion can be envisioned. In one extreme, as exemplified by the intense erosion of eastern Taiwan (Fig. 12–2), easily eroded mélange and accretionary-prism mudstones are rapidly uplifted in a humid, tropical, seismically active region, and erosion limits this steady-state mountain range to altitudes no higher than 1500 m despite intense tectonism (p. 98). At the other extreme, the Altiplano plateau of the central Andes in Bolivia and adjacent countries rose very rapidly in the last 12 million years, in a chronically arid region. Once it rose to heights of 4 to 6 km, the Altiplano plateau became even more cold and arid, and except for wind erosion, there is little or no denudation of the vast upland. Such a tectonic mass might become self-limiting by isostasy, for if nothing is removed, the orogen cannot continue to be uplifted indefinitely. It is interesting to postulate that the relief of all orogenic regions is bounded by these end members of erosion-limited and tectonic-limited conditions (Ahnert, 1988; Dahlen and Suppe, 1988).

SEQUENTIAL DEVELOPMENT OF FLUVIAL LANDSCAPES

The necessary foundation has now been laid for deducing the sequential development of a regional landscape under a climate that ensures year-round runoff by rivers that increase their discharge as they flow to the sea. In such a fluvial climate, continuous vegetational cover by grass and trees is assumed. It is necessary to deduce the development of individual river valleys and drainage basins prior to considering the regional landscape, because in fluvial landscapes networks of valleys are the basic landform units.

Sequential Development of River Valleys

As drainage networks expand over a subaerial landscape, most of the erosion is accomplished by the first-order gullies and streams (Fig. 9–1). Gradients are high, and as these streams grow headward and organize their subsidiary drainage basins, the landscape changes rapidly. New slopes shed waste directly into the first-order channels, which can deepen only if their streams can first move out the detritus supplied by mass wasting from their own valley walls. A wavelike front of intense erosion sweeps inland across the old landscape as new drainage networks expand.

If we could watch a particular area as an outward-expanding drainage network sweeps across it, we would see the initial slopes steepen toward the encroaching tributaries. The area would become a complex of small interfluves, side slopes, and gully floors. Then, as the network expands headward, the minor dissection would be consolidated into larger and less complex landforms, perhaps a single valley wall as a former first-order stream sends out its own headwater tributaries and becomes a higher-order trunk stream. At this stage, regional lowering of relief would become less rapid, although the local relief between interfluve and valley floor should continue to increase.

As a stream network expands, trunk streams and major tributaries change their functions from initial dissection to transportation. Downcutting may be deferred by an aggradation phase, as a former V-shaped gully is choked by sediment from its own expanding watershed. Quite early in the deduced sequence of events, we can expect the beginning of floodplains, because alluvium is temporarily or seasonally stored and weathered before it is removed downstream. Dams of alluvium may divert the streams against their banks to begin the process of spur trimming and valley-floor widening. By the time a stream has developed its own tributary system and is draining a watershed of 1 km², it is likely to be in a well-formed valley approximately 1 to 10 m deep, its low-water channel meandering or braiding around masses of poorly sorted colluvium and alluvium. Valley slopes will be at the angle of repose of the local bedrock or weathered debris and will extend down either to the channel banks or to fragments of floodplains.

Eventually, valley-side slopes will be trimmed back by stream erosion, and alluvium will be deposited on the valley floor so that the river flows between alluvial banks except at peak discharge. Midchannel bars may divert the main current to one or the other valley wall, so that for a time the stream has one bank of alluvium and one of bedrock. This process aids in valley widening and the development of a floodplain.

A significant event in the deduced sequential evolution of a stream valley is the development of a continuous floodplain. Depending on the grain size of the load and the complex interplay of hydraulic geometry variables, the stream channel will meander or braid in a layer of its own alluvium. At this stage, the mutual adjustment of channel shape, discharge, load, and slope becomes possible, and the stream is graded in an alluvial channel (p. 239). Erosion is not completed, but from this time onward, the river must mutually or sequentially readjust the entire complex of hydraulic geometry variables in response to changing conditions. Local perturbations such as landslides are less likely to affect a graded river, because any debris from the valley wall will accumulate as a cone or sheet on the edges of the floodplain and may weather there for years before the stream begins to rework it. As a consequence, valley-side profiles will change from initially straight slopes directly to the stream banks to complex curved forms, blending above into convex summits of divides and below into concave wash slopes that in turn blend into floodplains (Fig. 8–18).

When streams carry increased discharge, their longitudinal gradients decrease (Chap. 9). Therefore, in an expanding fluvial network such as that being used as the model for this deduction, an initial high-gradient, first-order tributary or gully will be replaced in time by successively lower-gradient valley segments. A net regional lowering of altitude must result. If the example of the Hydrographers Range is valid, with lower altitude and decreasing denudation rate, local slope gradient, slope length, and local relief will also decrease (Fig. 12–6). Slower mass wasting on more gentle slopes should permit more intense chemical weathering and thicker soils, with a corresponding reduction in the amount and grain size of sediment delivered to the river.

In general, then, the changes we observe in a downstream direction along present river valleys are those we can deduce for the temporal changes at any point along the river valley: a reduction in down-valley gradient, progressive growth of a continuous flood plain, removal of waterfalls and rapids except those on resistant structures, progressive widening of the cross-valley profile by spur trimming and adjacent valley-side retreat, and progressive loss of local first-order tributaries draining adjacent slopes.

Sequential Evolution of Regional Fluvial Landscapes

A regional fluvial landscape is simply an assemblage of individual drainage systems. Whereas the deduced development of a single river valley can assume a continued headward growth of the drainage network, on the regional scale there must come a time when independent drainage systems intersect, interfere, or compete. Deductions about the sequential evolution of regional fluvial landscapes must be strongly conditioned by the restraint of a finite limit for basin expansion.

There is no generalization possible about the initial appearance of a regional landscape just beginning to be dissected by rivers. The favorite initial stage postulated by deductive geomorphologists is the former floor of a shallow sea, newly emerged above sea level. The assumptions are that the sea floor is smooth and featureless and that it emerges quickly relative to erosional processes. Figure 12–8 suggests such a landscape, with scale omitted. The figure could represent 1 or 100 km². Any other constructional landscape—volcanic, tectonic, or depositional—could as well be taken as the initial form. Equally, or perhaps more probably, a landscape already undergoing fluvial dissection becomes the initial form for a new sequence of changes. It is well to recall the quotation (p. 218) that a landscape is best considered to have a heritage rather than an origin.

Regional landscapes require additional deductions about the sequential development of divide areas between adjacent or opposed drainage networks. Initially, the divide area is large relative to the expanding fluvial basins and networks. As the new generation of drainage basins intersect, however, the old land is converted to a series of fragments isolated between vigorous headward-eroding streams. Drainage divides become more sharply defined at this stage, but whether the divides become knife-edged ridges between adjacent steep valleys or smooth, convex hilltops depends on complex interrelations between structure, climate, and stream gradients.

As river valleys dissect a landscape, they are influenced by many subtleties of structural control. Weak-rock belts develop subsequent stream valleys. Divides become localized on the most resistant rocks, although superposition from cover strata onto a resistant structure may introduce complexities into the deduced evolutionary patterns.

Eventually, an evolving region will consist of a series of divides, valley sides, and valley floors, all mutually adjusted and graded with respect to each other. The drainage density is determined by structure and climate; drainage patterns reflect structural control or lack of it; divides are on the most resistant rock type between adjacent trunk streams; and the highest terrain is the region farthest inland, at the headwaters of the largest streams. A critical stage in regional landscape evolution is the final elimination of whatever initial forms were present. In significance, this stage of complete upland dissection is analogous to the development of a continuous floodplain along a segment of stream valley. In no way can the two events be considered simultaneous, however, because floodplains are constantly under development along a valley segment somewhere along each river.

Local, Available, and Critical Relief

Deductions concerning the development of regional fluvial landscapes subsequent to complete upland dissection depend on the interaction between mean regional elevation above sea level and the drainage density. With high initial elevation or closely spaced streams, valley-side slopes are likely to intersect as narrow divides while the valleys are still being deepened. With low regional elevation or widely

Figure 12-8. Hypothetical initial landscape, newly emerged from the sea. The initial surface is conformable to the sedimentary beds that underlie it. Drainage is consequent on the surface irregularities. (Based on drawings by W. M. Davis and C. A. Cotton.)

spaced streams, individual rivers will have completed their initial phase of rapid deepening and will have become the trunk streams of expanding networks while large interfluve areas are as yet unaffected by the new dissection. Glock (1932) defined **available relief** as the vertical distance between the initial upland flats and graded valley floors of dominant rivers and argued that if the available relief exceeded the local relief (the vertical distance from ridge crests to adjacent stream channels), total regional dissection of the oldland would be complete before floodplains began to accumulate on the local valley floors (Fig. 12–9a). If the available relief were less than the local relief, downcutting by trunk streams would be inhibited, because they would become lined with alluvium and graded, and upland denudation would be very slow (Fig. 12–9b). Glock envisioned a special case of **critical relief** such that regional disappearance of relict terrain on the interfluves would exactly coincide with the development of graded segments in the adjacent valleys. The concepts of available and critical relief need redefinition to incorporate the factor of drainage density (Chap. 9), but the basic concepts are fruitful nonetheless. With high available relief (or a high drainage density), regional landscapes will quickly be completely dissected to the stage called *ridge-and-ravine topography* by Hack (1960, p. 89). Subsequently, until alluvial floodplains extend upstream into the region, a long interval of time might pass, and a great amount of regional denudation might occur, without any significant sequential changes in the appearance of the landscape (Fig. 12–10). Competing streams might capture or be captured, resistant structures might develop as regional divides, and initially buried structures might be exposed and exert new controls on sequential development. But unless climate or structure change radically, the drainage pattern and local relief will seem "timeless." Not until regional denudation has brought the terrain low enough so that trunk streams into the region can become graded on alluvial plains can valley cross profiles soften, summits become rounded, and lower valley walls become concave wash slopes.

The Later Stages of Sequential Development

Accepting the premise that a landscape is at a finite altitude when tectonic uplift eventually ceases, we can deduce the later stages of sequential landscape development. The deduction incorporates the concept of *cyclic time,* measured in millions of years, during which the potential energy if the fluvial system gradually declines. Within the long time span of the cycle, on the shorter time scale of *graded time* (hundreds or thousands of years) the system will appear to be in equilibrium (Schumm and Lichty, 1965). When graded, a river floodplain widens only slowly, because the meandering channel only rarely encounters the bedrock of the valley wall. The loss of local relief in the headwater regions will decrease the sediment load and probably the grain size of the load, so all the hydraulic variables of the stream system must continue to change, although in mutually compatible ways. The denudation rate will probably decline from 1 to 10 cm/1000 yr to 0.1 to 1 cm/1000 yr. The net effect must be a gradual lowering of the longitudinal valley profile as the river slowly shifts laterally, trimming the valley flat beneath the floodplain. As relief is lowered, the slopes in interstream areas become more gentle. As floodplains become significant parts of the total landscape, the proportion of stream-dissected upland terrain must necessarily decrease. The observations that large quantities of the upland soil and slope-derived sediment that enter fluvial systems are not delivered to the sea are important to deductions about the later stages of landscape development. If sediment storage within fluvial systems is a normal part of their behavior and not the result of Holocene climate change or human activity (p. 306),

Figure 12-9. Complexities in comparing the deduced development of individual valleys with the development of the regional landscape. (a) With high available relief, upland is totally dissected but streams are vigorously downcutting. (b) With low available relief, trunk streams are not downcutting, and upland dissection is very slow.

36°00′N

35°30′N

109°00′E   109°30′E   110°00′E   110°30′E

0   10   20   30   40   50 km
Scale (approx.)

Figure 12-10.   Complete regional fluvial dissection of loss, Shanxi Province, China. (NASA ERTS E-1525-02455.)

we must deduce that an old landscape will be largely covered with colluvium and alluvium, which low-gradient streams are relatively incompetent to transport. As the fluvial denudation rate declines, wind erosion and transport become significant, especially on landscapes where agricultural practices cause seasonally exposed bare soil (Kirkby, 1980; Trimble, 1988).

When summit convexities are graded to lower slope concavities, which in turn are graded to floodplains, which are graded down-valley to the sea, what else can happen? Surely, as long as solar energy is supplied and sediment is exported, the landscape must continue to evolve. Local relief will decrease, slope angles will decrease, chemical weathering will increasingly dominate mechanical weathering, mass wasting will become slower and slower, and each increment of change will take a longer time. The extrapolations provided by the denudation rates in the Hydrographers Range (Fig. 12–5) offer some clues to the later stages of landscape development. The exponential loss of relief with time is such that of the nearly 5 million years of subaerial denudation that is predicted for an initial elevation of 1000 m, nearly one-half of the time is required to remove the final 10 m of relief. Unless tectonic uplift or volcanism renew the available relief, fluvial landscapes seem doomed to a condition of low elevation above sea level and low local relief for most of their life histories. A corollary to Powell's dictum is that if mountains cannot long remain mountains, they are very likely to become low rounded hills, deeply mantled with weathered residuum and drained by low-gradient river networks that are thoroughly adjusted to the slight but still declining available energy.

PENEPLAINS AND OTHER DEDUCED ENDFORMS OF EROSION

The History of an Idea

During the nineteenth century, geomorphologists gradually accepted the ideas of Hutton and Playfair that vast amounts of subaerial denudation had preceded the shaping of the modern landscape. The enormity of geologic time was demonstrated by stratigraphy and faunal succession. The erosion of valleys by rivers and glaciers became an accepted uniformitarian concept, and the progressive changes of landscapes through extended time became the subject for geomorphic speculation and debate. By 1874, G. K. Gilbert could propose that the basin-range fault blocks were much younger than the folding and metamorphism of the rocks within the ranges (p. 46), and in 1876, J. W. Powell used the phrase "the great denudation" for this erosion interval. Powell's dictum (p. 98), which has been noted repeatedly in this chapter, illustrates the attitudes of the time.

By 1883, W. M. Davis had begun development of his *geographical cycle,* or *cycle of erosion,* inductively generalizing from his own field work and that of his contemporaries that fluvial landscapes evolve in a predictable genetic sequence, which he characterized by analogy with the biological terms *youth, maturity,* and *old age* (Davis, 1885, 1899, 1902). Present readers should note that Davis wrote during the peak of intellectual enthusiasm for Darwinian evolution. Determinism was a popular philosophical scheme, and Davis's use of the term "cycle" implied not so much a steady-state condition but a directed sequence of events from an assumed initial state to a deduced endform. It was cyclic only by the assumption that the initial state could be recreated by tectonism or volcanism.

Davis assumed that rapid uplift initiated a "cycle" of erosion. The prevailing tectonic theory of his time called for worldwide pulses of orogeny followed by stillstand, so his assumption was readily accepted. After uplift, he deduced a stage of youth, when rivers rapidly deepened and extended their drainage basins and valley systems, and regional landscapes were dissected. He chose as the criterion of maturity in valleys the development of free-swinging meanders in trunk streams, presumably on alluvial floodplains. For the regional landscape, total removal of the former land surface from interfluves was the primary criterion of maturity. Other criteria, such as total elimination of waterfalls and lakes from river valleys, sharp-crested divides, and high drainage density, were also deduced by Davis but are probably better omitted because they are usually determined by structure or climate.

Davis emphasized that old age is not separated from maturity by any specific criteria. As in organisms, where maturity is defined by the specific condition of reproductive capacity, in geomorphic analysis the twin criteria of meandering rivers and complete upland dissection are specific features that can be observed. Old age is less easily defined. In organisms it is only an extension of maturity, with a decrease in metabolic activity, loss of facilities, and an intangible decline of ability. As a criterion of old age in landscapes, Davis suggested that floodplains would become several times wider than the meander belt. No recent studies of fluvial mechanics give any evidence that meander belts develop higher-order "supermeanders" or that floodplains increase their width through time beyond the amplitude of the meanders. The regrettable use of maps of the lower Mississippi River floodplain to illustrate an old age landscape has confused deductions about the later stages of erosion. Most introductory students of geology and geography are taught that the very young, depositional landscape of the aggraded Mississippi Valley (Fig. 10–2) is representative of old, erosional landscapes! Also, Davis's old landscapes were deduced to be covered with broad sheets of alluvium over which the rivers sluggishly meandered. A more likely deduction is that low hills of deeply weathered rock and colluvium would form more of the regional terrain than Davis implied. The hills would be the drainage divides and would either be localized by structural controls or be at the maximum distance from the sea. As noted earlier (p. 305), the apparent contradictions between upland denudation rates and river sediment loads may lead, when resolved, to important new deductions about the later stages of erosional landscape evolution.

In an attempt to describe the form of landscapes that have undergone long-continued weathering and erosion in humid climates, Davis (1889) introduced the elegant word **peneplain.** He used as a root the word "plain," in the geographical context of a regional surface of very low relief near sea level. Realizing that the ultimate base level is the limit of subaerial erosion, which like a mathematical limit may be approached but might not be achieved, he prefixed to the word "plain" the Latin derivation "pene," meaning almost. Thus "peneplain" was introduced to the scientific literature as *a surface of regional extent, low local relief and low absolute altitude, produced by long-continued fluvial erosion.*

How can we define a landscape that is "almost" gone? Apparently, the peneplain that W. M. Davis had in mind was far from a mathematical plane, or even a "peneplane," as some have proposed to spell the word. He envisioned an erosionally dissected landscape, one that includes the drainage networks of several master streams that enter the sea, in which the total relief is no more than about 100 m at the regional divides. One story is that he described a peneplain as a surface over which a horse could draw a carriage at a trot in any direction. In our modern era of excessive horsepower, it is hard to envision such a gently sloping surface of erosion, but it would not be planar.

Recent measurements of denudation rates provide quantitative support for the peneplain concept. Even at very low denudation rates of 1 cm/1000 yr or less, fluvial landscapes would be lowered, especially by solution, as much as 10 m/million years. From the rates listed in Table 12–1, Judson and Ritter (1964) concluded that the entire conterminous United States could be removed to sea level in 11 to 12 million years, if no isostatic or tectonic uplift occurs. Schumm (1963) concluded that regional peneplanation might require 15 to 110 million years, allowing for isostatic compensation. Ahnert (1970) suggested that 11 to 18.5 million years would be required to reduce the relief of a large mid-latitude drainage basin to 10 percent of its original relief. Clearly, the concept of cyclic time fits comfortably within the late Cenozoic Era (Young and Saunders, 1986, p. 15) and large areas of anorogenic continental landscapes have had time to be reduced to low altitude and low local relief. A complication that is yet to be evaluated is the importance of epeirogenic vertical movements (Chap. 2) with neotectonic rates well in excess of probable denudation rates.

Ancient continental shields have persisted within a few hundred meters above and below sea level for substantial periods of Phanerozoic time (Chap. 2; Wise, 1974). The relief on these cratonic areas has probably been regraded repeatedly during the last billion years. The present low relief of the Canadian shield for example, attributed by some to intense and repeated continental glaciation (Chap. 17), is replicated by the sub-Paleozoic and sub-Cretaceous unconformities where those rocks lap onto the shield (p. 424). Apparently, the shield has not had significant relief for at least the last 400 million years.

The exhumed erosion surface that truncates Precambrian basement rocks of the Argentine Pampean Ranges (Fig. 3–13) is another example of an ancient erosion surface of very low relief. It shows comparable monotony whether it has been exhumed from beneath late Paleozoic, Mesozoic, or late Cenozoic cover strata (pp. 47, 273). However, because the cover strata are not marine, the Pampean surface cannot be proved to have formed near sea level. Fairbridge and Finkl (1980) have proposed that ancient cratonic shields are repeatedly exhumed and modified, and are maintained at low relief throughout geologic time by a variety of erosional process. They propose that the term *peneplain* be applied to these polygenetic surfaces, broadening its definition to remove the restrictive condition of cyclic fluvial erosion. It appears that after several decades of emphasis on the role of processes in geomorphology, renewed interest is developing in the logical deductions permitted by cyclic time. Perhaps "NeoDavisian" geomorphology will become a productive approach.

Other Concepts of Sequential Development and Endforms

Davis's cycle of erosion made only passing reference to tectonic movements other than initial uplift followed by stillstand and erosion. He did allow for interruptions to the cycle by renewed intermittent uplift, but these were treated only as temporary disturbances that would initiate new downcutting and develop river terraces and compound valley forms. The most notable attempt at formulating a systematic geomorphology that related tectonic movement to erosion can be found in the writings of Walther Penck (Penck, 1924; English transl., 1953). His major works, written during a terminal illness and published posthumously in fragmentary form, are rightfully known for their complex and obscure German prose. Nevertheless, they attracted much attention among geomorphologists as an alternative or even a challenge to Davis's cyclic concept. Davis himself wrote a critical review of Penck's proposals (Davis, 1932) that became the primary source for most English-speaking geomorphologists. Unfortunately, certain errors of translation and interpretation in Davis's review have been misattributed to Penck ever since (Simons, 1962). The impact of Walther Penck's contributions on American geomorphology can be best judged by the published papers and discussions of a 1939 symposium on the subject (von Engeln, 1940). One of the better reviews of Penck's theories was by O. D. von Engeln (1942, pp. 256–68).

Penck accepted the idea that landscapes could be reduced to endforms of low relief but maintained that these endforms never became the initial forms for new episodes of dissection. By his theory, uplift was not rapid and then zero, as Davis supposed, but always began slowly, waxed to a maximum, and then waned gradually to a stop. During the long stage of initial slow uplift, all prior forms were destroyed, and a new surface of low relief, adjusted to a balance between uplift and degradation, developed. This initial surface of low relief was called a **Primärrümpf** by Penck and was the basis for his deductions. Instead of hypothesizing sequential forms, he maintained that slopes, once developed, retreat parallel to themselves and persist until they intersect. During downcutting by rivers, such as might be caused by accelerating uplift, the initial slopes would be broad and gentle, but as the rate of uplift accelerated, inner valley slopes would be steeper. Because each increment of slope profile retreated from the valley axis and was followed by a lower slope element of steeper gradient, convex skyward slopes would be diagnostic of accelerated uplift. Similarly, straight slopes would indicate a constant rate of stream downcutting, and concave slopes would indicate a waning rate of entrenchment. Thus, as uplift died away, landscapes were theorized to be an assemblage of concave forms, to be converted to a *Primärrümpf* during the next episode of initial slow uplift.

No erosional sequence of forms was allowed by Penck's scheme, because each morphologic assemblage was related to a certain tectonic condition. His commendable primary goal was to use geomorphology to determine regional tectonic history. Penck denied any climatic control of geomorphic processes other than glaciation, believing that tectonics alone determined landform assemblages. Yet it is a curious fact that Penck's ideas of slope retreat have been found most analogous to processes that shape mountain fronts and pediments in semiarid climates (Chap. 13).

A later generation of geomorphologists have used the pediment landform as the basis for deductions about sequential evolution of landscapes. Maintaining that hillslopes retreat parallel to themselves and leave a residual low slope of transportation (the pediment) at their base, these authors propose that huge areas can be reduced to a series of coalescing pediments capped by residual rock knobs and that successive generations of coalescing pediments (or *pediplains,* by analogy with peneplains) can migrate across a landscape, each one consuming higher and older surfaces by parallel scarp retreat, while in turn being consumed by the scarp retreat of a lower and younger generation of pediments (p. 337). No sequence of landforms is admitted by this theory either, because once formed, the pediments are essentially immutable, stable landforms (King, 1953). The origin and significance of pediments and related landforms is considered in the next chapter, but it is worth noting that very large areas of the earth's surface are pediment-prone, and in the geologic past, before the evolution of vascular plants, the entire subaerial landscape may have been shaped by pediment-forming processes that we now associate with semiarid climates.

The emphasis on internal self-adjustment among complex variables in open physical systems, such as characterizes modern stream hydrology (Chap. 9), has led to a hypothesis of "steady-state" landscapes (Hack, 1960). Critics of Davis's cycle of erosion argue that when alluvial floodplains develop, even in small streams, the landscape is essentially graded and therefore deny the progressive development of graded master streams as a criterion of the stage of regional evolution. The only sequential change they envision is an early one from initial disequilibrium to subsequent equilibrium. Thereafter, as long as the energy in the open system remains constant, the landforms also remain constant, even though mass is removed from the eroding terrane. Hack (1960, p. 85) applied the name "dynamic equilibrium" to this concept of a time-independent, open-system landscape in a steady state. He regarded most landscapes that had previously been called "maturely dissected peneplains" as nothing more than erosionally graded steady-state surfaces that carry no connotation of sequential development. Only when the energy of the whole system ultimately declines, as when the available relief is consumed, will the landscape evolve to one of reduced local relief, decreased slope gradients, and blunted divides.

Hack's landscape in dynamic equilibrium could be accommodated in Davis's cyclic theory by assuming extreme available relief, during the reduction of which great volumes of rock must be removed by rivers whose spacing and gradient are climate- and structure-dependent and, therefore, time-independent. However, his avowed purpose was to explain landscapes in an essentially noncyclic conceptual framework, and Hack rightfully challenged some of the hypotheses of multiple residual peneplains that have been inferred from accordance of summits in the Appalachian region. Summit accordance and many other landscape features, such as gravel-covered interfluves and summit regions, can well be explained by long-continued erosion and progressive adjustment to structure. Multiple cycles of erosion, each one terminating in a peneplain that was uplifted and dissected, are not required (Morisawa, 1989).

One can very well agree with Hack that most landscapes are in an uneasy dynamic equilibrium between the energy available for work and the work being done, during any particular interval of "graded" time. However, if regional elevation, and therefore the energy available for geomorphic change, continues to decline at a measurable rate, no landscape can be said to represent an open system in a steady state on the scale of "cyclic" time. As in the analogy of a rotary cement kiln (Chap. 5), the landscape must be constantly evolving toward equilibrium with the declining energy. The major contrast between Hack's "steady-state" landscape and Davis's cycle is that Hack deduced total regional dissection and graded rivers as a nearly permanent condition rather than as the mature stage of an evolving series.

The choice between these models depends on the relative rates of regional denudation and late Cenozoic tectonism. If, as in the example of the Hydrographers Range, total denudation in 5 million years can be inferred, evolving landforms can be deduced. If, however, tectonics and climatic change invalidate the assumption of initial uplift or other constructional process followed by stillstand and landscape evolution, the dynamic equilibrium model, changing only from disequilibrium to equilibrium, is most suitable as a basis for interpreting the present landscape. Until late Cenozoic climate change and tectonic history, especially the role of neotectonic epeirogeny, are better known, there seems to be room for multiple hypotheses in the interpretation of fluvially eroded landscapes.

REFERENCES

AHNERT, F. 1970, Functional relationships between denudation, relief, and uplift in large mid-latitude drainage basins: Am. Jour Sci., v. 268, pp. 243–63.

———, Modeling landform change, *in* Anderson, M. G., ed., Modelling geomorphological systems: John Wiley & Sons, Ltd., Chichester, England, pp. 375–400.

ANDERSON, M. G., ed., 1988, Modelling geomorphological systems: John Wiley & Sons, Ltd., Chichester, England, 458 pp.

BELL, M., and LAINE, E. P., 1985, Erosion of the Laurentide region of North America by glacial and glaciofluvial processes: *Quaternary Res.,* v. 23, pp. 154–74.

BÜDEL, J., 1982, Climatic geomorphology (transl. by L. Fischer and D. Busche): Princeton Univ. Press, Princeton, N.J., 443 pp.

CHAPPELL, J., 1974, Geomorphology and evolution of small valleys in dated coral reef terraces, New Guinea: Jour. Geology, v. 82, pp. 795–812.

COTTON, C. A., 1963, A new theory of the sculpture of middle-latitude landscapes: New Zealand Jour. Geol. Geophys., v. 6, pp. 769–74.

CURRAY, J. R., and MOORE, D. G., 1971, Growth of the Bengal deep-sea fan and denudation in the Himalayas: Geol. Soc. America Bull., v. 82, pp. 563–72.

DAHLEN, F. A., and SUPPE, J., 1988, Mechanics, growth, and erosion of mountain belts, *in* Clark, S. P., Jr., Burchfield, B. C., and Suppe, J., eds., Processes in continental lithospheric deformation: Geol. Soc. America Spec. Paper 218, pp. 161–78.

DALRYMPLE, G. B., 1971, Potassium–argon ages from the Pololu Volcanic Series, Kohala Volcano, Hawaii: Geol. Soc. America Bull., v. 82, pp. 1997–2000.

DAVIS, W. M., 1885, Geographic classification, illustrated by a study of plains, plateaus, and their derivatives [abs.]: Am. Assoc. Adv. Sci. Proc., v. 33, pp. 428–32.

———, 1889, Topographical development of the Triassic formation of the Connecticut valley: Am. Jour. Sci., v. 37, pp. 423–34.

———, 1899, The geographical cycle: Geog. Jour., v. 14, pp. 481–504 (reprinted 1954 *in* Geographical essays: Dover Publications, Inc., New York, pp. 249–78).

———, 1902, Base-level, grade, and peneplain; Jour. Geology, v. 10, pp. 77–111 (reprinted 1954 *in* Geographical essays: Dover Publications, Inc., New York, pp. 381–412).

———, 1932, Piedmont benchlands and Primärrümpfe: Geol. Soc. America Bull., v. 43, pp. 399–440.

DOLE, R. B., and STABLER, H., 1909, Denudation: U.S. Geol. Survey Water-Supply Paper 234, pp. 78–93.

DOUGLAS, I., 1967, Man, vegetation, and the sediment yield of rivers: Nature, v. 215, pp. 925–28.

VON ENGELN, O. D., 1940, Symposium: Walther Penck's contribution to geomorphology: Assoc. Am. Geographers Annals, v. 30, no. 4, pp. 219–84.

———, 1942, Geomorphology: The Macmillan Company, New York, 655 pp.

FAIRBRIDGE, R. W., and FINKL, C. W., JR., 1980, Cratonic erosional unconformities and peneplains: Jour. Geology, v. 88, pp. 69–86.

GIBBS, R. J., 1967, Geochemistry of the Amazon River system: Part I: Geol. Soc. America Bull., v. 78, pp. 1203–32.

GILLULY, J., 1964, Atlantic sediments, erosion rates, and the evolution of the continental shelf: Some speculations: Geol. Soc. America Bull., v. 75, pp. 483–92.

GLOCK, W. S., 1932, Available relief as a factor of control in the profile of a landform: Jour. Geology, v. 40, pp. 74–83.

GRAMLICH, J. W., LEWIS, V. A., and NAUGHTON, J. J., 1971, Potassium–argon dating of Holocene basalts of the Honolulu volcanic series: Geol. Soc. America Bull., v. 82, pp. 1399–404.

HACK, J. T., 1960, Interpretation of erosional topography in humid temperate regions: Am. Jour. Sci., v. 258-A, pp. 80–97.

HOLEMAN, J. N., 1968, Sediment yield of major rivers of the world: Water Resources Res., v. 4, pp. 787–97.

JUDSON, S., 1968a, Erosion rates near Rome, Italy: Science, v. 160, pp. 1444–46.

———, 1968b, Erosion of the land, or what's happening to our continents?: Am. Scientist, v. 56, pp. 356–74.

———, and RITTER, D. F., 1964, Rates of regional denudation in the United States: Jour. Geophys. Res., v. 69, pp. 3395–401.

KING, L. C., 1953, Canons of landscape evolution: Geol. Soc. America Bull., v. 64, pp. 721–52 (discussion and replies, v. 66, pp. 1205–14).

KIRKBY, M. J., 1980, The problem, *in* Kirkby, M. J., and Morgan, R. P. C., eds., Soil erosion: John Wiley & Sons, Ltd., Chichester, England, pp. 1–16.

LAMARCHE, V. C., 1968, Rates of slope degradation as determined from botanic evidence, White Mountains, California: U.S. Geol. Survey Prof. Paper 352-I, pp. 341–77.

LANGBEIN, W. B., and LEOPOLD, L. B., 1964, Quasi-equilibrium states in channel morphology: Am. Jour. Sci., v. 262, pp. 782–94.

LANGBEIN, W. B., and SCHUMM, S. A., 1958, Yield of sediment in relation to mean annual precipitation: Am. Geophys. Un., Trans., v. 39, pp. 1076–84.

LARSON, W. E., PIERCE, F. J., and DOWDY, R. H., 1983, The threat of soil erosion to long-term crop production: Science, v. 219, pp. 458–65.

MARCHAND, D. E., 1971, Rates and modes of denudation, White Mountains, eastern California: Am. Jour. Sci., v. 270, pp. 109–35.

MATHEWS, W. H., 1975, Cenozoic erosion and erosion surfaces of eastern North America: Am. Jour. Sci., v. 275, pp. 818–24.

MATTHIAS, G. F., 1967, Weathering rates of Portland arkose tombstones: Jour. Geol. Education, v. 15, pp. 140–44.

MENARD, H. W., 1961, Some rates of regional erosion: Jour. Geology, v. 69, pp. 154–61.

MILLIMAN, J. D., and MEADE, R. H., 1983, World-wide delivery of river sediment to the oceans: Jour. Geology, v. 91, pp. 1–21.

MILLS, H. H., 1976, Estimated erosion rates on Mount Rainier, Washington: Geology, v. 4, pp. 401–6.

MORISAWA, M., 1989, Rivers and valleys of Pennsylvania, revisited, *in* Gardner, T. W., and Sevon, W. D., eds., Appalachian geomorphology: Geomorphology, v. 2, no. 1–3, pp. 1–22.

PAINE, A. D. M., 1985, 'Ergodic' reasoning in geomorphology: Time for a review of the term?: Progress in Phys. Geography, v. 9, pp. 1–15.

PENCK, W., 1924, Die Morphologische Analyse: J. Engelhorn's Nachfolger, Stuttgart, 283 pp.

———, 1953, Morphological analysis of landforms (transl. by H. Czech and K. C. Boswell): St. Martin's Press, New York, 429 pp.

PITTY, A. F., 1971, Introduction to geomorphology: Methuen & Co. Ltd., London, 526 pp.

PLAYFAIR, J., 1802, Illustrations of the Huttonian theory of the earth: Dover Publications, Inc., New York, 528 pp. (facsimile reprint, 1964).

RAHN, P. H., 1971, Weathering of tombstones and its relationship to the topography of New England: Jour. Geol. Education, v. 19, pp. 112–18.

RUXTON, B. P., and McDOUGALL, I., 1967, Denudation rates in NE Papua from K–Ar dating of lavas: Am. Jour. Sci., v. 265, pp. 545–61.

SAUNDERS, I., and YOUNG, A., 1983, Rates of surface processes on slopes, slope retreat and denudation: Earth Surf. Proc. and Landforms, v. 8, p. 473–501.

SCHUMM, S. A., 1963, The disparity between present rates of denudation and orogeny: U.S. Geol. Survey Prof. Paper 454-H, 13 pp.

———, and LICHTY, R. W., 1965, Time, space, and causality in geomorphology: Am. Jour. Sci., v. 263, pp. 110–19.

———, MOSLEY, M. P., and WEAVER, W. E., 1987, Experimental fluvial geomorphology: John Wiley & Sons, Inc., New York, 413 pp.

SIMONS, M., 1962, The morphological analysis of landforms: A new review of the work of Walther Penck: Inst. Brit. Geographers Trans., no. 31, pp. 1–14.

TRIMBLE, S. W., 1975, Denudation studies: Can we assume stream steady state?: Science, v. 188, pp. 1207–8.

———, 1977, The fallacy of stream equilibrium in contemporary denudation studies: Am. Jour. Sci., v. 277, pp. 876–87.

———, 1988, The impact of organisms on overall erosion rates within catchments in temperate regions, in Viles, H. A., ed., Biogeomorphology: Basil Blackwell Ltd., Oxford, England, pp. 83–142.

WILSON, L., 1973, Variations in mean annual sediment yield as a function of mean annual precipitation: Am. Jour. Sci., v. 273, pp. 335–49.

WISE, D. U., 1974, Continental margins, freeboard and the volumes of continents and oceans through time, in Burk, C. A., and Drake, C. L., eds., The geology of continental margins: Springer-Verlag, Berlin, pp. 45–68.

WOLDENBERG, M. J., ed., 1985, Models in geomorphology: Allen & Unwin, Inc., Boston, 434 pp.

YAIR, A., and ENZEL, Y., 1987, The relationship between annual rainfall and sediment yield in arid and semi-arid areas. The case of the northern Negev, in Ahnert, F., ed., Geomorphological models: Theoretical and empirical aspects: Catena Supplement 10, pp. 121–35.

YOUNG, A., and SAUNDERS, I., 1986, Rates of surface processes and denudation, in Abrahams, A. D., ed., Hillslope processes: Allen & Unwin, Inc., Boston, pp. 3–27.

ZEITLER, P. K., JOHNSON, N. M., NAESER, C. W., and TAHIRKHELI, R. A. K., 1982, Fission track evidence for the Quaternary uplift of the Nanga Parbat region, Pakistan: Nature, v. 298, pp. 255–57.

PART III

CLIMATIC MORPHOGENESIS

A **morphogenetic system** is "the assemblage of processes or agencies locally or regionally operating under a prevailing climatic condition and combining to produce a particular type of landscape" (Cotton, 1958, p. 125; see also Tricart and Cailleux, 1972).[1] A **morphogenetic region** is then characterized by certain climatic factors that produce a distinct combination of geomorphic processes, which in turn may produce a landscape distinct in appearance from other regions (Birot, 1968). The concept of morphogenetic regions dominated by processes subordinates the factors of structure and time in geomorphology. Followed to the extreme conclusion, it suggests that each climatic region produces a unique assemblage of landforms independent of time and structure. No geomorphologist would accept such an extreme view, but within the science there is abundant room for debate about the relative importance of climatically controlled landforms and, for instance, structurally controlled landforms.

The dominance of fluvial processes on the subaerial part of this planet is the justification for giving those processes such emphasis in Part II. In some regions, however, water has never been known to flow. Sandy deserts may show only landforms shaped by wind; many glaciated terrains show evidence of ice erosion alone. Even within regions dominated by fluvial processes, the range of climatic variables, such as temperature and the amount and seasonality of precipitation, validates the study of climatic morphogenesis. The importance of climate in weathering and soil formation has already been described in Chap. 6.

W. M. Davis (1905a, b) considered the cool, humid, mid-latitude climate as "normal" and wrote at length about two "climatic accidents" that could affect landscapes: glaciation and aridity. We now realize that subtropical arid belts and polar icecaps are very much a part of the late

---

[1] For citations, see References, Chapter 13.

Cenozoic planetary climatic zonation and are not accidents. Even more significantly, the theory of plate tectonics provides a mechanism to move continents long distances over the earth's surface during the Cenozoic Era. As a result of plate motions and orogenies, climates have changed. The Antarctic ice cap and the present arid climate of most of Australia, for example, are related to late Cenozoic plate motions (p. 445). The late Cenozoic orogenic belt along the southern margins of Eurasia and the western margins of North and South America have profoundly affected the climate of the continents, and probably of the entire world (p. 436).

Even without considering the perplexing problems of past climatic change and plate motions, the present diversity of terrestrial climates is impressive and significant to geomorphology. The following table shows that a significant percentage of land area falls in each of five major climatic regions. According to other compilations, about 10 percent of the land area is now ice-covered, 20 percent is underlain by perennially frozen ground, and 33 percent is characterized by drainage that does not reach the sea. Unique combinations of processes are imposed on each of these landscape types.

Approximate Proportions of the Land Surface Falling within Köppen's Various Types at the Present Time

Climatic Group	Percentage of Land Surface
A. Tropical rain climates	20
B. Arid and semiarid climates	26
C. (Warm) temperate rain climates	16
D. Boreal forest and snow climates	21
E. Cold snow climates (treeless)	17

Source: Modifed from Lamb (1972), p. 513.

Not so obvious is the length of time available for the present morphogenetic systems to have shaped regional landscapes. In many regions, such as the 20 percent of the land area that was glacial as recently as 15,000 to 20,000 years ago but is now ice free, most of the landforms are the work of former glaciers and periglacial processes, not of the rivers that now flow there. Many, perhaps most, landscapes are *palimpsests* (p. 16) of forms shaped during a succession of different climatic conditions. In areas of former glaciation, relict and contemporary forms are relatively easy to distinguish, because glaciation imposes totally new processes on fluvial landscapes. Quaternary climatic changes in the lower latitudes apparently changed only the relative intensity of processes, not their kind. Areas now perennially humid became seasonally dry, and vice versa. One of the next great challenges to geomorphologists is to recognize the late Cenozoic sequence and succession of processes and landforms in the tropics and subtropics (Tricart, 1972). Because most of the developing nations of the world are in these regions, the geomorphic information can be applied to problems of water supply, land use, and mineral resources.

The six chapters of Part III build on the processes and deductions of Part II but reorganize the information with systematic emphasis on processes that are unique to or uniquely intense in certain climatic regions. For each region, an inquiry is made as to whether the assemblage of landforms is also unique, that is, whether the concept of morphogenetic regions applies only to processes or to landforms as well.

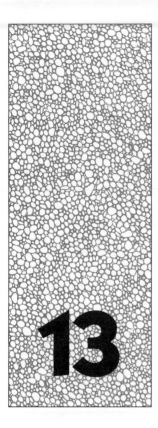

Arid, Semiarid, and Savanna Landscapes

13

DRY CLIMATES

The largest single identifiable morphogenetic region on earth is the dry region, where either seasonal or annual precipitation is insufficient to maintain vegetational cover and permit perennial streams to flow. Climatologists have devised various empirical formulas to define dry climates in terms of the ratio of precipitation to evapotranspiration. The temperature of a region must be incorporated into such formulas, because it is a major factor in evapotranspiration, but otherwise temperature is not a diagnostic parameter of arid climates. For instance, the Köppen system (Köppen and Geiger, 1936) defined the boundary between semiarid and humid climates with winter rains by the relationship

$$P \leq 20T$$

where P is the annual precipitation in millimeters and T is the mean annual temperature in Celsius degrees. The truly arid (desert) climate was defined as one in which precipitation is only one-half as great as the amount that locally separates the semiarid from the humid climate, or

$$P \leq 10T$$

Köppen and his assistants calculated other similar empirical formulas for climates in which precipitation is nonseasonal or is concentrated in the summer. Although apparently related only to mean annual precipitation and evaporation, Köppen's choice of boundaries was based primarily on vegetation. The parameters that he chose to define aridity also define areas in which insufficient moisture is available to support continuous vegetational cover of the ground. By Köppen's definitions, 26 percent of the land area is dry (see table, p. 328).

In another climatic classification, Thornthwaite (1948) attempted to correlate

329

potential evapotranspiration to other climatic variables, such as temperature and length of day, and thereby to define more closely the limits of dry climates as expressed in plant growth. His formulas apply primarily to the United States, where the minimum annual precipitation required for a humid climate increases southward from about 500 mm in North Dakota to about 750 mm in Texas. Less than 250 mm of annual precipitation produces a desert in almost any temperature range. By Thornthwaite's definitions, fully 36 percent of the land area is dry, even excluding the high-latitude frozen deserts.

The most recent attempt at classifying and delimiting arid regions was by Meigs (1952, 1953; Cooke and Warren, 1973). He used Thornthwaite's definitions but remapped areas based on better climatic data. Meigs' map (Fig. 13–1) shows 34 percent of the land area to be dry, including 4 percent defined as extremely arid, where no precipitation has fallen for at least 12 consecutive months at least once in the period of record and where rainfall lacks a seasonal rhythm.

By these various definitions, from one-fourth to more than one-third of our land is dry. Recent estimates by the United Nations expand the area of aridity to 43 percent of the land, owing in part to overgrazing and other unwise agricultural practices. Although not yet scientifically documented, these estimates suggest that aridity is expanding rapidly in our generation and emphasize the significance of the arid and semiarid morphogenetic regions.

In low latitudes, two belts of aridity coincide with the subtropical anticyclonic belts of high atmospheric pressure about 15 to 30° north and south of the equator. They are fringed by relatively narrow belts of semiarid transitional climates. In middle latitudes, dry climates are localized in the interior of the large continents. Temperature and evaporation are not as high in these regions as in the subtropics. Some precipitation may fall as snow. Mid-latitude dry climates are characterized by relatively large areas of semiaridity around smaller cores of true desert. Freezing temperatures during winter months are a common feature of mid-latitude semiarid climates. Small areas of dry climate also are on the west sides of the continents, where cold ocean currents flow offshore, and on the downwind, or rain shadow, side of mountain ranges. The polar deserts are excluded from consideration in this chapter, because they are either ice- or snow-covered, or underlain by perennially frozen ground (Chap. 15).

It should be noted that when climatologists attempt to define dry climates, they use the criterion of discontinuous vegetative cover. However defined, in arid regions erosion is not inhibited by a continuous cover of plants. Furthermore, rivers flow only seasonally at best, and even when flowing, they decrease in discharge downstream, in contrast to the streams in humid regions. Moisture deficiency causes groundwater to move very slowly and upward rather than downward in the soil, with the result that secondary minerals accumulate in soil profiles, and soil structure is strongly affected. For the geomorphologist, discontinuous flowing rivers, the lack of continuous ground cover, and the accumulation of secondary minerals in soil profiles define the dry morphogenetic region. Each has a significant effect on landscape evolution.

The tropical savanna climatic type was included with the tropical rainforest as the "tropical rain climates" by Köppen (p. 328). It belongs neither to the arid climates nor to the constantly wet tropical rainforest. Its diagnostic feature is the strongly seasonal precipitation. The tropical savanna climate is, in addition to being hot, dry for a significant part of each year, even though total rainfall exceeds the definitions used to define the dry regions by climatologists. In the savanna climate, which covers about 15 percent of the earth's land area, precipitation is strongly seasonal, usually coming during the hottest season. Toward the equator, the savanna grades into tropical rainforest, and poleward it grades into the subtropical dry belts. The savanna is a region of sparse open woodlands and thorny shrubs, with continu-

Figure 13-1. Dry regions of the world. (After Meigs, 1953.) Areas with mean annual temperature < 10°C are excluded. Definitions of degrees of aridity are discussed in the text.

Extremely arid regions

Arid regions

Semiarid regions

Sandy deserts

ous grass cover. Most photographs of savanna landscapes convey the impression of aridity, because most are taken during the dry season when the grass is brown and many trees have shed their leaves. During the wet months, however, very heavy rainfalls are common, and vast areas of savanna plains are under standing water. The year-round high temperatures and seasonally abundant water promote intense chemical weathering, even though rivers flow only during the rainy season and for a brief interval thereafter.

GEOMORPHIC PROCESSES IN DRY CLIMATES

The geomorphic significance of dry climates begins with the vegetation. Semiarid climates usually support a sparse grassland or **steppe** vegetation. In the United States, the boundary between humid and semiarid climates is approximated by the transition westward from medium-height grasses with a continuous turf or sod in the humid regions to short, shallow-rooted bunch grasses on otherwise bare ground in the semiarid regions. In arid regions, even the bunch grasses disappear, and the vegetation is, at best, widely spaced shrubs and salt-tolerant bushes. The maximum sediment yields have been reported from the arid/semiarid transition (Fig. 12–7), especially if the natural vegetation has been disturbed by the grazing of domestic cattle or by cultivated agriculture (Trimble, 1988). Wind assumes importance as an agent of erosion and transportation (Chap. 14).

In savanna regions, the seasonal rejuvenation of plant growth provides a ground cover during the rainy season, but barren ground is widespread during the dry months. Fires are common during the dry season; some are started by natural causes, but many are started by people in order to clear the land or to drive wild game. Some biogeographers suspect that savanna grassland has greatly expanded due to human activity. If so, this important morphogenetic region has become much more important in the later Cenozoic Era. At the earliest, savannas could not have existed prior to the early Tertiary evolution of grasses (Chap. 1). Grass pollen is common in Miocene continental sediments, and savannas probably had became widespread by then.

Soils and Weathering

The general rate of arid-climate weathering is very slow. The rare rainfalls and night dew apparently provide the moisture for hydrolysis and hydration. The puzzling fractured stones of hot deserts were described on page 123. Mechanical weathering becomes relatively important in dry regions as compared to chemical weathering, but many researchers believe that even in the driest deserts, chemical decomposition involving water dominates over mechanical disintegration.

With a deficiency of water, dry soils do not develop strong diagnostic horizons except for salt crusts and concretionary layers. The Mollisols of the humid temperate grasslands (Fig. 6–7 and Table 6–1) give way to the Vertisols, Aridisols, or Entisols of semiarid and arid regions. The definitive climatic criterion—precipitation less than evaporation—means that if soil moisture is available by lateral infiltration, as for instance in a desert basin flanked by more humid mountain slopes, the water is drawn upward through the soil by capillarity and evaporates either in the soil profile or at the ground surface. Dissolved mineral matter is precipitated when the soil water evaporates. Alkali salts (sodium and potassium carbonates), gypsum, and halite are found in desert soils. Under less extreme aridity, calcium carbonate is the most common mineral that accumulates, either as coatings on pebbles, as small concretions, or as massive cemented layers. The term "caliche" has been indiscriminately used for a variety of concretionary arid soils, but the informal adjective *calcic* is now recommended for soils that contain varying amounts of pedogenic carbonate minerals. **Calcrete** is a massive shallow accumulation of soil carbonate that may be a meter or more in thickness and may effectively control scarp formation. A series of six stages of calcium carbonate accumulation in arid and semiarid soils has been established, ranging from a few percent $CaCO_3$ that forms coatings on pebbles (Stage I), through layers of coalesced nodules (Stage III), to massive cemented surface layers 0.5 to 2 m in thickness in which 25 to 75 percent of the soil mass is pedogenic calcrete (Stages IV to VI) (Machette, 1985, p. 5). Stage I soils range from Holocene to late Pleistocene in age; Stage IV and older calcrete layers are found on land surfaces that range in age from 100,000 years to several million years. If the relative degree of calcification of soils on a series of river terraces or alluvial fans can be calibrated by radiometric dating of overlying lava flows or other means, it becomes very useful in reconstructing the late Cenozoic geomorphic history of the region (Chaps. 17 and 18). Most of the carbonate that has accumulated in dry soils is probably derived from windblown dust-size fragments of calcium-bearing minerals that were previously weathered by hydrolysis to calcium bicarbonate in solution (p. 130), then precipitated as calcium carbonate and blown by the wind. Lesser amounts are derived from weathering in place and from upward or laterally moving groundwater (Péwé et al., 1981, p. 182; Machette, 1985, p. 9).

Savanna soils are as yet little studied. The seasonal wet-and-dry climate, continuously warm, probably causes the maximum development of Ultisols and Oxisols (Table 6–1). Vertisols ("black cotton soils") are common in eastern Australia, India, and western Africa (Montgomery and Askew, 1983). Lateritic weathering (p. 142; Fig. 6–8) extends to a depth of 100 m or more, even though the seasonal runoff is not adequate to remove the weathered debris. The intensity of chemical weathering sets the savanna region apart from the arid and semiarid regions as a morphogenetic unit. Lateritic weathering, with its irreversible hardening of near-surface iron-rich layers, may have promoted the expansion of savanna grasslands after human beings cleared and burned the forests (Goudie, 1982, p. 41).

Rock Varnish

Exposed rock surfaces in arid and semiarid regions have long been known to darken with age, due to a surface accumulation of manganese and iron oxides called *desert varnish,* or more generally, **rock varnish.** Electron microscope and x-ray analyses have established that rock varnish is composed mostly of alkaline atmospheric dust, cemented by manganese and iron oxides that are mobilized and concentrated by manganese-oxidizing microbes (Dorn, 1981). These microbes are intolerant of acidic environments such as lichen-covered rocks or rocks with acid-producing endolithic cyanobacteria (p. 134) inhabiting their outer layers. They thrive where alkaline (especially Ca- and K-rich) dust maintains near-neutral pH values on the rock surface.

Rock varnish includes a fractional percentage of organic matter from the microbes as well as from airborne organic dust such as pollen grains, and radiocarbon ages can be determined for the time of varnish initiation. The techniques have been confirmed by dating rock varnish on various Holocene lava flows of known ages (Dorn et al., 1989). Also, the ratios of mobile cations such as K and Ca relative to stable elements such as Ti in the varnish decrease with time, and the cation ratios have been calibrated by radiocarbon and other dating methods to provide an alternate chemical dating method (Dorn, 1989). Rock-varnish dates provide only a minimum age for the exposure of a rock surface or stone tool, but the technique is one of the several important new means of dating land surfaces (p. 302). Of probable future significance is the fact that rock varnish is detectable by airborne or earth-orbiting spectrometers and can be used to estimate the relative age of alluvial fans, lava flows, and other barren rock surfaces by remote sensing.

Mass Wasting and Erosion

Soil creep in deserts is not the same as in humid climates, primarily because the surface layers are not laced together with a turf of interconnecting roots. Also, desert soils are likely to be broken into prisms of weakly cemented calcrete by vertical desiccation cracks, so that creeping masses are blocks rather than granular layers. Because of the coarse, angular debris, hillsides are likely to be steeper than those of humid regions, even in colluvium. Taluses are obvious at the bases of cliffs. Summit convexities develop on easily weathered rocks such as shale, but resistant strata, joints, dikes, and other structural units are clearly exposed and the landscape has an angular appearance with many scarps and benches.

In semiarid regions, slopes are at least partly stabilized by vegetation, usually shrubs (sagebrush, etc.) and grass. The association of plains (*steppe, prairie, pampas,* and *veld* are a few of the regional terms) with semiarid grasslands is notable and may be genetic. In hills or mountains, trees can be sheltered from sun, winds, and fire, but open plains offer no such protection. Fire, burning uninhibited across plains, may be the controlling factor of their vegetational patterns, even more important than climate (Wells, 1970).

Precipitation inadequate for trees is also typically erratic and variable from season to season or year to year. Semiarid regions are frequently cursed by droughts and floods, sometimes in the same year. The rate of landscape denudation is rapid under these conditions. Mountains in the semiarid climate have some of the most rapid rates of denudation found anywhere on earth, although denudation in mountainous tropical rainforests and wet periglacial mountain regions can be even more rapid (Tables 12–1 and 12–2).

Stream channels in arid and semiarid regions are likely to be flat floored, wide (to carry massive bed loads), and steep walled (because of the strong vertical or columnar jointing in dry soils). Their longitudinal profiles are steep. When dry, which is most of the time, they are floored with coarse alluvium. In Spanish-speaking regions, these distinctive channels are called **arroyos.** The equivalent Arabic term **wadi** (or **ouadi** in French) is widely used in North Africa and the Middle East.

The intensely seasonal precipitation of the tropical savanna briefly causes massive bed-load transport in braided channels that are dry for most of the year. Hillslopes are often bare joint-controlled rock, washed free of weathered debris. The savanna landscape has a unique assemblage of landforms that is described later in this chapter.

LANDFORMS OF DESERT REGIONS

Desert Plains and Plateaus

By far the greatest proportion of the world's arid regions are monotonous plains and plateaus in the interiors of large continents (Fig. 13–1). There is no compelling reason why dry regions should be nonorogenic, but nonorogenic regions are likely to form the large interior parts of continents where moist winds cannot penetrate, and nonorogenic regions also lack the relief necessary for orographic precipitation. In addition, internal drainage characteristic or even diagnostic of deserts causes alluviation that buries minor relief. Wind has been suggested as a leveling agent of desert plains as well, but evidence presented in Chapter 14 suggests that wind is more likely to increase desert relief by the excavation of basins and construction of major dune landforms.

Where alluvial fans merge in a desert basin, a saline lake (**playa, salar,** or **sabkha**) may form. Here, mud and salts accumulate to form the most level topographic surfaces developed on earth. Water less than a meter in depth may briefly accumulate, but when the water has evaporated, a great flat of polygonal-cracked mud or salt is all that remains. Residual brines may ooze from beneath the encrusted surface of the playas, too concentrated to evaporate further but liquid enough to engulf motor vehicles that are driven onto the surface. On other playas, hard crusts of salt and mud provide nearly perfect surfaces for racing high-speed cars and landing airplanes and space craft.

Dust and sand are swept from desert fans and floodplains, leaving behind a lag concentrate of gravel called a **reg,** or stony desert. A **desert pavement** may form a stable surface of closely fitted pebbles, each polished by sand blast and often faceted as well. If wind abrasion is not severe, rock varnish may coat the exposed surface of the pavement. If the rocks have been moved, their lighter under surfaces betray the movement. Animal trails and vehicle tracks across reg may remain visible for decades or centuries.

Approximately one-fourth of the Sahara and the Arabian desert are sandy (Chap. 14). The remainder is either **hammada** (barren rock) or reg. Rock ledges are pitted and polished by abrasive windborne sand, but the dominant relief is fluvial. Even in the driest desert plains, drainage networks cover the landscape. Some are used by present runoff, even if rain falls less than once in 12 months, but an unknown portion of the desert dissection is probably due to formerly greater runoff. The issue of climatic change (Chap. 18) is of major significance to theories of landscape evolution in arid regions. We do not yet know enough about either present processes or Quaternary climatic change in the dry regions to distinguish confidently between active and relict landforms. It has been argued that all but the most arid landscapes are essentially "fluvial" (Graf, 1988, p. 10).

Mountainous Deserts

The desert Basin and Range province of the southwestern United States and Mexico, the Pacific coastal desert of Chile and Peru, and the fold mountains adjacent to the Persian Gulf (Fig. 11–14) are examples of desert mountains (Fig. 13–2). Their area is small compared to the continent-sized deserts of Australia and the Sahara and the great desert plains of Arabia, trans-Caspian USSR, and western China. The Basin and Range province has been particularly significant in the development of the science of geomorphology, for it was the excellent analytic and descriptive literature about this area written in the late 19th century that became the basis for W. M. Davis's theory of landscape evolution. Europeans of that time also followed the exploration of the American west with great interest, so for a time the arid and semiarid territories of America became the leading topic of geomorphic thought. As noted in Chap. 3 (p. 45), the nearly unconscious association of fault-block tectonic relief with arid and semiarid landscapes has permanently biased geomorphic thought, especially in the United States (Graf, 1988, pp. 20–21).

Mountains in deserts are likely to be islands of nondesert climate rather than part of the desert. Orographic precipitation and sometimes snowmelt feed streams that dissect mountains in deserts much as they dissect mountains in other climatic zones. The contrast in morphogenesis is most notable at the base of desert mountains, where the streams discharge from the mountains onto the surrounding dry *piedmont* ("foot-of-the-mountain") slopes.

The dominant landform of intermontane deserts must be the alluvial fan (p.

Figure 13-2. Death Valley, California, northwest from Dante's View. Compare Figure 13-3. Note saline playa at the base of massive alluvial fans and the exposed regolith in the foreground. (Photo: U.S. Borax Co.)

261). Ephemeral rivers decrease in discharge as they enter desert basins and deposit their entire load as they evaporate or infiltrate. It is no exaggeration to say that desert mountains are progressively buried in their own waste. The great apron of coalescing fans at a mountain front is called a **bajada** (or bahada).

The alluvial fans of Death Valley (Fig. 13–2) have been analyzed as equilibrium, or steady-state, landforms (Denny, 1965, 1967). Each fan has an area equal to one-third to one-half of the mountain watershed that feeds it. Fans grow until their surface area becomes so large that the rate of sediment supply from the mountain is balanced by the rate of erosion of the fan surface by gullies that originate on the fan itself. If the catchment area is large and the depositional basin is small, bajadas will comprise nearly the entire piedmont area (Fig. 13–3). Frequent shifts in distributary channels and capture of aggrading, sediment-transporting channels by eroding fan channels ensure the symmetric evolution of fan forms (Chap. 10).

The best-known mountainous desert is the Sonoran region of the Basin and Range province in southwestern United States and adjacent Mexico. Much of the area is graded to sea level by intermittent streams and by tributaries to the lower Colorado River and Gila River. Although the region appears to consist of mountains rising above broad alluvial plains or fans, much of the intermontane surface is of planed rock, similar to that which composes the mountains. An estimated 20 percent of the region consists of mountains, 40 percent of barren or thinly veneered bedrock plains, and 40 percent of thick alluvium (McGee, 1897, p. 91). Early explorers were surprised to find that the gently sloping piedmont surfaces 5 to 8 km from the mountain front were cut across hard granitic rocks, with only a thin layer of alluvium on the surface: "so thin that it may be shifted by a single great storm" (McGee, 1897, p. 91). The wide extent of barren piedmont plains in the Sonoran district is probably permitted by the external drainage, which exports alluvium to the sea rather than permitting it to accumulate in closed intermontane basins.

Figure 13-3. Bajada and saline playa of Death Valley, California. Dante's view (Fig. 13-2) in right center. Rock varnish (p. 333) of variable darkness indicates the relative age of the fan surfaces. (Satellite image: SPOT Image Corp., Reston, VA © CNES, 1990, courtesy T. G. Farr.)

The significance and extent of such rock surfaces in the Sonoran desert and elsewhere has led to extensive further study. Bryan (1922, p. 88) formally adopted the elegant term **mountain pediment** for

> a plain which lies at the foot of mountains in an arid region or in headwater basins within a mountain mass. The name is applied because the plain appears to be a pediment upon which the mountain stands. A mountain pediment is formed by erosion and deposition of streams, usually of the ephemeral type, and is covered with a veneer of gravel in transit from higher to lower levels. It simulates the form of an alluvial fan.

In classic architecture, a pediment is the triangular end of a low-pitched gable roof, on which a frieze is often carved. The term *mountain pediment* was chosen because in profile, when viewed from a distance, each mountain peak seems to stand at the crest of a low conical plain, triangular in profile (Fig. 13–4). Today, the landform is referred to simply as a **pediment.** It has no relation to the term "pedistal," with which it is frequently confused, or to the very general descriptive term "piedmont," which simply means "at the foot of the mountain." The true statement that pediments are piedmont landforms has confused generations of beginning geomorphology students.

The slope on pediments, as on alluvial fans, increases toward the mountain front. Typical gradients range from 1 to 7°. Headward portions of pediments may penetrate into a mountain mass, isolating segments of the range and possibly intersecting with pediments from the other side of the range at pediment passes (Howard, 1942). The pediment form is an excellent example of a water-spreading wash

Figure 13-4.   Mountains near Maricopa, Arizona, surmounting a pediment. The boulder-controlled mountain front has an abrupt junction angle with the pediment. (Photo: R. E. Moeller.)

slope (Fig. 8–20). Functionally, it is an ideal surface to distribute and dissipate a mass of water and sediment introduced at its apex. By original definition and conventional use, the pediment is a landform of dry regions. However, some authors have unfortunately extended the term to include all water-eroded plains.

As first used by Gilbert (1882, p. 183), the term "pediment" was casually applied to the surfaces of the alluvial fans that encircle desert mountains. Bryan's definitive study, following McGee's observations in the same general area, emphasized that pediments are slopes of transportation cut on bedrock, usually covered with a veneer of alluvium in transit from high to lower levels. In form and function, a pediment is similar to an alluvial fan, the difference being that a pediment is an erosional landform and a fan is constructional.

The origin of pediments is closely related to morphogenetic processes of dry regions. McGee's dramatic eye-witness account of a **sheetflood** in the Sonoran desert (McGee, 1897, pp. 100–101) was especially influential in establishing the sheetflood as an important geomorphic process.

The shower passed in a few minutes and the sun reappeared, rapidly drying the ground to whiteness. Within half an hour a roar was heard in the foothills, rapidly increasing in volume; the teamster was startled, and set out along the road up the valley at best speed; but before he had gone 100 yards the flood was about him. The water was thick with mud, slimy with foam and loaded with twigs, dead leaflets and other flotsam; it was seen up and down the road several hundred yards in either direction or fully half a mile in all, covering the entire surface on both sides of the road, save a few islands protected by exceptionally large mesquite clumps at their upper ends. The torrent advanced at race-horse speed at first, but, slowing rapidly, died out in irregular lobes not more than a quarter of a mile below the road; yet, though so broad and tumultuous, it was nowhere more than about 18 inches and generally only 8–12 inches in depth. . . . For perhaps five minutes the sheetflood maintained its vigor, and even seemed to augment in volume; the next five minutes it held its own in the interior, though the advance of the frontal wave slackened and at length ceased; then the torrent began to disappear at the margin, the flow grew feeble in the interior, the water shrank and vanished from the margin up the slope nearly as rapidly as it had advanced, and in half an hour from the advent of the flood the ground was again whitening in the sun, save in a few depressions where muddy puddles still lingered.

McGee's description led to a general assumption that sheetfloods erode weathered bedrock surfaces to the perfection of pediments. Other geomorphologists proposed that lateral corrosion by streams (or **streamfloods,** episodic high-intensity discharges confined to channels) is the dominant process of pediment formation (Johnson, 1932). The controversies between proponents of these and other hypotheses of pediment evolution were summarized by Hadley (1967).

Field observations of desert storms are sufficiently rare that processes of erosion and deposition are poorly known. Rahn (1967) described examples of flood activity on piedmont slopes in Arizona and concluded that the floods on pediments were streamfloods, confined to channels but exhibiting high velocity and capable of impressive erosion and deposition. Only when rain fell directly onto bajadas were sheetfloods produced. Rahn concluded that pediments are likely to be "born dissected" by channeled flow near mountain fronts and later smoothed by weathering and sheetfloods as the mountain front retreats.

Apparently, pediments widen by the lateral migration of rills and ephemeral channels that periodically impinge on the mountain front and undercut taluses and cliffs. They also expand by the normal mass-wasting retreat of cliffs and taluses and by abrasion of the rock-cut pediment surface by sheetfloods. The relative importance of the three processes seems to vary from region to region. Analysis of these details is difficult, because all erosion is slow in dry regions, and some pediments probably formed under a different climate and are now being slowly dissected. A strong contemporary opinion is that many, if not all, pediments were first formed by more humid morphogenetic processes, were buried, and are now being exhumed (Tuan, 1959; Oberlander, 1974). Evidence of climatic change in presently dry climates is reviewed in Chapter 18.

As pediments expand into the mountain slopes, the lower part of the pediments in turn may be progressively buried by the rising bajada alluvium (Fig. 13-5). For this reason, pediments in truly arid regions, where the local base level is a playa, are generally only narrow rock-cut fringes between the mountain fronts and the bajada. The pediments plunge downslope beneath the thickening alluvial fill toward the center of the **bolson,** or intermontane basin. By those who insist on precise definitions, the smoothly graded wash slope at such a mountain front would be called a *pediment* in its upper part, where the entire thickness of alluvium is in transit, and a *bajada* in its lower part, where the alluvium has permanently come to rest

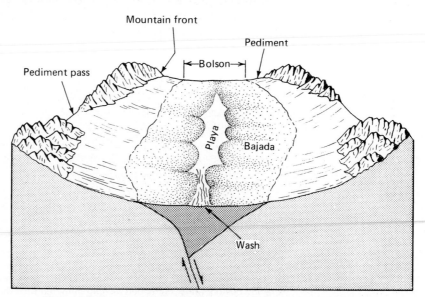

Figure 13-5. Typical assemblage of mountainous desert landforms.

in the bolson. The boundary between pediment and bajada is then defined by the thickness of alluvium that can be episodically reworked by fluvial processes.

The geometry of desert basins and mountains is such that as a basin is filled, its surface area increases, whereas the area of the adjacent mountain decreases by erosion. Therefore, successive increments of mass eroded from the mountain cause a decreasing rate of vertical accretion in the receiving basin and the development of a convex-skyward buried bedrock surface, even though the piedmont surface is always concave-upward (Fig. 13–6). Lawson (1915) referred to the convex buried rock surface as the **suballuvial bench** and to the concave exposed portion as the *subaerial bench*. The latter term is synonymous with pediment. Wells on the margins of many bolsons in the Great Basin demonstrate that their subsurface geometry is as Lawson inferred. The original fault scarp of the basin-range structure is now far out under the bajada, and large portions of pediments truncate older sedimentary formations or indurated fan gravels. Figure 13–6 illustrates that a slight lowering of base level by tectonism or climatic change could extend and regrade large areas of a pediment across former bajada sediments.

Pediments that form on weak-rock lowlands at the base of mountains carved from resistant rock are especially likely to be complex, multiple surfaces. Weak-rock lowlands are rapidly eroded by rivers from the mountain front, leaving broad, fan-shaped pediments heading at major canyon mouths. These are cut across erodible rocks, but are protected from deep dissection by an alluvial veneer of coarse gravel transported from the mountain canyons (Fig. 13–7). Their coarse alluvial bed-load requires the pediment streams to maintain steep gradients to the local basin floor. If, however, the alluvial cover is breached, gullies form in the underlying weak rocks and erode rapidly headward. These gullies carry primarily silt and clay from the erodible substrata, and the pediment cover gravel is not abundant enough to choke the gullies. Therefore, they grow headward with steep headwalls and the pediment cover gravel slumps down their flanks from interfluve remnants. Lateral shifting of the gullies in this manner can introduce multiple levels of pediment remnants without any tectonic movement or climate change.

Tectonic uplift and climatic change can compound the development of multiple, "nested" pediment remnants (Oberlander, 1989, p. 57). In the Sierras Pampeanas of northwestern Argentina, fission-track dating of volcanic ash interbedded in thin pediment alluvium has documented at least five pediment levels that range in age from 2.5 to 0.3 million years. While some may be the result of multiple dissection episodes on the pediment slopes, active tectonism is certainly involved in the older, most dissected pediments (Strecker et al., 1989).

Figure 13-6. Lawson's theory for the origin of a convex-upward suballuvial rock bench. Equal volumes of alluvium cause successively smaller increments of aggradation in the widening bolson. Minor continued faulting demonstrates that the original fault-block boundary is now well away from the mountain front. Vertical scale is exaggerated.

Figure 13-7. Evolution of topography where a stream flows from resistant rocks across deformed weak rocks. (Adapted from Rich, 1935, Fig. 3.)

The most extensive pediments are in regions that have sufficient precipitation to allow at least occasional direct surface runoff to the sea. Under these conditions, which are technically semiarid rather than arid, weathered sediment can be transported to the sea instead of filling bolsons and progressively burying the lower edges of developing pediments. The Sonoran Desert in Arizona and in adjacent Mexico, the classic region of pediment development in North America, has pediments, alluvium-floored plains, and river channels all graded to sea level in the Gulf of California. Water rarely flows in the channels. The entire Sonoran piedmont landscape that so impressed early explorers may be relict from times of greater precipitation.

The interior plateaus of large parts of Africa are great, gently sloping pediments, and large parts of the dry continent, Australia, are also pediment prone. The arid interior of Asia is less well known than other dry regions, but there, too, pediments can be predicted to be among the dominant landforms. Considering that 26 to 36 percent of the land area is arid or semiarid, the pediment may be the most common erosional landform on earth.

If there is a landscape characteristic of the arid morphogenetic region, it should consist of barren rock or gravel-armored plains scarred by braided ephemeral channels and arroyos. Eolian processes produce deflation basins and dune fields (Chap. 14). Mountains are fringed with cliffs, taluses, pediments, fans, and playas. Large areas of the earth do have such landform assemblages, and the arid zone is one of the more widely accepted morphogenetic regions. Even so, the degree to which the forms were shaped by processes that were formerly more intense has yet to be resolved. Quaternary climates formerly more humid than now can be demonstrated in many desert regions (Chap. 18).

LANDFORMS OF SEMIARID REGIONS

The semiarid *steppes, prairies, veld,* and *pampas* of the world are characterized by (1) grass cover; (2) annual precipitation up to twice the amount that locally defines the desert boundary; and (3) rivers that are graded externally to the region, usually to the sea. Slopes and drainage networks are integrated regionally, and erosional landforms evolve toward sea level rather than the rising base level of a closed basin.

Steppe regions are generally plains or low dissected plateaus. Large areas of grassland, such as the Great Plains of North America, are the constructional surfaces of huge piedmont alluvial fans (Fig. 10–11). The Pampas region of Argentina is a featureless plain of windblown silt or *loess* (Chap. 14) that covers an area of more than 1 million km². The dissected plains of north central China and the interior plains of the southern Soviet Union are even larger semiarid steppes on wind-blown loess. In other regions, especially Australia and southern Africa, grassy plains are either pediments or depositional surfaces. Most have now been converted to agricultural use.

SAVANNA LANDFORMS

The savanna landscape of Africa was first made known through the work of German geomorphologists early in this century (Bornhardt, 1900; Passarge, 1928). They called it an *Inselberglandschaft,* or **inselberg landscape,** with reference to the steep-sided, isolated hills and mountains of barren rock that rise like islands from nearby flat plains (Fig. 13–8). During the wet months, wide areas of the plains are flooded, and some of the hills are truly islands.

The term *inselberg* subsequently was extended to the residual knobs on pediments in arid and semiarid regions, and the distinction between morphogenetic processes in truly dry regions and the savanna wet-and-dry region became blurred. In an effort to restore precision, Willis (1934; King, 1948) proposed that the barren, domelike "sugar-loaf" hills without taluses so common in savanna regions should be called *bornhardts,* after the pioneer geomorphologist. Twidale (1982, pp. 15, 124) maintained this usage by distinguishing the most common granite inselbergs from unspecified other types, and designating the most abundant of the granite inselbergs as bornhardts. He defined **bornhardt** as domical inselbergs developed on a single massive joint block. He designated two other minor types of inselbergs on granite rocks; **castle koppies,** angular and castellated hills (like the small turrets and slotted walls of a medieval castle), and **nubbins,** knolls of block- or boulder-strewn bedrock exposure. In his usage, bornhardts, castle koppies, and nubbins are structurally controlled landforms (Chap. 11), analogous to corestones and tors (Figs. 11–3 and 11–4). As such, they are asserted to be azonal or non-morphogenetic landforms shaped by the structural properties of granitic rocks (including such factors as mineralogy and fractures) (Twidale, 1982, p. 24). Nevertheless, the unique regional landscape of inselbergs, especially the most common variety of granite bornhardts, rising like islands above surrounding flat plains, require a special set of morphogenetic conditions that are most often found in the tropical savanna climate. These include warm temperatures, at least seasonally humid conditions, ac-

Figure 13-8.   Inselberg near Einasleigh, Queensland, Australia. (Photo: Ian Douglas.)

tive subsurface weathering by groundwater, and slow or only episodic fluvial incision (Twidale, 1982, p. 136). Tectonically stable cratonic plains also seem to be required, with their tendency to remain near sea level and undergo very slow denudation (p. 321).

In the savanna climate, with high year-round temperatures and heavy seasonal rainfall, chemical weathering extends to great depths (Fig. 6–9; Faniran and Jeje, 1983). With only seasonal runoff, however, the cratonic savanna-zone rivers cannot remove the massive sediment loads furnished by weathering. Rather than entrenching their valleys, rivers braid or flood across extremely flat **wash plains.** Runoff that does not evaporate infiltrates to the water table. The result of rapid chemical weathering by groundwater and inhibited stream erosion is that wash plains are underlain by deep zones of altered rock (Büdel, 1957; Cotton, 1961). Although the ground surface is a monotonous plain of alluvium and weathered bedrock, at a depth of perhaps 100 m an irregular surface is in the process of formation at a "weathering front" in the zone of seasonally fluctuating water table. Small masses of fresh rock that are bypassed by the subsurface weathering are eventually exhumed as tors and corestones (Fig. 11–3 and 11–4); such small forms are commonplace on jointed granitic and similar rocks in all humid and formerly humid climates. Under the optimal climate conditions of the tropical savanna, and usually in a non-orogenic tectonic setting of cratonic crystalline rocks with irregular joint spacing that includes some very large unfractured monoliths, the inselberg landscape evolves. The huge size of some savanna inselbergs leads to the conviction that they are of considerable age, possibly having survived several episodes of rejuvenation and lowering of the adjacent wash plains. When the weathering front has bypassed a large monolith, its preservation and eventual expression as an inselberg seem assured. Thus structural control and climatic morphogenesis combine to produce one of the more striking landscapes of the earth. The process has been referred to as a "double surface of leveling" *("doppelten Einebnungsflächen")* by Büdel (1957). Figure 13–9 outlines Büdel's hypothesis of inselberg development.

Fractures and other structural controls are the only obvious ways of creating irregularities on the subsurface weathering front and potential inselbergs. Thomas (1974, p. 196) listed 12 lines of evidence to support Büdel's hypothesis, including a description of a large excavation in the weathered layer that artificially exposed a

Figure 13-9. Hypothesis of a "double surface of leveling" in the tropical savanna. (a) Wash plain of seasonal flooding is as much as 100 m above the weathering front. Pediments fringe the wash plain. (b) Wash plain is lowered by rejuvenation or climatic change. Inselbergs and marginal pediments are exhumed or regraded to the lowered wash plain. (From Büdel, 1957, Figs. 5 and 7.)

typical inselberg. It is possible that purely chance events at the surface, such as a landslide, might strip the regolith off an emerging rock mass, allow it to be washed clean, and thereby preserve it from future deep chemical weathering. The development of a benched landscape on the Sierra Nevada without structural control has been attributed to similar randomness of exposure (Wahrhaftig, 1965).

Some inselbergs are encircled by shallow moats at their bases, probably due to concentrated subsurface weathering by runoff and infiltration at the base of the inselberg (Twidale, 1968, pp. 105–24; 1976, pp. 7–17; Twidale and Bourne, 1975). As wash plains are lowered, these moats and weathered notches in the flanks of steep-sided inselbergs (Fig. 13–10) are exhumed to record former ground levels. Other inselbergs have narrow pediments around their bases (Fig. 13–9), which led to the confusing extension of the word inselberg to residual hills that are surrounded by much larger pediments in other types of dry regions. The savanna pediments are narrow and of low gradient, 3° being the maximum reported in Ghana (Clayton, 1956). The minor development of pediments and the great extent and flatness of wash plains distinguish the savanna inselberg landscape from the pedimented landscapes of dry regions. The contrast is due to the intensity of chemical weathering in the savanna climate, by which exposed rock is so weakened that it is easily eroded, and to the heavy seasonal precipitation, by which the detritus is distributed across the wash plains.

There seems to be adequate evidence to postulate that the inselberg landscape defined by Bornhardt is a valid example of a morphogenetic landscape assemblage unique to the savanna climate. Certainly its best expression is in the seasonally wet-and-dry tropics. On the wetter, equatorial side of the savanna regions, the wash plains become forest covered, stabilized, and eroded by perennial streams. On the dry poleward side of the savanna belt, desert landforms replace the inselbergs and wash plains, because chemical weathering becomes less effective. However, both climatic boundaries of the savanna zone have shifted during the Quaternary, and some undetermined amount of savanna landscape is found relict in both adjacent regions (Chap. 18).

Figure 13-10. Steepened margin of a granite inselberg caused by multiple episodes of subsurface weathering and subsequent exposure of the weathering front, Pildappa Hill, South Australia. (Photo: C. R. Twidale.)

REFERENCES

BIROT, P., 1968, Cycle of erosion in different climates (transl. by C. I. Jackson and K. M. Clayton): Univ. of California Press, Berkeley, 144 pp.

BORNHARDT, W., 1900, Zur Oberflächengestaltung und Geologie Deutsch-Ostafrikas; D. Reimer, Berlin, 595 pp.

BRYAN, K., 1922, Erosion and sedimentation in the Papago country, Arizona: U.S. Geol. Survey Bull. 730-B, pp. 19–90.

BÜDEL, J., 1957, Die "doppelten Einebnungsflächen" in den feuchten Tropen: Zeitschr. für Geomorph., v. 1, pp. 201–28.

CLAYTON, R. W., 1956, Linear depressions (Bergfüssniederungen) in savanna landscapes: Geog. Studies, v. 3, pp. 102–26.

COOKE, R. V., and WARREN, A., 1973, Geomorphology in deserts: Univ. of California Press, Berkeley, 394 pp.

COTTON, C. A., 1958, Alternating Pleistocene morphogenetic systems: Geol. Mag., v. 45, pp. 125–36.

———, 1961, Theory of savanna planation: Geography, v. 46, pp. 89–101.

DAVIS, W. M., 1905a, Complications of the geographical cycle: Internat. Geog. Cong., 8th, Washington, 1904, Repts., pp. 150–63 (reprinted 1954 *in* Geographical essays: Dover Publications, Inc., New York, pp. 279–95).

———, 1905b, Geographical cycle in an arid climate: Jour. Geology, v. 13, pp. 381–407 (reprinted 1954 *in* Geographical essays: Dover Publications, Inc., New York, pp. 296–322).

DENNY, C. S., 1965, Alluvial fans of the Death Valley region, California and Nevada: U.S. Geol. Survey Prof. Paper 466, 62 pp.

———, 1967, Fans and pediments: Am. Jour. Sci., v. 265, pp. 81–105.

DORN, R. I., 1981, Microbial origin of desert varnish: Science, v. 213, pp. 1245–47.

———, 1989, Cation-ratio dating of rock varnish: A geographic assessment: Progress in Phys. Geography, v. 13, pp. 559–96.

———, JULL, A. J. T., DONAHUE, D. J., LINICK, T. W., and TOOLIN, L. J., 1989, Accelerator mass spectrometry radiocarbon dating of rock varnish: Geol. Soc. America Bull., v. 101, pp. 1363–72.

FANIRAN, A., and JEJE, L. K., 1983, Humid tropical geomorphology: Longman, Inc., New York, 414 pp.

GILBERT, G. K., 1882, Contributions to the history of Lake Bonneville: U.S. Geol. Surv. 2nd Ann. Rept., pp. 167–200.

GOUDIE, A., 1982, The human impact: Man's role in environmental change: MIT Press, Cambridge, Mass., 316 pp.

GRAF, W. L., 1988, Fluvial processes in dryland rivers: Springer-Verlag, Berlin, 346 pp.

HADLEY, R. F., 1967, Pediments and pediment-forming processes: Jour. Geol. Education, v. 15, pp. 83–89.

HOWARD, A. D., 1942, Pediment passes and the pediment problem: Jour. Geomorphology, v. 5, pp. 2–31, 95–136.

JOHNSON, D. W., 1932, Rock fans of arid regions: Am. Jour. Sci., v. 223, pp. 389–420.

KING, L. C., 1948, A theory of bornhardts: Geog. Jour., v. 112, pp. 83–87.

KÖPPEN, W., and GEIGER, R., 1936, Handbuch der Klimatologie, v. 1, part C: Gebrüder Borntraeger, Berlin, 44 pp.

LAMB, H. H., 1972, Climate: Present, past and future, v. 1: Methuen & Co. Ltd., London, 613 pp.

LAWSON, A. C., 1915, The epigene profiles of the desert: Calif. Univ. Pub. Geol. Sci., v. 9, pp. 23–48.

MACHETTE, M. N., 1985, Calcic soils of the southwestern United States *in* Weide, D. L., ed., Soils and Quaternary geology of the southwestern United States: Geol. Soc. America Spec. Paper 203, pp. 1–21.

McGee, W J, 1897, Sheetflood erosion: Geol. Soc. America Bull., v. 8, pp. 87–112.

Meigs, P., 1952, Arid and semiarid climatic types of the world: Internat. Geog. Cong., 17th, Washington 1952, Proc., pp. 135–38.

———, 1953, World distribution of arid and semiarid homoclimates, in Reviews of research in arid zone hydrology. UNESCO, Paris, pp. 203–10 and maps.

Montgomery, R. F., and Askew, G. P., 1983, Soils of tropical savannas, in Bourlière, F., ed., Ecosystems of the World 13: Tropical Savannas: Elsevier Scientific Publishing Co., Amsterdam, pp. 63–78.

Oberlander, T. M., 1974, Landscape inheritance and the pediment problem in the Mohave Desert of southern California: Am. Jour. Sci., v. 274, pp. 849–75.

———, 1989, Slope and pediment systems, in Thomas, D. S. G., ed., Arid zone geomorphology: John Wiley & Sons, Inc., New York (a Halsted Press Book), 372 pp.

Passarge, S., 1928, Panoramen afrikanischer—Inselberglandschaften: Dietrich Reimer, Berlin, 15 pp., 25 plates.

Péwé, T. L., Péwé, E. A., Péwé, R. H., Journaux, A., and Slatt, R. M., 1981, Desert dust: Characteristics and rates of deposition in central Arizona, in Péwé, T. L., ed., Desert dust: Origin, characteristics, and effects on man: Geol. Soc. America Spec. Paper 186, pp. 169–90.

Rahn, P. H., 1967, Sheetfloods, streamfloods, and the formation of pediments: Assoc. Am. Geographers Annals, v. 57, pp. 593–604.

Rich, J. L., 1935, Origin and evolution of rock fans and pediments: Geol. Soc. America Bull., v. 46, pp. 999–1024.

Strecker, M. R., Cerveny, P., Bloom, A. L., and Malizia, D., 1989, Late Cenozoic tectonism and landscape development in the foreland of the Andes: Northern Sierras Pampeanas (26°–28°S), Argentina: Tectonics, v. 8, pp. 517–34.

Thomas, M. F., 1974, Tropical geomorphology: John Wiley & Sons, Inc., New York (a Halsted Press book), 332 pp. (in U.K. by The Macmillan Press Ltd.).

Thornthwaite, W., 1948, An approach toward a rational classification of climate: Geog. Rev., v. 38, pp. 55–94.

Tricart, J., 1972, Landforms of the humid tropics, forests and savannas (transl. by C. J. K. de Jonge): Longman Group Ltd., London, 306 pp.

———, and Cailleux, A., 1972, Introduction to climatic geomorphology (transl. by C. J. K. de Jonge): Longman Group Ltd., London, 295 pp.

Trimble, S. W., 1988, The impact of organisms on overall erosion rates within catchments in temperate regions, in Viles, H. A., ed., Biogeomorphology: Basil Blackwell Ltd., Oxford, England, pp. 83–142.

Tuan, Yi-Fu, 1959, Pediments in southeastern Arizona: Univ. Calif. Pub. Geog., no. 13, 140 pp.

Twidale, C. R., 1968, Geomorphology, with special reference to Australia: Thomas Nelson (Australia) Ltd., Melbourne, 406 pp.

———, 1976, Analysis of landforms: John Wiley & Sons Australasia Pty. Ltd., Sydney, Australia, 572 pp.

———, 1982, Granite landforms: Elsevier Scientific Publishing Co., Amsterdam, 372 pp.

———, and Bourne, J. A., 1975, Episodic exposure of inselbergs: Geol. Soc. America Bull., v. 86, pp. 1473–81.

Wahrhaftig, C., 1965, Stepped topography of the southern Sierra Nevada, California: Geol. Soc. America Bull., v. 76, pp. 1165–89.

Wells, P. V., 1970, Postglacial vegetational history of the Great Plains: Science, v. 167, pp. 1574–82.

Willis, B., 1934, Inselbergs: Assoc. Am. Geographers Annals, v. 26, pp. 123–29.

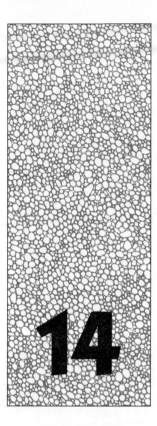

14

Eolian Processes and Landforms

ENERGETICS OF EOLIAN PROCESSES

In Chapter 5 it was noted that a substantial amount of the solar energy received at the earth's surface is converted into thermal and density contrasts in the atmosphere, which produce winds. Perhaps as much as 0.7 percent of the thermal energy exchange is accomplished by wind (Fig. 5–2), which, considering the enormity of the solar energy available for geomorphic processes, should give eolian processes and landforms a significant place in geomorphology. Yet the direct effectiveness of eolian processes is rated quite low (p. 209), less than rivers, glaciers, and wind-driven waves. The present general opinion of the minor role of eolian geomorphic processes is especially surprising in view of the very substantial proportion of earth's land area that is arid and semiarid, where vegetation is restricted in its ability to cover protectively the weathered detritus at the ground surface.

Wind is considered to be a comparatively minor agent of geomorphic change primarily because of the low density of air as compared to rock and water. The relatively low ratio of the densities of immersed rock and water $[(2.65-1.0)/1.0 \simeq 1.6:1]$ means that most of the fluvial sediment load can be carried in suspension, buoyed by the turbulence of the moving and often muddy water. However, rock fragments are 2000 times heavier than the air they displace. Only very fine dust particles are moved in suspension except at extremely high wind velocities. Also because of its low density, moving air exerts a relatively low pressure against obstructions, whereas water flowing downhill can accomplish significant erosion by simple hydraulic force against submerged rock ledges.

Eolian processes and landforms are most obvious in arid regions, where lack of vegetation not only makes wind erosion, transportation, and deposition possible but enables the resulting landforms to be observed. Eolian landforms are safely regarded as part of the arid morphogenetic system if due provision is made for those restricted areas in other climatic regions where sediment supply and exposure to winds also permit eolian landforms to develop. For example, belts of coastal dunes are on sandy coastlines in all climatic regions, although their frequency is less in the humid tropics (p. 510). The braided channels made by glacial meltwater are notable sources for windblown dust and sand owing to their variable discharges and their extensive nonvegetated exposed bars and abandoned channels. The massive sheets of loess (Fig. 14–17) downwind from outwash channels in glaciated areas attest to the effectiveness of meltwater streams to provide a renewable supply of fine-grained alluvium for subsequent wind transport. Dry periglacial regions are especially susceptible to eolian processes. Fine-grained soils in any climatic region are likely to have minor eolian landforms such as blowouts and dunes, especially if they have been disturbed by inappropriate agricultural practices. As noted in Chapter 12, wind may prove to be a more significant denudation agent in humid regions than heretofore believed (p. 319).

WIND EROSION

Wind erosion occurs either by **deflation,** the process of removing loose sand and dust, or by **abrasion,** using the entrained sand grains as tools against rock surfaces or other grains. The widespread industrial use of sandblasting by an air jet with entrained fine sand illustrates the potent abrasive capacity of windblown sand (Suzuki and Takahashi, 1981). It has been shown experimentally that abrasion by impact, in contrast to abrasion by rubbing, can be accomplished by particles that are softer than the surface they strike. Even ice crystals at −10 to −25°C, with a hardness of between 2 and 3.5 on the Mohs scale of hardness used by geologists, have been observed to break edges and corners from cleavage rhombs of minerals with hardness of 3 to 5.5 (Dietrich, 1977). Rock surfaces are fluted, grooved, pitted, or polished within decades after exposure (Hickox, 1959; Whitney and Dietrich, 1973, p. 2571). Sand-sized angular quartz grains become rapidly rounded by impact with each other in experimental apparatus, and produce large quantities of silt-sized fragments in the process. The implication is that significant amounts of dust can be derived from sand by abrasion, even though desert sand is notable for its excellent sorting and general lack of silt and clay particles. The finer fractions are clearly transported far from a sandy desert. For example, dust from central Asia is seasonally collected in Hawaii (Whalley et al., 1987; Schneider et al, 1990).

Cobbles and pebbles that show wind abrasion are called **ventifacts** (Fig. 14–1). If undisturbed, a ventifact develops a facet facing the effective wind direction and sloping down to windward. Most ventifacts have multiple facets that intersect along sharp keel-like ridges, testifying to more than one direction of effective abrasion. Large wind-eroded boulders and talus blocks, unlikely to have moved, may be used to infer one or more effective wind directions. Smaller cobbles and pebbles are likely to be moved by a great variety of processes, so their facets have little significance for determining effective winds (Sharp, 1949). Bricks and blocks of easily abraded gypsum cement placed on a plot in a California desert were undermined by creep and wind scour and tilted, rotated, and overturned, sometimes within a few months (Sharp, 1980). The detailed fluting, pitting, and rill marks on some ventifacts have been attributed to abrasion by dust rather than by sand (Whitney and Dietrich, 1973).

Undercut, mushroom-shaped pedestal rocks in desert areas are commonly attributed to wind abrasion. However, experiments (p. 355) prove that the median height of windblown sand transport is only about 1 cm over a layer of similar-sized grains and little more than 10 cm over hard-rock surfaces. Most authorities now doubt that wind abrasion could cut the notches or overhangs on desert cliffs or large boulders that are 1 to 2 m in height. More likely, desert weathering processes (p. 348) involving exfoliation or spalling on sheltered surfaces are the cause of the un-

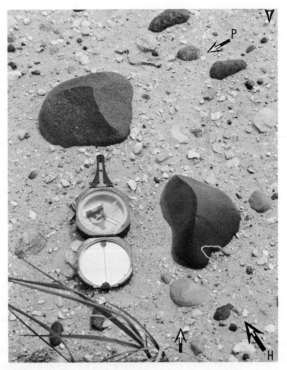

Figure 14-1.   Ventifacts on lag gravel between lake-shore dunes, Sleeping Bear, Leelanau County, Michigan. North is at the bottom. P, prevailing winds; H, high-velocity winds; unlabeled arrow at bottom indicates other effective winds. (Photo: M. I. Whitney.)

dercut surfaces. Some contain rock paintings or are encrusted with gypsum crystals, neither of which could survive if abrasion were an active process.

Minor eolian abrasional landforms include the chutes or flumes that notch low escarpments around playas and other desert plains. They are aligned with the locally effective winds and are probably scoured by windborne sand being swept through small gullies. They are usually cut in weakly indurated sediments and are separated by tonguelike or teardrop-shaped streamlined, wind-abraded ridges called **yardangs** (Blackwelder, 1934; Ward and Greeley, 1984). Deflation of weakly indurated sediment is at least as important as is abrasion in shaping yardangs (McCauley et al., 1977). The shape of yardangs probably originates from initial surface irregularities such as gullies and their interfluves at a scarp on the edge of a deflation basin, and evolve toward relatively stable streamlined bodies with a width/length ratio of 1/4 and a tapered downwind end, considered to be the shape of least resistance of a solid body immersed in a moving fluid (Ward and Greeley, 1984, p. 832). However, yardangs are dynamic, small landforms that are probably removed as fast as new ones form by differential erosion.

On the high, cold, dry Altiplano plateau of the central Andes in Argentina, Chile and Bolivia, volcanic ignimbrite sheets that cover thousands of square kilometers are systematically eroded by southeast-trending parallel *quebradas* (gullies) that are aligned with the powerful prevailing winds from the northwest (Fig. 4–19). They are probably eroded in part by fluvial action along cooling joints in the ignimbrite plateaus, with wind abrasion preferentially elongating and extending those open joints that are aligned with the wind, and perhaps with drifting sand filling those joints that trend obliquely to the wind. Aligned quebradas in one area in Chile are typically spaced 30 to 200 m apart and range in depth from a few meters to 30 m (Guest, 1969). They may be several km in length and commonly trend oblique to the local slope. They are a characteristic feature of many ignimbrite sheets of the

Altiplano and are a useful criterion for identifying areas of ignimbrite on space images (Fielding et al., 1986).

Deflation hollows of various sizes are known to form quickly by wind erosion on exposed erodible terranes. When the vegetation on stabilized dunes is destroyed by fire or overgrazing, *blowouts* develop within months. When excavated below root depth, a blowout can grow as deep as the effective winds can lift sand (Fig. 14–2). Dry creep keeps the adjacent slopes at the appropriate angle of repose. The limiting depth for blowouts is usually the water table, because moist sand resists deflation, and plants may grow on the floor of the basin as well. Over the more arid sections of the American Great Plains, thousands of small basins mark sandy districts. They were termed "buffalo wallows" by early explorers because animals frequented the ponds found in them. Although primarily the result of deflation during dry years, buffalo and domestic cattle surely aid deflation by loosening the sun-dried soil in the floors of the blowouts as they seek water. After being farmed for decades, some of the Great Plains blowouts were observed to deflate actively during the dry, fallow years of the mid-1930s.

Larger closed basins in arid regions, some of them many kilometers in diameter and over 100 m in depth, have variously been attributed to solution, differential compaction, faulting, and deflation. Tectonic causes can usually be affirmed or eliminated by geologic mapping or drilling, but the other possibilities are difficult to separate. Big Hollow, near Laramie, Wyoming, is about 5 km wide, 15 km long, and about 100 m deep. It is considered an inactive blowout, developed on fine-grained Mesozoic sedimentary rocks. The enormous Qattara Depression (Fig. 14–3) in the western Egyptian Sahara is estimated to have an excavated volume of 3200 km³ (Peel, 1966, p. 21). Its floor reaches 134 m below sea level, less than 100 km inland from the Mediterranean coast. Annual precipitation ranges from 25 to 50 mm on the northern rim to less than 25 mm in the western part of the basin, although torrential winter rains have been reported (Albritton et al., 1990). The Qattara Depression may have originated as a Miocene river valley that was abandoned and segmented by deep karst solution during the late Miocene "Messinian crisis" when the Mediterranean basin became a desert at least 2.5 km below sea level (Hsü, 1983; Norman and Chase, 1986). A cavern in Eocene limestone was penetrated by a water

Figure 14-2.   Active blowout in coastal dunes on Cape Cod, Massachusetts. Vegetated blowout in background is at the water table.

well 228 to 246 m below the western floor of the depression, strengthening the evidence that paleokarst processes may have helped deepen the basin (Albritton et al., 1990, p. 959). When the Mediterranean basin refilled in Pliocene time, the sea again transgressed over western Egypt, depositing thin limestone. Subsequently, the Plio-Pleistocene and Miocene limestone caprock was breached, and a combination of mass wasting and slope retreat on easily eroded mudstones that underlie the surrounding limestone escarpments and wind deflation of the resulting sediment from the basin floor have been the primary processes involved in deepening the basin (Ball, 1927, 1933). If the Qattara Depression is assumed to have formed by deflation in the last 2 million years, the eolian denudation rate is 9 cm/1000 yr, impressively comparable to reported rates of fluvial denudation (Chap. 12). The possible role of limestone solution and exhumed paleotopography in the origin of the basin await further research.

Many other Saharan depressions have been discovered by analyses of the wartime aerial photographs of the region. Numerous very large depressions in Pliocene continental sediments of the Mongolian Desert, the P'ang Kiang Hollows described by Berkey and Morris (1927, pp. 336–37, 347), also are confidently attributed to a combination of Pleistocene deflation and arid scarp retreat. In a general review of desert landscapes, Peel (1966) urged renewed attention to the origin of large desert basins, suggesting that wind erosion, rather than tending to reduce the relief of desert landscapes, might actually increase it. Only the water table can limit the depth of deflation if structures do not produce a gravel lag, and wind strength is adequate. The water table might actually be lowered as basins develop (Murray, 1952). It is notable that erosional desert landscapes, unlike fluvial landscapes, do not evolve toward the ultimate base level of the sea. Although tectonic desert basins seem to be sediment traps (pp. 336 and 340), it is possible that some basins could deepen almost without limit by deflation alone.

Figure 14-3. The Qattara Depression, a desert basin with an area below sea level of 19,500 km², within 56 km of the Mediterranean coast. The depression has been suggested as a site for hydroelectric power generation by siphoning water from the Mediterranean Sea and letting it fall through turbines to the floor of the basin, where it would evaporate. (Redrawn from Ball, 1933.)

THE DYNAMICS OF EOLIAN EROSION AND TRANSPORT

Wind is always turbulent. All of the equations for turbulent flow in water (Chap. 9) also apply to air, although the values of the parameters are very different. As in turbulent flow in water or any other fluid, above a thin boundary layer where the velocity is zero, the velocity initially increases rapidly, then less rapidly, with height (Fig. 14–4a). In air, the thickness of the stagnant boundary layer over a flat granular bed is equal to about $\frac{1}{30}$ of grain diameter. Above the stagnant layer, wind velocity is proportional to the logarithm of the height (Fig. 14–4b), from which the important parameter $V*$, the **drag velocity,** was defined as being proportional to the rate of increase of wind velocity with the logarithm of height (Bagnold, 1941, p. 51). By definition,

$$V* = \sqrt{\tau/\rho}$$

where τ is the surface shear stress and ρ is the density of air. Although $V*$ has the dimensions of a velocity, it really represents the vertical velocity gradient near a rough surface. When it exceeds a threshold level ($V*_t$) equal to the vertical settling velocity of a particle of a certain size, that particle will begin to oscillate at the surface beneath the wind, then suddenly leave the surface in a nearly vertical direction (Greeley and Iverson, 1985, p. 70). Bagnold (1941) demonstrated that the threshold velocity is proportional to the square root of grain diameter for grains larger than about 0.1 mm (Fig. 14–5a). The relationship is similar to the threshold of erosion in water defined by Hjulström's curve (Fig. 9–10a). Figure 14–5b is simplified from Figure 9–10a and redrawn with a linear vertical axis for a direct comparison of the threshold of erosion in air and water. However, Figure 14–5a uses the calculated values of the threshold velocity $V*_t$ in air as a vertical axis, whereas Figure 14–5b uses Hjulström's measured average velocity in flowing water at least 1 m in depth, so the curves are not exactly analogous. In both curves, the threshold of erosion is linear for larger grains, but below a minimum grain size of about 0.08 mm in air and about 0.2 mm in water, the threshold velocity for erosion rises steeply. In water, this is explained in part by cohesion and in part by the relative thickness of the laminar-flow boundary layer (Fig. 9–10b). In air, small particles on a flat surface lie entirely within a stagnant boundary layer. The Shields curve for air is similar to that for water for sand and larger grains, but rises much more steeply for silt and clay-size particles. A layer of dry portland cement powder will not move on the floor of a wind tunnel even when $V*_t$ exceeds 100 cm/s, a wind strong enough to move fine pebbles with diameters of 4.6 mm (Bagnold, 1941, p. 90). This phenomenon should be obvious to anyone who has observed the very localized plume of dust that trails a moving vehicle on a dry dusty road on a windy day. Immediately behind the disturbance, the dust is again immobile.

For sand particles exposed to wind, there is also a lower *impact threshold* of motion, at a velocity less than the fluid threshold, caused by other sand grains already in motion striking particles at rest. If smaller grains have already been set in motion, their impacts will cause larger grains to move at velocities only about 80 percent of the threshold velocity predicted for the particular grain size (Fig. 14–5a).

Eolian sediment transport can be compared to fluvial transport only to a limited degree. In both air and water, some particles are suspended, some bounce or roll at the bottom of the moving fluid, and some are dragged along in contact with the substrate. Because of the great density contrast between air and rock fragments, only very fine detritus, with an effective diameter of less than 0.2 mm and usually less than the minimum size of very fine sand (0.0625 mm), can be maintained in suspension by the upward component of atmospheric turbulence. This is *dust* that grades into extremely fine particulate matter called *haze, smoke,* or *aerosols,* and

(a)

(b)

Figure 14-4. Velocity distribution for two different wind strengths on (a) linear and (b) logarithmic height scales. The slope of the lines in Fig. 14-4b is V∗, the velocity gradient near the surface. (Bagnold, 1941, Fig. 15a and b, © Chapman & Hall Ltd., London, 1973.)

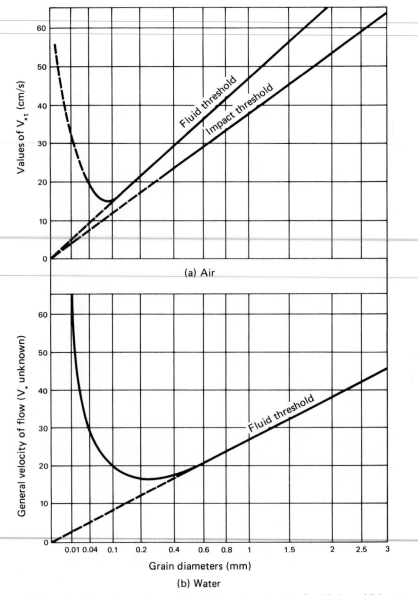

Figure 14-5. Variation of the threshold velocity with grain size for (a) air and (b) water. Grain size drawn on a square root scale to show the relation $V \propto \sqrt{d}$ for larger grains. (Bagnold, 1941, Figs. 28 and 29, © Chapman & Hall Ltd., London, 1973.)

can be carried entirely around the world by planetary winds before settling out or being washed out of the air.

Loose sand particles with a diameter of 0.2 mm or larger can be stirred by a steady light wind of 5 m/s but do not remain in suspension. They roll or slide along the ground or bounce, transferring momentum to other grains by collisions. Bagnold (1941) made a thorough laboratory and field study of the physics of wind-blown sand, and by ingenious experiments first demonstrated that most sand grains transported by wind do so in short asymmetric trajectories, never far from the ground. The process is termed **saltation.** A falling grain strikes one or more other grains and makes a visible dimple or crater in the sand surface. Although the disturbed particles are scattered in various directions, as they rise into the airstream

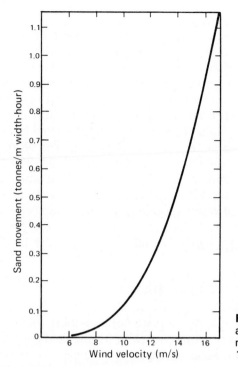

Figure 14-6. Relation between the flow of average dune sand and the wind velocity as measured at a standard height of 1 m. (Bagnold, 1941, Fig. 22.)

and fall again they are swept forward. Just prior to impact they are usually approaching the ground at an angle of only 10 to 16° (Bagnold, 1941, p. 19). Above some threshold velocity, the inertia transferred from the wind to saltating grains is sufficient to maintain continuous forward transport of saltating sand. A lesser amount of sand is driven forward over the ground surface without saltating. This component Bagnold called **surface creep.** It is roughly analogous to the traction or bed load transported by rivers.

Bagnold was able to describe the geometry of the characteristic path or trajectory of a saltating sand grain in terms only of its initial upward velocity (derived from prior impact) and the vertical velocity gradient, V_*. As wind velocity increases, so does the velocity gradient near the ground, where most saltating grains move (Fig. 14–4). The effect of increased wind speed is to impart a higher forward velocity to saltating grains, which in turn produces a rapid increase in the amount of sand in transport (Fig. 14–6). Most of the sand saltating over a surface of similar-sized grains does not rise higher than 1 or 2 m, regardless of wind speed. Because of the logarithmic decrease in wind velocity close to the ground, the median height of the trajectories of all saltating sand grains is only about 1 cm above the ground.

Saltating sand grains absorb the kinetic energy of moving air and convert it to kinetic energy of additional moving sand. The sharp increase in sand transport with increased wind speed is a remarkable illustration of the tendency for all forms of sediment transport to increase sharply during times of high energy inputs, absorbing much of the energy. As in the case of river alluvium during a flood, or beach sand during a coastal storm, a sheet of sand on a desert absorbs energy during windstorms by moving downwind. If an upwind supply of new sand is available, the desert looks just the same after the storm, even though the depositional landforms are composed of sand that formerly rested somewhere upwind.

The impact of a saltating sand grain can move a much larger grain forward by surface creep. Bagnold (1941, pp. 33–37) estimated that significant creep occurs in grains six times the diameter of saltating grains. However, creeping grains move only a few millimeters with each impact, whereas saltating grains have characteristic trajectories a meter or more in length. Here is a potent method of size sorting, in which the size fraction moving by saltation greatly outdistances coarser grains that move only by surface creep. During strong winds over poorly sorted alluvium, the finest fraction is completely exported from the region as dust; a selected size fraction (perhaps including three-fourths of the total volume moved) is driven downwind by saltation; a comparatively small amount of coarse sand (perhaps one-fourth of the total sand volume) is driven downwind by creep; and a gravel lag is left behind. Wind passing from a sheet of moving sand onto a barren rock surface is no longer slowed by the drag of loose, saltating sand. The abrupt increase in surface wind velocity sweeps the hard surface clean, but when any swept-up grains again pass onto a sand surface, the wind velocity is checked and deposition results. Deserts thus consist of various combinations of barren, windswept, and sandblasted exposed rocks *(hammada),* gravel or stony lag deposits *(reg),* and sand sheets or dune fields *(erg)* (Chap. 13). The boundaries between these various terranes are abrupt.

DEPOSITIONAL LANDFORMS OF EOLIAN SAND

A patch of sand on a desert plain has a curiously definitive size limitation. Under gentle winds, barely strong enough to stir the sand grains, the patch is spread downwind and grows thinner or disperses. No new sand is delivered to the patch from upwind, because the available sand there is sheltered by bedrock or gravel irregularities. Under strong winds, however, a preexisting sand patch larger than a critical size grows in thickness and in size as its upwind border extends still farther upwind. The change in behavior is related to wind-transport dynamics described in the preceding paragraph. Strong winds create surface velocities over hard surfaces that can transport sand grains downwind. Over sand, however, the wind energy near the surface is absorbed by saltating sand, and the new sand derived from upwind is deposited in increasing amounts until an equilibrium is reached with the amount leaving the patch. For net accumulation to occur on an exposed sand patch, it must have a minimum initial downwind length of 4 to 6 m, several times the length of the saltation trajectory (Bagnold, 1941, p. 183).

A smooth sheet of sand is inherently unstable under gentle wind. The saltation method of sand movement results in most grains of a given size following trajectories of similar length. At the upwind edge of a sand patch, where transport first begins, a lag of coarser grains is left behind. The saltating grains moving downwind from that lag area impact at the mean distance of saltation and produce another lag area. By this process, **sand ripples** develop, transverse to the wind direction. Their typical wavelength is 1 m or less, very similar to the characteristic flight length of saltating grains. Sand ripples are characterized by a lag concentrate of coarse grains at their crests where the fine grains have been removed, and finer grains in their troughs. Because of the rather low angle of impact of saltating grains, each ripple crest receives more than the average number of saltation impacts, and creates a shadow zone of fewer impacts in its lee.

Under stronger winds, the surface of a sand patch loses its rippled form, because less size sorting takes place. Then the entire surface layer is sheared forward by saltation and surface creep. However, under such conditions, new sand is also trapped from upwind sources, and the sand sheet is not eliminated but increases in area and thickness as long as an upwind sand supply is available.

Large accumulation landforms of windblown sand are called **dunes.** The dynamics just described make a sharp distinction between ephemeral ripples on a sand surface, which form under gentle winds, and dunes, which are distinctly larger

Figure 14-7. Grain size (the coarse twentieth percentile) of sand in ripples, dunes, and draas of three Saharan ergs. (Wilson, 1972, Fig. 2.) This clear relationship between grain size and the size of the landform has not been found in other ergs (Lancaster, 1988b).

and stable or growing landforms (Fig. 14–7). Even larger-scale complex or multiple dunes are called **draas** (Wilson, 1972). To survive and grow, dunes must be at least 4 to 6 m in downwind length and are commonly much larger. It is common for the sand on dunes to be rippled, just as large swells on the ocean surface have small wind waves on their surfaces in turn. Although ripples have the coarsest sand on their crests, on dunes the coarse grains creep or roll to the troughs and the finest sand remains on the crests (Lancaster, 1989b, p. 75).

Wilson (1972) found a distinct correlation between the size of the coarse twentieth percentile of sand grains and the landforms from which they were collected in three northern Saharan ergs (Fig. 14–7). Ripples with the coarsest grains had wavelengths approaching those of dunes with very fine sand, but even those dunes with coarse sand did not have wavelengths as long as draas composed of fine sand. Wilson found no transitional forms. The general application of these relationships between grain size and landforms has not been confirmed by subsequent studies in Namibia, the Kalahari desert, and the Gran Desierto in Mexico. However, in a given erg, the spacing of some dune types can be correlated with the grain size of the coarse twentieth percentile, and other types show no correlation (Lancaster, 1988b, p. 973).

Bagnold (1941, p. 188) classified sand accumulations under five headings, described next.

Sand Shadows and Sand Drifts

Wherever wind velocity is checked by a fixed obstruction, sand may strike the obstruction and then fall at its windward base or be swept into the lee of the obstruction and accumulate as a streamlined mound. Both landforms are classed as **sand shadows.** They are fixed in form and size by the size and aerodynamic shape of the obstruction. For instance, a small tree raised above the ground on a narrow trunk will not accumulate a shadow, whereas a tuft of grass or a ground-hugging shrub may produce a substantial lee shadow. Some vegetation will trap sand and grow upward within the accumulating mound, forming a crude hemispherical mound of sand held in a mesh of roots.

Sand drifts are the accumulations in the lee of a local high-velocity zone between obstacles. Most commonly they accumulate downwind from a place where wind is funneled through a gap at a higher-than-average velocity. Another important site of sand drifts is at the foot of a downwind cliff or escarpment. Sand blown off the plateau surface sinks into the still air in the lee of the cliff. When such sand drifts grow large enough, they begin to trail off downwind from the base of the cliff.

On images of the Martian surface, and on photos and electronic images of terrestrial deserts, sand shadows and drifts are collectively called *wind streaks*.

Dunes

True dunes can best be considered as deformable obstructions to airflow. They are free to move, divide, grow, and shrink, but they are not dependent on a fixed obstruction for their maintenance. Two principles that have been previously emphasized are critical to dune formation. First is the principle that a free sand patch larger than 4 to 6 m is likely to increase in size and thickness and thus become a mound or hill. The second principle is that like a cliff, a dune develops a lee-side pocket of low-velocity air that promotes deposition there. As soon as a mound of sand grows to moderate size, it develops the classic dune asymmetry of a gentle windward slope and a steep angle of repose *slip face* to leeward (Figs. 14–8 and 14–14). The sand on the windward slope is firm and compact, but as it blows over the crest and cascades down the slip face, it may become loose enough to make walking difficult. Note that the internal structure of a dune is fixed by the original deposition and slight compaction; whereas the dune *form* may move without changing shape, the internal *structure* is immobile. Eroded edges of slip-face bedding are commonly exposed on windward slopes (Fig. 14–9). Less commonly, gently dipping sheetlike accumulations of windward-slope deposits are preserved in dune structures.

Dunes have a wide range of forms, some of which were unknown until aerial

Figure 14-8. Cross section of an active dune. Much of the slip-face deposition finally results from sand flow and slumping.

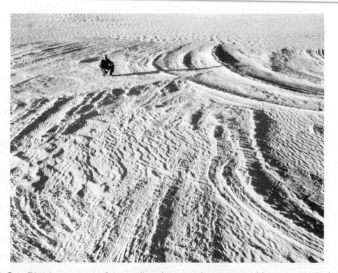

Figure 14-9. Dipping strata of an earlier dune cycle truncated by cross winds. Barchan dune area, White Sands, New Mexico. (Photo: Tad Nichols.)

photography and space images became available for large areas of hitherto unexplored sand regions in the Sahara and Arabia. Bagnold recognized only two basic dune forms, the *barchan* or *crescentic dune,* and the *seif* or *longitudinal dune.* He maintained that *transverse dunes,* with their long axes at right angles to the effective wind, are inherently unstable, and that they break up into shorter segments as barchans or become elongate in the direction of the wind as longitudinal dunes (Bagnold, 1941, p. 205). Other geomorphologists recognize additional dune forms (Fig. 14–10 and 14–11). Transverse dunes are actually very common in areas of copious sand supply and weak winds, despite Bagnold's deemphasis (Hack, 1941; Wilson, 1972). In semivegetated areas, blowouts are paired with downwind *blowout dunes* or *parabolic dunes.* Very large *star dunes* (Fig. 14–12) may be several hundred meters in height and several kilometers in diameter, with radiating sinuous ridges culminating at a common crest. They accumulate under effective winds that blow from several different directions (Breed and Grow, 1979, p. 275; Lancaster, 1989a). At various seasons, they may show slip faces on various sides of their multiple radiating arms. They do not fit in the classification scheme of Fig. 14–11. They seem to grow in height rather than migrate, as though they are at some focus of depositional winds.

Longitudinal or **linear dunes** may be huge landforms extending hundreds of kilometers in length, a kilometer or more in width, and several hundred meters in height. They are especially well developed in the heart of the tradewind deserts, where the wind is either from a constant direction or varies seasonally. Longitudinal dunes cover approximately 30 percent of the total area of eolian deposition (Tsoar and Møller, 1986). Their great size and abundance, and the fact that many of them are now relict, has lead to much speculation about their origin. Bagnold (1941) suggested that they were built by seasonal shifting winds that preferentially elongated one arm of a large barchan dune until it became a wavy, elongate longitudinal dune. Livingstone (1986) and Tsoar (1989) demonstrated that longitudinal dunes extend parallel to the dominant wind direction, but that other effective winds, at an angle to the dominant flow, change direction as they cross the dune crest, to blow nearly parallel to the dune crest. These winds may either erode or deposit, because of velocity changes, but always with a longitudinal transport component. Great systems of longitudinal dunes cross the arid interiors of Australia and South Africa (Fig. 14–13), but most are now covered by sparse vegetation and soil and seem to be relict from one or more earlier periods of greater aridity. Active longitudinal dunes are found in areas nearly devoid of vegetation, where winds are persistent for many months and sand supply is irregular. The dune ridges are separated by gravel-armored reg or hammada.

Transverse dunes (Fig. 14–14) are associated with massive quantities of sand and relatively ineffective winds. They are not common in the great sandy deserts but are common along coastlines (Cooper, 1958, 1967) and on zones of erodible sandy bedrock or alluvium. Although their crests are ridges roughly perpendicular to the effective wind, in detail the crests have sinuous or "fish-scale" outlines or break up into fields of isolated barchan or longitudinal dunes. The *barchanoid ridge* is a common transitional type (Fig. 14–10).

Barchan dunes (Fig. 14–14) are the classic eolian landforms. Their crescentic form consists of a gently inclined windward slope and a steep lee side around which the horns or cusps of the dune project downwind, making the slip face concave to the downwind direction. Barchans are isolated dunes that migrate freely across rock or gravel desert plains, usually downwind from some other dune form (Inman et al., 1966). They have a nearly constant sand mass, and are large enough to maintain themselves as they migrate. They represent a remarkable balance between accumulation, transportation, and erosion.

LINEAR DUNES. Arrows show probable dominant winds.

STAR DUNES. Arrows show effective wind directions.

REVERSING DUNES. Arrows show wind directions.

BLOWOUT DUNES.

DOME DUNES.

PARABOLIC DUNES. Arrow shows prevailing wind direction.

BARCHAN DUNES. Arrow shows prevailing wind direction.

BARCHANOID RIDGE. Arrow shows prevailing wind direction.

TRANSVERSE DUNE. Arrow shows prevailing wind direction.

Figure 14-10. Basic dune forms, with probable wind directions where applicable. (From McKee, 1979, pp. 11–13.)

Figure 14-11. Dune forms as a response to three semiquantified variables: wind effectiveness, vegetative cover, and sand supply. Wind direction is assumed to be nearly constant. (Modified from Hack, 1941, Fig. 19.)

Parabolic dunes and irregular **blowout dunes** are characteristic of partially stabilized sandy terranes that develop blowouts (Fig. 14–2). The crescentic shape of parabolic dunes has a very different orientation from that of barchan dunes, likely to confuse a novice. On the lee side of an elongate blowout a dune may accumulate, its windward slope rising from the blowout and the slip face *convex* downwind in plan view. A parabolic dune may become elongated (a "hairpin" or "upsiloidal" dune) by downwind migration of a blowout until it is split into two minor longitudinal ridges paralleling the long axis of the blowout. These dunes are usually much longer and narrower than barchan dunes and are always associated with a central blowout. They should not be confused with barchans, especially on aerial photographs, because the resulting interpretation of effective wind direction will be reversed.

Figure 14-12. Saharan erg near 30°10′N, 8°15′E, dominated by closely spaced star dunes. (Photo: U.S. Air Force, from collection of H.T.U. Smith.)

Figure 14-13. Dunes of the northern Namib Sand Sea, Namibia. The oceanic Benguela Current transports sand northward to this arid coast, where it is blown inland and northward. Dune patterns include barchans and crescentic dunes near the coast, and complex linear (longitudinal) dunes that break into fields of star dunes near ephemeral river channels in the east. (Landsat MSS 40394-08264: courtesy N. Lancaster.)

Whalebacks and Zibar Dunes

Where longitudinal dunes have migrated downwind, the coarser sand is left behind as a prominent low, rolling ridge of coarse sand, lacking slip faces. These **whalebacks** or **zibar dunes** (Greeley and Iverson, 1985, p. 171) attest to the stability of longitudinal dunes in form and spacing over long periods of time. Either very large longitudinal dunes migrated a long distance downwind or longitudinal dunes repeatedly followed the same tracks across the desert. Either case implies former wind effectiveness, sand supply, or both, far in excess of present conditions. The whalebacks described by Bagnold (1941, p. 230) in the Egyptian desert are 1 to 3 km in width and 50 m high and extend for 300 km. Modern longitudinal dunes may surmount these great ridges but are much smaller and segmented. Sometimes as many as four chains of modern parallel longitudinal dunes surmount a single whaleback (for similar Arabian forms, see Holm, 1960). These forms represent one of the more intriguing aspects of the effect of climatic change on landforms (Chap. 18).

Undulating Fixed Sand Sheets

Large areas of semiarid steppe or prairie (Chap. 13), especially in the middle latitude interiors of the large continents, are underlain by sandy sediments, have a

Figure 14-14. Transverse or barchanoid coastal dunes breaking inland into barchans. Guerro Negro dune field, Baja California, Mexico. View northward, Laguna Manuela in background. Dunes in foreground are about 6 m high. (Photo: D.L. Inman.)

sparse grass cover, and have intermittent rain or winter snow. As the result of either drought or fire, local areas of shallow deflation basins develop, alternating with subdued parabolic dune forms. Winds are not sufficiently effective to create dunes with active slip faces, but the combination of deflation and deposition makes a landscape of undulating or "billowing" hills, somewhat analogous to the top of a cloud cover as viewed from an airplane (Fig. 14–15). During wet years or seasons, ponds accumulate in hollows between the hills. High permeability and low precipitation prevent stream dissection even when the water table is near the surface. The Sand Hills region of Nebraska (Smith, 1965; Ahlbrandt and Fryberger, 1980) is a typical example. Large areas of the Russian steppe and Iran are also of this origin. As a transitional landscape between arid and subhumid climates, these large regions can be expected to show many relict landforms of conditions both more moist and more arid than now experienced. Buried soil profiles, sand-filled drainage systems, polycyclic drainage networks, and buried sites of human occupance attest to the mobility of sandy terranes under marginal precipitation.

Sand Sheets (Ergs)

In contrast to the fixed dunes of grassy steppes, from one-third to one-fourth of the true deserts consist of vast "sand seas" or **ergs** (McKee, 1979). Their surfaces consist of a variety of dune forms, ranging from low mounds to huge star dunes. These largest of all the sandy landscapes are fixed primarily by lithologic factors. Sand is thick and continuous enough for the surface to include the full range of eolian landforms; ripples, dunes, and draas (defined by Wilson, 1973, as ridges about 1 to 3 km apart). There may be barren rock areas and reg; Wilson (1973, p. 78) suggested a preliminary definition of an erg as "an area where wind-laid sand deposits cover at least 20 percent of the ground, and which is large enough to contain draas." The minimum size ranges from 1 to 40 km². At least 30 named ergs exceed 120,000 km²

Figure 14-15. The Sand Hills region of Nebraska. Stabilized dunes along Nebraska Highway 97, south of the Dismal River. (Photo: W. J. Wayne.)

(Wilson, 1973, Table 1). Those in Africa, Arabia, and Australia are on or downward from easily weathered, sandy sedimentary rocks, probably deposited in Cenozoic or Mesozoic marine embayments or freshwater lakes. They are more nearly "timeless" residuals than evolving landforms, either constructional or erosional, although sand is known to both enter and leave even the largest ergs (Lancaster, 1988a, 1989b). Wind transport and deposition constantly shift the surface layers, but the regional landscape is one of awesome monotony (Fig. 14–16).

Figure 14-16. Dunes in the Rub' al Khali, an erg covering 600,000 km² of southeastern Saudi Arabia. (Photo: Arabian American Oil Company.)

DUST AND LOESS

It was noted earlier that loose silt- and clay-size particles ("dust") can be extremely resistant to wind shear and entrainment. However, if stirred into suspension by vehicles, animals, impacting sand grains, or vortical winds ("dust devils", "whirlwinds"), dust can be lifted hundreds or thousands of meters into the air and carried long distances. Dust that has been transported for distances of less than 100 km has grain sizes in the silt range (approximately 5 to 50 μm). Finer dust stays in suspension as an aerosol and is only precipated by rainfall; it is mostly clay size (2 to 10 μm) (Péwé, 1981, p. 3). It can travel around the earth on high-altitude winds. Fine windblown dust is the major component of deep-sea sediments; the Sahara is the major source for the equatorial Atlantic Ocean and central Asia contributes most of the dust to the north Pacific (Schneider et al., 1990).

Thick deposits of homogeneous unstratified silt are draped over or bury other landforms in the middle latitudes. The extremely well-sorted, fine-grained sediment is generally known by the German term **loess,** meaning simply "loose," or unconsolidated. Loess has an eolian origin, derived from glacial outwash and other alluvium (Fig. 15–10) or deserts, although minor water transport, mass wasting, and pedogenic processes are accepted as part of the overall process of loess accumulation (Pye, 1987, pp. 198–200). Deposits similar to loess but showing evidence of water retransport, soil formation, and other alteration are often referred to as "loess-like." The extremely good size sorting during long-distance transport of windblown silt accounts for most of the peculiar and significant properties of loess. It is a material, not a kind of landscape, but where thick enough to bury the underlying landscape, its surface deserves consideration as a constructional landform. Furthermore, the soil-forming and erosional properties of loess terranes also give them special geomorphic interest.

Loess and similar deposits form a cover 1 to 100 m thick over an estimated 10 percent of the land area (Pécsi, 1968). Thirty percent of the United States has an eolian cover (Ruhe, 1974, p. 152). In detailed studies, loess sheets in Europe and North America have been shown to consist of multiple depositional units, each correlative with a glacial age and separated by unconformities with soil profiles representing interglacial conditions. In Europe, archeologic sites are found in the buried soils. Loess deposits thicken exponentially toward source areas, usually outwash plains or the undulating fixed sand sheets that were probably barren during glacial ages (Fig. 14–17).

Loess thicknesses in excess of 300 m have been reported in parts of China, but more typically thicknesses are in the range of 20 to 30 m. Buried soils demonstrate that deposition was intermittent, probably controlled by the availability of a renewable source such as the braided floodplain of a glacial meltwater river. Fossil snail shells in the Mississippi Valley loess are of species that inhabit moist woodlands. Probably the thick loess along the downwind sides of glacial outwash rivers was trapped by trees growing adjacent to the floodplains (Fig. 14–17). Grass is also an effective dust trap, from which periodic rains wash the accumulated dust down to the soil surface. Winter dust blowing from barren fallow fields commonly discolors the snow downwind of the source areas. When the snow melts, the resulting loess would show no stratification, even though it was a seasonal accumulation.

The largest known region of loess and loess-like sediments is in north central China, where they cover at least 635,000 km² (Liu, 1988, p. 13). The Chinese loess is probably not glacially derived. It becomes coarser toward the northwest and grades into sandy loess and the sand of the Gobi desert of Mongolia (Derbyshire, 1983). Loess deposition began about 2.4 million years ago, according to paleomagnetic correlations and other dating techniques (Kukla et al., 1988; Liu, 1988, p. 52). Deposition continues today at a rate of several millimeters per year (Derbyshire,

Figure 14-17. Loess thickness in Illinois (compare with Fig. 15-10). (From Willman and Frye, 1970, Plate 3.)

1983). It has numerous soil profiles developed within it, and its history is intimately involved with global Quaternary climate change (Chap. 18).

Most of the world's loess is dominantly silt-size quartz (mostly 20 to 60 μm; Smalley and Smalley, 1983), but reflects the lithology of the source terranes and may be highly calcareous. The abundance of quartz indicates a source-area weathering environment in which quartz was stable but minerals such as feldspars, mica and ferromagnesian minerals were chemically weathered. Because detrital quartz is usually sand-sized, the abundance of quartz in loess has been puzzling. Glacial abrasion (Chap. 17) to silt size is clearly indicated by the angular, fractured grains of quartz in the loess of formerly glaciated areas. Experimental evidence of eolian abrasion of quartz to silt-size fragments seems to have resolved former concerns about the origin of the silt-sized quartz in non-glacial loess (Whalley et al., 1987).

The loess of the Argentine Pampas is derived from weathered volcanic rocks and contains a high percentage of pyroclastic glass shards and plagioclase indicative of an Andean source (Teruggi, 1957). Its origin and age must be related to the chronology of eolian denudation in the high volcanic regions of the Andes (p. 349). Deposition began in the late Pliocene and continues today.

Because of its fineness, carbonate content, and homogeneity, loess consolidates readily and forms stable steep cliffs, often with a "case-hardened" surface of secondary calcium carbonate. Most of the calcium carbonate in loess is derived from weathering in the source area, not from weathering or groundwater transport after deposition (p. 333). No more than 3 to 10 percent of the carbonate content of the Chinese loess is estimated to have been derived from local weathering by atmosphere CO_2 or by groundwater (Liu, 1988, p. 4). The depositional surface is featureless, although parabolic dunes of loess-like silt and clay (**lunettes**) accumulate on the lee sides of dry lake beds in southern Australia (Hills, 1940; Stephens and Crocker, 1946). Most often, the uniform deposition results in topography that shows strong inheritance from preloess conditions (Lugn, 1935, p. 159). Loess erodes readily in humid climates and forms a characteristic high-density network of gullies with steep or vertical banks controlled by the columnar soil structure typical of loess. The loess sheets associated with the last glaciation are relatively undissected and unweathered, but beyond the drift border of the last ice sheets, older loesses have much more developed erosional landscapes. Quaternary loess sheets can be dated by stratigraphic correlation with tephras and glacial drift and by paleomagnetic correlation with other Quaternary deposits. In the late Quaternary, radiocarbon dates on fossil snails and interbedded peat are useful.

REFERENCES

AHLBRANDT, T. S., and FRYBERGER, S. G., 1980, Eolian deposits in the Nebraska Sand Hills: U.S. Geol. Survey Prof. Paper 1120-A, 24 pp.

ALBRITTON, C. C., JR., BROOKS, J. E., ISSAWI, B., and SWEDAN, A., 1990, Origin of the Qattara Depression, Egypt: Geol. Soc. America Bull., v. 102, pp. 952–60.

BAGNOLD, R. A., 1941, The physics of blown sand and desert dunes: Methuen & Co. Ltd., London, 265 pp.

BALL, J., 1927, Problems of the Libyan desert: Geog. Jour., v. 70, pp. 21–38, 105–28, 209–22.

———, 1933, Qattara depression of the Libyan desert and the possibilities of its utilization for power-production: Geog. Jour., v. 82, pp. 289–314.

BERKEY, C. P., and MORRIS, F. K., 1927, Geology of Mongolia: Natural history of central Asia, v. 2, Am. Museum Nat. Hist., New York, 475 pp.

BLACKWELDER, E., 1934, Yardangs: Geol. Soc. America Bull., v. 45, pp. 159–66.

BREED, C. S., and GROW, T., 1979, Morphology and distribution of dunes in sand seas observed by remote sensing, *in* McKee, E. D., ed., A study of global sand seas: U.S. Geol. Survey Prof. Paper 1052, pp. 253–302.

COOPER, W. S., 1958, Coastal sand dunes of Oregon and Washington: Geol. Soc. America Mem. 72, 169 pp.

————, 1967, Coastal dunes of California: Geol. Soc. America Mem. 104, 131 pp.

DERBYSHIRE, E., 1983, Origin and characteristics of some Chinese loess at two locations in China, *in* Brookfield, M. E., and Ahlbrandt, T. S., eds., Eolian sediments and processes: Developments in sedimentology 38, Elsevier Science Publishers, Amsterdam, pp. 69–90.

DIETRICH, R. V., 1977, Impact abrasion of harder by softer materials: Jour. Geology, v. 85, pp. 242–46.

FIELDING, E. J., KNOX, W. J., JR., and BLOOM, A. L., 1986, SIR-B radar imaging of volcanic deposits in the Andes: IIEE Trans. on Geoscience and Remote Sensing, v. GE-24, pp. 582–89.

GREELEY, R., and IVERSON, J. D., 1985, Wind as a geological process on Earth, Mars, Venus and Titan: Cambridge Univ. Press, Cambridge, England, 333 pp.

GUEST, J. E., 1969, Upper Tertiary ignimbrites in the Andean Cordillera of part of the Antofagasta Province, northern Chile: Geol. Soc. America Bull., v. 80, pp. 337–62.

HACK, J. T., 1941, Dunes of the western Navaho country: Geog. Rev., v. 31, pp. 240–63.

HICKOX, C. F., JR., 1959, Formation of ventifacts in a moist, temperate climate: Geol. Soc. America Bull., v. 70, pp. 1489–90.

HILLS, E. S., 1940, The lunette, a new landform of eolian origin: Aust. Geographer, v. 3, pp. 15–21.

HOLM, D. A., 1960, Desert geomorphology in the Arabian peninsula: Science, v. 132, pp. 1369–79.

HSÜ, K. J., 1983, The Mediterranean was a desert: Princeton Univ. Press, Princeton, New Jersey, 197 pp.

INMAN, D. L., EWING, G. C., and CORLISS, J. B., 1966, Coastal sand dunes of Guerrero Negro, Baja California, Mexico: Geol. Soc. America Bull., v. 77, pp. 787–802.

KUKLA, G., HELLER, F., MING, L. X., CHUN, X. T., SHENG, L. T., and SHENG, A. Z., 1988, Pleistocene climates in China dated by magnetic susceptibility: Geology, v. 16, pp. 811–14.

LANCASTER, N., 1988a, On desert sand seas: Episodes, v. 11, pp. 12–17.

————, 1988b, Controls of dune size and spacing: Geology, v. 16, pp. 972–75.

————, 1989a, Star dunes: Progress in Phys. Geography, v. 13, pp. 67–91.

————, 1989b, The Namib Sand Sea: Dune forms, processes and sediments: A. A. Balkema, Rotterdam, The Netherlands, 180 pp.

LIU TUNGSHENG, 1988, Loess in China, 2nd ed.: Springer-Verlag Berlin Heidelberg, 224 pp.

LIVINGSTONE, I., 1986, Geomorphological significance of wind flow patterns over a Namib linear dune, *in* Nickling, W. G., ed., Aeolian geomorphology: Allen & Unwin, Inc., Boston, pp. 97–112.

LUGN, A. L., 1935, Pleistocene geology of Nebraska: Nebraska Geol. Survey, Bull. 10, Ser. 2, 223 pp.

McCAULEY, J. F., GROLIER, M. J., and BREED, C. S., 1977, Yardangs, *in* Doehring, D. O., ed., Geomorphology in arid regions: Publications in Geomorphology, State University of New York in Binghamton, pp. 233–69.

McKEE, E. D., ed., 1979, A study of global sand seas: U.S. Geol. Survey Prof. Paper 1052, 429 pp.

MURRAY, G. W., 1952, The water beneath the Egyptian western desert: Geog. Jour., v. 118, pp. 443–52.

NORMAN, S. E., and CHASE, C. G., 1986, Uplift of the shores of the western Mediterranean due to Messinian desiccation and flexural isostasy: Nature, v. 322, pp. 450–51.

PÉCSI, M., 1968, Loess, *in* Fairbridge, R. W., ed., Encyclopedia of geomorphology: Reinhold Book Corp., New York, pp. 674–78.

PEEL, R. F., 1966, The landscape in aridity: Inst. Brit. Geographers Trans., no. 38, pp. 1–23.

PÉWÉ, T. L., 1981, Desert dust: An overview, *in* Péwé, T. L., ed., Desert dust: Origin, characteristics, and effect on man: Geol. Soc. America Spec. Paper 186, pp. 1–10.

PYE, K., 1987, Aeolian dust and dust deposits: Academic Press Inc. (London) Ltd., London, 334 pp.

RUHE, R. V., 1974, Geomorphology: Houghton Mifflin Co., Boston, 246 pp.

SCHNEIDER, B., TINDALE, N. W., and DUCE, R. A., 1990, Dry deposition of Asian mineral dust over the central North Pacific: Jour. Geophys. Res., v. 95, no. D7, pp. 9873–78.

SHARP, R. P., 1949, Pleistocene ventifacts east of the Big Horn Mountains, Wyoming: Jour. Geology, v. 57, pp. 175–95.

———, 1980, Wind-driven sand in Coachella Valley, California: Further data: Geol. Soc. America Bull., v. 91, pp. 724–30.

SMALLEY, I. J., and SMALLEY, V., 1983, Loess material and loess deposits: Formation, distribution and consequences, *in* Brookfield, M. E., and Ahlbrandt, T. S., eds., Eolian sediments and processes: Developments in sedimentology 38, Elsevier Science Publishers, Amsterdam, pp. 51–68.

SMITH, H. T. U., 1965, Dune morphology and chronology in central and western Nebraska: Jour. Geology, v. 73, pp. 557–78.

STEPHENS, C. G., and CROCKER, R. L., 1946, Composition and genesis of lunettes: Royal Soc. S. Australia Trans., v. 70, pp. 302–12.

SUZUKI, T., and TAKAHASHI, K., 1981, An experimental study of wind abrasion: Jour. Geology, v. 89, pp. 23–36.

TERUGGI, M. E., 1957, The nature and origin of Argentine loess: Jour. Sed. Petrology, v. 27, pp. 322–32.

TSOAR, H., 1989, Linear dunes—forms and formation: Progress in Phys. Geography, v. 13, pp. 507–28.

——— and MØLLER, J. T., 1986, The role of vegetation in the formation of linear sand dunes, *in* Nickling, W. G., ed., Aeolian geomorphology: Allen & Unwin, Inc., Boston, pp. 75–95.

WARD, A. W., and GREELEY, R., 1984, Evolution of yardangs at Rogers Lake, California: Geol. Soc. America Bull., v. 95, pp. 829–37.

WHALLEY, W. B., SMITH, B. J., MCALISTER, J. J., and EDWARDS, A. J., 1987, Aeolian abrasion of quartz particles and the production of silt-size fragments: Preliminary results, *in* Frostick, L. E., and Reid, I., eds., Desert sediments: Ancient and modern: Geological Society Special Publication no. 35, pp. 129–38.

WHITNEY, M. I., and DIETRICH, R. V., 1973, Ventifact sculpture by windblown dust: Geol. Soc. America Bull., v. 84, pp. 2561–81.

WILLMAN, H. B., and FRYE, J. C., 1970, Pleistocene stratigraphy of Illinois: Illinois Geol. Survey, Bull. 94, 204 pp.

WILSON, I. G., 1972, Aeolian bedforms—their development and origins: Sedimentology, v. 19, pp. 173–210.

———, 1973, Ergs: Sedimentary Geol., v. 10, pp. 77–106.

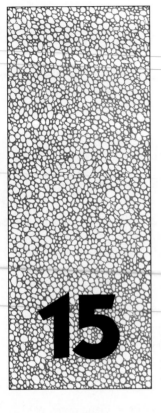

15 Periglacial Morphogenesis

DEFINITIONS AND EXTENT

As originally defined and generally used early in this century, the term **periglacial** referred to the region in Europe that was peripheral to the margins of the Pleistocene glaciers where frost action was inferred to have been particularly active in the past. The traditional emphasis was on past climatic conditions now more equitable, but the term applies equally well to the tundra regions of today, in high latitudes primarily in the northern hemisphere. "Ideally, it is a zone of permanently- [preferred usage is 'perennially'] frozen subsoil (permafrost), seasonally-thawed topsoil (active layer), frequent changes of temperature across the freezing point, and an incomplete vegetation cover of herbaceous plants and dwarf species of trees, for example dwarf birch and willow" (Sparks and West, 1972, p. 98).

The modern tendency is to consider the periglacial environment as near-glacial in the sense of either location or intensity of conditions. No unique climate definition of the periglacial region has been formulated. Essential to all definitions is the concept of intense frost action and bare ground for at least part of the year (thus excluding glacier-covered regions, which are a separate topic). Permafrost is not required by most definitions (Washburn, 1980; Péwé, 1983). Features described on subsequent pages, including gelifluction and patterned ground, are commonly regarded as periglacial phenomena, but are known to form in areas that lack permafrost. Nevertheless, almost all periglacial regions are underlain by permafrost, and fossil evidence of permafrost in soil structures is taken as proof of a former periglacial environment (Péwé, 1983, p. 157). Because saline ocean water will not freeze at 0°C, extensive areas of relict permafrost extend seaward in the Arctic continental shelves of North America and Asia. It has been preserved by subzero, but unfrozen, seawater since it formed in times of lower sea level during the last ice age. For practical reasons, the modern periglacial zone is regarded here as coincident with the region of perennially frozen ground, or **permafrost** (Fig. 15–1), but excluding the glacier-covered areas. Some synonyms include *cryergic* or *cryogenic* processes and morphogenetic regions.

The definitive feature of periglacial morphogenesis is intense freezing and seasonal thawing. Gelifluction (p. 375) and nivation (p. 379) are also characteristic processes. Some authors have included in the morphogenetic region all areas that experience seasonal frost, but this definition is too broad to be useful. By including in the definition the criterion of perennially frozen subsoil (permafrost), the modern periglacial morphogenetic region becomes much more coherent and identifiable (Péwé, 1969; 1983). Also, perennially frozen ground produces soil and sediment structures and landforms that remain after the climate changes, so relict periglacial conditions can be identified and the affected areas can be shown on maps.

Figure 15-1. Distribution of permafrost in the northern hemisphere. Isolated areas of alpine permafrost not shown on the map exist in high mountains and outside the map area in Mexico, Hawaii, Japan, and Europe. (Péwé, 1983, Fig. 9-1.)

No single climatic region corresponds to the periglacial zone. Low-latitude plateaus and mountain tops are included, as well as the cold, dry continental interiors of Asia and North America and the maritime boreal coasts of Iceland, Canada, Scandinavia, and Alaska. In some periglacial regions, the mean annual temperature is close to 0°C, and the annual range varies only slightly because of maritime influences. In other regions, a short summer interval may be distinctly hot, with 20 hours or more of sunshine, but winters are dark and intensely cold and dry.

Neither can the former periglacial regions peripheral to the Pleistocene ice sheets be compared directly with the arctic tundras of today. Relict low-altitude periglacial soil structures have been discovered at latitudes at least as low as 40°N in the central United States (Johnson, 1990, Fig. 1). Even if the periglacial interval was brief, its record at a latitude more than halfway toward the equator from the north pole should remind us not to extrapolate modern climatic conditions to the Pleistocene record. In the central plains of the United States at the edge of a Pleistocene ice sheet, the summer sun at noon was within 17° of the zenith. Days and nights must have been of nearly the same length as present ones and not at all like the seasonal alternation of continuous daylight and darkness that characterizes the tundra regions north of the Arctic Circle.

Permafrost now underlies 22 to 25 percent of the world's non-glacier-covered land area, including about 82 percent of Alaska, about 50 percent of Canada and the Soviet Union, and 20 percent of China (Fig. 15–1). All that is required to produce permafrost is a mean annual soil temperature below freezing. Because soil temperature at a depth of a meter or two closely approximates mean annual air temperature, the latter measure is often used to define the permafrost region. However, differences in soil moisture, porosity, vegetation cover, dissolved components, and groundwater pressure have marked effects on the thickness of the frozen layer, and subcategories of *alpine, discontinuous,* and *continuous* permafrost are used. Some permafrost may have persisted since the last glaciation; in other areas, the permafrost has formed in postglacial time. As with glaciers, the permafrost limit seems to be shrinking during the last century. The maximum known depth of permafrost is 1600 m, but over wide areas of the arctic, a thickness of several hundred meters is commonplace (French, 1976, p. 56). Only 1 or 2 m at the surface is subject to seasonal melt. This is the **active layer.**

Despite the vagueness of its definition, the periglacial morphogenetic region is impressive in size and geomorphic significance. The table on page 328 lists 38 percent of the earth's land area under either a cold, treeless, snow climate or a boreal forest and snow climate. At least half of the combined areas of the two regions is underlain by permafrost. At various times during the Quaternary Period, recently enough so that some relict landforms display periglacial morphogenesis, perhaps as much as 40 percent of the land area has been subjected to periglacial conditions (Tricart, 1970, p. 28). Büdel (1982, pp. 109, 119) argued that fully 95 percent of the landforms of the mid-latitude northern hemisphere are relict from former periglacial conditions. He considered subtropical savanna and periglacial morphogenetic processes as the two most important sets of landscape-forming processes on earth. In reviewing Büdel's viewpoint, Cotton (1963, p. 769) suggested that it is possible that most of the landforms in mid-latitude regions, regarded by W. M. Davis and others of his time as the result of the "normal" processes of fluvial erosion (p. 210), may be relict from past periglacial conditions. It would indeed be an odd trick of fate if the landforms of the northeastern United States and western Europe, which provided the observational evidence for the deduced "normal cycle of erosion," were actually relict periglacial landforms. Geomorphologists with field experience in present-day periglacial regions are steadily demonstrating the impact of former periglacial morphogenesis in regions of presently temperate climates. Both for its present and past importance, periglacial morphogenesis is an important current topic of geomorphology.

PERIGLACIAL PROCESSES

The periglacial processes are characterized more by the criterion of degree or intensity than of kind. Weathering and soil formation are controlled by the intensity and frequency of freezing and thawing and by the generally frozen subsoil. Mass wasting is dominated by solifluction of the active layer over permafrost (gelifluction). Rock falls and rock glaciers (p. 385) are common in mountainous regions. Fluvial erosion is inhibited by the brevity of seasonal stream flow and the massive loads of sediment fed into streams by gelifluction. Wind becomes a relatively important agent of transport and deposition, with loess sheets creating major downwind constructional landforms. In the review that follows, most of the processes have been described in previous chapters and will be only briefly noted. The emphasis in this chapter is on the ways the processes interact to produce a landscape recognizable as the result of periglacial morphogenesis.

Weathering

Freezing and Frost Cracking. Despite continued uncertainty as to the exact process by which rocks are fractured by freezing and thawing of interstitial water (pp. 124–26), there is no doubt that regions now in the periglacial environment are characterized by great quantities of angular, fractured rock detritus (Fig. 15–2). The size of the fragments is largely a function of lithology, pore size, permeability, degree of microcracking, freezing intensity, and water content (Lautridou, 1988). *Frost scaling* causes thin flakes to break from rock surfaces; *frost splitting* is thought to be a true breaking of previously uncracked rock; and *frost wedging* exploits preexisting

Figure 15-2. Wright Valley, South Victoria Land, Antarctica. The mean annual temperature is about −20°C. No soil develops on the steep taluses. (Photo: C. Bull, Institute of Polar Studies, Ohio State University.)

fractures. Laboratory experiments suggest that repeated freezing and thawing can disaggregate mineral grains as small as 1 μm in diameter (McDowall, 1960; Lautridou, 1988, p. 43). Water is a necessary but little-evaluated component in the process, but precipitation records for periglacial regions do not give much useful information about the availability of water within the ground. Freezing must be slow enough to permit pore water to migrate toward the freezing front. Certainly, dry rocks will not be disrupted by repeated temperature fluctuations through 0°C. Even for saturated rocks, the critical temperature for fracturing is probably well below 0°C. A range of −4 to −15°C, reached by continuous slow cooling at a rate of less than −0.1 to −0.5°C/hr, is calculated to be optimal for crack growth in granite and marble (Walder and Hallet, 1985). Rocks with medium-sized pores and moderate permeability show an increase in shattering down to temperatures of −8 to −10°C, but little change at colder temperatures. However, a temperature below −10°C has been reported to be more effective in fracturing chalk (Lautridou, 1988). Air-temperature fluctuations do not affect rocks to a depth of more than a few centimeters, so probably the annual temperature range is the significant climatic factor in mechanical weathering by freezing (Cook and Raiche, 1962). Nevertheless, with all these qualifications and uncertainties, alternate freezing and thawing is generally considered the dominant weathering process in the periglacial zone.

Chemical Weathering.   Freezing has been so closely associated with periglacial morphogenesis that other weathering processes have been slighted. Hydration, hydrolysis, and oxidation will occur whenever aerated water contacts minerals, but at low temperatures and with few microorganisms, chemical processes proceed only very slowly. Crusts of hydrated salts form on rocks at McMurdo Sound, Antarctica, and might be important to rock weathering there. In other coastal periglacial regions, where sea salts might collect by evaporation under low atmospheric humidity, similar weathering might also be expected.

An instructive illustration of weathering in a polar desert was provided by Kelly and Zumberge (1961). They analyzed a sequence of quartz diorite specimens from a single exposure at Marble Point, on McMurdo Sound, Antarctica. The samples ranged in degree of weathering from fresh rock to completely disaggregated, iron-stained grus. Although the authors assumed that a sequence of chemical changes would be illustrated by the samples, in reality no chemical weathering was found, except for oxidation of iron-rich biotite and accessory minerals. Bulk chemical and mineralogic composition did not change. No clay minerals formed. The disintegration was caused not by oxidation or hydration but by frost cracking and the subsequent crystallization of salts in pores and fractures. The site is exposed to sea spray during the brief summer, but no rain falls. This study provided substantial quantitative support for the general but previously unsubstantiated opinion that mechanical weathering predominates over chemical weathering in cold, dry climates. The question of limestone solution in cold regions is discussed in Chapter 7.

Mass Wasting

Frost heaving is the process of lifting surface particles by the formation of ice in soil. As ice nuclei form in soil, free water migrates toward these nuclei, which grow as veins or lenses of ice while adjacent sediments are dewatered (Everett, 1961). The grain size and pore size of sediments determine the rate of ice-crystal segregation and water movement. Silt is the optimal grain size for ice-crystal growth, and frozen silt may contain 50 to 150 percent ice by weight (French, 1976, p. 76). Ice masses as thick as 40 m have been reported from Canada (Washburn, 1980, p. 43).

Needle ice consists of clusters of slender ice crystals up to several centimeters in length that form overnight in frosty wet soils. Although the phenomenon is much less impressive than the formation of segregated ice masses in perennially frozen ground, needle ice plays an important role in loosening surface soils and in lifting small stones to the surface of fine sediments. On hill slopes, frost heave and needle ice promote frost creep by lifting particles at right angles to the sloping surface, then permitting them to settle vertically on remelting.

Gelifluction and Frost Creep.   Long before the term "periglacial" had been proposed, European geologists were aware that thick sheets of poorly sorted angular debris mantled their landscape beyond the limits of glaciation. The material was given quaint colloquial terms such as "head," or "coombe rock," and had been hypothesized to be the result of cold-climate, but not glacial, processes. During an expedition to the Falkland Islands, J. G. Andersson observed fluid sheets of debris moving down gentle slopes over perennially frozen ground and producing deposits similar to the "head" of European authors. He named the process **solifluction** (p. 183 and references there), later modified to **gelifluction** to specify flow over a frozen substrate. The process is generally accepted as being responsible for the deposits of "head," and, in fact, massive deposits of relict gelifluction debris are accepted as evidence of a former periglacial climate (Cotton and Te Punga, 1955).

Because of the perennially frozen substrate, mass wasting is not a deep phenomenon in periglacial regions. Gelifluction occurs only in the active layer, rarely more than 2 m thick. Resistant rock ledges are exposed to freeze and thaw as the active layer flows downhill, and they stand out as tors and low structurally controlled ridges on debris-covered slopes. Gelifluction occurs even on very low slope angles, perhaps as low as 1° on rocks that break down to clay-rich debris (Fig. 15–3). On bouldery debris, gelifluction slopes may be as steep as 20° (Tricart, 1970, pp. 105–6).

During the brief high-latitude summer thaw, the active layer, a meter or so in thickness, composed of tundra peat, rock rubble, and other weathered debris, may

Figure 15-3.   Tundra landscape of extremely low relief and gentle convex slopes above river-bank cliffs on the Colville River near Umiat, Alaska. The bentonitic shale is prone to gelifluction and other mass-wasting processes. (Photo: T. L. Marlar, U.S. Army Cold Regions Research and Engineering Lab.)

flow down slopes of almost negligible gradient, because meltwater saturates the active layer but cannot penetrate the frozen ground beneath. The mass of gelifluction debris may flow with a kind of rolling motion like the endless tread of a tracked vehicle. Arcuate ridges and troughs mark the toe, or lower part of the mass. The rate of downhill motion in gelifluction lobes in East Greenland is several times more rapid than the nonlobate creep on wet slopes of similar gradient (Fig. 8–2) (Washburn, 1967, pp. 94–95). Although the annual rate of solifluction movement may be little more than 100 mm, the entire movement takes place during the brief summer season when net movement may be several mm/day on slopes of less than 10° (Fig. 15–4).

Washburn accurately surveyed lines of conical targets driven into the soil across Greenland hillsides, and by repeated resurveys that extended over a period of nine years, he was able to separate the components of gelifluction and **frost creep** (soil creep due to the ratchetlike motion of particles during alternating freeze–thaw cycles) (Fig. 15–5). His periodic resurveys proved that frost creep was quantitatively slightly more important on most slopes than gelifluction. An unexpectedly large component of *retrograde motion* due to cohesion caused his targets to move uphill during the summer thaw season, with respect to the vertical plane through the targets at the time of maximum frost heave. The retrograde motion partly or entirely cancels downslope movement by gelifluction. Points P_1 to P_4 in Figure 15–5 trace a typical path of a soil particle on a Greenland hillside during an annual cycle. The diagram should be examined step by step, from the "jump" (P_1–P_2) during frost heave, through the downhill movement (P_2–P_3) during thaw and gelifluction, to the interaction between retrograde motion and gelifluction (P_3–P_4) that determine the net annual downslope motion. In no instance was the net motion uphill. Although retrograde motion in some years largely negated the role of gelifluction and frost creep, it cannot exceed them. By freezing insulated trays of saturated silt tilted at various low angles, Higashi and Corte (1971) were able to reproduce in a laboratory the various combinations of frost creep, gelifluction, and retrograde motion reported by Washburn.

Figure 15-4. Large gelifluction lobe, Mesters Vig district, northeast Greenland. (Photo: A. L. Washburn.)

Along with the specific processes of gelifluction and frost creep, a variety of other mass movements are common features of periglacial morphogenesis. Snow and rock avalanches are frequent, especially during the springtime. Mud flows and debris flows break out of gelifluction lobes and sheets and are really only an extension of the gelifluction process. On clay-rich soils, debris flows in channels with levees have been described (Fig. 15–3) (Anderson et al., 1969; Harris, 1987; French, 1988).

Ground Ice.   The annual cycle of freeze and thaw in the active layer produces large lateral stresses in the mixture of rock fragments, ice, and water. On level ground, slabs and cobbles may be lifted by frost heaving, but fine sediment fills in beneath them so that the layer becomes sorted, with the coarsest fragments at the surface. Even more commonly, cells of differential motion mutually interfere to form a variety of polygonal patterns (Fig. 15–6). The patterns may consist of either sorted or unsorted debris. On slopes steeper than 3 percent, polygonal patterns become attenuated downslope (Fig. 15–7). The driving forces that produce these and other patterns are freezing and thawing combined with gravity. During freezing, pore water migrates toward centers of ice nucleation in frost-susceptible soil in the active layer, and the bulk density of the mass is greatly reduced by the presence of up to 70 percent ice. During thawing, which begins at the surface and migrates downward into the active layer, dewatering increases the bulk density. The density contrast is believed to be adequate to promote free convection of sediment in the active layer, with the formation of polygonal patterns (Hallet and Prestrud, 1986). A related process of convective circulation of water within the active layer, caused by the anomalous greater density of water at 4°C at the surface of the active layer

Figure 15-5.   Diagram of frost creep, gelifluction, and retrograde components of movement by a survey target on a perennially frozen hillside in northeast Greenland. (Washburn, 1967, Fig. 5.)

compared to the less dense water at 0°C at the permafrost contact beneath the active layer, may also promote free convection and polygonal patterns (Gleason et al., 1986; Krantz et al., 1988). While water convection in the density gradient of 0 to 4°C is far too weak to transport sediment, it can transfer heat to the base of the active layer, creating a corrugated relief that then promotes differential sorting in the overlying active layer. The specific landforms are described in a subsequent section of this chapter.

The most prominent polygonal patterns are in areas of extremely low mean annual temperatures, probably averaging −6 to −8°C. Permafrost is normally several hundred meters thick. In such regions, networks of **ice-wedge polygons** cover many thousands of square kilometers (Fig. 15–6). Each polygon, from 3 to 30 m in diameter or more, is outlined by shallow troughs, beneath each of which is a wedge-shaped foliated mass of ice, a meter or more broad at the top and extending downward from 3 to 7 m or more (Lachenbruch, 1962). Most authorities accept the contraction theory of ice-wedge polygon formation (Leffingwell, 1915; Lachenbruch, 1962) as the best explanation, although as noted in Chapter 6 and previously in this chapter, the exact dynamics of fracture in freezing solids is by no means fully understood. As outlined by Lachenbruch (Fig. 15–8), the contraction theory is based on seasonal cracks that develop in totally frozen ground during winter. Ice and soil, like other solids, contract on cooling, and may explosively develop vertical cracks several millimeters wide and a meter or more in depth. In the spring, water from the active layer seeps down the crack and refreezes in the cold permafrost. During the following summer, warming and expansion (but not thawing) of the frozen layer causes sediments to be upturned and deformed against the ice wedge.

Figure 15-6. Ice-wedge polygons, northeastern Mackenzie River delta, Canada. Dark high-centered polygons are 10 to 20 m in diameter. (Photo: J. R. Mackay.)

Each winter the freeze-cracking is repeated along the initial plane of weakness, and each summer a new lamella of ice is introduced. After several centuries, large ice wedges have formed, their spacing and intersections outlining ice-wedge polygons. **Sand wedges** can form by a similar process in arid cold regions, where blowing sand fills surface cracks that open during the coldest season. Figure 15–8 should be studied carefully to note that ice wedges form below the active layer, due to the thermal contraction of solid soil and ice. The stress generated by the 9 percent expansion of water on freezing is an entirely different process, unrelated to ice-wedge formation (Worsley, 1985). Because they can be replaced by later sediments as ice-wedge casts, ice wedges are among the few features that are preserved as valid criteria of former permafrost in regions not now periglacial. However, the identification of fossil ice wedges is difficult and controversial (Black, 1976; Péwé, 1983; Johnson, 1990).

Erosion and Transport Processes

Nivation.   A subtle progression of processes can be observed beginning with weathering under a snow bank, to meltwater erosion from beneath a snow bank, to the downhill erosive creep of water-saturated or refrozen snow, to glacial erosion. The first three processes are grouped under the heading of **nivation,** the intensification of weathering and erosion associated with immobile patches of snow that persist into the summer season in the periglacial environment. Nivation, along with freezing and thawing and gelifluction, is one of the three groups of processes that characterize periglacial morphogenesis (St. Onge, 1969). The definition is complex because it includes both weathering and erosion, and describes only the intensification of processes that are also active in nonperiglacial regions (Thorn, 1988).

Snow patches that persist in sheltered positions on hillslopes below the altitude of permanent snowfields have been observed to produce **nivation hollows** after only a few seasons. Weathering is intensified in the saturated ground beneath the snow. The saturated snowmass produces a flow of water toward its edges, especially

Figure 15-7.   Relict patterned ground and stone stripes on Snake River basalt, north of Glenns Ferry, Idaho (Malde, 1964). Plateau gradient is 2 percent southwestward; colluvium gradient is 20 percent southeastward. Light-colored areas are grassy; dark areas are basalt fragments. (Vertical aerial photo: U.S. Dept. of Agriculture.)

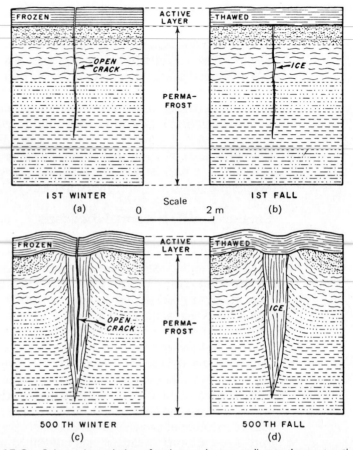

Figure 15-8. Schematic evolution of an ice wedge according to the contraction-crack theory. (From Lachenbruch, 1962, Fig. 1.)

on the downhill side. Water percolating to the base of the snowbank carries away fine particles from the rock or soil beneath. Freezing and thawing may be intensified at the edges of the shrinking snowbank as well, although the snow insulates the rock and soil beneath it. Nivation hollows tend to be self-enlarging; a hollow once produced will catch more snow and shelter a larger snowbank in successive years (Fig. 15–9). Periglacial hill slopes commonly are lined with shallow nivation hollows, especially on the shaded and downwind side of ridge crests where snow accumulates and is shaded from sun for the longest time. Permafrost promotes widening rather than deepening of the hollows, because only the active layer is removed by nivation processes. In the conterminous United States, the favored location for nivation hollows is on the northeastern side of ridges, where snow is blown by westerly winds and is sheltered from sunlight in early summer. In the western Great Basin, nearly 80 percent of the nivation hollows face toward the northeast (Dohrenwend, 1984).

There is a tendency to equate nivation hollows with incipient glacial cirques (p. 422) and to imply a genetic progression from residual snow patches in sheltered hollows to permanent snowfields in basins, to cirques from which alpine glaciers flow (Chap. 17). Although small nivation hollows have been observed to form and enlarge over a period of several years, their ultimate size and function has not been determined. Most of them are found in areas of former glaciation, and the initial hollows may have been relict glacial forms. During the onset of glaciation on a mountain range the snowline is progressively lowered, so that nivation hollows in a high ridge are likely to be buried by permanent snow as new ones form at lower elevations. Conversely, during deglaciation and a rising snowline, nivation hollows are likely to be superimposed on preexisting relief at progressively higher levels, rather than to enlarge in place (Thorn, 1988, p. 22).

The originator of the term *nivation* (Matthes, 1900) found no evidence that seasonal snow patches moved or abraded their beds. Subsequently, other researchers have either asserted or denied that slow creep in the snow associated with compaction, or actual slip of the snowbank over bedrock, contributed to bedrock erosion. A study of the nivation process in the Snowy Mountains of Australia (Costin et al., 1964) proved conclusively that the downhill creep and slide of snowbanks dragged basal rock fragments over bedrock and produced visible, although minor, abrasion marks on both the transported fragments and the subsurface. In a subsequent study (Costin et al., 1973), steel rods notched to various depths were embedded in rock in nivation hollows, and their deformation or breakage under the force of moving snow was calibrated by laboratory tests. The pressure generated ranged from 160 to 3790 kPa (1.6 to 37.9 bar), equal to or greater than the shear stresses calculated for the base of valley glaciers (p. 395). Clearly, seasonal snow patches in nivation hollows, with floors that slope at angles of 10 to 27°, do move downhill and abrade their floors with bed-load fragments. Because one of the definitive features of glaciers is evidence of motion (Chap. 16), proof of snow-patch abrasion makes the distinction between glacial erosion and nivation even less clear. In areas of possible former periglacial or glacial conditions, abraded surfaces have been cited as proof of glaciation. Such evidence may need to be reevaluated. At least it should be used with great caution.

Fluvial Erosion.   Rivers in the periglacial zone have extremely complex hydraulic geometry. Runoff, controlled by snow melting, is temperature-dependent rather than precipitation-dependent. Infiltration is zero over permafrost. If there is rainfall, the direct runoff is increased even more by additional snowmelt. Seasonal

Figure 15-9.   Nivation hollows on cryoplanation terraces cut into a greenstone ridge on Indian Mountain near Hughes, Alaska. (Reger and Péwé, 1976, Fig. 1. Photo: T. L. Péwé.)

and diurnal fluctuations are great. High-gradient mountain streams and large rivers on plains may continue to flow under ice cover throughout the winter. Even so, discharge is greatly decreased because of the lack of overland flow in winter. Small, low-gradient rivers may freeze solid and cease to flow entirely. Because the periglacial zone in general is semiarid or arid, river discharges are never high except during the melting season.

Springtime is the season of violent fluvial action. When meltwater begins to flow into headwater streams, the streams swell and burst their ice cover or flow over frozen river beds, progressively developing a flood wave that moves downstream. If the thaw season begins in the headwaters and progresses toward the river mouth, as it generally does in the larger south-to-north-flowing rivers of arctic North America and Siberia, huge ice jams and floods are seasonal events that are part of the streams' regimen. Briefly, the rivers move much larger loads and much coarser sediment than their mean annual discharges would suggest. Initially, the floods flow over basal ice on those rivers that were frozen solid, so bank erosion is much more extensive than is channel deepening.

The discharge of small periglacial streams, especially mountain streams, follows a distinct daily cycle during the summer. Shortly after the hottest part of the day, the rivers reach peak discharge as overland runoff reaches the channels. Before dawn after a freezing night, runoff will have nearly ceased. Sediment transport and deposition goes through a daily cycle under such conditions, with alluvium deposited among shifting, braided channels. The sediment load delivered to streams also fluctuates with the daily cycle of gelifluction and other forms of mass wasting. Most periglacial streams, other than mountain torrents, flow on aggraded beds. Mass wasting seems to deliver sediment faster than periglacial rivers can move it to the sea. Aggradation results, but it is not yet clear how much of the aggradation is modern and how much is relict from Pleistocene times.

Eolian Transport. Low precipitation, diurnal fluctuations in discharge, freeze-drying of exposed sediment, and strong winds combine to make the periglacial zone a region of strong eolian action. Clouds of sand and dust can be seen rising from the floodplains of rivers as the spring flood subsides (Fig. 15–10). Dunes are common features on the downwind banks of rivers. Sand sheets, less well sorted than dunes and sometimes stratified, are widespread in northern Europe and are attributed to windblown sediment settling on snow and being prevented from forming dunes. A previously unknown relict erg of 7000 km² area has been discovered on the Arctic coastal plain of Alaska (Carter, 1981). Very large areas of present and former periglacial regions are mantled with loess (p. 382; Fig. 14–17).

PERIGLACIAL LANDFORMS

Major Landforms

The combination of intense frost action, permafrost, gelifluction, nivation, strong winds, incomplete vegetation cover, and seasonal flowing rivers imparts a strong morphogenetic imprint on landscapes. Periglacial plains are wastelands of patterned ground, rubble-strewn hillslopes, gelifluction lobes and sheets, and seasonal bogs or shallow lakes of surface water unable to infiltrate because of the permafrost layer. The background of Figure 15–3 gives a good impression of the relief of periglacial plains on nonresistant rocks. Gelifluction causes mass wasting on very low slopes, producing broad convex summits. Stream action is so brief and ineffective that small valleys cannot entrench but are choked by gelifluction debris. Large areas are covered with patterned ground. Mountainous periglacial regions add

Figure 15-10. Clouds of silt transported by wind from the Delta River floodplain, central Alaska. Compare with Figure 14-17. (Photo: U.S. Navy, courtesy T. L. Péwé.)

block fields, rock glaciers, massive taluses, and a variety of terraced or benched slopes shaped by nivation and gelifluction. Many well-illustrated descriptions of periglacial scenery have been published (Muller, 1947; Hopkins et al., 1955; Wahrhaftig, 1965; Ferrians et al., 1969; Péwé, 1975). Only the major landforms and some of the more spectacular minor forms are noted here. For more detailed descriptions of periglacial phenomena, books by Péwé (1969), Brown (1970), Tricart, (1970), French (1976), Washburn (1980), and Williams and Smith (1989) are recommended, especially to American readers. An enormous literature in Russian, Polish, French, and the Scandinavian languages is readily available, as well.

Asymmetric Valleys. Some periglacial valleys show notable asymmetry in cross profile, even though structural control can be eliminated. The most typical condition is for north-facing slopes (in the northern hemisphere) to be steeper than the south-facing slopes (French, 1976, p. 178). Where valleys trend north or south, the asymmetry disappears. The theory is that gelifluction is more intense on the valley side facing the sun for the longest time each day, and the shaded side stays colder, less prone to gelifluction, and steeper (Currey, 1964). Also, the copious colluvium from the south-facing slopes forces rivers to erode and steepen their north-facing valley walls preferentially. Relict valleys with such asymmetry are common in southern Indiana (latitude 38°N) adjacent to the Wisconsin glacial margin (Wayne, 1967). By contrast, Tricart (1970, p. 137) suggested that near the southern limit of the arctic periglacial regions, permafrost, gelifluction, and gentle slopes might persist on cold, north-facing valley walls even as drier, warmer, nonpermafrost slopes maintain steeper gradients on the south-facing valley sides.

Cryoplanation Terraces and Cryopediments. Flattened summit areas and benched hillslopes in Alaska were attributed to an assemblage of periglacial weathering, erosional, and depositional processes collectively termed **altiplanation** (Eakin, 1916, pp. 78–82). Current authors prefer the equivalent term, **cryoplanation**

(Priesnitz, 1988). By implication, the composite of cryoplanation processes could result in a regional periglacial surface of low local relief, not controlled by any base level. The deduced or implied penultimate landform of periglacial regions, the "cryoplain" or "altiplain," has never been seriously proposed as a climatic analog of a peneplain or a "pediplain" (pp. 321 and 323). However, **cryoplanation terrace** (or the Russian equivalent, **goletz terrace**) is a widely accepted term for the typical benchlike landforms (Fig. 15–9).

Cryopediments were first recognized as relict periglacial landforms in eastern Europe in the 1950s (for English-language reviews, see French, 1976, p. 155, and Washburn, 1980, p. 237). The term has been extended to present-day periglacial regions. In shape and size they are similar to cryoplanation terraces, but are on lower piedmont slopes backed by steep mountain slopes, analogous to pediments in arid and semiarid regions (p. 337). It has been proposed that both cryoplanation terraces and cryopediments begin as aligned nivation hollows or a nivation bench on a periglacial hillside, after which frost weathering maintains a steep slope or cliff at the back of the bench while gelifluction maintains a smooth, gentle "slope of transportation" that, like a semiarid pediment, is primarily an erosional surface even when covered by gelifluction debris in transit across it (Te Punga, 1956; Reger and Péwé, 1976).

Block Fields, Rock Streams, and Rock Glaciers.   Masses of angular rock rubble on summits (**block fields** or **felsenmeer**) or streaming down mountainsides (**rock streams**) are striking features of modern periglacial landscapes, and their relict equivalents are also well known. Relict rock streams in the Appalachian Mountains consist of large, subangular boulders and are now devoid of vegetation. They extend as much as 1000 m downhill from a cliff or exposed ledge (Fig. 15–11). Although not now in motion, the Appalachian rock streams were almost certainly formed by periglacial mass wasting (Potter and Moss, 1968; Clark and Ciolkosz, 1988). Gelifluction and frost creep are the likely processes, because although the rock streams now lack the water-retaining fine sediments in the upper few meters that would be required to maintain motion, the large blocks have interstitial finer

Figure 15-11.   Hickory Run boulder field, Carbon County, Pennsylvania. Boulders reach 2 m in length. (Photo: W. Bolles, Pennsylvania Dept. of Education.)

sediments between them at depth. Hickory Run boulder field, one of the larger of the relict Appalachian rock streams, has a surface gradient of only 1° (Smith, 1953).

Rock glaciers were once thought to be moribund valley glaciers but are now recognized as distinct landforms, transitional between periglacial and glacial in origin (Giardino et al., 1987). White (1971a, p. 44) defined rock glaciers as "accumulations of unsorted, till-like, coarse to fine rock debris, with an ice core and/or interstitial ice, having a glacier shape and spreading downvalley." Active rock glaciers are either tongue-shaped along valley axes or lobes on valley walls. They have steep frontal slopes, often with active taluses, and arcuate pressure ridges on their surfaces (Fig. 15–12). All known rock glaciers are derived either from glacial debris downvalley from an active or inactive glacier, or from avalanche debris, sliderock, or colluvium on a valley wall (Johnson, 1987, p. 191; Barsch, 1988, p. 97; Humlum, 1988).

If a mass of poorly sorted colluvium has sufficient ice packed within its pore spaces, the plastic deformation of the ice can produce downhill motion similar to but slower than glacier flow. The arbitrary difference between a rock glacier and a glacier is that the former is mostly rock debris with interstitial ice, whereas the latter is mostly ice with included rock fragments. Several rock glaciers in the central Rocky Mountains are ice-cored and are either transitional upslope to true valley glaciers or are degenerating former glaciers (White, 1971a, b, 1976; Potter, 1972; Benedict et al., 1986). Larger rock glaciers in Alaska are ice-cemented rock fragments, similar to glaciers in all but the relative proportion of ice and rock (Wahrhaftig and Cox, 1959). Rock glaciers move forward by internal deformation and debris slides at their termini, resulting in a rolling or continuous-tread form of transport. Although they have not been proved to erode the rock surfaces over which they move, their movement is clearly transitional to glacial processes.

Figure 15-12.   Vertical aerial view of the Arapaho (Colorado) rock glacier in late summer. Direction of motion (northeasterly) is toward the lower left of the picture. Principal lobe is about 200 m wide (White, 1971b). (Photo: J. B. Benedict.)

Minor Landforms

The landscape of periglacial regions is marked by a variety of patterns, many of which barely deserve the name "landform." Much of the terrain is depositional rather than erosional because of the contrast between intense mass wasting and inefficient stream erosion. On the alluvial plains and other surfaces of low relief, permafrost overlain by the active layer contorts, erupts, or collapses the land surface into a maze of polygonal patterns, bogs or shallow lakes, ice-cored domes, and pits. Only the more striking forms are illustrated here.

Patterned Ground. The relationship of ice-wedge polygons to permafrost has been described previously (p. 378). Less conspicuous polygons, outlined by open cracks or shallow depression but not underlain by ice wedges, are called frost-crack polygons (Washburn et al., 1963). They are not diagnostic of permafrost and can form in any seasonally cold region if snow cover is thin during a cold interval. Some frost cracks may be preserved by windblown silt packing into them, but for the most part, the patterns are visible only during cold weather and are not preserved.

Polygons may be made of sorted or unsorted sediment and may have raised centers or raised rims (Fig. 15–6) (Washburn, 1980; Krantz et al., 1988, and references therein). Generally, ice-wedge polygons are made of unsorted sediments. Sorted polygons involve more severe rearrangement of rock fragments, with coarser particles lifted to the surface or thrust outward to the edges of the polygons. Convection within the sorted debris, or in water in the active layer, is thought to be involved in the sorting process (p. 377). A higher proportion of fine-grained sediment probably promotes sorting, because water is held longer without refreezing in fine sediments, and ice masses may segregate from the freezing soil to disrupt and heave the original stratification, if any. Sorted polygons may be preserved because seasonal-flowing water permeates the coarse network of polygonal boundaries without eroding the fine sediments from the polygon cores. Sorted polygons are not diagnostic of the permafrost zone, although they are found there. They seem to be more abundant on wet, cold mountain plateaus and other places with intense freeze–thaw activity.

Sorted polygons, when occurring on slopes, are distorted laterally by gravity to **stone stripes** (Fig. 15–7), parallel segregations of coarse and fine sediment trending downhill. Like sorted polygons, stone stripes require active freeze–thaw processes but are not restricted to regions of permafrost. Once formed, the coarse stripes will channel runoff from the hillside. The percolating water probably erodes fine sediment from beneath the coarse debris and aids in lowering the stripes of coarse debris into trenches down hillsides. The excellent drainage provided by stone stripes enables hillslopes to remain stable at unusually steep angles, perhaps to 30°, without slumping or sliding.

A variety of *frost mounds* and *palsas,* less than 10 m in height, form from periglacial groundwater processes. Some form within a single winter season by refreezing of the active layer (Pollard, 1988; Seppälä, 1988). A palsa in the traditional Lappland terminology was a peat hummock with a frozen core of ice and peat. In wide usage, it now includes mounds with cores of mineral sediment as well. Many of these minor landforms are associated with the discontinuous permafrost region (Fig. 15–1) where winter freezing is severe but groundwater is mobile for an extended summer season and peat bogs are abundant.

The many bogs of subarctic periglacial regions have a striking ribbed pattern of turf stripes between elongate pools of water. These are the **string bogs** that cover a large proportion of the subarctic taiga or forested region. Permafrost is probably not involved in their formation, so they might be excluded from the periglacial morphogenetic region except for their widespread occurrence at the margins of the periglacial zone. Their present extent can be related to the extensive glacial derangement of rivers in formerly glaciated regions. Their origin is hypothesized to be due to lines of organic foam and debris that wash from the spruce forests and lodge as low dams across cold, slush-filled creeks or flooded areas (Thom, 1972). According to that hypothesis, string bogs form progressively from the downstream edge, each line of turf and grass forming at right angles to the slow flow of water, where it traps more organic debris until the next ridge forms above it.

Pingos.   Under certain hydrologic conditions, laccolithic intrusions (Chap. 3) of ice arch the overlying ground into a fractured dome or blister called a **pingo** (Fig. 15–13). Pingos occur at the edges of alluvial plains or old lake beds and at the bases of long slopes. Probably the hydrostatic head in groundwater is intensified by pressure generated at the base of the permafrost or in unfrozen layers within permafrost. At some low point on an aquifer, pressure builds up sufficiently to arch the overlying permafrost and build up a lenticular mass of ice in the core of the mound. These are called "open system" pingos because groundwater continues to flow toward them and they can grow. Other pingos are thought to be of a "closed-system" type, wherein the progressive inward and downward freezing of an old lake or bog generates pressure in the enclosed saturated area sufficient to drive pore water upward, where it freezes as a segregated mass in the core of a mound (Mackay, 1962; Holmes et al., 1963; Pissart, 1988). Fossil pingos, low circular mounds enclosing poorly drained basins, are reported from many areas of western Europe. A few similar forms in Illinois (Flemal et al., 1973) and Pennsylvania (Marsh, 1987) have been described, but they seem less common in North America than in Europe (Flemal, 1976). Like fossil ice wedges, fossil pingos are excellent evidence of former permafrost, and their presence near the southern limits of former glaciers gives important information about the former extent of the periglacial zone.

Thermokarst.   The global warming trend of the last two centuries has caused a reduction in area of permafrost regions. Some areas that were formerly underlain by continuous permafrost now have discontinuous or only sporadic occurrences. As permafrost melts, the reduction of total volume and the collapse of ground over former segregated ice masses produces a pitted karstlike terrain, to which the term

Figure 15-13.   Pingo, Wollaston Peninsula, Victoria Island, Northwest Territories, Canada. (Photo: A. L. Washburn.)

thermokarst has been applied (Czudek and Demek, 1970). Many areas of permafrost are preserved only because of the insulating effects of forests or tundra vegetation. If land is cleared for roads, railways, airports, or pipelines, the collapse of permafrost can lead to destruction of the installations. An enormous literature of permafrost engineering has grown in recent years as arctic oil fields and pipelines have been developed and built.

The tundra is a delicate ecosystem that exhibits a distinct periglacial landform assemblage. The trans-Alaskan oil pipeline provoked violent controversy about its effects on tundra life, scenery, and the frozen substrate. It has been built, but with greater awareness of the regional environment than if it had been initially authorized without full examination of its impact. Lessons learned there can lead to better preservation and management of other tundra regions (Walker et al., 1987).

REFERENCES

ANDERSON, D. M., REYNOLDS, R. C., and BROWN, J., 1969, Bentonite debris flows in northern Alaska: Science, v. 164, pp. 173–74.

BARSCH, D., 1988, Rocky glaciers, *in* Clark, M. J., ed., Advances in periglacial weathering: John Wiley & Sons Ltd., Chichester, England, pp. 69–90.

BENEDICT, J. B., BENEDICT, R. J., and SANVILLE, D., 1986, Arapaho rock glacier, Front Range, Colorado, U.S.A.: A 25-year resurvey: Arctic and Alpine Res., v. 18, pp. 349–52.

BLACK, R. F., 1976, Periglacial features indicative of permafrost: Ice and soil wedges: Quaternary Res., v. 6, pp. 3–26.

BROWN, R. J. E., 1970, Permafrost in Canada: Univ. of Toronto Press, Toronto, 234 pp.

BÜDEL, J., 1982, Climatic geomorphology (transl. by L. Fischer and D. Busche): Princeton Univ. Press, Princeton, N.J., 443 pp.

CARTER, L. D., 1981, A Pleistocene sand sea on the Alaskan Arctic coastal plain: Science, v. 211, pp. 381–83.

CLARK, G. M., and CIOLKOSZ, E. J., 1988, Periglacial geomorphology of the Appalachian highlands and interior highlands south of the glacial border—a review: Geomorphology, v. 1, pp. 191–220.

COOK, F. A., and RAICHE, V. G., 1962, Freeze–thaw cycles at Resolute, N.W.T.: Geog. Bull., no. 18, pp. 64–78.

COSTIN, A. B., JENNINGS, J. N., BAUTOVICH, B. C., and WIMBUSH, D. J., 1973, Forces developed by snowpatch action, Mt. Twynam, Snowy Mountains, Australia: Arctic and Alpine Res., v. 5, pp. 121–26.

COSTIN, A. B., JENNINGS, J. N., BLACK, H. P., and THOM, B.G., 1964, Snow action on Mt. Twynam, Snowy Mountains, Australia: Jour. Glaciology, v. 5, pp. 219–28.

COTTON, C. A., 1963, A new theory of the sculpture of middle-latitude landscapes: New Zealand Jour. Geol. Geophys., v. 6, pp. 769–74.

——— and TE PUNGA, M. T., 1955, Fossil gullies in the Wellington landscape: New Zealand Geographer, v. 2, pp. 72–75.

CURREY, D. R., 1964, A preliminary study of valley asymmetry in the Ogotoruk Creek area, northwestern Alaska: Arctic, v. 17, pp. 84–98.

CZUDEK, T., and DEMEK, J., 1970, Thermokarst in Siberia and its influence on the development of lowland relief: Quaternary Res., v. 1, pp. 103–20.

DOHRENWEND, J. C., 1984, Nivation landforms in the western Great Basin and their paleoclimatic significance: Quaternary Res., v. 22, pp. 275–88.

EAKIN, H. M., 1916, The Yukon–Koyukuk region, Alaska: U.S. Geol. Survey Bull. 631, 88 pp.

EVERETT, D. H., 1961, Thermodynamics of frost damage to porous solids: Faraday Soc. Trans., v. 57, pp. 1541–51.

FERRIANS, O. J., KACHADOORIAN, R., and GREENE, G. W., 1969, Permafrost and related engineering problems in Alaska: U.S. Geol. Survey Prof. Paper 678, 37 pp.

FLEMAL, R. C., 1976, Pingos and pingo scars: Their characteristics, distribution, and utility in reconstructing former permafrost environments: Quaternary Res., v. 6, pp. 37–53.

————, HINKLEY, K. C., and HESLER, J. L., 1973, DeKalb mounds: A possible Pleistocene (Woodfordian) pingo field in north-central Illinois, in Black, R. F., Goldthwait, R. P., and Willman, H. B., eds., The Wisconsinan Stage: Geol. Soc. America Mem. 136, pp. 229–50.

FRENCH, H. M., 1976, The periglacial environment: Longman Group Ltd., London, 309 pp.

————, 1988, Active layer process, in Clark, M. J., ed., Advances in periglacial weathering: John Wiley & Sons Ltd., Chichester, England, pp. 151–98.

GIARDINO, J. R., SHRODER, J. F., JR, and VITEK, J. D., eds., 1987, Rock glaciers: Allen & Unwin, Inc., Winchester, Mass., 355 pp.

GLEASON, K. I., KRANTZ, W. B., CAINE, N., GEORGE, J. H., and GUNN, R. D., 1986, Geometrical aspects of sorted patterned ground in recurrently frozen soil: Science, v. 232, pp. 216–20.

HALLET, B., and PRESTRUD, S., 1986, Dynamics of periglacial sorted circles in western Spitsbergen: Quaternary Res., v. 26, pp. 81–99.

HARRIS, C., 1987, Mechanisms of mass movement in periglacial environments, in Anderson, M. G., and Richards, K.S., eds., Slope stability: John Wiley & Sons Ltd., Chichester, England, pp. 531–59.

HIGASHI, A., and CORTE, A. E., 1971, Solifluction: A model experiment: Science, v. 171, pp. 480–82.

HOLMES, G. W., FOSTER, H. L., and HOPKINS, D. M., 1963, Distribution and age of pingos in interior Alaska: Permafrost Internat. Conf. Proc., NAS-NRC Pub. 1287, pp. 88–93.

HOPKINS, D. M., KARLSTROM, T. N. V., BLACK, R. F., WILLIAMS, J. R., PÉWÉ, T. L., FERNOLD, A. T., and MULLER, E. H., 1955, Permafrost and ground water in Alaska: U.S. Geol. Survey Prof. Paper 264-F, pp. 113–46.

HUMLUM, O., 1988, Rock glaciers appearance level and rock glacier initiation line altitude: A methodological approach to the study of rock glaciers: Arctic and Alpine Res., v. 20, pp. 160–78.

JOHNSON, P. G., 1987, Rock glacier: glacier debris systems or high-magnitude low-frequency flows, in Giardino, J. R., Shroder, J. F., Jr., and Vitek, J. D., eds., Rock glaciers: Allen & Unwin, Inc. Winchester, Mass., pp. 175–92.

JOHNSON, W. H., 1990, Ice-wedge casts and relict patterned ground in central Illinois and their environmental significance: Quaternary Res., v. 33, pp. 51–72.

KELLY, W. C., and ZUMBERGE, J. H., 1961, Weathering of a quartz diorite at Marble Point, McMurdo Sound, Antarctica: Jour. Geology, v. 69, pp. 443–46.

KRANTZ, W. B., GLEASON, K. J., and CAINE, N., 1988, Patterned ground: Sci. American, v. 259, no. 6, p. 68–76.

LACHENBRUCH, A. H., 1962, Mechanics of thermal contraction cracks and ice-wedge polygons in permafrost: Geol. Soc. America Spec. Paper 70, 69 pp.

LAUTRIDOU, J.-P., 1988, Recent advances in cryogenic weathering, in Clark, M. J., ed., Advances in periglacial weathering: John Wiley & Sons Ltd., Chichester, England, pp. 33–47.

LEFFINGWELL, E. DEK., 1915, Ground-ice wedges: The dominant form of ground-ice on the north coast of Alaska: Jour. Geology, v. 23, pp. 635–54.

MACKAY, J. R., 1962, Pingos of the Pleistocene Mackenzie delta area: Geog. Bull., v. 18, pp. 21–63.

MALDE, H. E., 1964, Patterned ground in the western Snake River Plain, Idaho, and its possible cold-climate origin: Geol. Soc. America Bull., v. 75, pp. 191–207.

MARSH, B., 1987, Pleistocene pingo scars in Pennsylvania: Geology, v. 15, pp. 945–47.

MATTHES, F. E., 1900, Glacial sculpture of the Bighorn Mountains, Wyoming: U.S. Geol. Survey, 21st Ann. Rept., pt. 2, pp. 167–90.

McDOWALL, I. C., 1960, Particle size reduction of clay minerals by freezing and thawing: New Zealand Jour. Geol. Geophys., v. 3, pp. 337–43.

MULLER, S. W., 1947, Permafrost or permanently frozen ground and related engineering problems: Edwards Bros., Ann Arbor, Mich., 231 pp.

PÉWÉ, T. L., ed., 1969, The periglacial environment, past and present: McGill–Queen's Univ. Press, Montreal, 487 pp.

——, 1975, Quaternary geology of Alaska: U.S. Geol. Survey Prof. Paper 835, 145 pp.

——, 1983, The periglacial environment in North America during Wisconsin time, *in* Porter, S. C., ed., Late-Quaternary environments of the United States, v. 1, The Late Pleistocene: Univ. of Minnesota Press, Minneapolis, pp. 157–89.

PISSART, A., 1988, Pingos: An overview of the present state of knowledge, *in* Clark, M. J., ed., Advances in periglacial weathering: John Wiley & Sons Ltd., Chichester, England, pp. 279–97.

POLLARD, W. H., 1988, Seasonal frost mounds, *in* Clark, M. J., ed., Advances in periglacial weathering: John Wiley & Sons Ltd., Chichester, England, pp. 201–29.

POTTER, N., JR., 1972, Ice-cored rock glacier, Galena Creek, northern Absaroka Mountains, Wyoming: Geol. Soc. America Bull., v. 83, pp. 3025–57.

——, and Moss, J. H., 1968, Origin of the Blue Rocks block field and adjacent deposits, Berks County, Pennsylvania: Geol. Soc. America Bull., v. 79, pp. 255–62.

PRIESNITZ, K., 1988, Cryoplanation, *in* Clark, M. J., ed., Advances in periglacial weathering: John Wiley & Sons Ltd., Chichester, England, pp. 49–67.

REGER, R. D., and PÉWÉ, T. L., 1976, Cryoplanation terraces: Indicators of a permafrost environment: Quaternary Res., v. 6, pp. 99–109.

SEPPÄLÄ, M., 1988, Palsas and related forms, *in* Clark, M. J., ed., Advances in periglacial weathering: John Wiley & Sons Ltd., Chichester, England, pp. 247–78.

SMITH, H. T. U., 1953, Hickory Run boulder field, Carbon County, Pennsylvania: Am. Jour. Sci., v. 251, pp. 625–42.

SPARKS, B. W., and WEST, R. G., 1972, The ice age in Britain: Methuen Co. Ltd., London, 302 pp.

ST. ONGE, D. A., 1969, Nivation landforms: Canada Geol. Survey Paper 69–30, 12 pp.

TE PUNGA, M. T., 1956, Altiplanation terraces in southern England: Biul. Peryglac., v. 4, pp. 331–38.

THOM, B. G., 1972, Role of spring thaw in string-bog genesis: Arctic, v. 25, pp. 237–39.

THORN, C., 1988, Nivation: A geomorphic chimera, *in* Clark, M. J., ed., Advances in periglacial weathering: John Wiley & Sons Ltd., Chichester, England, pp. 3–31.

TRICART, J., 1970, Geomorphology of cold environments (transl. by E. Watson): Macmillan and Co. Ltd., London, 320 pp.

WAHRHAFTIG, C., 1965, Physiographic divisions of Alaska: U.S. Geol. Survey Prof. Paper 482, 52 pp.

—— and Cox, A., 1959, Rock glaciers in the Alaska Range: Geol. Soc. America Bull., v. 70, pp. 383–436.

WALDER, J., and HALLET, B., 1985, A theoretical model of the fracture of rock during freezing: Geol. Soc. America Bull., v. 96, pp. 336–46.

WALKER, D. A., WEBBER, P. J., BINNIAN, E. F., EVERETT, K. R., LEDERER, N. D., NORDSTRAND, E. A., and WALKER, M. D., 1987, Cumulative impacts of oil fields on northern Alaskan landscapes: Science, v. 238, pp. 757–61.

WASHBURN, A. L., 1967, Instrumental observation of mass-wasting in the Mesters Vig District, northeast Greenland: Meddelelser om Grønland, v. 166, no. 4, 296 pp.

——, 1980, Geocryology: A survey of periglacial processes and environments: John Wiley & Sons, Inc., New York, 406 pp.

——, SMITH, D. D., and GODDARD, R. H., 1963, Frost cracking in a middle-latitude climate: Biul. Peryglac., v. 12, pp. 175–89.

WAYNE, W. J., 1967, Periglacial features and climatic gradient in Illinois, Indiana, and western Ohio, east-central United States, *in* Cushing, E. J., and Wright, H. E., Jr., eds., Quaternary paleoecology: Yale Univ. Press, New Haven, Conn., pp. 393–414.

WHITE, S. E., 1971a, Rock glacier studies in the Colorado Front Range, 1961 to 1968: Arctic and Alpine Res., v. 3, pp. 43–64.

———, 1971b, Debris falls at the front of Arapaho rock glacier, Colorado Front Range, U.S.A: Geograf. Annaler, v. 53, ser. A, pp. 86–91.

———, 1976, Rock glaciers and block fields, review and new data: Quaternary Res., v. 6, pp. 77–97.

WILLIAMS, P. J., and SMITH, M. W., 1989, The frozen earth: Fundamentals of geocryology: Cambridge University Press, Cambridge, England, 306 pp.

WORSLEY, P., 1985, Periglacial environment: Progress in Phys. Geography, v. 9, pp. 391–401.

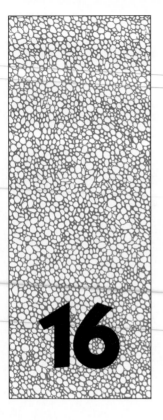

Glaciers as Morphogenetic Agents and Landforms

16

Glaciers are masses of ice (with included rock debris and air) on land, which show evidence of present or past internal deformation and motion. Water in the solid state has thermal and physical properties very different from those in the liquid state. The transition from one state to the other is abrupt. Therefore, landscape development under the glacial morphogenetic system is very different from the processes and forms of the fluvial system that are dominated by liquid water, and the boundaries between the glacial morphogenetic region and other regions are unusually sharp.

Of all the water near the surface of the earth, only about 2 percent is on the land, in the solid state as glacier ice (Table 5–1; Fig. 5–5). Even this small fraction is sufficient to cover entirely one continent (Antarctica) and most of the largest island (Greenland) with ice to an average thickness of 2.2 km over Antarctica and 1.5 km over Greenland. At the present time, about 10 percent of the earth's land area is ice covered. An additional 20 percent has been ice covered repeatedly during the glaciations of the Pleistocene Epoch, much of it as recently as 15,000 years ago. Glacial morphogenesis has been the dominant factor in shaping the present landscape of North America northward of the Ohio and Missouri Rivers and of Eurasia northward of a line from Dublin eastward through Berlin to Moscow and beyond the Urals (Fig. 16–1). In addition, mountains and plateaus have been glaciated to an altitude 1000 to 1500 m lower than their present snowline.

Glaciers, especially continent-sized ice sheets, pose unique problems for geomorphic description. In one sense, the surface of an ice sheet is a constructional landform, built of layers of snow compressed to ice. However, ice is an easily deformed plastic solid and assumes a dome-shaped or elliptical cross section (Figs. 16–9, 16–10, and 16–12) that must be defined as being structurally controlled. Erosion plays only a minor role in shaping the surface of a glacier, yet glaciers are powerful eroding agents of other rocks. An estimated 7 percent of all modern erosion is being done by glaciers (p. 209), which makes them a poor second to rivers in total erosional effectiveness. Nevertheless, considering their limited area today, they

Principal areas presently
covered by glacier ice

Principal areas formerly
covered by glacier ice

Figure 16-1.   Existing large glaciers and principal areas formerly covered by glacier ice,
northern hemisphere. An additional area on the floor of the Barents Sea between northern
Norway and Spitzbergen may also have been glacier covered. (Redrawn from Flint, 1971,
Figs. 4-8 and 4-9.)

make a respectable contribution to continental denudation. Glaciation increases
the erosion rate on a mountain by at least an order of magnitude over the rate on a
comparable, but unglaciated, mountain (Chap. 12).

We must study glaciers in three ways: first, as a peculiar but abundant monomineralic rock with unusual thermal and rheologic properties; second, as a mobile,
diverse, dynamic landscape shaped by unusual geomorphic processes such as rheid
flow, brittle fracture, and melting; and third, as a powerful geomorphic agent that is
shaping and has shaped many terrains. Three distinct scientific disciplines are involved in glacial morphogenesis. The first is **glaciology,** the science of ice. Next is
glacier morphology, the description and explanation of ice surfaces or "ice-scapes."
These two subjects are covered in this chapter. The third scientific discipline is **glacial geomorphology,** which describes and explains the geomorphic work of glaciers,
past and present. It is a large enough subject to be considered separately in Chapter
17. British authors use "glacierized" to describe a landscape covered by glaciers,
and distinguish that term from "glaciated," meaning a landscape now ice-free, but
with the diagnostic landforms and sediments of former glaciers. No geomorphologist can appreciate the work of ancient glaciers in shaping some of the world's most

dramatic landscapes unless he or she understands the physical and thermal properties of ice. But all the theoretical studies in the world cannot replace a walk onto the surface of a modern glacier for the comprehension of glacial morphogenesis.

GLACIOLOGY

Glaciology is the scientific study of ice. It includes the study of ice crystals in high clouds, hail, and snow; frozen lake, river, and ocean water; and glacier ice. The Voyager spacecraft exploration of Jupiter, Saturn, and Uranus and their moons has broadened the interest of glaciology to include the ices of ammonia (NH_3) and methane (CH_4), for these, along with H_2O ice, are probably the materials of which many of the moons of the outer planets are composed (Krass, 1984; Jankowski and Squyres, 1988). Glaciologists can be metallurgists, meteorologists, physicists, or geologists. Ice can be studied either as an easily deformed crystalline solid with a structure to be analyzed by metallurgical techniques or as a geomorphic agent that erodes valleys and transports massive loads of rocky debris. What we know about glacier ice has been learned from a wide field of experiment and observation (Lliboutry, 1965; Glen, 1974, 1975; Hutter, 1983; Weertman, 1983).

Flow in Glaciers

Plastic Deformation (Creep).   When an ice specimen is subjected to a low directional or *shear stress* (pressure or tension) and then released, it elastically recovers its original shape. When the shear stress is increased to above a certain threshold called the *elastic limit* or *yield strength,* irreversible deformation or *strain* results (Fig. 16–2). If steady-state shear stress is maintained, the sample deforms by continuous plastic strain or *creep.* Although there are several high-pressure polymorphs of H_2O ice, only one form is known to occur naturally on planet Earth. It is known as ice I, crystallizes in the hexagonal crystal system, and is characterized by three intermolecular glide axes 120° apart on a plane perpendicular to the *c* optical axis (Fig. 16–2). Deformation of an ice crystal is easy along any of the glide axes,

Figure 16-2.   Single crystal of ice, 13 mm² cross section, *c* optical axis 45° to specimen axis deformed by sustained tension of 0.4 kg (= 0.3 MPa). (Photo: J. W. Glen, reproduced from the Journal of Glaciology by permission of the International Glaciological Society.)

and as vectors they define a *basal glide plane* on which single ice crystals easily creep under very low shear stress. The basal plane is the orientation of the incomplete hexagonal skeletal ice crystals we call snowflakes. Under a low sustained pressure of 0.3 MPa (3 bar), an ice crystal can be drawn out into a thin ribbon in only a few days (Fig. 16–2) (Glen, 1952). If an ice grain is constrained from the easy deformation on the basal glide plane, as for example if it is completely surrounded by other grains of ice and held in an unfavorable orientation relative to the applied shear stress, it will still deform, but much more slowly. The creep rate is 500 times more rapid for a given shear stress in the "easy glide" basal plane of a single crystal than in various "hard glide" crystal orientations or in polycrystalline aggregates (Fig. 16–3) (Weertman, 1983, p. 224). The cluster of experimental and field-determined strain-stress relationships in the lower central part of Figure 16–3 are regarded as the best ones to use in calculations or modeling of glacier flow. They demonstrate that, in general, a mass of polycrystalline ice subjected to a shear stress of 0.3 MPa (3 bar) deforms by an amount equal to its original length in a few years.

Polycrystalline ice deforming under shear stress behaves very much as any ceramic a few tens of degrees Celsius below its melting point, or in somewhat the same way that iron deforms when it is heated to a bright red color. Either in single crystals or in polycrystalline masses, at only slightly increased stresses the strain rate of ice becomes very rapid. By the simplifying assumption that ice is a perfect plastic, which does not deform whatsoever at stresses less than 0.1 MPa (1 bar) but deforms infinitely rapidly at a slightly larger stress, theoretical models of ice flow have been developed that closely approximate the flow of valley glaciers and ice caps. The approximation is sometimes referred to as the "one-bar flow law." It would plot on Figure 16–3 as a vertical line at a shear stress of 0.1 MPa (1 bar). Several lines of evidence suggest that shear stresses at the base of glaciers are never far from the range 0.05 to 0.15 MPa (0.5 to 1.5 bar), so the simplifying assumption of the one-bar flow law has proved useful.

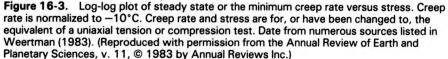

Figure 16-3. Log-log plot of steady state or the minimum creep rate versus stress. Creep rate is normalized to −10°C. Creep rate and stress are for, or have been changed to, the equivalent of a uniaxial tension or compression test. Date from numerous sources listed in Weertman (1983). (Reproduced with permission from the Annual Review of Earth and Planetary Sciences, v. 11, © 1983 by Annual Reviews Inc.)

More precisely, the plastic deformation of ice compiled from laboratory experiments and measurements in ice tunnels and boreholes (Fig. 16–3) fits a power-function equation of the form

$$\dot{\epsilon} = A\tau^n$$

where $\dot{\epsilon}$ is the creep rate, or rate of permanent deformation measured in percent change of length per time, τ is the shear stress, and A and n are constants. This equation is called the "Glen flow law" in recognition of the original experimenter (Glen, 1952; Paterson, 1981, pp. 26, 85). Experimental values of n for both single crystals and polycrystalline ice generally have a mean of about 3. When plotted on a log-log graph, the value of n of about 3 is illustrated by a thousand-fold increase in creep rate for each 10-fold increase in shear stress (Fig. 16–3). Although the absolute creep rates vary greatly between single crystals in the easy glide orientation and polycrystalline glacier ice with varying amounts of impurities, the power function increase in creep rate with increased shear stress is similar in most experiments. The constant A, but not n, is strongly temperature dependent, so that for a given shear stress, the strain rate is only one-fifth as great in ice at $-25°C$ as in ice at $-10°C$ (Paterson, 1981, p. 28).

It is important to note that confining, or hydrostatic, pressure does not significantly affect the strain rate of ice. Only shear, or directional, stress causes plastic deformation. As with rock and other solids, confining pressure only prevents brittle fracture and thereby permits creep to be measured. Confining pressure does not make the ice at the bottom of a glacier any more "plastic" than the ice at the surface. Many serious errors in interpreting glacier flow have been made because of this misconception.

On the basis of the equation of plastic flow, it can be shown that a slab of ice resting on an inclined rock surface will deform downslope by laminar flow as a solid, without any melting and refreezing. The rate of movement at the ice surface is proportional to the thickness and surface slope of the ice (Paterson, 1981, p. 86). Theory and observation of actual glaciers agree that the motion is most rapid at the surface and progressively less rapid with depth (Fig. 16–4). Theory and observation also agree that flow is controlled by the surface slope of the glacier, not by the slope of the rock floor beneath it, as long as the ice thickness is significantly greater than the bedrock relief. It is quite possible for basal ice to move into and out of an ice-filled basin or over an obstruction, controlled only by the slope on the surface of the ice above the basin or obstruction (p. 418).

Theoretical models that assume ice to be a plastic solid with rheidity (p. 13) defined by the Glen flow law give excellent predictions of the shape, thickness, and rate of flow of glaciers and of floating ice shelves, but the results are not perfect. The obvious reason is that real glaciers are complex masses of ice, with internal temperature gradients and impurities, that move over or between rock surfaces along discontinuities that cannot be modeled theoretically. Also, during movement, deforming ice crystals in a polycrystalline mass grow and reorient themselves in the stress field by a process of solid-state recrystallization. Thus the grain size and orientation, which greatly affect the rate of deformation, change during deformation. Ice under high shear stress may fracture or form small, interlocking grains, but if the stress is low and uniform, these small grains "anneal" to form oriented grains as wide as 10 cm. Ice recovered by deep boreholes in glaciers usually show grains several cm across, oriented with their c optical axes parallel and vertical; in this configuration, horizontal "easy glide" flow is facilitated, and the deformation rate increases by a factor of 4 over the rate for randomly oriented grains (Fig. 16–3) (Weertman, 1983, p. 229).

Figure 16-4. Distribution of velocity within the Saskatchewan Glacier, Alberta, Canada at a measured cross section, as viewed from below, with the ice moving toward the right front of the block diagram. About 30 percent of the maximum surface motion is due to basal sliding. (From Meier, 1960, Fig. 42.)

Basal Sliding and Regelation Slip. The Glen flow law shows that the strain rate of glacier ice is typically proportional to the third power of the shear stress. Near the base of a glacier, where bedrock irregularities protrude into the ice, local shear stresses increase, and local deformation is correspondingly very greatly increased. This *enhanced plastic flow* is proportional to the size of the obstruction, so the larger the obstruction, the more important is the enhanced plastic flow.

If ice is at or very close to its melting temperature, an additional form of basal flow can occur, called **regelation slip.** This is not a solid-state phenomenon but involves pressure-dependent melting and freezing (Fig. 16–5). If the ice is at the melting temperature for the appropriate confining pressure, on the upglacier side of an obstruction the ice presses against rock, locally increasing pressure, depressing the melting temperature, and producing water *at the same temperature as the surrounding ice.* The water migrates along grain boundaries within the ice or over the rock surface to regions of less pressure, especially on the downglacier side of the obstruction where ice is being drawn away from the rock. Here the water refreezes as bubble-free clear blue *regelation ice* (Fig. 16–6). Regelation is essentially an isothermal process. The only thermal gradients are those produced by the pressure gradients around the obstruction. The heat released by refreezing on the downstream side of the obstruction migrates upstream, primarily through the obstruction, to be absorbed by the melting at the upstream side. The heat transfer is faster through small obstacles, so the smaller the obstruction the more important is regelation slip.

Mathematical models of basal sliding are imperfect, because it is essentially the result of boundary conditions between unlike materials and phases. The controlling factor, assuming that the temperature is at or very near the melting point, is the bed roughness, or size of the limiting obstruction (Paterson, 1981, p. 112; Drewry, 1986, p. 10). Large obstructions are more efficiently passed by enhanced plastic flow, but smaller obstructions are more efficiently passed by regelation slip. Theory suggests that the controlling wavelength of bedrock obstruction, which would permit the optimal contributions by both plastic deformation and regelation

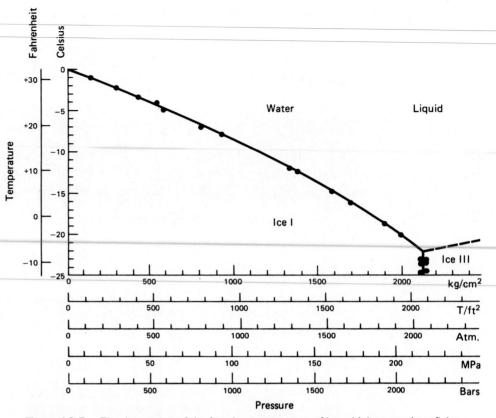

Figure 16-5. The depression of the freezing temperature of ice with increased confining pressure. At confining pressures greater than 207 MPa, more dense phases of ice crystallize (p. 124).

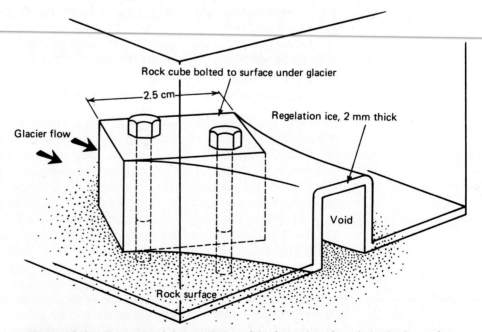

Figure 16-6. Experimental demonstration of the formation of regelation ice beneath a temperate glacier. (From Peterson, 1970.)

slip, is about 0.5 m (Kamb and LaChapelle, 1964; Paterson, 1981, p. 119). However, obstructions of this size are not common on glaciated rock surfaces. Streamlined glaciated landforms about 10 times as large are common and are thought to be stable forms under moving ice, somewhat like ripple marks and dunes under moving water and air. The lack of agreement between theory and observation illustrates the imperfections of present understanding of basal sliding.

Enhanced plastic flow and regelation slip are typically associated with a film of water at the ice–rock interface. Because of this layer, there is very low shear stress at a flat rock surface. Only obstructions that protrude upward into the ice can resist the force of flow parallel to the average bed surface. Basal sliding is then controlled by factors such as the size and spacing of roughness elements on the bed (p. 417), the cross-section size and shape of a glaciated valley, and the amount and grain size of rock fragments within the basal ice. As shown by borehole measurements (Fig. 16–4), basal sliding (including regelation slip) can account for 50 percent or more of the total observed surface motion of valley glaciers. At the extreme of 100 percent basal sliding, the movement becomes similar to a rock slide or slump along a basal surface (Chap. 8).

Brittle Fracture; Faults and Crevasses. In their thin downstream or marginal regions, glaciers also "flow" by movement along discrete shear planes such as thrust faults. Pressure of moving ice against marginal ice that is either frozen fast to rock, or too full of rock debris to move, or blocked by a large obstruction, causes a thrust fault to form, dipping upglacier. Actively moving ice shears forward and over the stagnant basal portion (Schytt, 1956). Thrust faults of this sort may reach entirely through the glaciers and are one means by which rock debris from the base of the ice is brought upward to the surface.

Brittle failure also contributes to surface topography. Wherever a glacier moves over an obstruction at shallow depths or drops down a buried cliff, the surface ice is fractured by a maze of **crevasses,** mostly vertical and extending tens of meters deep (Figs. 16–7 and 16–13). These are tension cracks that heal at the depth where the stresses can be compensated by plastic deformation. The deepest known crevasses, in the cold brittle ice of Antarctica, are about 50 m deep. Crevasses often become bridged by drifting snow and are one of the major dangers for explorers. A prominent arcuate crevasse, called the **bergschrund,** is at the extreme head of most valley glaciers (Fig. 17–3). It is caused by compaction of snow into ice and ice motion away from the rock face.

Surges. A small percentage of glaciers are known to move suddenly forward at their termini, advancing at a rate of kilometers per year instead of the average rate of a few meters per year (Paterson, 1981, Chap. 13). These **surging glaciers** typically flow normally for 10 to 100 years, then suddenly surge forward for 1 to 3 years. They are not randomly distributed; over 6 percent of the 2356 glaciers in the St. Elias Mountains on the Alaska-Canada border have surged, and other concentrations are in mountains in Asia, the Andes, Iceland, eastern Greenland, and Svalbard (Clarke et al., 1986). Surges rarely last more than a few months to a year, but they create a chaotic pattern of crevasses on the glaciers (Kamb et al., 1985). They were first reported on a group of Alaskan glaciers early in this century (Tarr, 1909) and were attributed to an excessive accumulation of snow from avalanches produced by an earthquake in 1899. This idea has not been supported by subsequent observations, but after the major Alaskan earthquake of 1964, a close watch was kept on the glaciers of the Alaska Range in hopes of observing the initial phases of surging.

Events leading up to the well documented 1982–1983 surge episode of Variegated Glacier in Alaska had been monitored for a decade (Kamb et al., 1985; Ray-

Figure 16-7. Yentna Glacier, Alaska Range. Late summer view; equilibrium line is visible toward the left rear. Rhythmic ridges on the left are *ogives,* formed by seasonal melting as the glacier descends a buried ridge. Lateral moraines of tributary glaciers become medial moraines in foreground, but maintain their individual identity. (Photo: Austin Post, U.S. Geological Survey.)

mond and Harrison, 1988). The ice in the middle and upper parts of the glacier had been thickening since 1973, and the velocity of flow had been increasing, even though the terminus remained stagnant. Until 1978, most of the increased rate of flow was during the summer, but subsequently the ice began to move more rapidly in winter as well. This indicated that basal sliding was accelerating even in winter. In 1982 the headward part of the glacier began moving forward at rates of 2 to 15 m/day. The surge then propagated rapidly downglacier, increasing from a pre-surge flow rate of 0.2 m/day or less to 40 to 60 m/day. The surface of the glacier was intensely crevassed by the rapid motion. A borehole to the base of the glacier confirmed that 95 percent of the surging motion was by basal sliding. The zone of water saturation in the ice had risen to within 90 percent of the ice thickness, approaching the level at which the ice was floating. As the water in the ice migrated down to the terminus and drained away in a spectacular flood, the surge stopped. The excess water that accumulated within the glacier was only indirectly correlated with air temperature and precipitation. Apparently, it is a characteristic of Variegated Glacier to accumulate water over an interval of about 20 years, then rapidly drain during a surge (Kamb et al., 1985).

A growing hypothesis is that an ice sheet might surge along a restricted segment of its perimeter, independent of the classic theory of radial-spreading plastic flow (p. 409). If so, the glacier margin could rapidly advance in one region even while shrinking elsewhere (Wright, 1973; Fulton, 1989, pp. 187–88). The possibility has important implications for Pleistocene deglaciation chronology (Chap. 17).

Flow Lines within Glaciers.　　Glaciers originate in regions where snow accumulation exceeds loss, and they flow outward or downward to regions where losses exceed accumulation (Fig. 16–8). The *terminus,* or down-glacier extremity, represents the line where losses by all causes (melting, sublimation, erosion, and calving into water—collectively termed **ablation**) equal the rate at which ice can be supplied by accumulation and forward motion. In the *zone of accumulation,* vectors of particle motion are downward into the ice mass as each year's snowfall adds a new surface layer to the glacier. Erosion at the glacier bed is intensified. In the *zone of ablation,* vectors of particle motions point toward the ice surface, because a surface layer is annually removed, exposing progressively deeper ice (Fig. 16–8). Net sediment deposition is probable. This is the predominant method of bringing sediment to the surface of a glacier, more important than thrust faulting. At the cross section of the glacier where upglacier net accumulation and downglacier net ablation are in exact balance, internal flow vectors are parallel to the glacier surface. The equilibrium cross section of a glacier is located at the **equilibrium line,** or down-glacier edge of net annual accumulation. Theoretical models of glacier flow fit observations at the equilibrium line especially well.

Within the zone of accumulation of a valley glacier, velocity generally increases downglacier. This is called *extending flow,* because an imaginary mass within the glacier becomes longer and thinner as it moves downvalley (assuming that the valley sides are parallel). Within the zone of ablation, longitudinal velocity generally decreases, perhaps to zero at the toe of the glacier. This is called *compressing flow,* because an imaginary mass within the glacier becomes shorter and thicker as it flows (again assuming no change of the valley width). The equilibrium cross section is therefore the place where the flow rate is at the maximum. Locally within a glacier, other areas of extending and compressing flow also can be identified (Fig. 16–8). For example, flow may be compressive upstream from a bed obstruction, with the ice thickening and the glacier surface bowing upward as a reflection of the buried obstacle. Extending flow is common where ice passes over a buried sill and drops rapidly. At such places the ice may fracture into a crevasse field (Chap. 17) or simply thin rapidly by extending flow. In plan view, compressing and extending flow can be expected respectively on the inside and outside of curves. Surface fracture patterns on glaciers are diagnostic of regions of extending and compressing flow (Fig. 16–7). Beneath the ice, regions of extending and compressing flow can control regelation and enhanced plastic flow, and determine whether the glacier erodes its bed or deposits sediment (Chap. 17). The surging part of a glacier is dominated by extending flow, but the lower end is in compressing flow until the surge reaches it, when it rapidly thins and surges forward (McMeeking and Johnson, 1986).

Flow lines in an ice cap or ice sheet are less easy to predict. The ice must spread radially in plan view from the area of accumulation, as well as outward along any cross section. However, the general flow lines (Fig. 16–9) are probably much as inferred by early workers: downward beneath the zone of accumulation near the center and then outward to the zone of ablation. Obstructions such as enclosing mountain ranges obviously complicate the flow patterns. Paterson (1981, p. 161) reviewed the theories of flow in ice sheets and concluded that the horizontal component of ice motion in the central part of a large ice cap is very low. For a hypothetical ice sheet 2000 km wide, with a central thickness of 4700 m, a parabolic profile, and net accumulation rate of 15 cm of ice per year, Paterson calculated that it would take a particle of ice 75,000 years to travel to the edge from a starting point on the surface 50 km from the center. About 60 percent of the time would be spent in moving the first 250 km downward and outward from the starting point. The dimensions and rates are reasonably appropriate for the existing Antarctic and

Figure 16-8. Trajectories of particle motion in a valley glacier. The large area of the zone of accumulation is not well shown in this longitudinal cross section. Sections labeled (+) and (−) are local areas of compressing and extending flow.

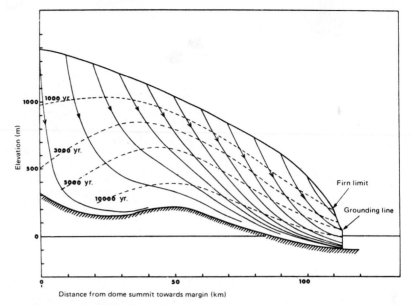

Distance from dome summit towards margin (km)

Figure 16-9. Particle paths and calculated ages of ice in a cross section of Law Dome, a small independent ice dome on the edge of the Antarctic ice sheet in Wilkes Land (111°E longitude). (From Budd and Morgan, 1973, Fig. 4.)

Greenland ice sheets and for the former North American and Scandinavian ice sheets.

For various reasons, excluding hydrostatic pressure, shear strain (the vertical rate of change of velocity) in an ice cap should be most rapid near the base. Ice at higher temperatures deforms more rapidly, and the highest temperatures should be near the base of an ice sheet because of the poor thermal conductivity of ice and the geothermal heat flow. A possible contribution of frictional heat must also be considered. Also, the greatest shear stress is at the base of the ice, although this factor does not vary greatly. If any regelation slip occurs, it too will be at the base. Therefore, reasonable models of flow in ice caps are obtained by assuming that all the shear takes place in a basal layer only a few meters thick, and the overlying ice is passively transported. Under such assumptions, flow rate is proportional only to the surface slope of the ice, which develops an elliptical or parabolic profile (Figs. 16–10 and 16–12). An excellent approximation of the thickness of an ice sheet within a few hundred kilometers of its margin is given by the equation

$$h = 4.7\sqrt{d}$$

where h is the thickness of the ice in meters on a horizontal base at distance d in from the margin of the ice sheet in meters. This equation is derived from the Glen flow law (p. 396) with an assumed basal shear stress of 0.1 MPa (1 bar). A few minutes with a calculator will demonstrate how well this equation approximates actual surface profiles of ice sheets and ice caps (Fig. 16–10). However, no theoretical model fits either the central or near-marginal parts of any ice sheet very well. Neither does any account for surges or other disproportionate rates of flow along certain radii while the remainder of the ice sheet remains in equilibrium.

Surface Flow in Valley Glaciers. Among the oldest observations of glacier flow were reports that lines of stakes emplaced straight across valley glaciers would bulge progressively downvalley year after year, proving that the center part of the ice surface moved faster than the margins. Many subsequent experiments have confirmed the fact (Fig. 16–4). Because most of the differential shear is very near

Figure 16-10. Surface profiles of (a) the ice sheets of Greenland and Antarctica, and (b) smaller ice sheets and ice caps. In both figures, bedrock relief is slight relative to ice thickness. (Robin, 1964, Fig. 5.)

the base of a glacier, the mean velocity of the surface ice is within a few percent of the mean velocity of the entire glacier cross section (Paterson, 1981, p. 105). As a useful result of this approximation, glaciologists can estimate the annual discharge of ice through the equilibrium-line cross section of a glacier by measuring surface velocities only. It is not necessary to drill boreholes and measure velocities within the ice. In this respect especially, the laminar flow of a glacier is much simpler than the turbulent flow of a river, for which an entire velocity cross section must be measured to determine mean discharge. Shear, or the lateral rate of change of velocity, is also high near the edges of valley glaciers. Most crevasses develop there, in orientations diagnostic of extending or compressing flow.

Thermal Classification of Glaciers

From the foregoing discussion of ice deformation and glacier flow, two categories of glaciers can be visualized. One category is of glaciers that are cold from top to bottom, with the entire thickness of ice well below the pressure-melting temperature. The other category is of glaciers that are at or very close to the pressure-melting temperature, so that small pressure variations can produce melting and regelation slip. Actually, this thermal classification is better applied to parts of glaciers rather than entire valley glaciers or ice caps, because the thermal status of one part of a glacier may be different from that of another part. A discussion of the thermal status of glaciers and parts of glaciers uses terminology first proposed over 50 years ago (Lagally, 1932; Ahlmann, 1933, 1935), but the examples and conclusions are much more recent. The thermal condition of glaciers has a very important bearing on both their form and their ability to shape landscapes.

Cold, Polar, or Dry-Base Glaciers.   A 1387-m borehole through the Greenland ice sheet at Camp Century, about 100 km from the west edge of the ice sheet,
was completed in 1966 (Figs. 16–11 and 16–12). At the time, it was the deepest
borehole made in ice and the first to penetrate entirely through a polar ice cap. The
temperature profile illustrates well the characteristics of a **cold,** or **polar,** glacier.
The ice temperature at a depth of 10 m was −24°C, very close to the mean annual
air temperature. A minimum temperature of −24.6°C was reached at a depth of
154 m; then the temperature rose steadily to −13.0°C at the base of the ice. The entire thickness of ice at the Camp Century borehole is far below the pressure-melting
temperature. No water phase can exist, regelation slip cannot occur, and the rather
slow surface motion of only 3.3 m/yr must be due entirely to plastic deformation
and minor brittle fracturing near the surface.

Cold, or polar, glaciers are also called **dry-base** glaciers, because the basal ice
must be frozen firmly to the rock beneath. Experiments have demonstrated that the
surface bond between rock-forming minerals and ice is considerably stronger than
the yield strength of ice, so that any motion of a cold glacier must be by internal
plastic deformation. Just one consequence of this fact is that cold glaciers cannot be
expected to produce striations and grooves by sliding over bedrock. Most of the
motion is presumed to be concentrated in the basal, relatively "warm" ice, however, and differential shear in the basal zone might break off rock projections or
fracture larger rock fragments in transport.

Temperate, or Wet-Base, Glaciers.   As the confining pressure due to the
weight of the overlying ice increases downward in a glacier, the melting temperature
is depressed (Fig. 16–5). Drawn in Figure 16–11 is the theoretical graph of the temperature of ice at its pressure-melting point with depth, which is depressed in a glacier at a rate of about −0.7°C/km. If the mean annual surface temperature of the
accumulating snow and ice is close to 0°C, then with progressive burial, pressure
melts a minute amount of ice, and the ice and water phases coexist at progressively
lower temperatures with depth. This is the definition of a **temperate,** or **wet-base**
glacier. Regelation slip and enhanced basal sliding are much more rapid than solid-
state plastic flow, even in a solid as deformable as ice, and flow rates in temperate
glaciers are likely to be at least twice as fast as in polar glaciers. Striated, grooved
basal rock surfaces are possible, as are numerous other significant forms of glacier
transport and erosion (Chap. 17), because the ice slides over rock or compact sediment on water. In addition, the sub-ice water adds its erosional, transportational,
and depositional work to that being done by the ice (p. 417).

An important characteristic of a temperate glacier derives from the negative
slope of the pressure-melting curve. Because the temperature at the base is less than
at the surface, no geothermal heat can flow upward as it does in a polar glacier. The
geothermal heat that reaches the base of a temperate glacier cannot be conducted
upward but must melt basal ice. This geothermal heat, plus an increment of heat
generated by sliding friction that is also trapped at the base of the ice, assures that a
temperate glacier must move on a wet base (Weertman, 1966). The nature of the
water film is not certain. Most of it is probably a very thin film, only 1 or 2 mm in
thickness, but it may fill pockets and tunnels in the ice in the lee of rock obstructions, or gather in deeper basins (p. 417). Certainly, permafrost cannot exist under a
temperate glacier (Werenskiold, 1953).

Intermediate Categories.   Ahlmann (1933, 1935) subdivided polar glaciers
(*arctic* glaciers in his original terminology) into two classes, *sub-polar* and *high-
polar*. Although both are cold at depth, summer melting on the sub-polar glacier
might produce a surface layer 10 to 20 m deep, which is at the melting temperature. Ahlmann's high-polar glacier would be typified by the Greenland icecap at

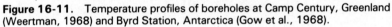

Figure 16-11. Temperature profiles of boreholes at Camp Century, Greenland (Weertman, 1968) and Byrd Station, Antarctica (Gow et al., 1968).

Figure 16-12. Profiles and estimated temperatures of the Greenland ice sheet (the Inland Ice). Vertical exaggeration 50:1. (Weidick, 1975, Fig. 4.)

Camp Century (Fig. 16-11) in which not even a surficial layer of snow melts in the summer.

Lagally (1932), in a classification otherwise almost identical to Ahlmann's polar and temperate glaciers, proposed an *intermediate* type that would be cold or polar at the top but be at the pressure-melting temperature near the base. The intermediate category of Lagally was accepted as an interesting theoretical possibility but was generally overlooked until 1968, when the first borehole was drilled to bedrock at Byrd Station in Antarctica (Fig. 16–11) (Gow et al., 1968). At a depth of 2164 m, the bottom of the ice was reached, but at the basal contact was a film of water at least 1 mm in thickness, under enough pressure to rise 50 m up the drill hole and displace the antifreeze solution that had been used to keep the hole open. At the base of the ice sheet, the calculated melting temperature is −1.6°C. The surface temperature is −28.8°C. The steeper-than-average thermal gradient in the lower half of the Byrd Station borehole suggests a geothermal heat flow about 50 percent greater than the world average. Because West Antarctica is an ice-covered volcanic archipelago, the higher heat flow is reasonable. The significance of the borehole data to the history and nature of flow of the Antarctic ice sheet has yet to be fully evaluated. Subsequent radar depth soundings of the Antarctic ice sheet have indicated that large areas of the basal ice are wet (Oswald and Robin, 1973; Radok, 1985). The effectiveness of the wet-base condition is demonstrated by the fact that the Byrd Station drill bit was unable to obtain a sample of the sub-ice material, because the ice moved laterally several centimeters per day and constantly pinched the drilling tool in the borehole. The substrate may be either rock or saturated, clay-rich basal sediment.

GEOMORPHOLOGY OF GLACIER SURFACES

The simplest morphologic subdivision of glaciers is to distinguish those that flow between confining rock walls, or **valley glaciers,** and those that flow unconfined by virtue of their great thickness, the **ice caps** and **ice sheets** (Sharp, 1988). Although the morphology of all glaciers is controlled by the rheology of ice, important differences are to be noted whether flow is confined or unconfined.

Valley Glaciers

The essential component for any valley glacier is an upland area in which annual snow accumulation exceeds all ablation losses. As the accumulating snow layers thicken and compact, the yield stress of ice is exceeded, and glacier flow begins.

The conversion of snow to glacier ice follows a regular sequence. Fresh snow is full of entrapped air and may have a bulk specific gravity even lower than 0.1. Snowflakes readily sublime, so old snowflakes lose their frilly margins and become more globular. Meltwater often freezes onto snowflake nuclei, so aging also tends to increase the size of the ice granules. Old snow, such as remains on sheltered mountain slopes in early summer, has the texture of very coarse sand. This is the late-season *corn snow* or *buckshot snow,* familiar to avid skiers. The loose granular mass is about one-half ice and one-half entrapped air and has a bulk specific gravity of about 0.5.

Snow that has survived a summer melting season is called **firn** in German or **névé** in French. Both terms are widely used in English. Firn, or névé, is an intermediate step in the conversion of snow to glacier ice. It is granular and loose unless it has formed a crust. It represents the net positive balance between winter accumulation and summer losses.

As successive annual layers accumulate, the deep firn is compacted. Individual ice grains freeze together. By definition, when the grains are frozen together so that the mass is impermeable to air, firn becomes glacier ice. The bulk specific gravity is usually about 0.8 by this stage of consolidation. Most glacier ice is a polycrystalline mass of frozen water plus a variable amount of air. Other components are dust and rock fragments that have fallen, washed, or been blown onto the ice surface, and rock that has been eroded from beneath the glacier. The bulk specific gravity of glacier ice ranges from about 0.8 to 0.9, which is approximately the specific gravity of pure, gas-free ice.

In a mountainous terrain, the colder temperatures and frequently higher precipitation at higher altitudes produce ice fields at the heads of glaciers. This is the zone of accumulation (Fig. 16–8), and as a rule of thumb, about 70 percent of the surface area of a valley glacier is in this zone. The zone of accumulation is separated from the zone of ablation by the equilibrium line (p. 401). The annual discharge of ice through the cross section of the equilibrium line is an important measure of the glacier's regimen. The equilibrium line is approximated by the **firn limit,** or the lowest line of surviving accumulation at the end of summer, and by the annual **snow line** on the glacier and adjacent mountain slopes. The firn limit on a glacier is always a little lower in altitude than the equilibrium line, because the motion of the glacier carries the firn forward. The common summertime sight of a fringe of dirty gray firn ice downglacier from the edge of new snow is not evidence that the climate is warming and the snowline is rising.

Downglacier from the equilibrium line, the glacier has net annual loss of ice. Melting is the major factor, but sublimation, wind ablation, and iceberg calving into lakes or the ocean also contribute. A glacier is in equilibrium only if annual discharge through the equilibrium line is equal to the net annual accumulation upglacier and the net annual ablation downglacier. Otherwise, it will grow or shrink. Ablation is at a maximum toward the downglacier terminus, and the terminus itself marks the line where ablation is complete. Some valley glaciers have steep ice faces at their termini, but others end deeply buried in transported rock debris, so that the exact location of the terminus is uncertain.

Many large valley glaciers are fed by more than one *tributary glacier* (Fig. 16–7). The surfaces of merging glaciers join at a common level, because ice is easily deformed and responds to the lateral stresses imposed by an entering tributary. Tributary glaciers do not turbulently mix with the trunk glacier, as rivers do. Because of the nature of flow in ice, the tributary glacier retains its identity as a separate stream or surface ribbon of ice (Fig. 16–7), although it may become progressively attenuated or enfolded into the trunk glacier.

Glaciers have much higher surface gradients than rivers, commonly averaging 10 percent (6°). On ice cliffs or ice falls steeper than 45°, ice avalanches are nearly continuous. Ledges buried beneath the ice are reflected by crevasse fields and ice pinnacles *(séracs)* on the ice surface (Figs. 16–7 and 17–3). The long profile of a glacier surface is likely to be marked by level areas, steep drops, and even local reversals of slope, much in contrast to the surface gradient of a river.

On the surface of a temperate valley glacier in the ablation zone, a variety of ephemeral landforms develop due to melting, collapse of ice tunnels, and erosion by streams flowing on, in, or along the glacier. Many features are analogous with karst features and may be collectively called **glacier karst. Thaw lakes** and swallow holes (**moulins**) mark the ice surface. Moulins serve as conduits for surface meltwater to flow down into the glacier, eventually joining subglacial channel systems (p. 417). They are inclined as much as 45° toward the glacier margin because the water tries to flow along the internal water pressure gradients, perpendicular to the equipotential lines (Fig. 7-4) in the saturated zone of the glacier (Holmlund, 1988). Segregations of debris-rich ice stand as ice-cored mounds or ridges above the debris-covered surface (Chap. 17). Lakes may drain and then abruptly form again when a crevasse or moulin opens at depth, and then clay-rich sediments seal it closed again. Ice tunnels within or at the base of the glacier discharge sediment-laden water from the terminus.

Ice Caps and Ice Sheets

Unconfined glaciers are so large that they form their own orographic precipitation pattern. The accumulation zone is in the central region of the dome, and the ablation zone is on the periphery, below the equilibrium line. The regimen of an ice cap is very sensitive to slight climatic changes, because a slight raising or lowering of the equilibrium line on a convex-skyward dome can have an enormous effect on the relative areas of accumulation and ablation. Because the climate has warmed since they formed, numerous small or moderate-sized ice caps on Iceland, Scandinavia, and the Canadian Arctic islands could not re-form today if they once fell below the firn limit; the present sustaining orographic precipitation is a consequence of the ice caps themselves.

The radial profiles of continent-sized ice sheets and the somewhat smaller, but otherwise similar, ice caps differ very little from each other (Fig. 16–10). Theories of radial flow that assume shear stresses of only 20 to 30 kPa (0.2 to 0.3 bar) at the horizontal base of ice sheets, and either plastic flow concentrated just above the base or basal sliding, predict parabolic or elliptical radial profiles very similar to those actually observed (p. 403). The equations of the predicted profiles are not very sensitive to either internal temperature or rate of accumulation, two factors that strongly affect the steepness of valley glaciers. Note the considerable vertical exaggeration on Figures 16–10 and 16–12. At true scale, the thickness of the ice could barely be shown by a slight thickening of the horizontal base lines. Published maps and radar profiles (Walker et al., 1968; Drewry, 1983; Bogorodsky et al., 1985) give an excellent idea of the surface relief, bottom relief, and thickness of large ice sheets.

Most of the Greenland ice sheet perimeter is a series of broad, low ramps. Along some parts of the perimeter, where the terminus is on land, the glacier ends as a vertical ice cliff as much as 100 m in height. The cliffs may be due to a combination of factors, such as low temperatures, compass orientation relative to a low sun angle that causes sublimation directly from the cliff face, and cold, brittle ice almost free of enclosed sediment except at the base. The Greenland ice sheet (known as the *Inland Ice*) rests on a basin-shaped interior plateau surrounded by mountain ranges (Fig. 16–12) (Reeh, 1989). Except for a few places, the basal temperatures are below 0°C. Around much of its margin, the ice moves between exposed or slightly buried mountain peaks as **outlet glaciers.** These have many of the geomorphic properties of valley glaciers. At the head of each outlet glacier is an area of shallow concavity sloping toward the outlet, with many wide, deep crevasses. This is the "drawdown" area, or "ice shed" of that particular outlet glacier.

Much of the perimeter of the Antarctic ice sheet is also a great cliff, but that cliff ends at a **grounding line** in the ocean (Fig. 16–9) from which enormous tabular icebergs calve off (Swithinbank, 1988). A few **nunataks,** peaks surrounded by ice, rise above the surface (Fig. 16–13). The west Antarctic sector, in the longitudal zone that faces northward toward the Pacific Ocean between New Zealand and South America, is grounded 1 to 2 km below sea level in two large tectonic basins from which ice flows into the Ross Sea and Weddell Sea ice shelves and numerous other smaller ice shelves (Fig. 16–13). The West Antarctic Ice Sheet is in a precarious status. It is wet based and grounded far below sea level except at its discharge areas, where it is held back by *ice rises,* buried hills over which it rides in the transitional zone where outlet glaciers become floating ice shelves (Fig. 16–13). Continued global warming and rising sea level (Chap. 17) could cause the West Antarctic Ice Sheet to float over the ice rises and massively surge. The resulting rise of global sea level of about 6 m within the next few centuries could have disastrous consequences to a large portion of the world's population (Van Der Veen and Oerlemans, 1987).

Figure 16-13. Oblique aerial photograph up the Fleming outlet glacier, Graham Land, Antarctica. Buffer Ice Rise (foreground) deflects and fractures the floating Wordie Ice Shelf. (Photo: U.S. Navy for U.S. Geological Survey, TMA 1835 F33, courtesy R. S. Williams, Jr.)

The surface of the Antarctic ice sheet is a monotonous dome with only minor areas of exposed bedrock, mostly near the coast (Drewry, 1983; Radok, 1985). Flow lines define ice divides and drainage basins, but surface gradients are very low, and little relief is apparent. Much of the East Antarctic Ice Sheet has precipitation of less than 50 mm/yr of water equivalent, making it one of the driest regions on earth. Surface temperatures are the coldest known on earth, yet large areas are shown by radar reflection profiling to have water at the base. The Cenozoic history of the Antarctic ice is reviewed in Chapter 17.

REFERENCES

AHLMANN, H. W., 1933, Scientific results of the Swedish–Norwegian Arctic Expedition in the summer of 1931, Part VIII, Glaciology: Geograf. Annaler, v. 15, pp. 161–216, 261–95.
––––– 1935, Contribution to the physics of glaciers: Geog. Jour., v. 86, pp. 97–113.
BOGORODSKY, V. V., BENTLEY, C. R., and GUDMANDSEN, P. E., 1985, Radioglaciology: D. Reidel Publishing Company, Dordrecht, The Netherlands, 254 pp.
BUDD, W. F., and MORGAN, V. I., 1973, Isotope measurements as indicators of ice flow and paleo-climates, *in* van Zinderen Bakker, E. M., ed., Paleoecology of Africa and the surrounding islands and Antarctica: A.A. Balkema, Cape Town, South Africa, v. 8, pp. 5–22.
CLARKE, G. K. C., SCHMOK, J. P., OMMANNEY, C. S. L., and COLLINS, S. G., 1986, Characteristics of surge-type glaciers: Jour. Geophys. Res., v. 91, pp. 7165–80.

DREWRY, D. J., ed., 1983, Antarctica: Glaciological and geophysical folio: Scott Polar Research Institute, Cambridge, England, 9 map sheets with text.

————, 1986, Glacial geologic processes: Edward Arnold (Publishers) Ltd., London, 276 pp.

FLINT, R. F., 1971, Glacial and Quaternary geology: John Wiley & Sons, Inc., New York, 892 pp.

FULTON, R. J., co-ordinator, 1989, Quaternary geology of the Canadian Shield, in Fulton, R. J., ed., The Quaternary geology of Canada and Greenland: Geol. Soc. America, The geology of North America, v. K-1, pp. 177–317.

GLEN, J. W., 1952, Experiments on the deformation of ice: Jour. Glaciology, v. 2, pp. 111–14.

————, 1974, Physics of ice: U.S. Army Corps of Engineers, Cold Regions Res. and Eng. Lab., Cold Regions Sci. and Eng. Mono. II-C2a, 81 pp.

————, 1975, Mechanics of ice: U.S. Army Corps of Engineers, Cold Regions Res. and Eng. Lab., Cold Regions Sci. and Eng. Mono. II-C2b, 43 pp.

GOW, A. J., UEDA, H. T., and GARFIELD, D. E., 1968, Antarctic ice sheet: Preliminary results of first core hole to bedrock: Science, v. 161, pp. 1011–13.

HOLMLUND, P., 1988, Internal geometry and evolution of moulins, Storglaciären, Sweden: Jour. Glaciology, v. 34, pp. 242–48.

HUTTER, K., 1983, Theoretical glaciology: D. Reidel Publishing Company, Dordrecht, The Netherlands, 510 pp.

JANKOWSKI, D. G., and SQUYRES, S. W., 1988, Solid-state ice volcanism on the satellites of Uranus: Science, v. 241, pp. 1322–25.

KAMB, B., and LaCHAPELLE, E., 1964, Direct observations of the mechanism of glacier sliding over bedrock: Jour. Glaciology, v. 5, pp. 159–72.

KAMB, B., RAYMOND, C. F., HARRISON, W. D., ENGELHARDT, H., ECHELMEYER, K. A., HUMPHREY, N., BRUGMAN, M. M., and PFEFFER, T., 1985, Glacier surge mechanism; 1982–1983 surge of Variegated Glacier, Alaska: Science, v. 227, pp. 469–79.

KRASS, M. S., 1984, Ice on planets of the solar system: Jour. Glaciology, v. 30, pp. 259–74.

LAGALLY, M., 1932, Zur Thermodynamik der Gletscher: Zeitschr. für Gletscherkunde, v. 20, pp. 199–214.

LLIBOUTRY, L., 1965, Traité de glaciology (2 vols.): Masson et Cie, Paris, 1040 pp.

McMEEKING, R. M., and JOHNSON, R. E., 1986, On the mechanics of surging glaciers: Jour. Glaciology, v. 32, pp. 120–22.

MEIER, M. F., 1960, Mode of flow of Saskatchewan Glacier, Alberta, Canada: U.S. Geol. Survey Prof. Paper 351, 70 pp.

OSWALD, G. K. A., and ROBIN, G. D., 1973, Lakes beneath the Antarctic ice sheet: Nature, v. 245, pp. 251–54.

PATERSON, W. S. B., 1981, The physics of glaciers, 2nd ed.: Pergamon Press, Elmsford, N.Y., 380 pp.

PETERSON, D. N., 1970, Glaciological investigations on the Casement Glacier, southeast Alaska: Ohio State Univ. Inst. of Polar Studies, Rept. no. 36, 161 pp.

RADOK, U., 1985, The Antarctic ice: Sci. American, v. 253, no. 2, pp. 98–105.

RAYMOND, C. F., and HARRISON, W. D., 1988, Evolution of Variegated Glacier, Alaska, U.S.A., prior to its surge: Jour. Glaciology, v. 34, pp. 154–69.

REEH, N., 1989, Dynamic and climatic history of the Greenland Ice Sheet, in Fulton, R. J., ed., The Quaternary geology of Canada and Greenland: Geol. Soc. America, The geology of North America, v. K-1, pp. 795–822.

SCHYTT, V., 1956, Lateral drainage channels along the northern side of the Moltke Glacier, northwest Greenland: Geograf. Annaler, v. 38, pp. 64–77.

SHARP, R. P., 1988, Living ice: Understanding glaciers and glaciation: Cambridge University Press, Cambridge, England, 225 pp.

SWITHINBANK, C., 1988, Satellite image atlas of glaciers of the world: Antarctica: U.S. Geol. Survey Prof. Paper 1386-B, 278 pp.

TARR, R. S., 1909, The Yakutat Bay region, Alaska: U.S. Geol. Survey Prof. Paper 64, 183 pp.

VAN DER VEEN, C. J., and OERLEMANS, J., eds., Dynamics of the West Antarctic Ice Sheet: D. Reidel Publishing Company, Dordrecht, The Netherlands, 368 pp.

WALKER, J. W., PEARCE, D. C., and ZANELLA, A. H., 1968, Airborne radar sounding of the Greenland ice cap: Flight 1: Geol. Soc. America Bull., v. 79, pp. 1639–46.

WEERTMAN, J., 1966, Effect of a basal water layer on the dimensions of ice sheets: Jour. Glaciology, v. 6, pp. 191–207.

———, 1968, Comparison between measured and theoretical temperature profiles of the Camp Century, Greenland, borehole: Jour. Geophys. Res., v. 73, pp. 2691–700.

———, 1983, Creep deformation of ice: Ann. Rev. Earth Planet. Sci., v. 11, pp. 215–40.

WEIDICK, A., 1975, Review of Quaternary investigations in Greenland: Grønlands Geologiske Undersøgelse Miscell. Papers, no. 145; distributed as Ohio State Univ. Inst. of Polar Studies, Rept. no. 55, 161 pp.

WERENSKIOLD, W., 1953, Extent of frozen ground under the sea bottom and glacier beds: Jour. Glaciology, v. 2, pp. 197–200.

WRIGHT, H. E., JR., 1973, Tunnel valleys, glacial surges, and subglacial hydrology of the Superior Lobe, Minnesota, *in* Black, R. F., Willman, H. B., and Goldthwait, R. P., eds., The Wisconsinan Stage: Geol. Soc. America Mem. 136, pp. 251–76.

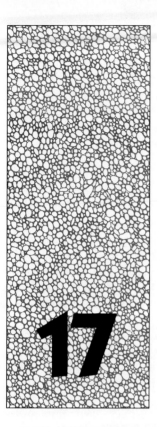

17 Glacial Morphogenesis

The glacial climatic zone is defined by the combination of temperature and precipitation that permits glaciers to cover the rocky landscape. The glacial morphogenetic region is sharply circumscribed by the edge of a glacier. This is one of the few examples that can be cited where geomorphic processes and forms change abruptly at a sharp boundary. Only at a coast is there such an equally abrupt process-controlled change in landforms.

The difficulty of describing glacial morphogenesis is that most of the processes go on beneath a layer of ice, which is properly regarded as a common rock type at the earth's surface. Therefore, strictly speaking, glacial morphogenesis is a subsurface activity, and the true glacier landscape is on top of the ice, not beneath it. These are the forms described in the preceding chapter. In this chapter we consider the geomorphic work of glaciers in shaping the surface morphology of other rocks. However, these are not landscapes until they are exposed by the disappearance of the glaciers. In this sense, all glaciated landscapes are relict even if they were exposed by the previous year's retreat of a glacier terminus.

The rheid behavior of glacier ice under various conditions of temperature and pressure was described in Chapter 16. In this chapter, much of the same terminology is used, but here the emphasis is on glacial erosion, transport, and deposition of rock and sediment. Ice is considered as a geomorphic agent rather than as a deformable rock on which peculiar landscapes are to be found.

Because of the effectiveness of glacial erosion, transport, and deposition, and of the associated work of water and wind, glaciated landforms preserve their morphogenetic origins long after glaciers cease to shape them. In very large areas of Europe and North America, relict landforms of glacial origin totally dominate the few landforms developed in postglacial time, even though the areas have been deglaciated for as long as 18,000 years. Much of the total postglacial landscape evolution may have been produced during a brief phase of periglacial morphogenesis immediately after deglaciation (p. 372), but the impact of these late-glacial processes has yet to be firmly established. Clearly, glacial processes are able to shape landscapes so powerfully that a much longer period of fluvial erosion is required to remove the evidence of glaciation. The oldest glacial deposits of the central United States, assigned to the Nebraskan glaciation, are deeply weathered and dissected, but they still control the regional topography although they are over 2 million years old (p. 437).

GLACIAL EROSION

Processes

Crushing and Fracture.   The crushing strength of most rocks greatly exceeds the combination of normal and shear stresses that can be applied to the bed by overriding glacier ice. However, at a point contact between a rock particle embedded in the ice and the rock substrate, pressure is adequate to chip or flake both the particle and the substrate. A wide range of distinctive fractures can be seen on glaciated rock surfaces, including *chattermarks* and *crescentic fractures* (Drewry, 1986, p. 34). These are made by hard particles rolling at the ice–rock interface that elastically compress the rock surface in front of them and stretch the rock behind them to create rhythmic fractures that are reproduced by rolling a hard steel ball over glass under a heavy load. Large rock obstructions that protrude upward into ice can also be broken by the excess ice pressure on their upglacier sides, even though temperate ice will attempt to flow around such an obstruction by intensified plastic flow and regelation. Because surface rocks are usually weathered and full of microfractures and pores, fracture may occur at stresses lower than the predicted strength of the rocks.

Plucking.   During regelation slip, water migrates to the downglacier side of a sub-ice obstruction and refreezes (p. 397). Fractured rock can be incorporated into the regelation layer and carried away by the glacier. This process is called glacial **plucking.** It is not clear whether regelation ice has enough tensile strength to actually pull away lee-side pieces of rock or whether pieces already loosened by the stress of the moving ice are simply frozen in and carried away (Boulton, 1974, pp. 69–75). Whatever the process, plucking is much more effective than abrasion on fractured rocks. Jahns (1943) reconstructed the sheeting joints on crystalline metamorphosed rocks near Boston, Massachusetts (p. 120), to show that plucking had removed several times the volume of rock from the lee sides of hills than abrasion had removed from the upglacier, or **stoss** (German for *push*) sides.

Abrasion.   Table 17–1 lists the hardness, by the standard Mohs' scale, of ice at various temperatures. From the table it is clear that some soft minerals or weathered rock surfaces can be abraded if very cold ice slides over them. However, the very low yield stress of ice is more likely to cause plastic deformation of the ice pressing against a rock surface rather than abrasion of the rock. Furthermore, sliding will be trivial or nonexistent without a water film at the ice–rock interface, at which temperature ice is a very soft mineral.

The abraded scratches and grooves that characterize glaciated rock surfaces are undoubtably caused by rock fragments in the ice rather than by the ice itself (Boulton, 1974, p. 43). If the glacier is sliding over its base on a water film, any fragment larger than the thickness of the water film will be rolled or dragged against the rock surface. Glaciated surfaces show a range of abraded microrelief forms from hairline *striations* to *grooves* a meter or more in depth and width. Careful study of glacial striations and grooves can sometimes reveal the direction of ice motion, as

TABLE 17–1

Brittle Hardness of Ice at Various Temperatures (Mohs' Scale)

Temperature (°C)	Hardness
0	1.5
−15	2–3
−30	3–4
−40	4
−78.5	6 (hardness of orthoclase)

Source: Shumskii (1964), pp. 39–40.

when a piece of rock is chipped off and dragged downglacier over the surface, or when a resistant projection protects less resistant rock in its lee from abrasion (Fig. 17–1). Regelation slip (p. 397) must be an effective process of abrasion, because if a sand grain is embedded in basal ice, it will move into contact with the rock surface as the enclosing ice melts. Abrasion is ineffective beneath polar glaciers, where there is no sliding at the ice-rock interface (p. 405) (Boulton, 1974, p. 83).

Not only bedrock surfaces are abraded; the rock fragments carried in glaciers also are subjected to abrasion. A significant proportion of glacially transported pebbles and cobbles have planocurved *facets* on one or more sides, showing that they were held in contact with other rocks as they ground past each other. Two of the strongest proofs that a region has been glaciated by a wet-base glacier are (1) a regional pattern of striated and grooved bedrock surfaces, and (2) abraded stones of exotic origins in the detritus overlying the surfaces.

Abrasional effectiveness depends on the relative hardnesses of the "tool" held by the ice and the rock substrate, the flux of abrasive particles in contact with the bed, and the effective force with which the particles are held against the bed (Hallet, 1979). At the base of a temperate glacier, the normal stress (ice thickness) is equally applied to the bed and the tools in the basal ice, so abrasional effectiveness is not dependent on glacier thickness. Increased force against the bed decreases the rate of movement of the particle, and may actually cause it to *lodge* immoble, so that abrasion is in no way directly proportional to the force on the particle. The flux of abrasive particles in contact with the bed is reduced by a debris concentration above a certain relatively low value. Glaciers with significant basal sliding will abrade their beds until the basal load becomes large and the sliding rate decreases, when deposition will begin (Hallet, 1981).

Figure 17-1. Glacially or subglacial fluvially abraded surface of limestone, with a protruding chert concretion, a streamlined tail of limestone, and a shallow marginal moat. (Photo: A. Meike.)

Some polished and striated rock surfaces show patterns of flow that diverge from the regional direction in minute, complex ways (Fig. 17–2). Local reversals of flow direction that give eddylike patterns, helical flow along shallow troughs, and potholes suggest a highly plastic abrasive substance (Gjessing, 1966). Possibly, the base of some temperate and intermediate glaciers (pp. 405–07) is underlain by a sheared mass of water-saturated, probably clay-rich, sediment under high confining pressure and shearing stress (Alley, 1989; Alley et al., 1989). A basal layer of impermeable, shearing sediment was detected by seismic reflection techniques beneath Variegated Glacier in Alaska prior to the 1982–83 surge (p. 399). It may have helped confine water near the bed and build the water pressure to a level at which the glacier was nearly floating during the surge (Richards, 1988). The sediment moves generally in the same direction as the overlying ice but sensitively responds to the locally intensified stress gradients around obstructions. Whether it is capable of abrading the rock surface beneath it, or whether it absorbs all the shearing motion of the glacier within itself, is not known. Such saturated basal sediment has been exposed by rapid retreat of some glacier termini and has been observed in tunnels dug along the base of valley glaciers. It also makes constructional landforms when it squeezes up into crevasses near the margins of glaciers.

The combination of abrasion and plucking exploits structural details of rocks, especially joints. A variety of streamlined landforms, such as *stoss-and-lee topography, roches moutonnées,* and *crag-and-tail* hills are variants of the combined abrasion and plucking process. They range from a few meters to about 1 km in length. They may form equally well under ice sheets or valley glaciers, but a wet-base or temperate thermal status is necessary. All are small hills with abraded up-glacier sides, but the crag-and-tail variant has a streamlined tail of compact sediment or erodible bedrock sheltered behind a more resistant crag, whereas the other two forms have steepened, plucked lee faces.

Figure 17-2. Glacially scoured metamorphic rock suggesting a highly plastic abrasive medium, near Portør, Norway, on the Skagerrak coast. (Photo: Just Gjessing.)

Fluvial Erosion Beneath Glaciers

With the recognition of the widespread occurrence of wet-base glaciers, both at the present time and during former ice ages, the erosional role of flowing water beneath glaciers has received new emphasis. The basal film of water observed beneath temperate and intermediate glaciers is not capable of mechanical erosion, although by virtue of its high confined pressure, high dissolved CO_2 content, and low temperature it can chemically corrode carbonate rocks (Drewry, 1986, pp. 73–82). The most important effect of the water layer is to reduce the resistance to basal sliding. Bed roughness is an important factor in controlling glacier flow (p. 397). Roughness typically includes a wide range of dimensions, as shown by fractal analysis (Fig. 1–2). If the water film covers and negates the smaller part of the roughness spectrum, the flowing ice "sees" only the larger irregularities, and will slide much faster (Weertman and Birchfield, 1983, p. 22). By carefully mapping the glacial erosional features on a recently deglaciated limestone surface, Walder and Hallet (1979) concluded that at least 20 percent of the surface area had been separated from the base of the glacier by water-filled channels, with the remainder covered by a thin water film.

Cavities with dimensions of several meters may be opened on the downglacier side of rock knobs that protrude into temperate glaciers. These become water filled, especially during the summer melt season, and may coalesce into a system of water-filled channels. Where water flows turbulently in discrete channels or tunnels at high velocity and with a significant sediment load, it can actively erode the sub-ice rock in ways fully analogous with channel erosion by surface streams. Channels may close by plastic flow during the winter season, and reopen or form in new areas during subsequent meltwater seasons. A complex set of sub-ice tunnels has been observed under an outlet glacier from a small Norwegian ice cap that is used for hydroelectric power generation (Hooke et al, 1985). When the meltwater season begins and the previous tunnels have closed, high water pressures are observed at the base of the ice. The relatively warm and muddy meltwater can erode both the base of the ice and the bedrock beneath the ice. *Potholes* in bedrock eroded by vortices in subglacial rivers are well-known glacial erosional landforms. Very large *tunnel valleys* were eroded into unconsolidated sediments by subglacial meltwater along some margins of the late Pleistocene ice sheets. Those in central Minnesota are up to 1 km in width and 30 km in length (Mooers, 1989). Whether the tunnels entrench into the substrate or melt upward into the ice, and whether they close by ice deformation or become filled with sediment to become eskers (p. 430), is a complex function of the amount of meltwater, the basal water pressure, ice plasticity, sediment load, and many other variables.

Erosion of subglacial sediment and its transporation to the ice margin by subglacial and englacial ("within the glacier") streams is claimed to be the source of almost all of the outwash in front of the Malaspina Glacier in Alaska (Gustavson and Boothroyd, 1987). An extreme view is that episodic subglacial floods from beneath the southern part of the North American Laurentide Ice Sheet were responsible for most of the glacial erosional landforms on both bedrock and sediment, including polished and abraded streamlined forms (Sharpe and Shaw, 1989; Shaw, 1989). A continuing debate concerns the relative importance of subglacial water and highly plastic basal sediment in producing the polished, striated, and fluted surfaces on hard rocks so diagnostic of glacial erosion (Figs. 17–1 and 17–2). Evidence that a thick layer of plastic or fluidized debris at the base of a glacier may inhibit rather than enhance erosion has forced the realization that "fluvial" erosion may be important beneath wet-base glaciers.

Erosional and Residual Landforms

Troughs.   The erosional landforms of valley glaciers rank with the most spectacular scenery on earth. The European Alps are so thoroughly dissected by valley glaciers that the entire assemblage is sometimes simply referred to as *alpine topography* (Figs. 17–3, 17–4, and 17–5). The fundamental landform component is the **glacial trough,** the channel of a former valley glacier. The cross profile of a glacial trough is a catenary curve, not unlike the cross profile of a river channel but quite different from the profile of an entire river valley. Troughs are on the order of 1 km or more in width, and commonly hundreds of meters or more in depth. Concave slopes dominate, rising from the valley floor to steep cliffs or stepped-back, ice-abraded shoulders (Figs. 16–7, 17–5, and 18–7). In long profile, glacial troughs may be a series of basins separated by cross-valley ledges of more resistant rocks (Fig. 17–4) (Gutenberg et al., 1956). The long profile is structurally controlled by intense glacial plucking and abrasion on more highly fractured or easily eroded rocks to create the basins. Alternating segments of compressing and extending flow may differentally erode the glacier bed and enhance initial structurally controlled irregularities. Typically, deep-basin erosion is localized by valley-wall constrictions, and sills are associated with widening of the valley walls (Shoemaker, 1986). If not filled with sediment, the basins form chains of lakes, each with a waterfall over the downstream ledge to the next lake. These delightful chains are called **paternoster lakes,** from the resemblance to shining beads on a rosary. It is helpful to visualize a glacial trough as the analog to a river channel at bankfull discharge (Chap. 9). Just as a river channel may have alternating pools and riffles, the basal ice can flow up and down over rocks of varying resistance as long as the surface gradient provides the necessary downvalley gravitational potential.

Trough Lakes and Fiords.   In a glacial trough that has either a rock sill or a massive end moraine blocking its lower end, a **trough lake** will drown the valley floor. Trough lakes are usually long, steep-sided, deep, and cold, with a minimal inflow of sediment because of the clean abraded valley walls and the upstream settling basins on most tributary streams. The Swiss lakes, the Como region of northern Italy, the Lake District of England, the Finger Lakes of central New York (Fig. 17–6), and many other regions of large trough lakes have become justly famous as scenic tourist areas. Hydroelectric projects make use of the large storage volumes and abrupt gradients associated with trough lakes in mountainous regions.

The side walls of glacial troughs are straightened by abrasion and plucking. The interlocking spurs that interrupt views along a river valley are cut away, giving a series of **ice-faceted spurs** that align between lateral valleys (Fig. 16–7). Where tributary glaciers joined the trunk ice stream, smaller glacial troughs form, but their floors are usually well above the floor of the main valley. Waterfalls cascade down the main valley wall from these **hanging troughs.** The accordance of valley floors, so important to Playfair's law, is absent in alpine glacial scenery.

Fiords are glacial troughs that have been entered by the sea (Fig. 20–6). Their form is no different from that of any other trough except that near the outer coast, many fiords shoal rapidly to only a fraction of the depth between more restricting walls farther inland, where multiple basins may be 1 km or more in depth. In some fiords, the sill at the mouth is an end moraine, but usually it is bedrock and is attributed to a reduction in intensity of vertical erosion as the valley glacier either began to float (Crary, 1966) or spread laterally as a piedmont lobe on the land that would have then been exposed by the worldwide lowering of sea level during glacial intervals. The analysis of extending and compressing flow in glaciers (p. 401) assumes that longitudinal flow is confined between parallel walls so that no lateral ex-

1. Valley glacier
2. Hanging glacier
3. Regenerated glacier
4. Firn basin (Accumulator)
5. Glacier tongue (Ablator)
6. Glacier cap
7. Bergschrund
8. Marginal crevasses
9. Transverse crevasses
10. Ice fall with seracs
11. Pressure arches (Ogives, Forbes' Bands)
12. Glacier snout with stream from ice cave
13. Crevasses
14. Hummocky sands
15. Glacier tables
16. Glacier mills
17. Dead ice buried beneath kettle moraine
18. Lateral moraine
19. Terminal moraine
20. Medial moraine
21. Old moraines
22. Glacier stream
23. Incipient avalanche glacier
24. Ice apron
F.L. Firn line separating Accumulator from Ablator

Figure 17-3. Glacier landforms in an Alpine region. (Streiff-Becker, 1947. Reproduced by courtesy of the International Glaciological Society from *Journal of Glaciology* vol. 1, No. 2, p. 64–65.)

THE SAME REGION AFTER DISAPPEARANCE
OF THE GLACIERS

1. Glaciated valley
2. Cirque or corrie
3. Cirque or corrie lake
4. Rock barrier
5. Upper limit of ice erosion
6. Rock wall
7. Rock shoulder
8. Glacial striations
9. Valley steps
10. Glacier bed
11. Drumlins
12. Roches moutonnées
13. Basal or ground moraine
14. Terminal moraine (of advance)
15. Recessional moraine
16. Glaciofluvial gravels
17. Contemporaneous river channel
18. Eskers (Osers)
19. Kames

Figure 17-4. Glacial landforms following deglaciation of an Alpine region. (Streiff-Becker, 1947. Reproduced by courtesy of the International Glaciological Society from *Journal of Glaciology* vol. 1, no. 2, p. 64/65.)

Figure 17-5. View northeastward over Mount Cook (3800 m) up the Tasman Glacier, New Zealand. Horns and arêtes are prominent. (Photo: New Zealand National Publicity Studios.)

Figure 17-6. The Finger Lakes of central New York. All the lakes drain north to Lake Ontario (top) via the Oswego River. Cayuga Lake, the easternmost of the two longest lakes, is 61 km long. A light snow cover enhances the relief. (NASA ERTS E-1243-15244.)

pansion or contraction occurs. Where their valley walls widen, as on a piedmont plain, glaciers spread laterally and thin abruptly. Basal erosion decreases dramatically and the trough floor rapidly rises to a terminal sill (Shoemaker, 1986).

Cirques.   At the head of each trough, in the accumulation area of the glacier that eroded the valley, are steep-headed, semicircular basins, or **cirques.** The bowl shape is a consequence of intense frost action and nivation processes adjacent to the ice field plus the flow lines in the zone of accumulation (Figs. 17–3 and 17–4) that carry surface rock debris downward toward the base of the glacier, where the bed is heavily abraded. A cirque has much the shape of a human heel print in sand and is often "down-at-the-heel," that is, more deeply eroded as a basin behind an abraded rock sill at the lip of the cirque. A cirque lake, or **tarn,** dammed by the rock sill, occupies the basin in many cirques.

Cirques are a unique and diagnostic feature of alpine glaciation and have been given many local names, indicating their recognition as landforms. In addition to cirque (French), other terms such as *cwm* (Welsh), *corrie* (Scottish), *kar* (German), *botn* (Norwegian), and *nisch* (Swedish) are encountered in the geomorphic literature. The association of active cirques with the accumulation areas of modern valley glaciers led to the recognition of similar relict forms very early in the history of geomorphology. By nature of their position high on mountain slopes, many cirques that fed Pleistocene valley glaciers still carry small cirque glaciers or greatly reduced valley glaciers in them. Many cirques are compound, with a nested series of progressively smaller and higher forms representing stages in the reduction of icefield size in late-glacial and postglacial time.

Cols, Arêtes, and Horns.   If a plateau or mountain range is lightly but incompletely dissected by glacial troughs and cirques, the former upland surfaces remain as remnants between the steep, concave glaciated forms (Fig. 17–7). This condition is well shown in the central Rocky Mountains and the Big Horn Mountains, where it has been described as a **scalloped upland,** or **biscuit-board** topography. The latter term describes the landscape well, by analogy to a thick layer of biscuit dough on a cutting board, with arcuate pieces removed by cirque erosion.

It is not hard to deduce the sequential development of a scalloped upland into a **fretted upland,** completely dissected by cirques. As two steep-headed cirques intersect, U-shaped notches cut the divide between their headwalls. Such a gap is called a **col** in alpine terminology. A saw-toothed divide consisting of a series of cols and intervening ridge segments is an **arête.** Where three or more cirques intersect, a higher pyramidal or triangular-faceted peak may rise above the general level of the arête. These peaks are the **horns** (or *Matterhorn peaks,* after the type example) of alpine scenery. A summit area of arêtes, cols, and horns represents the complete destruction of the preglacial surface (Fig. 17–5), and is analogous to the mature dissection of a regional fluvial landscape. The most prominent horns are at the intersection of three or more radiating arêtes.

Through Valleys and Breached Divides.   If mountain glaciation increases in intensity until a mountain ice cap develops, former regional divides may be breached by glacial erosion, and major diversions of postglacial stream flow result. The Brooks Range of Alaska was glaciated more intensely on the southern, snowier side of the regional divide, whereas only short valley or piedmont glaciers formed on the arid northern Arctic slope. Headward erosion by the south-flowing glaciers breached the preglacial regional divide in several places, and major ice streams that originally flowed southward on relatively gentle profiles were diverted down the steeper northward troughs. Breached divides of this sort become strongly abraded

Figure 17-7.   A scalloped upland on a broad, gentle divide (foreground) evolving by intersection of opposing cirques into a fretted upland of arêtes, horns, and cols at higher altitudes (background).

and low. In the Brooks Range, the divide at Anaktuvuk Pass, between the Yukon and Arctic drainage, is only slightly over 600 m above sea level and is on a swampy valley floor among moraines (Porter, 1966).

Similar breached divides in the glaciated northern Appalachian Mountains were produced by multiple advances of the Pleistocene ice sheets that overrode the preglacial divides between the St. Lawrence–Great Lakes drainage and the southerly draining Susquehanna, Delaware, and Ohio Rivers. Powerful erosion through constricted valleys at the base of the ice sheets produced the Finger Lakes troughs (Fig. 17–6) on the north side of the divide but left flat-gradient, swampy **through valleys** between north- and south-flowing tributaries (Tarr, 1905). The present divide is inferred to be close to the preglacial position in most areas, but valley forms have been modified into a series of intersecting troughs "like the intersecting passages of some unroofed ants' nest" (Clayton, 1965, p. 51; see also Linton, 1963; Coates, 1966). Undoubtedly, much of the erosion of the through valleys can be attributed to meltwater overflowing into the southerly drainage from large ice-dammed lakes north of the divide at times when the ice margin blocked the St. Lawrence lowland. Each ice advance must have been preceded and followed by an intense period of fluvial erosion in meltwater channels across the divides.

Regional Denudation

Valley Glaciers.   Table 12–2 presents some evidence that the regional denudation rate in mountains with glaciers approaches or exceeds 1m/1000 yr, the most rapid of any morphogenetic region. The combination of high relief, steep slopes, intense nivation, and glaciers must indeed have a powerful impact on landscapes. The share of the work borne by glaciers alone is hard to evaluate, because glaciers are just one aspect of geomorphic processes in high snowy mountains. It seems certain that alpine glaciation, in which most of the upland is buried under a mountain ice cap from which valley glaciers radiate, must be one of the most rapid means of destroying mountains. Deductions concerning an alpine "cycle of erosion" were never satisfactory, because after the "fretted upland" stage of mature dissection, any further erosion must lower the ice fields toward or below the regional equilibrium line, reduce the accumulation areas, and thus reduce the effectiveness of glacial processes before a deduced endform can be reached. The **gipfelflur,** or "sea of peaks" all at about the same elevation, is a well-known feature of alpine scenery (Fig. 17–5), but whether the horns that rise above mountain ice fields represent remnants of an ancient erosion surface or are nothing more than the mean level of intersection of adjacent cirques and troughs cannot be determined. Also, in the Holocene Epoch (a postglacial or interglacial interval), glaciers all over the earth have greatly receded, and the high modern denudation rates of mountainous regions are at least partly due to the many oversteepened slopes relict from glacial erosion and deposition but now exposed to mass-wasting and fluvial erosion (Church and Ryder, 1972). Some authors have proposed a hypothesis of glacial protection, in which glaciation preserves a landscape by burying it and stopping fluvial erosion. That hypothesis might be considered for high-polar, dry-base ice sheets or ice caps (Andrews and LeMasurier, 1973, p. 80) but is hardly tenable for valley glaciers.

Ice Caps and Ice Sheets. An extreme range of views can be found concerning the effectiveness of large ice sheets in denuding landscapes. White (1972) argued that the Laurentide ice sheets of eastern North America completely stripped up to 1 km of Paleozoic and younger sedimentary rocks off large areas of the Canadian shield. Other authors (Ambrose, 1964) have noted that the buried relief of the crystalline shield, where it is unconformably overlain by lower Paleozoic or Cretaceous sedimentary rocks, is very similar to the local relief on the shield today. By this line of reasoning, multiple Pleistocene glaciations have done little more than to remove the weathered surface layers and exploit structural weaknesses by glacial plucking (Prest, 1971, p. 697). The oldest known glacial deposits in the central United States contain igneous and metamorphic rocks derived from the Canadian shield, proving that the shield was exposed at least in part prior to the onset of late Quaternary glaciation (Gravenor, 1975). The view of minimal regional denudation by ice sheets is supported by the rather small volume of glacial deposits on the periphery of the Laurentide ice sheet and by the fact that most of the glacial sediment was derived from relatively near the ice margin, especially on the southern quadrant in the United States. From the volume of glacial sediment on the North American continent, many authors have argued that a total of no more than 20 m was eroded from the Canadian shield during the Quaternary ice ages. Only 1 to 2 m of resistant shield rock was removed by the latest glaciation, although as much as 6 m might have been removed by the earliest ice sheets, which eroded the preglacial regolith (Kaszycki and Shilts, 1987, p. 158). Such rates are of the same order as the rates of fluvial erosion expected for shield terranes (Fig. 12–2). They imply that ice sheet erosion is not intense, and that ice sheets could even be regarded as "protection" from fluvial erosion. A less confident estimate is that a total of 120 m has been removed from the glaciated portion of North America by Pleistocene glacial erosion, and that most of the volume is now in the deep sea, not on land. This estimate is based on many assumptions of deep-sea sediment volumes, grain-size distribution, and age (Bell and Laine, 1985). Even so, it suggests that ice sheets may lower a landscape by a factor of 5 or 6 times faster than fluvial processes. Although not necessarily providing protection from fluvial processes, ice sheets do not seem to be dramatic erosive agents.

Discoveries of bauxite pebbles in glacial deposits on Martha's Vineyard, Massachusetts (Kaye, 1967), and deep weathering on Long Island, New York that is characteristic of the humid tropics (Liebling, 1973; Blank, 1978), support the inference that in preglacial times, definitely in the late Mesozoic Era and possibly in the late Tertiary Period, subtropical or tropical weathering affected the northeastern United States. Such weathering is known to affect rocks to depths of 100 m or more (Fig. 6–9), so it is possible that glacial erosion has accomplished little more than to remove the previously weathered rock from northeastern North America. The very large crystalline **erratics** derived from the Precambrian shield, but widely scattered in southern Canada and the glaciated United States east of the Rocky Mountains, need not represent erosion into fresh, unweathered rock; they could well be corestones (Fig. 11–3) from the preglacial weathered layer (Feininger, 1971). The control of stoss-and-lee landforms around Boston, Massachusetts by sheeting joints (Jahns, 1943), referred to earlier in this chapter, is also consistent with the hypothesis that glacial erosion by the Laurentide ice sheets removed much weathered rock but did not fundamentally reshape the bedrock relief. The nearly total lack of rock debris in the tabular icebergs derived from Antarctica suggests that glacial erosion is now inactive there (Warnke, 1970). However, the thick deposits of glacier-derived submarine sediments around Antarctica suggest that at some time, probably in the late Cenozoic Era, the Antarctic ice sheet eroded a considerable mass off the continent.

GLACIAL TRANSPORT AND DEPOSITION

Glaciers carry rock debris on, in, and beneath the ice. Valley glaciers probably derive most of their sediment loads from intense weathering and mass wasting on adjacent slopes. Rock debris that composes lateral and medial moraines seems to stay fairly high in the glacier cross section (Fig. 16–7). An unknown amount is actually eroded by the moving glacier. Flow lines in a valley glacier carry surface debris downward into the glacier in the zone of accumulation and upward to the surface in the zone of ablation (Fig. 16–8), so debris derived by mass wasting or by subglacial erosion can be thoroughly mixed. Fluvial sediment transport from beneath the margin of wet-base glaciers has largely replaced other mechanisms as the process most favored to deliver sediment to the margins of glaciers (Evanson and Clinch, 1985; Gustavson and Boothroyd, 1987). More and more, glacial and fluvial processes are being studied in conjunction with each other as a "glacial-fluvial system" analogous to the fluvial system introduced in Chapter 9 (Davidson-Arnot et al., 1982; Ashley et al., 1985; Gurnell and Clark, 1987).

Ice sheets that bury landscapes cannot gain sediment by mass wasting, because nothing is exposed above the ice. The Antarctic and Greenland ice sheets are notably free of any sediment other than minute amounts of windblown dust, except very near their bases. Neither ice sheet appears to be transporting much rock debris, but they may not be good models for the former ice sheets of Europe and North America, which expanded and shrank repeatedly during the Pleistocene Epoch.

Several traditional terms have been formalized to describe rock debris that is both in transport on modern glaciers and is inferred to have been transported by former glaciers. The oldest word is **moraine,** used to describe the hummocky or ridged depositional landscapes beyond the margins of alpine glaciers and also to describe the rock debris in and on glaciers today (Figs. 17–3 and 17–4). Some confusion results if the word moraine is used to describe both the material and the landforms, so only the latter meaning is used here. Moraine, then, is simply a collective term for depositional landforms of direct glacial origin. One possible classification of moraine and related glaciofluvial forms is outlined on the following pages.

For centuries before the glacial theory of Agassiz (1840), the poorly sorted, unconsolidated surficial deposits of northern Europe were supposed to have been the result of the biblical flood of Noah's time. In Germany, the term *diluvium* was widely used. In the British Isles, the same material was called *drift,* a term also used to describe marine sediment in transport along coasts. With the acceptance of the glacial theory to explain these deposits, **glacial drift** (or simply **drift** if no ambiguity is possible) has come to be the general English term for all sediment transported and deposited by glaciers or associated rivers. It is important not to confuse landforms and materials when studying glacial morphogenesis, so *drift* is used consistently to describe glacier-transported sediment, and *moraine* is used as a general term for glacier-depositional landforms.[1] The value of the distinction should become apparent in the following pages.

Glacial Drift

The following brief glossary defines some of the more common terms for categories of glacial drift. Details of composition, sorting, stratification, and genesis are left to specialized texts (Flint, 1971; Price, 1973; Embleton and King, 1975; Jopling and McDonald, 1975; Sugden and John, 1976; Evenson et al., 1983; Eyles, 1983; Drewry, 1986), but this vocabulary is necessary to appreciate the genesis and morphology of morainic and related glaciofluvial landscapes.

1. **Till** Nonsorted, nonstratified glacial drift, deposited directly from ice without reworking by meltwater. Called *boulder clay* by some British workers.
   a. *Basal,* or *lodgment,* till—Compact, tough, dense till, often clay-rich, deposited at the base of a temperate or wet-base glacier. Stones may show preferred orientation parallel to former flow. Basal till can also accumulate by melting at the glacier base (*melt-out till;* Drewry, 1986, pp. 130–33).
   b. *Ablation* till—Less consolidated than basal till, often lacking the finer grain sizes. Carried on the surface of melting ice and "let down" without being overridden. Also used for the cover of debris on the ablation zone of modern glaciers, although "ablation moraine" is also applied to the sheetlike form of that debris.
2. **Ice-contact stratified drift** Drift modified by meltwater during or after deposition. Considerable stratification and sorting is usual. Although water-modified, the large range of sizes, chaotic sedimentary structures, slump structures, faults, surface topography, and inclusions of till all confirm an origin in contact with, or in close proximity to, melting ice.
3. **Outwash** Alluvium deposited by meltwater not in close proximity to melting ice. Sorting and stratification are more regular than in ice-contact stratified drift. Similar to other coarse alluvium, except as influenced by the extreme variability of discharge from glacial sources.
4. **Lacustrine sediment** Lake-bottom deposits, usually fine grained, formed in settling basins on or near melting ice. Often forms silt and clay plains of broad extent, with distinctive erosional and soil-forming properties. Sometimes *varved,* or composed of rhythmic silt–clay couplets each representing an annual cycle of deposition.
5. **Eolian sediment** Sand dunes, sand sheets, and loess derived from other types of glacial drift. Genetically undifferentiated from other eolian sediments except by regional and temporal association with other drift.

---

[1] See the discussion on page 12 concerning the use of the terms *terrane* and *terrain.*

6. **Glacial marine sediment** Clay-rich, but often stony, marine deposits on shelves and deep-ocean floor. Characterized by angular, mechanically crushed fragments and anomalous coarse particles embedded in fine mud. Coarse fraction attributed to ice rafting.

Moraines and Other Constructional Drift Landforms

Some moraines and constructional glaciofluvial landforms are genetically related to a specific kind of drift. Examples are *till plains,* which by definition are low-relief areas of eroded or deposited till; *eskers,* which by origin in an ice tunnel or channel must be made of ice-contact stratified drift; or *outwash plains,* which are depositional plains of outwash. Other terms for constructional glacial landforms do not indicate the nature of the drift from which they are built. *End moraines* range in composition from entirely till to entirely ice-contact stratified drift: Almost all moraines contain some mixture of these materials.

In an attempt to sort out the nomenclature of moraines, Prest (1968) proposed a classification based on the orientation of the landforms with respect to ice movement (Table 17–2). The classification is intended for landforms built by continental ice sheets but includes forms built by valley glaciers as well. The scheme is vaguely similar to the genetic classification of sand dunes (Figs. 14–10 and 14–11) in that longitudinal (parallel to ice flow), transverse (perpendicular to ice flow), and nonoriented forms are included. The transverse forms are mostly ice-marginal deposits, built during the last recession of the ice margin, but include some forms that are shaped beneath ice during ice advance. The forms that parallel ice flow, by contrast, are nearly all produced by actively moving ice, although some form in stagnant, melting ice along channels and crevasses parallel to former flow directions. The nonoriented forms are commonly built of a mixture of ablation till and ice-contact stratified drift, during a time of regional melting with insignificant ice motion. Longitudinal forms are likely to predate transverse and nonoriented forms in the same region. For instance, younger end moraines or kame moraines may be

TABLE 17–2
Classification of Moraine[a]

Transverse to Ice Flow (Controlled Deposition)	Parallel to Ice Flow (Controlled Deposition)	Nonoriented (Uncontrolled Deposition)
Ground moraine—corrugated.	Ground moraine—fluted and drumlinized; includes drumlin, drumlinoid ridge, crag-and-tail hill.	Ground moraine—hummocky (low relief); includes some ablation moraine.
End moraine; includes terminal, recessional, and push moraine.		Disintegration moraine—hummocky and/or pitted (mainly high relief); includes dead ice, stagnation, collapse, and ablation moraine.
Ice-thrust moraine.	Marginal and medial moraine.	
Ribbed moraine.		
De Geer moraine; includes cross-valley moraine.	Interlobate and kame moraine.	
Interlobate and kame moraine.	Linear ice-block ridge; includes ice-pressed and ice-slump ridge, and crevase filling.	Interlobate and kame moraine.
Linear ice-block ridge; includes ice-pressed and ice-slump ridge, and crevasse filling.		Irregular ice-block ridge and rim ridge; includes ice-pressed and ice-slump ridge, and ablation slide moraine.

Source: Prest (1968).

[a] Included are landforms constructed from till or other drift that was deposited directly from, or in association with, glacier ice.

banked against, or draped over, drumlins and fluted ground moraines. Some of the morainic terms apply equally well to transverse, longitudinal, and nonoriented forms. For example, drift pressed into the base of a crevassed and melting ice sheet may produce ridges oriented in several directions or produce a nonoriented maze of ridges and mounds. Drift is a notoriously mixed sediment; within a single moraine (Fig. 17–8), deltaic structures, lodgment till, and deformed ice-contact stratified drift may be found together.

Regional patterns are helpful in interpreting glacial landforms, because the terrane over which the glacier moved and the ablation history largely determine the regional character of both the drift and the landforms. Many morainic forms can be assembled in a genetic sequence that was called a *glacial series* (Fig. 17–8) by early workers (Penck and Brückner, 1909; von Engeln, 1921). The principle is a good one. At any one time during the uncovering of a landscape by ice, an end moraine may be forming at the ice margin, fronted by an outwash plain. Ice tunnels and crevasses behind the moraine may be the sites of eskers, crevasse fillings, and kames. Meltwater may be traced through an eroded channel across the end moraine to a fan or delta that coalesces with the outwash plain in front of it. The ice-contact, or *proximal,* part of the end moraine is of till or ice-contact stratified drift. On the *distal* or outwash side of the end moraine, the drift is mostly outwash, with inclusions of slumped ice-contact stratified drift or till. The entire assemblage represents contemporary events and, when understood as a unit, gives a good concept of the activity along a receding glacier terminus.

Morainic Landforms Constructed Primarily of Till. Many of the ice-parallel morainic landforms are streamlined, and when developed in clusters give a vivid impression of ice motion. Similar erosional forms, such as roches moutonées and crag-and-tail hills have been described previously. Best known of the streamline ice-molded forms are **drumlins** (Muller, 1974). These are elliptical or ovoid hills, blunt on the upglacier end, with an elongate downglacier tail (Fig. 17–9). They occur in clusters, most frequently in areas of clay-rich or sandy basal till. Internal composition is not a criterion for calling a streamlined hill a drumlin, and known examples range from those that are nearly all rock with a thin veneer of till to those that are

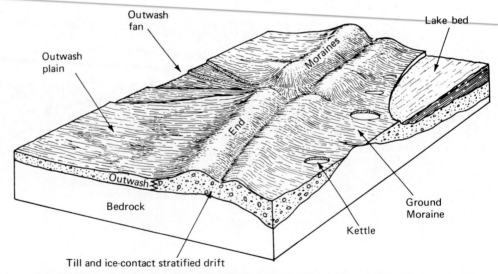

Figure 17-8. The concept of a glacial series. A group of morainic and related glaciofluvial landforms are either synchronous or sequential in time and are genetically related.

entirely till with no rock core. Classic examples are to be seen in northern Ireland, New Brunswick, eastern Massachusetts, southern Ontario and central New York, and southern Wisconsin. In each region, hundreds or thousands of drumlins give the landscape a distinct "grain," or linear trend. The drumlins tend to be of comparable size and shape within each region and show evidence of harmonic spacing (Reed et al., 1962). Table 17–2 lists as variants such forms as fluted (ridged) ground moraine and elongate "drumlinoid" ridges.

An hypothesis of "dilatant flow" within water-saturated basal till has been proposed to explain drumlin formation (Smalley and Unwin, 1968; Smalley, 1981). A plastic mixture of rock debris, ice, and water, similar to the hypothesized plastic scouring substance beneath wet-base glaciers (p. 416), could flow as a viscous fluid as long as it retains water. If, however, an increased pressure region upglacier of a bedrock obstruction or even of a large stone within the mass were to cause local dewatering, the mixture would stiffen and lodge. The initial obstruction would be enlarged by the mass of rigidified till and continue to grow to some equilibrium streamlined form or until another obstruction in the vicinity caused a similar response. The hypothesis has the advantage of explaining the size, shape, compositional variation, and spatial distribution of drumlins. The process requires a wet-base glacier, which is also consistent with the usual abundance of ice-contact stratified drift in drumlin fields. Local pressure and temperature gradients at the base of the ice, associated with extending and compressing flow, may localize drumlin formation in certain zones broadly parallel to the ice margin. One stimulating hypothesis is that drumlins are the streamlined erosional remnants of subglacial meltwater floods, analogous to the mid-channel bars of giant paleofloods (p. 250) (Shaw, 1989).

Figure 17-9.   Drumlin field southeast of Lake Athabaska, Saskatchewan, Canada. Ice flowed from the northeast. Also shown are numerous kettle lakes and a system of anastomosing eskers. Low-angle illumination is from the southeast. If the drumlins appear inverted, view the picture upside down. (Photo: Surveys and Mapping Branch, Dept. of Energy, Mines, and Resources, Canada.)

Glacial Landforms Constructed Primarily of Ice-Contact Stratified Drift. A great variety of distinctive landforms are made of ice-contact stratified drift. Oriented and nonoriented forms are both common. The terminology is complicated, because various terms can be applied to either single landforms or groups of forms. Some landforms are parts of moraines; many are not.

Kame is representative of the complexities of this category of landforms. By derivation, a kame is an isolated hill or mound of stratified drift deposited in an opening within or between ice blocks (Fig. 17–10). If the opening was between ice and a valley wall, the resulting form can be called a **kame terrace. Kame deltas** are flat-topped hills with an ice-contact proximal face and a semicircular plan view, deposited in standing water at an ice margin. A **moulin kame** is believed to form under stagnant ice at the base of a *moulin* or glacier-karst swallow hole (p. 408). It is clear that a variety of environments adjacent to melting glaciers can produce kames.

Another important landform of ice-contact stratified drift is the **esker** (Fig. 17–9). An esker is a sinuous or meandering ridge of stratified sand and gravel, commonly with very well-rounded pebbles and cobbles. Eskers form in open channels on glaciers and in tunnels in or beneath them. If large quantities of sediment-laden meltwater are discharging through a karstlike hydrologic system within a glacier, the sediment may be deposited on the floors of various tunnels or channels while the ice roof or walls are widened or raised by melting. Eskers show sinuosities comparable to the dimensions of rivers and clearly show by internal structures that they are water-laid. However, they cross bedrock ridges, rising and falling as they are traced across moraine landscapes. Confined flow must be invoked to permit them

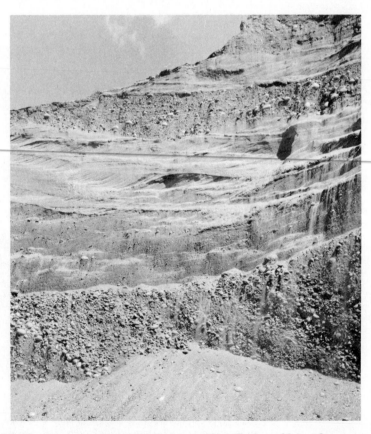

Figure 17-10. Internal structure of a kame near West Durham, Maine. Coarse gravel foreset beds at bottom were deposited by a prograding delta among ice blocks. Overlying glaciofluvial sand and gravel were probably deposited by an aggrading meltwater river.

to be deposited on uphill gradients. The esker systems of northeastern United States, eastern Canada, and southern Finland and Sweden are noteworthy. Most trend broadly perpendicular to major end moraines and parallel to former ice flow, probably because water flowed under hydrostatic head toward the ice margin (Shreve, 1985).

A **kettle** is a basin of nondeposition within morainic terrain formed where an ice block was buried by till or ice-contact stratified drift and was then melted. Many

Figure 17-11. Hypothetical sequence of events in the evolution of a flat-topped kame. (a) Cover of ablation till over stagnant ice. (b) Till slumps and washes into deepening crevasses, exposing ice on a chaotic ablation surface. (c) Glaciofluvial sediment washed into a depression when ice melted, later became the crest of a kame, surrounded by a rim of till. Other ice-contact stratified drift deposits are slumped and tilted by final melting.

Figure 17-12. View upstream over the braided valley train of the Tasman River, South Island, New Zealand. Tasman Glacier, mostly covered with ablation drift, in right distance. Mount Cook is high peak on the left. (Photo: V. C. Brown & Son, Photographers.)

areas of ground moraine are chaotic mazes of kettles among kames and **crevasse fillings,** the latter representing sediment castings made in ice molds which then melted. Extremely complex internal structures, representing multiple episodes of melting and drift deposition, characterize many areas of kames and kettles (Fig. 17-11). Small kettles develop angle-of-repose side slopes and circular outlines very similar to those of collapse dolines (Fig. 7-10). Larger ones are complexly segmented (Fig. 17-9). Kettle lakes are favorite scenic attractions and a fisherman's paradise. Thoreau's Walden Pond may be the best known.

       Glacial Landforms Constructed Primarily of Outwash.   Meltwater alluvium aggrades the floors of glaciated valleys downstream from melting glaciers. Meltwater streams have high gradients and braided patterns appropriate for their massive bed loads and fluctuating discharges. Where outwash is confined between valley walls downstream from a glacier, the aggraded surface is called a **valley train** (Fig. 17-12). On the margin of an ice sheet in areas of low relief, broad **outwash plains** form from confluent alluvial fans that emerge from multiple meltwater valleys through end moraines. The Icelandic term **sandur** (plural: sandr) is sometimes applied to outwash plains. The Icelandic sandr are notable for major episodic floods that inundate hundreds of square kilometers during events called *jökulhlaups* ("glacier bursts"), when ice dams are floated by subglacial water pressure and release masses of impounded water from the margins of mountain ice caps. Active valley trains and outwash plains are excellent source areas for loess (p. 365), because they carry massive sediment loads, but may nearly dry up at night or during winter, when melting is slower.

INDIRECT MORPHOGENETIC EFFECTS OF GLACIATION

Postglacial Isostatic Recovery

Sound theory and abundant evidence (Fig. 17-13) support the conclusion that ice caps larger than 50 km in diameter isostatically depress the earth's crust (p. 27). The response time of the earth to the removal of an ice sheet is slow enough so that, following deglaciation, the land surface continues to rise for more than 10,000 years at an exponentially decreasing rate until isostatic adjustment is complete. The Laurentide and Scandinavian ice sheets may have been 3000 m thick at their maximum (Fig. 17-14), so the potential isostatic recovery is an impressive 1000 m. However, depending on the rate and nature of deglaciation, much of the recovery may have been completed prior to the land being uncovered by ice (Andrews, 1970; Fulton, 1989). Uplift continues today at a maximum known rate of 9 mm/yr near the north end of the Baltic Sea (Fig. 17-15).

    The most obvious geomorphic effects of postglacial isostatic uplift are to be found on coasts (Chaps. 19 and 20), where emergence so dominates that no shoreline has a chance to develop large erosional or depositional forms before it is uplifted beyond the reach of waves (Blake, 1975). River gradients must be affected by the tilting associated with uplift, and lakes or swamps may have formed or drained, but the extreme deranged drainage patterns of glaciated regions make the subtleties of tilting hard to recognize. Rivers flowing away from the center of uplift should show simultaneous rejuvenation as their gradients are increased by tilting; rivers flowing toward the center of maximum uplift, as do many of the rivers that drain into Hudson Bay, should have their gradients reduced by the greater differential uplift of their lower reaches; this contributes to the abundance of lakes and swamps in glaciated regions. The differential tilt on postglacial shorelines is typi-

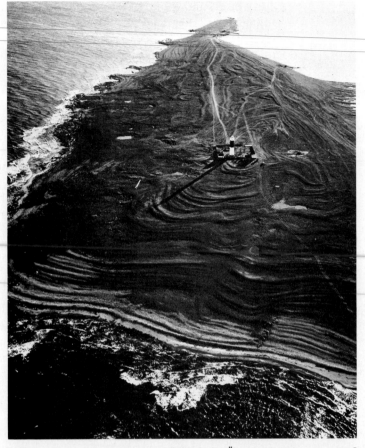

Figure 17-13. Emerged Holocene beach ridges on Östergarnsholm, eastern Gotland, Sweden. The contemporary uplift rate is about 2 mm/yr (Fig. 17-15). (Photographer and copyright holder: Arne Philip, Visby, Sweden. Courtesy IGCP Project Ecostratigraphy.)

cally of the order of 1 m/km, or 0.1 percent, which is comparable to the gradient of many river segments, so the effect on stream hydrology cannot be neglected.

The Great Lakes of North America form a natural laboratory for studying postglacial uplift and tilting (Teller, 1987). Abandoned shorelines and contemporary water-level gages prove that the lakes are deepening on their southern shores and shoaling on the northern. For instance, Lake Michigan and Lake Huron now form a continuous water level through the very deep Straits of Mackinac. The

Figure 17-14. Estimated surface profile of the Fennoscandian ice sheet along a line eastward from Trondheim, Norway, to Helsinki, Finland. Note the very large area of ice that was grounded below sea level in the Baltic basin. (Robin, 1964, Fig. 7.)

Figure 17-15. Contemporary emergence rates in the Baltic basin and vicinity (mm/yr). Emergence is relative to sea level, which is also rising 1 mm/yr or more. (Flint, 1971, Fig. 13-2.)

northeastern parts of the combined lakes are now rising about 30 cm/100 yr relative to their southwestern parts. Furthermore, the outlet of the combined lakes near Detroit is rising at about 7.5 cm/100 yr relative to a former glacial overflow at Chicago. The discharge of Lake Michigan will be diverted naturally from the St. Lawrence system to the Mississippi in the next 3200 years if control dams in the Chicago Sanitary and Ship Canal do not prevent the changes (Clark and Persoage, 1970, p. 632). The magnitude of the potential change suggests that widespread regional drainage diversions are possible by postglacial uplift. However, it should be emphasized that the uplift is a *recovery*, or a return to a previous condition, and not the equivalent of epeirogenic uplift.

Glacial Eustatic Fluctuations of Sea Level

The volume of water abstracted from the sea at times of ice-sheet expansion and returned to the sea during interglacial intervals of the Pleistocene Epoch has had a worldwide effect on morphogenesis. Global changes of sea level are called **eustatic,** meaning that the entire ocean is affected; changes due to fluctuating ice volume are called **glacial eustatic.** Estimates of the maximum glacial eustatic lowering of sea level during the latest full glacial interval, as compared to today, vary from 80 to 160 m, with the best estimate being about 120 m (Fairbanks, 1989). The lowering was enough to lay bare about one-half of the continental shelf area so that all rivers had to extend and deepen their valleys to the lower base level. At present, all rivers enter the sea either in estuaries or across thick deltaic fills. Both features are a direct consequence of glacial eustatic sea-level fluctuations but are so much a part of the Quaternary fluvial geomorphic system that we find it hard to deduce what river mouths would be like if sea level had been stable for the last few million years

(Chap. 10). The nickpoints produced during glacial low sea level have migrated up-stream various distances and are generally buried under postglacial alluvium, so they are not part of the modern valley morphology. However, it is a useful mental exercise to deduce the hypothetical morphology of any major river valley during glaciation, with the assumption that sea level was 120 m lower than present (compare with Fig. 10–2). During glacial intervals, large paired terraces must have flanked the gorgelike lower valleys of many rivers.

Glacial eustatic fluctuations of sea level have affected all the coastlines of the earth, with much greater impact than postglacial uplift which principally affected only the deglaciated areas. The rising sea level of the last 15,000 years has drowned all coasts except those with rapid glacial-isostatic or other tectonic uplift. During the most rapid phase of deglaciation, from about 10,000 to 7000 years ago, sea level may have risen at a rate of 10 mm/yr. The consequences of this submergence to coastal morphology are considered at length in Chapters 19 and 20. It can be noted here that the sea now appears to be only a few meters below its level of the last inter-glacial, about 120,000 to 140,000 years ago. Many coastal landforms that on first glance seem to be in equilibrium with modern processes have been shown to be rel-ict from the last interglacial and only recently reoccupied by the sea. Many mor-phologic features of lower river valleys may be equally suspect of having been inherited from an earlier time.

CHRONOLOGY OF LATE CENOZOIC GLACIATION

To appreciate the impact of glacial morphogenesis on the earth, a review of the chronology of glaciations is required. Most of the earth's landforms have taken shape during late Cenozoic time, when ice sheets had formed in Antarctica and pos-sibly elsewhere in mountainous regions.

Evidence of glaciers in West Antarctica can be traced back at least 27 million years. Hyaloclastites (p. 79) of that age are exposed above the ice plateau of Marie Byrd Land but were erupted under ice. Thus an ice cap probably covered the vol-canic archipelago of West Antarctica as early as Late Oligocene time and perhaps earlier (LeMasurier, 1972). By 25 to 30 million years ago, Australia had crossed 50°S latitude as it moved toward the equator (p. 487), the Circum-Antarctic Cur-rent had developed (Kennett et al., 1974), and the large East Antarctic plateau was probably ice-covered. At that time, the modern global atmospheric and oceanic cir-culation patterns began to develop.

Between about 14 and 7 million years ago, the Andes Cordillera of South America rose, forming a powerful meridianal barrier to southern hemisphere at-mosphere circulation. In the last 10 million years, tectonic uplift of western North America and southern Asia has created an atmospheric circulation pattern condu-cive to major climatic change in the northern hemisphere, with the mid-latitude winter westerly winds dipping south over the Atlantic and Pacific Oceans and swinging far to the north around the cold high pressure cells over the Asian and North American continental interiors (Ruddiman et al., 1990). In addition, the ris-ing Himalaya Mountains and the Tibetan Plateau climatically isolated southern Asia from the continental interior, increasing the intensity of the seasonal wet-dry monsoonal climate of the Indian Ocean region (Ruddiman et al., 1989).

The Greenland ice sheet may have formed 3 million years ago in part because of the emergence of the isthmus of Panama and the development of the Gulf Stream (Keigwin, 1982). Microfossil assemblages and oxygen-isotope ratios of deep-sea sediments in the northeastern Atlantic Ocean show that between about 3.4 and 2.0 million years ago, the water changed from warm-temperate to subpolar in a series of steps (Loubere and Moss, 1986). Glaciation is documented in Iceland at

3.1 million years ago, but major horizons of ice-rafted sand in a deep-sea core south of Iceland and west of Ireland are abundant only during the last 2.4 million years. Isotope compositions show that the earliest documented Pliocene glaciation around the northern Atlantic had a magnitude fully comparable to the better known middle and late Pleistocene glaciations (Shackleton et al., 1984).

The chronology of the late Cenozoic glaciations of North America and Europe is not established, although rapid progress is now being made by fission-track and potassium–argon dating on volcanic tephras that are interstratified with drift (p. 92). In particular, several fine-grained tephras known collectively as "Pearlette Ash" are widespread on the Great Plains of the United States from South Dakota to Texas. Formerly thought to be a single ash bed of late Kansan or early Yarmouth age (Flint, 1971, pp. 549–51), the tephras at numerous localities give ages that cluster in three groups. The oldest dates of about 1.9 million years are from tephra that presumably is younger than the Nebraskan glaciation, the oldest of the named glaciations in central North America. Other ash beds with ages of 1.2 million years and 0.6 million years probably bracket the second oldest, or Kansan, glaciation. All three groups of ashes predate the Illinoian glaciation (Naeser et al., 1973; Boellstorff, 1974). Early Pleistocene outwash terraces in the Rhine valley of Europe are also overlain by volcanic ash about 0.6 million years old.

The radiometric dates summarized above demonstrate that substantial regions experienced glacial morphogenetic conditions well before the Pleistocene Epoch (the traditional "Ice Age") began, about 1.6 million years ago. The traditional subdivision of the Pleistocene Epoch into four glacial ages with three intervening interglacial ages is obsolete, and will certainly be greatly revised in the near future (Richmond and Fullerton, 1986).

A major glaciation, in many regions the most extensive of all, is called the Illinoian in North America and the Riss or Saale in Europe. It began less than 0.5 million years ago and consisted of several advances separated by local retreats but is not yet adequately dated.

There are numerous botanical, pedological, and geological lines of evidence that the next youngest recognized event in the glacial-interglacial sequence was an interglacial, called the Sangamon in North America and the Riss/Würm or Eem in Europe. Sea level was a few meters above present, spanning or recurring during the interval from at least 140,000 to 120,000 years ago. Numerous coral reefs grew above present sea level at that time, and the eroded bases of these reefs are the foundation of modern reefs (Chap. 19). In areas of strong tectonic uplift, prominent coastal terraces record this latest interglacial interval (Fig. 2–3). The continents were somewhat warmer than present. Deciduous forests extended into what is now the conifer zone north of the Great Lakes, for instance. The required warmth is consistent with less glacier volume than today and with a sea level a few meters higher than present. Therefore, part or all of the interval of 140,000 to 120,000 years ago, dated primarily from emerged coral reefs, is widely and confidently correlated with the last interglacial.

Sometime slightly less than 120,000 years ago, the earth's climate cooled sharply, and the most recent major glaciation began. Sea level fell in a series of pulses, separated by brief intervals of high sea level that were nevertheless well below present sea level (Fig. 17–16). The entire interval from about 115,000 to about 10,000 years ago is now called the Wisconsin glaciation in North America and the Weichsel or Würm glaciation in Europe. It consisted of short periods of ice advance, called **stadials,** separated by periods of retreat but colder than an interglacial, called **interstadials.** The interstadials are dated by coral reef growth that produced terraces on rising orogenic coasts (Figs. 2–3 and 17–16) (Bloom et al., 1974) and in glaciated regions by radiocarbon ages of organic deposits between the younger drifts. Each stadial seems to have been more severe than the previous, cul-

Figure 17-16. Estimate of the interstadial high sea levels of the last 105,000 years, based on an assumed sea level of +6 m at 124,000 years ago. (Bloom and Yonekura, 1985, Fig. 6-4 and references therein.)

minating in the maximum advance of the northern hemisphere ice sheets only about 15,000 to 18,000 years ago. The outermost moraines of the last glaciation around the periphery of the Laurentide and Scandinavian ice sheets all fall within that time range.

The worldwide cooling during the Wisconsin glacial maximum was not extreme. The average change in summer sea-surface temperature 18,000 years ago was only −2.3°C (CLIMAP Project members, 1976; Gates, 1976). However, the North Atlantic surface water was as much as 18° colder, the western North Pacific was 10°C colder, and the equatorial Pacific was 6°C colder than present summer temperatures (Fig. 17–17). The extreme cooling of the North Atlantic correlates well with the morphogenetic evidence of glacial and periglacial conditions in eastern North America and northwestern Europe (p. 372).

Deep ice cores from the central domes of the East Antarctic Ice Sheet have been isotopically analyzed to give estimates of full-glacial conditions there (Jouzel et al., 1989). The mean annual surface temperature was about 9°C colder than the present −53 to −55°C. Precipitation may have been only 50 percent of the present very low amount. The ice sheet may have thinned by a few hundred meters. More atmospheric dust was trapped in the slowly accumulating ice, perhaps reflecting more arid conditions on nearby continents and wide continental shelves exposed by lower sea level. The CO_2 content of air bubbles trapped in the ice is 80 parts per million less than that of the modern atmosphere (Barnola et al., 1987), a natural change comparable in magnitude but opposite in sign to the increase of the last two centuries, which is attributed to the combustion of fossil fuels during the industrial revolution. The Antarctic ice cores hold a priceless record of past climate change by which human beings can measure their impact on the earth's climate.

The marginal retreat of the northern hemisphere Wisconsin/Weichsel ice sheets was dramatic. By 11,000 years ago, the ice margins were fluctuating but receding north of the Great Lakes (Teller, 1987) and in the Baltic Sea. The Gulf Stream had diverted back toward northern Europe from its glacial-maximum track toward Portugal. By 9000 years ago, the Scandinavian ice sheet was separating into small highland residuals in northern Sweden, and the sea was entering Hudson Bay, calving the thickest part of the complex Laurentide ice sheet and radically changing the flow directions of the ice that remained on the highlands east and west of the Bay (Fulton, 1989). By 6500 years ago, the remnant Laurentide ice sheets were gone, sea level was within a few meters of its present level, and the climate was approximately as today.

In the past 6000 years there have been several climatic oscillations that have affected geomorphic and pedologic processes, but they are not comparable to the glacial-interglacial changes. Perhaps they have involved a temperature range of 1°C in the middle latitudes, in contrast to the 8 to 10°C difference between glacial and postglacial climates in the same regions. Between 6000 and 3000 years ago, most areas were warmer and drier than today. This was the **Hypsithermal Interval** (Deevey and Flint, 1957). Most of the cirque glaciers in the United States and Canadian Rockies and the European Alps were gone, as were most of the small mountain ice caps in Norway. The valley glaciers along the Alaskan coast retreated far up their valleys. Then, in a cooling trend with a series of minor oscillations, the climate approached present conditions, and valley glaciers expanded somewhat. The last century has seen a dramatic retreat of valley glaciers from the moraines built during the last times of expansion between A.D. 1650 and 1850. Systematic geomorphic study of glaciers has been done almost entirely during an interval of glacier retreat and negative glacier budgets. We know very little about the events that accompany glacial advances (except when caused by surges).

The causes of glaciations and of climatic change in general are unknown. Certain periodicity, as in Figures 17–16 and 18–1, suggests that planetary orbital parameters might be involved. Certainly, glaciation in middle latitudes was accompanied by lower temperatures over the entire earth (Fig. 17–17). As is discussed in Chapter 18, there is less information about climatic change in the low latitudes or nonglaciated regions than in glaciated regions.

The boundaries of glaciated regions can be determined to within a kilometer or less, so it has been relatively easy to define the glacial morphogenetic region. Not at all as simple are the tasks of defining the former extent of the periglacial zone and of identifying the effects of alternating or changing morphogenetic regions outside of the glacial limits. There is abundant evidence that climates have changed during the late Cenozoic Era while the modern landscape has taken shape. We use the glacial chronology, as imperfect as it is, as a framework for the chronology of morphogenetic change in other regions, but many mistakes have been made, and many assumptions have been proved false. An urgent need now is to date and measure the magnitude of the past changes in the earth's climate as a base line for monitoring the effects of human intervention. The possibility that human-induced climatic warming could melt enough glacier ice to raise sea level a few more meters, and thereby trigger a catastrophic surge of the West Antarctic Ice Sheet over the next few centuries (p. 409) is now the focus of a massive international research effort.

REFERENCES

AGASSIZ, L., 1840, Études sur les glaciers: privately published, Neuchâtel, Switzerland, 346 pp.

ALLEY, R. B., 1989, Water-pressure coupling of sliding and bed deformation: II. Velocity-depth profiles: Jour. Glaciology, v. 35, pp. 119–29.

———, BLANKENSHIP, D. D., ROONEY, S. T., and BENTLEY, C. R., 1989, Water-pressure coupling of sliding and bed deformation: III. Application to Ice Stream B, Antarctica: Jour. Glaciology, v. 35, pp. 130–39.

AMBROSE, J. W., 1964, Exhumed paleoplains of the Precambrian shield of North America: Am. Jour. Sci., v. 262, pp. 817–57.

ANDREWS, J. T., 1970, Geomorphological study of post-glacial uplift with particular reference to Arctic Canada: Inst. Brit. Geographers Spec. Pub. 2, 156 pp.

——— and LEMASURIER, W. E., 1973, Rates of Quaternary glacial erosion and corrie formation, Marie Byrd Land, Antarctica: Geology, v. 1, pp. 75–80.

ASHLEY, G. M., SHAW, J., and SMITH, N. D., eds., 1985, Glacial sedimentary environments: Soc. Econ. Paleontologists and Mineralogists Short Course No. 16, 246 pp.

BARNOLA, J. M., RAYNAUD, D., KOROTKEVICH, Y. S., and LORIUS, C., 1987, Vostok ice core provides 160,000-year record of atmospheric CO_2: Nature, v. 329, pp. 108–14.

BELL, M., and LAINE, E. P., 1985, Erosion of the Laurentide region of North America by glacial and glaciofluvial processes: Quaternary Res., v. 23, pp. 154–74.

BLAKE, W., JR., 1975, Radiocarbon age determinations and postglacial emergence at Cape Storm, southern Ellesmere Island, Arctic Canada: Geograf. Annaler, v. 57, ser. A, pp. 1–71.

BLANK, H. R., 1978, Fossil laterite on bedrock in Brooklyn, New York: Geology, v. 6, pp. 21–24.

BLOOM, A. L., BROECKER, W. S., CHAPPELL, J. M. A., MATTHEWS, R. K., and MESOLELLA, K. J., 1974, Quaternary sea level fluctuations on a tectonic coast: New $^{230}Th/^{234}U$ dates from the Huon Peninsula, New Guinea: Quaternary Res., v. 4, pp. 185–205.

BLOOM, A. L. and YONEKURA, N., 1985, Coastal terraces generated by sea-level change and tectonic uplift, in Woldenberg, M. J., ed., Models in geomorphology: Allen & Unwin, Inc., Winchester, Mass., pp. 139–54.

BOELLSTORFF, J., 1974, Fission-track ages of Pearlette family ash beds: Comment: Geology, v. 2, p. 21.

BOULTON, G. S., 1974, Processes and patterns of glacial erosion, in Coates, D. R., ed., Glacial geomorphology: State Univ. of New York Publications in Geomorphology, Binghamton, N.Y., pp. 41–87.

CHURCH, M., and RYDER, J. M., 1972, Paraglacial sedimentation: A consideration of fluvial processes conditioned by glaciation: Geol. Soc. America Bull., v. 83, pp. 3059–72.

CLARK, R. H., and PERSOAGE, N. P., 1970, Some implications of crustal movement in engineering planning: Canadian Jour. Earth Sci., v. 70, pp. 628–33.

CLAYTON, K. M., 1965, Glacial erosion in the Finger Lakes region (New York State, U.S.A.): Zeitschr. für Geomorph., v. 9, pp. 50–62 (discussion, v. 10, pp. 475–77).

CLIMAP PROJECT MEMBERS, 1976, Surface of the ice-age earth: Science, v. 191, pp. 1131–37.

COATES, D. R., 1966, Discussion of K. M. Clayton, "Glacial erosion in the Finger Lakes region (New York State, U.S.A.)": Zeitschr. für Geomorph., v. 10, pp. 469–74.

CRARY, A. P., 1966, Mechanism for fiord formation indicated by studies of an ice-covered inlet: Geol. Soc. America Bull., v. 77, pp. 911–29.

DAVIDSON-ARNOTT, R., NICKLING, W., and FAHEY, B. D., eds., 1982, Research in glacial, glacio-fluvial, and glacio-lacustrine systems: Geo. Books, Norwich, England, 318 pp.

DEEVEY, E. S., and FLINT, R. F., 1957, Postglacial Hypsithermal interval: Science, v. 125, pp. 182–84.

DREWRY, D., 1986, Glacial geologic processes: Edward Arnold (Publishers) Ltd., London, 276 pp.

EMBLETON, C., and KING, C. A. M., 1975, Glacial geomorphology: John Wiley & Sons, Inc., New York, 573 pp.

VON ENGELN, O. D., 1921, The Tully glacial series: N.Y. State Museum Bull. 227–228, pp. 39–62.

EVENSON, E. B. and CLINCH, J. M., 1985, Debris transport mechanisms at active alpine glacier margins: Alaskan case studies, in Kujansuu, R., and Saarnisto, M., eds., INQUA till symposium, Finland 1985: Geol. Survey Finland, Spec. Paper 3, pp. 111–36.

EVENSON, E. B., SCHLÜCHTER, C. H., and RABASSA, A., eds., 1983, Tills and related deposits: A.A. Balkema, Rotterdam, 454 pp.

EYLES, N., ed., 1983, Glacial geology: An introduction for engineers and earth scientists: Pergamon Press, Elmsford, N.Y., 409 pp.

FAIRBANKS, R. G., 1989, A 17,000-year glacio-eustatic sea level record: Influence of glacial melting rates on the Younger Dryas event and deep-ocean circulation: Nature, v. 342, pp. 637–42.

FEININGER, T., 1971, Chemical weathering and glacial erosion of crystalline rocks and the origin of till: U.S. Geol. Survey Prof. Paper 750-C, pp. 65–81.

FLINT, R. F., 1971, Glacial and Quaternary geology: John Wiley & Sons, Inc., New York, 892 pp.

FULTON, R. J., ed., 1989, Quaternary geology of Canada and Greenland: Geol. Soc. America, The geology of North America, v. K-1, 839 pp.

GATES, W. L., 1976, Modeling the ice-age climate: Science, v. 191, pp. 1138–44.

GJESSING, J., 1966, On "plastic scouring" and "subglacial erosion": Norsk Geografisk Tidsskrift, v. 20, pp. 1–37.

GRAVENOR, C. P., 1975, Erosion by continental ice sheets: Am. Jour. Sci., v. 275, pp. 594–604.

GURNELL, A. M., and CLARK, M. J., eds., 1987, Glacio-fluvial sediment transfer: An alpine perspective: John Wiley & Sons Ltd., Chichester, England, 524 pp.

GUSTAVSON, T. C., and BOOTHROYD, J. C., 1987, A depositional model for outwash, sediment sources, and hydrologic characteristics, Malaspina Glacier, Alaska: A modern analog of the southeastern margin of the Laurentide Ice Sheet: Geol. Soc. America, v. 99, pp. 187–200.

GUTENBERG, B., BUWALDA, J. P., and SHARP, R. P., 1956, Seismic explorations on the floor of Yosemite Valley, California: Geol. Soc. America Bull., v. 67, pp. 1051–78.

HALLET, B., 1979, A theoretical model of glacial abrasion: Jour. Glaciology, v. 23, pp. 39–50.

———, 1981, Glacial abrasion and sliding: Their dependence on the debris concentration in basal ice: Ann. Glaciology, v. 2, pp. 23–28.

HOOKE, R. L., WOLD, B., and HAGEN, J. O., 1985, Subglacial hydrology and sediment transport at Bondhusbreen, southwest Norway: Geol. Soc. America Bull., v. 96, pp. 388–97.

JAHNS, R. H., 1943, Sheet structure in granites: Its origin and use as a measure of glacial erosion in New England: Jour. Geology, v. 51, pp. 71–98.

JOPLING, A. V., and McDONALD, B. C., eds., 1975, Glaciofluvial and glaciolacustrine sedimentation: Soc. Econ. Paleontologists and Mineralogists Spec. Pub. 23, 320 pp.

JOUZEL, J., RAISBECK, G., BENOIST, J. P., YIOU, F., LORIUS, C., RAYNAUD, D., PETIT, J. R., BARKOV, N. I., KOROTKEVITCH, Y. S., and KOTLYAKOV, V. M., 1989, A comparison of deep Antarctic Ice Cores and their implications for climate between 65,000 and 15,000 years ago: Quaternary Res., v. 31, pp. 135–50.

KASZYCKI, C. A., and SHILTS, W. W., 1987, Glacial erosion of the Canadian shield, *in* Graf, W. L., ed., Geomorphic systems of North America: Geol. Soc. America, Centennial Special Vol. 2, pp. 155–61.

KAYE, C. A., 1967, Fossiliferous bauxite in glacial drift, Martha's Vineyard, Massachusetts: Science, v. 157, pp. 1035–37.

KEIGWIN, L., 1982, Isotopic paleoceanography of the Caribbean and east Pacific: Role of Panama uplift in late Neogene time: Science, v. 217, pp. 350–53.

KENNETT, J. P., HOUTZ, R. E., ANDREWS, P. B., EDWARDS, A. R., GOSTIN, V. A., HAJOS, M., HAMPTON, M. A., JENKINS, D. G., MARGOLIS, S. V., OVENSHINE, A. T., and PERCH-NIELSON, K., 1974, Development of the Circum-Antarctic Current: Science, v. 186, pp. 144–47.

LeMASURIER, W. E., 1972, Volcanic record of Antarctic glacial history: Implications with regard to Cenozoic sea levels, *in* Price, R. J., and Sugden, D. E., eds., Polar geomorphology: Inst. Brit. Geographers Spec. Pub. no. 4, pp. 59–74.

LIEBLING, R. S., 1973, Clay minerals of the weathered bedrock underlying coastal New York: Geol. Soc. America Bull., v. 84, pp. 1813–16.

LINTON, D. L., 1963, The forms of glacial erosion: Inst. Brit. Geographers Trans., v. 33, pp. 1–28.

LOUBERE, P., and MOSS, K., 1986, Late Pliocene climatic change and the onset of northern hemisphere glaciation as recorded in the northeast Atlantic Ocean: Geol. Soc. America Bull., v. 97, pp. 818–28.

MOOERS, H. D., 1989, On the formation of the tunnel valleys of the Superior Lobe, central Minnesota: Quaternary Res., v. 32, pp. 24–35.

MULLER, E. H., 1974, Origins of drumlins, *in* Coates, D. R., ed., Glacial geomorphology: State Univ. of New York Publications in Geomorphology, Binghamton, N.Y., pp. 187–204.

NAESER, C. W., IZETT, G. A., and WILCOX, R. E., 1973, Zircon fission-track ages of Pearlette family ash beds in Meade County, Kansas: Geology, v. 1, pp. 187–89.

PENCK, A., and BRÜCKNER, E., 1909, Die Alpen im Eiszeitalter: Tauchnitz, Leipzig, 3 v., 1199 pp.

PORTER, S. C., 1966, Pleistocene geology of Anaktuvuk Pass, Central Brooks Range: Arctic Inst. of North America Tech. Paper 18, 100 pp.

PREST, V. K., 1968, Nomenclature of moraines and ice-flow features as applied to the glacial map of Canada: Canada Geol. Survey Paper 67–57, 32 pp.

———, 1971, Quaternary geology of Canada, *in* Geology and economic minerals of Canada, Economic Geology Rept. no. 1, 5th ed.: Dept. Energy, Mines and Resources, Ottawa, Canada, pp. 676–764.

PRICE, R. J., 1973, Glacial and fluvioglacial landforms: Oliver & Boyd Ltd., Edinburgh, 242 pp.

REED, B., GALVIN, C. J., and MILLER, J. P., 1962, Some aspects of drumlin geometry: Am. Jour. Sci., v. 260, pp. 200–210.

RICHARDS, M. A., 1988, Seismic evidence for a weak basal layer during the 1982 surge of Variegated Glacier, Alaska, U.S.A.: Jour. Glaciology, v. 34, pp. 111–20.

RICHMOND, G. M., and FULLERTON, D. S., 1986, Summation of Quaternary glaciations in the United States of America: Quaternary Sci. Rev., v. 5, pp. 183–96.

ROBIN, G. D., 1964, Glaciology: Endeavour, v. 23, pp. 102–7.

RUDDIMAN, W. F., PRELL, W. L., and RAYMO, M. E., 1989, Late Cenozoic uplift in southern Asia and the American west: Rationale for general circulation modeling experiments: Jour. Geophys. Res., v. 94, pp. 18,379–91.

SHACKLETON, N. J., and sixteen others, 1984, Oxygen isotope calibration of the onset of ice-rafting and history of glaciation in the North Atlantic region: Nature, v. 307, pp. 620–23.

SHARPE, D. R., and SHAW, J., 1989, Erosion of bedrock by subglacial meltwater, Cantley, Quebec: Geol. Soc. America Bull., v. 101, pp. 1011–20.

SHAW, J., 1989, Drumlins, proglacial meltwater floods, and ocean responses: Geology, v. 17, pp. 853–56.

SHOEMAKER, E. M., 1986, The formation of fjord thresholds: Jour. Glaciology, v. 32, pp. 65–71.

SHREVE, R. L., 1985, Esker characteristics in terms of glacier physics, Katahdin esker system, Maine: Geol. Soc. America Bull., v. 96, pp. 639–46.

SHUMSKII, P. A., 1964, Principles of structural glaciology (transl. by D. Kraus): Dover Publications, Inc., New York, 497 pp.

SMALLEY, I. J., 1981, Conjectures, hypothesis, and theories of drumlin formation: Jour. Glaciology, v. 27, pp. 503–5.

—— and UNWIN, D. J., 1968, Formation and shape of drumlins and their distribution and orientation in drumlin fields: Jour. Glaciology, v. 7, pp. 377–90.

STREIFF-BECKER, R., 1947, Glacierization and glaciation: Jour. Glaciology, v. 1, pp. 63–65.

SUGDEN, D. E., and JOHN, B. S., 1976, Glaciers and landscape: John Wiley & Sons, Inc., New York (a Halsted Press Book), 376 pp.

TARR, R. S., 1905, Drainage features of central New York: Geol. Soc. America Bull., v. 16, pp. 229–42.

TELLER, J. T., 1987, Proglacial lakes and the southern limit of the Laurentide Ice Sheet, *in* Ruddiman, W. F., and Wright, H. E., Jr., eds., North America and adjacent oceans during the last deglaciation: Geol. Soc. America, The geology of North America, v. K-3, pp. 39–69.

WALDER, J., and HALLET, B., 1979, Geometry of former subglacial water channels and cavities: Jour. Glaciology, v. 23, pp. 335–62.

WARNKE, D. A., 1970, Glacial erosion, ice rafting, and glacial-marine sediments: Antarctica and the Southern Ocean: Am. Jour. Sci., v. 269, pp. 276–94.

WEERTMAN, J., and BIRCHFIELD, G. E., 1983, Basal water film, basal water pressure, and velocity of traveling waves on glaciers: Jour. Glaciology, v. 29, pp. 20–27.

WHITE, W. A., 1972, Deep erosion by continental ice sheets: Geol. Soc. America Bull., v. 83, pp. 1037–56.

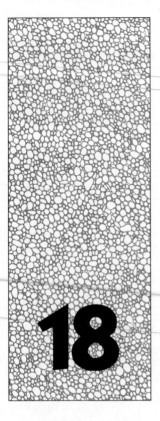

18

Alternating Quaternary Morphogenetic Systems

THE CONCEPT OF ALTERNATING QUATERNARY MORPHOGENETIC SYSTEMS

In the preceding five chapters, it has not been possible to describe morphogenetic systems without reference to examples of relict Quaternary landforms. Glacial moraines of several ages are 1000 m or more lower on mountains than any valley glaciers now reach, or 2000 km nearer the equator than the latitude of any present ice sheet. Some have asserted that the entire landscape that formed the basis of W. M. Davis's "normal cycle of erosion" is relict periglacial scenery (p. 444). Deserts are dissected by river networks in which no water has flowed in human history; longitudinal dune systems are now soil- and grass-covered. Any attempt to classify morphogenetic regions is confronted immediately by the need to separate presently evolving landforms from those that are relict from earlier Quaternary time. The problems are so difficult that only a start has been made in solving them, and numerous contradictions and inconsistencies among the morphogenetic models and examples are cited in this chapter. All that can be done is to outline the general concepts of Quaternary climatic change and to illustrate by selected examples the impact of alternating morphogenetic systems on the present landscape. Relicts of Tertiary or pre-Cenozoic morphogenetic systems, claimed for tectonically inactive regions especially in Africa and Australia, may be due either to latitudinal shifts in the positions of the continents (Frakes and Kemp, 1972; Kennett et al., 1974) or to pre-Quaternary climatic changes. If there are difficulties in recognizing Quaternary changes of climate and their impact on landscapes, the problems of even older events and forms are at present insuperable. However, there is real promise that eventually the positions of the continents in pre-Quaternary times will be worked out with sufficient accuracy so that the effects of plate tectonics and climatic change can be distinguished. Thus far, the record is primarily from submarine sediment cores; the corresponding events on land are only dimly perceived.

Although the causes of climatic change are unknown, it is clear that the earth has experienced a series of climatic alternations, each of comparable magnitude, during at least the last 2.4 million years. The chronology of these events insofar as it is now known was reviewed at the end of Chapter 17. The most obvious effect of the changes was the repeated expansion of the glacial climates on the larger and more humid parts of the northern hemisphere continents, with a corresponding expansion of the periglacial morphogenetic region as well. Alternating with the glacial episodes have been a series of interglacial (and "interperiglacial") intervals with morphogenetic regions approximating their present distribution. The time scale for the full cycle from glacial to interglacial conditions has been about 100,000 years. Within the last glaciation and presumably during earlier ones as well were a series of lesser climatic fluctuations (Fig. 17–16) more nearly glacial than interglacial in character. Quite possibly, interglacial intervals may also have had lesser climatic fluctuations. Certainly, the Holocene has been marked by such lesser fluctuations, some of which have been called "little ice ages," or "neoglaciations."

In lower latitudes, especially in the seasonally wet-and-dry climatic zones that border the great arid regions, Quaternary climatic change was expressed more as changes in effective precipitation than as temperature changes. Although we have abundant reasons to believe that the climatic changes of the Quaternary Period were worldwide, the resulting changes in morphogenesis varied from region to region. We know more about the strongly contrasting glacial–interglacial morphogenetic systems than we know about any of the others, and in the following sections, the chronology of these contrasting systems is used as a temporal framework within which other alternating morphogenetic systems can be placed. It must be kept in mind that the time scale of 10^3 to 10^5 years, in which most of the changes that are described here have occurred, is very short in terms of the evolution of landscapes as deduced in Chapter 12. No regional landscape can be described without considering the impact of Quaternary climatic changes.

Tertiary and Early Quaternary Morphogenetic Changes

Evidence in the northeastern United States of preglacial deep weathering of a sort usually associated with tropical or subtropical climates (p. 424) hints at morphogenetic changes far greater than the alternating systems of Quaternary glacial–interglacial times. However, the older relict forms and deposits are so rare that they are only recorded, not organized into a morphogenetic chronology. Similarly, during a long interval of erosion in the late Mesozoic and early Cenozoic Eras, much of Australia was reduced to low relief, with a deeply weathered landscape on which ferricrete crusts (p. 143) were widespread. The climate is assumed to have been warm and humid for as long as 60 to 200 million years. The separation of Australia from Antarctica by lithospheric plate motions, beginning about 25 to 30 million years ago (p. 436), initiated epeirogenic movements in Australia that rejuvenated rivers and dissected the ancient landscape into plateaus and mesas. Conditions favorable for laterite and bauxite landscape armor became less favorable in the late Cenozoic Era, and had practically ended by Quaternary time (Twidale, 1983). Much of Australia is now either too dry or too cool for comparable weathering. Late Cenozoic plate motions and orogenies disrupted former climatic patterns and intensified latitudinal climatic gradients. Possibly the alternating Quaternary morphogenetic systems are just one result of the tectonic-produced pattern of continent orientation that developed progressively during the late Cenozoic Era. Interpretation is complicated by the effects of biological evolutionary changes, especially the first appearance of grasses (p. 14), and accelerating human modification during the last 6000 years.

For at least the last 800,000 years, there is reasonable evidence for alternating climatic morphogenesis, recorded in deep-sea stratigraphy. The isotopic chemistry of pelagic and benthonic foraminifera is a record of changes in ice volumes and, to a lesser extent, oceanic surface temperatures (Fig. 18–1). Deep-sea cores that record these changes are dated by radioactive decay or by reference to paleomagnetic reversals with an assumption of constant sedimentation rate throughout the length of each core (Shackleton and Opdyke, 1973, 1976). About 22 isotope stages, each representing a significant change in terrestrial ice volumes and therefore of temperature, are recorded in the last 800,000 years. Some changes are of interglacial–glacial status, whereas other isotope stages represent glacial stadia or interstadials. The complex cyclic oscillations that have taken place on a scale of 10^5 years should warn us to use great caution in interpreting any landscape as the result of a single morphogenetic system. The amplitudes of the larger climatic oscillations shown by Figure 18–1 are equivalent to the contrast between full glacial and interglacial morphogenetic conditions on the land.

Temperature and Precipitation Changes in Late Quaternary Time

During the last glacial maximum, approximately 15,000 to 20,000 years ago, the present morphogenetic systems were strongly displaced (Figs. 17–17, 18–2, and 18–3). Mean annual temperature was lowered 8 to 10°C in the mid-latitude continental interiors and a few degrees in the surface waters of the tropical oceans. The cooling caused the greatest latitudinal shift of climatic zones in the middle latitudes of the northern hemisphere and lesser displacement of tropical zones. According to one model (Fig. 18–2), the zones of tropical forest and humid savanna actually expanded slightly as a result of decreased evaporation and cooler mean temperature. Another model (Fig. 18–3) does not show an expansion of the tropical humid regions. Neither of Figures 18–2 and 18–3 was intended as anything more than an idealized meridianal transect of average climatic conditions. As such, both are still very useful. However, Figure 17–17 demonstrates that during the last glacial maximum, the temperature changes were not the same on all meridians but were very complex. We assume, but have little proof, that the cooling associated with the previous Pleistocene glaciations was similar.

A large research team has attempted to compare computer-generated general circulation models of the earth's atmosphere with geologic records of climatic con-

Figure 18-1. Oxygen-isotope ratios from a planktonic foraminiferal species in a tropical Pacific deep-sea sediment core. Delta (δ) is related to ice volumes on land in such a way that isotope stage 1 (uppermost 22 cm of core) represents the Holocene, stage 2 is the last full glacial interval (from 13,000 to 32,000 years ago), and so on. The Brunhes–Matuyama paleomagnetic reversal, which occurred 730,000 years ago, is recorded at a depth of 1200 cm in the core. Isotope stages 1 to 5 are similar to the record of sea-level changes inferred from emerged coral reefs (Fig. 17-16). (Shackleton and Opdyke, 1973, Fig. 9.)

Figure 18-2. The present climatic zones along a meridianal belt from the equator to the north pole through Europe, compared with the hypothetical climatic zones during the last glaciation. (Büdel, 1957, Fig. 1.)

ditions at various places during the last 18,000 years (COHMAP members, 1988). Although most of the factual data about climate change since the maximum of the last ice age comes from sites in the mid-latitude northern hemisphere, enough tropical and southern hemisphere sites have been studied to provide an initial test of the validity of the current computer models. The boundary conditions for computer simulations include (Fig. 18–4): (1) Incident solar radiation at 18,000 years ago similar to the present, but seasonality greatly increased between 15,000 and 9000 years ago by known cyclical changes in the tilt of the earth's axis and the time of year when the earth is nearest the sun (Berger, 1978); (2) ice-sheet area at 100 percent of full glacial conditions until 15,000 years ago, then rapid ablation to present conditions by 7000 years ago; (3) sea-surface temperatures averaging 4°C colder than present until 15,000 years ago, then warming to present averages by 9000 years ago; and (4) atmospheric CO_2 content much reduced during the last ice age (until 15,000 years ago), and then rising to the level known prior to the industrial revolution. Given such changes, global climates at 3000-year intervals of late-glacial and post-glacial time were modeled and compared to known facts. Many of the examples of climate change described later in this chapter were used as tests for the computer models.

Figure 18-3. Model of the effective precipitation in the northern hemisphere during a glaciation and today (or during an interglacial interval). (From Flohn, 1953, Fig. 1.)

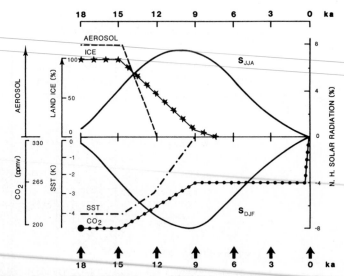

Figure 18-4. Boundary conditions for the COHMAP simulation with the Community Climate Model for the last 18 ka (thousand years ago). External forcing is shown for northern hemisphere solar radiation in June through August (S_{JJA}) and December through February (S_{DJF}) as percent difference from the radiation at present. Internal boundary conditions include land ice (Ice) as percent of 18-ka ice volumes, global mean annual sea-surface temperatures (SST) in degrees K as departures from present, excess glacial-age aerosol (Aerosol) with arbitrary scale, and atmospheric CO_2 concentration (CO_2) in parts per million by volume. The arrows mark the times of seven sets of simulation experiments. The one for 18 ka included the lowered CO_2 concentration (200 ppmv, large filled circle): the others had the CO_2 concentrations of the control case (330 ppvm) rather than a stepwise increase. Experiments incorporating increased glacial-age aerosol loading have not been completed. (COHMAP members, 1988, Fig. 2.) (Copyright by the AAAS.)

A major generalization for mid-latitude northern hemisphere ice-age sites is that annual temperatures as much as 10°C colder than present were common along the southern margin of the North American and Scandinavian ice sheets, with powerful global westerly winds, as well as easterly winds which poured down off the edges of the ice sheets. Forests were generally absent from Europe and permafrost was widespread. Mediterranean lowlands were treeless and somewhat drier than now, although the lower temperatures and less evaporation permitted some pluvial lakes to expand. In North America, the southward deflection of the westerly storm track and lower temperatures allowed lakes in the Great Basin to expand, and woodlands to cover larger parts of the American southwest.

In low latitudes, the steepened seasonal temperature gradients between 12,000 and 9000 years ago increased the intensity of the seasonal *monsoon* circulation, bringing more moisture to the Sahara, Arabia, and southern Asia. Conditions in the southern hemisphere were cooler than present at 18,000 years ago, but are generally less well known.

The worldwide cooling of the last glaciation is well established, but the corresponding effects on precipitation are not at all clear. Most authorities reason that overall cooling must have reduced total evaporation and therefore total precipitation; values commonly suggested are about 80 percent of the present values shown by Figure 5–5 and Table 5–2 (Flint, 1971, p. 417). The former notion that glaciations were also times of increased total precipitation ("pluvials") is very likely wrong.

Even though temperature and total precipitation may have decreased during a glacial age, the *effective* precipitation may have increased because of reduced evaporation. In particular, the large areas of subtropical aridity (Fig. 13–1) that characterize the modern planetary circulation may have been reduced in size and displaced toward the equator (Figs. 18–2 and 18–3). The Mediterranean steppe regions on the poleward sides of the arid belts were widened as winter rains brought moisture to areas such as the northern Sahara and the southern edge of the Australian desert. The shifts on the equatorial side of the arid regions are still uncertain. Aridity may have displaced part of the savanna region, or the humid tropics may have stayed approximately fixed and simply compressed the dry zones. Both hypotheses are illustrated in Figures 18–2 and 18–3. Case studies from Brazil, Africa, and Australia, presented later in this chapter, illustrate the difficulties of interpretation.

Part of the evidence for inferring climates during glacial ages is morphogenetic. At present, the snowlines on maritime west coasts rise to the east as the available moisture is lost. The altitude of glacial-age cirque floors, although 1000 to 1500 m lower than the present snowline, also rise inland at about the same rate as the modern snowline (Fig. 18–5a). The parallelism of the two lines suggests that the cause of the glacial-age lowering of the snowline was cooler temperature rather than increased precipitation (Flint, 1971, pp. 63–73). If precipitation had decreased inland away from the coast more rapidly during glaciation than at present, coastal cirques would be lowered by an amount proportional to the cooling, but cirque-floor altitudes would rise more steeply inland and converge with the modern snowline

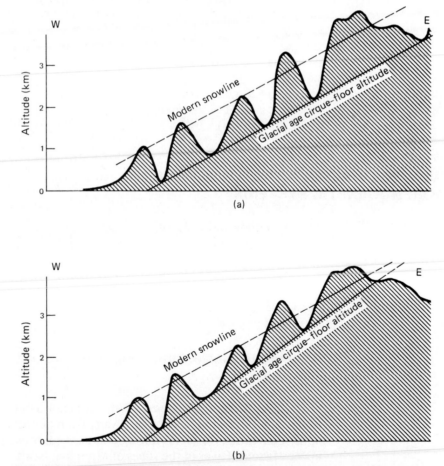

Figure 18-5. Theory of glacial snowline lowering as applied to a maritime west coast mountain range. In the most usual condition, (a), the snowline of glacial time (as indicated by cirques) was lowered approximately parallel to the modern snowline. If inland convergence were found, (b), cooling and either a reduction of precipitation inland or an increase in precipitation near the coast could be inferred.

(Fig. 18–5b). Such convergence has been found in the Cascade Range of Washington, where west-wind moisture was blocked by a large glacier lobe in the Puget Sound lowland (Porter, 1964).

Morphogenetic Effects of Late Quaternary Climatic Change

To illustrate the hypothetical morphogenetic effects of the climatic changes during a glacial age, Figure 6–9 should be compared with Figures 17–17, 18–2, and 18–3. In Chapter 6 the zonal or latitudinal changes in weathering processes were described with climate as a major factor. If the climatic zones have shifted, the nature and intensity of weathering and soil-forming processes must also have changed. Similar deductions can be made for mass wasting and the fluvial morphogenetic system.

As an indirect effect of glaciations, sea level repeatedly fell at least 100 m (Fig. 17–16). This had the effect of rejuvenating major rivers and steepening both their longitudinal and cross-valley profiles in coastal regions. Also, in regions with wide continental shelves, present maritime climates became more continental as the shoreline withdrew 100 km or more seaward of its present position. Superimposed on the colder and possibly drier glacial-age climates, the increased continentality probably contributed to the morphogenetic contrasts of these regions. For example, southern Korea today has a humid subtropical climate, with temperature extremes greatly reduced by the moderating effect of the seas around the peninsula. During full-glacial lowering of sea level, the entire Yellow Sea was dry land and the Korea peninsula was incorporated into the Asian mainland. Massive periglacial gelifluction deposits on the south coast of Korea (35°N), presumably dating from the most recent glacial age, testify to the extreme morphogenetic changes that occurred there, mostly as a result of the change from a maritime to a continental climate (Guilcher, 1976, pp. 655–58).

EXAMPLES OF ALTERNATING QUATERNARY MORPHOGENETIC SYSTEMS

The magnitude and chronology of Quaternary climatic changes are the subjects of active research and are certain to be revised. However, the concept forms a framework for a series of examples that illustrate the effects of alternating morphogenetic systems on a variety of landscapes. Inconsistencies and outright contradictions will be discovered by comparing the various examples, yet they illustrate the powerful impact of Quaternary climatic change on landscapes. Eventually, a theory must be developed that can mutually reconcile all of these and many other examples.

Alternating Glacial (and Periglacial) and Interglacial Systems

Upper Mississippi Basin.   With its major tributaries, the Missouri and Ohio Rivers, the Mississippi carried meltwater from almost the entire southern quadrant of the Laurentide ice sheets during at least the last two glaciations (Fig. 16–1). The upper Missouri River in the states of Montana, North Dakota, and northern South Dakota was created during the Pleistocene Epoch from segments of ancestral rivers that flowed northeastward down the regional slope of the Great Plains toward Hudson Bay (Fig. 18–6). Successive ice-sheet advances from the Hudson Bay lowland moved southwestward up the gentle slope of the Great Plains, damming the ancestral rivers and forcing them to drain southeastward along the ice margin until they joined the Mississippi (Flint, 1971, pp. 232–35; Bluemle, 1972). The modern trenchlike valley of the upper Missouri River trends almost perpendicular to the regional slope and the ancestral, drift-filled valleys. Most of the diversions are correlated with the Illinoian glaciation (p. 437), but some are Wisconsinan in age. Almost all of the tributaries of the upper Missouri River enter from its right side, and are segments of the former Hudson Bay drainage network. The reservoirs of the large dams that have been built along the Missouri River (Oahe Dam, South Dakota; Garrison Dam, North Dakota; Fort Peck Dam, Montana) reproduce the ice-dammed lakes that occupied those valleys during late Quaternary glaciations.

The Ohio River had a comparable but older origin. In preglacial time, the western Appalachian drainage probably flowed northwest but was directed more nearly westward by one of the earliest ice sheets (Tight, 1903; Teller, 1973; Swadley, 1980). A buried valley comparable in size to the present Ohio River valley has been traced westward across Ohio, Indiana, and Illinois. The valley of the ancestral river, named the *Teays Valley,* has been nearly buried by several subsequent drifts (Fig. 18–6). At least two glacial-lacustrine deposits in buried valleys of the Teays system have reversed magnetic polarity, and are interpreted as having been deposited because of ice damming prior to 730,000 years ago (Eschman, 1985). The western part of the Teays system, known as the Mahomet bedrock valley in Illinois, may be somewhat younger but contains the record of several pre-Illinoian glaciations (Kempton, 1983). The early ice sheets that buried the Teays drainage system may have spread southward across Michigan prior to the erosion of the modern Great Lakes basins (Johnson, 1985), which have been progressively deepened by succeeding glaciations.

During each glacier advance southward across the central lowlands of the United States, north-flowing rivers were blocked by ice and diverted southward to the Mississippi system. A sequence of events can be deduced, with an initial phase of high meltwater discharge but primarily suspended load, because the coarser drift was trapped in ice-marginal lakes. Then, as the ice margin crossed the divide be-

Figure 18-6. Preglacial drainage networks in the upper Mississippi drainage basin. Present rivers are shown in bold lines.

tween northward-flowing and Mississippi drainage, as at the southern edges of the Great Lakes, outwash poured down the Mississippi. The previously eroded deep trench, cut to a glacially lowered sea level (Fig. 10–2), began to aggrade. Later, when the ice margin had again retreated north of the ice-marginal lakes on the regional divide, only water and suspended load were carried by the Mississippi. The modern meandering channel developed on fine-grained, floodplain alluvium. Postglacial isostatic uplift (p. 433) has aided the upstream reaches to entrench and leave outwash terraces on valley sides, but the downstream part of the river has aggraded a modern floodplain to rising sea level, deeply burying the outwash sand and gravel of glacial times. The alternation of glacial and interglacial hydrologic regimes has left flights of loess-capped terraces and many buried channels throughout the Mississippi River drainage basin. In general, the regional relief has been decreased by thick sheets of glacial drift with a subdued morainal topography.

Southward from the maximum advances of the Pleistocene ice margins, tundra vegetation extended at least 300 km on the Appalachian Plateau (Maxwell and Davis, 1972; Jacobson et al., 1987). Periglacial patterned ground is known from many regions of central and east-central United States (pp. 372 and 384). Periglacial morphogenesis must have influenced the development of the Mississippi Basin, but the effects have not been separated from the massive effects of glaciation. Although tundra-like, the late-glacial periglacial zone of the United States was not like the present high-latitude tundra, for the sun remained visible throughout winter days and was high overhead during the summer.

Northwestern Europe.   The present climate of Europe grades from cool maritime forests on the Atlantic seaboard (1) eastward to continental steppe grasslands in Poland and the Soviet Union, and (2) southward to Mediterranean subhumid conditions. During each glaciation, most of nonglaciated Europe north of the Alps became a periglacial region. The present maritime influence was greatly decreased by the width of the exposed continental shelf and by the southward diversion of the Gulf Stream (Fig. 17–17), which then did not affect the European climate north of Spain (Ruddiman and McIntyre, 1973). The Alpine ice cap that formed synchronously with each expanding Scandinavian ice sheet blocked warm air and moisture from the south.

Glacial–interglacial alternating climatic morphogenesis seems to have followed a repeated pattern (West, 1968; van der Hammen et al., 1971). With the onset of glaciation, the climate grew cold, but initially remained fairly humid in western Europe. Intense gelifluction characterized the wet, cold, periglacial interval. Then, during each of the glacial maxima, the climate became cold and dry because pack ice and lower sea level blocked the northward flow of the warming Gulf Stream. Thick loess deposits mantled frost-churned residual soils and the gelifluction deposits of the wetter phase. As the ice margins retreated again, the climate warmed and became more humid, but with several sharp fluctuations. The pollen stratigraphy in western European bogs and lakes records a late-glacial tundra environment changing to birch and pine forest about 8500 years ago. The climate then was cold and rather dry. About 5000 years ago, the climate became maritime and moist and perhaps 2°C warmer than the average of today. This was the Hypsithermal Interval in Europe. Heavy deciduous forests covered northwestern Europe, and peat bogs were widespread on uplands and lowlands. About 3000 years ago, the forests began to degenerate, perhaps owing to a slightly cooler, drier climate but more likely owing to the beginning of Neolithic agriculture. People have strongly affected subsequent morphogenesis in western Europe by cutting the forests and planting crops.

The highlands of Scandinavia and the Alps support small ice caps and valley glaciers today, although all but a possible few have reformed after the Hypsithermal

Interval and are not the remnants of the last glaciation. The drift plains of northern Europe have low relief, dissected by abandoned ice-marginal valleys. Loess is abundant, especially in the intermontane basins of eastern Europe. The cereal-producing steppe lands of Austria, Poland, Hungary, and the Ukrainian SSR are all loess mantled, with silt derived from large outwash rivers such as the Danube, Dneiper, and Don.

South of the glacial limit, western European landscapes are dominated by the multiple episodes of periglacial morphogenesis. Slopes are mantled with thick gelifluction sheets. Valleys are heavily aggraded. Cave stratigraphy and rock paintings prove that Paleolithic people hunted the tundra animals during the last glacial age and witnessed the periglacial processes (Valladas et al., 1986).

High-Altitude New Guinea. One of the most striking results of alternating glacial and interglacial morphogenetic systems is the landscape of high-altitude equatorial regions. On the island of New Guinea, 3 to 6° south of the equator, only three of the highest mountain peaks, all above 4600 m in height, now support glaciers (Verstappen, 1964). From an undetermined time prior to 22,000 years ago until as recently as 9000 years ago, mountain ice caps or valley glaciers shaped at least 20 mountains, covering a total area of 2000 to 2200 km^2 (Hope and Peterson, 1976, p. 174). The firn limit was about 1000 m lower than present, and the treeline, as well as vegetational zones in the lower rainforests, were comparably lowered (Galloway et al., 1973; Hope and Peterson, 1975; Bowler et al., 1976). The result is that areas now in high-altitude grasslands, just above the treeline, have U-shaped glaciated troughs, paternoster lakes, hanging valleys, and other typically alpine landforms (Fig. 18–7). Moraine ridges cross valley floors and dam lakes. Soils are shallow. Weakly developed cirques dissect mountain ridges into arêtes. Areas between 3200 and 2200 m, now in montane rainforest, were covered by alpine grasslands until about 10,000 years ago. Valley forms change from broad, open troughs in the deglaciated highlands to steep, deep gorges across the forested lower mountain ridges. Entire assemblages of landforms, each appropriate to a modern climatic and vegetational zone on the mountain sides, are found relict about 1000 m lower. The entire shift is consistent with a climatic cooling of about 5 to 6°C, with no significant change in precipitation (Löffler, 1972; 1977, pp. 63–68). The climatic interpretation is complicated by the influence of human agriculture for the last several thousand years but is nevertheless very similar to the history of montane equatorial regions in Africa and South America.

Alternating Pluvial and Nonpluvial Systems

The term **pluvial** ("rainy") implies greater total precipitation but in fact is more accurately defined as greater *effective* precipitation, the excess of precipation over evapotranspiration. The climatic cause can be greater precipitation, reduced evaporation, more effective seasonal distribution of precipitation, or any combination of these. Many regions now arid or semiarid show evidence of former pluvial conditions. In some of the regions, times of pluvial morphogenesis correlate with glacial ages; in others, the glacial ages were nonpluvial.

Figure 18–3 summarizes the hypothesis that during a glacial age, pluvial conditions prevailed in the zones between latitudes 25 and 35°, where there is widespread aridity or semiaridity today (Fig. 13–1). Three examples of alternating pluvial–nonpluvial morphogenesis are offered to demonstrate that in this latitudinal zone, pluvial and glacial ages are correlative. Subsequently, three other examples of relict pluvial morphogenesis demonstrate a correlation with interglacial ages. The latter three examples are from tropical regions, not from the extratropical latitudinal belts.

Figure 18-7. Pindaunde Valley, a glaciated trough with rock-basin lakes on the flank of Mount Wilhelm, Papua New Guinea. The building at the lower end of the lakes is at an altitude of 3500 m. Glaciated landforms extend down to 3200 m, below which steep V-shaped valleys are covered with tropical rainforest. (Photo: G. S. Hope.)

Southwestern United States. About 140 closed tectonic basins mark the semiarid or arid Basin and Range province of southwestern United States (Fig. 1–6). Most of them contained lakes during the last glaciation, ranging from shallow playas to Lake Bonneville, which was more than 330 m deep at its maximum. A few of the lakes were fed by meltwater from valley glaciers on the adjacent ranges, but most of them developed only because of a more favorable precipitation/evaporation ratio. Various lines of evidence suggest that the pluvial lakes resulted from a combination of cooler temperatures, more precipitation, especially from winter storms, and reduced evaporation related to cooler conditions and perhaps greater cloudiness. Estimates vary widely, with temperature estimates ranging from 3 to 16°C cooler and precipitation estimates ranging from only 50 percent to almost 200 percent of present amounts (Smith and Street-Perrott, 1983; Spaulding et al., 1983; Dohrenwend, 1987; Benson and Thompson, 1987). One widely quoted source estimated that the average pluvial climate of Nevada was about 3°C cooler, with 68 percent more precipitation than present and evaporation reduced by 10 percent (Mifflin and Wheat, 1979). Typically, this would be the equivalent of the present climate in the relatively cool and moist northwestern corner of the state spreading to the presently much drier and warmer southern sections. Secondary fluctuations in the levels of the pluvial lakes may have been related to the position of the jet stream along the southwestern margin of the Laurentide Ice Sheet, and therefore be broadly synchronous with glacial advances and retreats.

In two pluvial lake basins, end moraines that were built at the mouths of valleys by glaciers from adjacent mountain ranges are cliffed by high-level lake shorelines. Outwash correlated with recessional moraines built deltaic terraces at the level of the pluvial shorelines as well. In only these two lakes, pluvial Lake Russell in the Mono Basin and pluvial Lake Bonneville in the Great Salt Lake Basin, is it possible to prove by such morphologic evidence that glaciation and pluvial conditions were synchronous. However, a growing number of radiocarbon-dated sediment cores and shoreline deposits from various playas prove that they were pluvial during at least the last glaciation and probably during earlier glacial intervals as well (Oviatt and Nash, 1989; Waters, 1989). Radiocarbon dates and isotopic analyses of rock varnish (p. 333) on alluvial fans in Death Valley, California demonstrate at least three episodes of fan building in late Pleistocene time, prior to the onset of Holocene aridity (Dorn et al., 1987, 1989). The fans aggraded during brief pluvial intervals and were entrenched by streams during subsequent more lengthy arid intervals. In the present extremely arid climate of Death Valley the extensive areas of fans darkened by rock varnish (Fig. 13–3) are relict.

From 24,000 to 12,000 years ago, at least 70 percent of the lake basins in the Great Basin for which radiocarbon dates are available were high (defined as being between 70 and 100 percent full relative to overflow level). Lake levels fell rapidly in the last 10,000 years, and many basins were even drier than present in the interval between 10,000 and 5000 years ago (Fig. 18–8). The chronology fits the history of continental glaciation (p. 438) very well.

The most obvious morphologic effects of a pluvial climate are the shoreline features of the former lakes. Shore platforms, wave-cut cliffs, spits, bars, and deltas all trace the ancient shorelines (Fig. 18–9). River terraces are graded to former lake levels. Basin floors are blanketed with lacustrine sediments, including salt layers and buried soils that record nonpluvial conditions. Sand dunes are associated with

Figure 18-8. Relative percentages of pluvial basins in the western United States containing dated materials at elevations indicative of low, intermediate, or high stages versus their radiocarbon ages. (Smith and Street-Perrott, 1983, Fig. 10-6.)

Figure 18-9. View southeast over Payson, Utah. Large gravel spit in foreground was build northward by longshore currents in a succession of pluvial lakes in the Great Salt Lake Basin. Benches were eroded during falling lake levels. (Photo: H. J. Bissell.)

ancient shorelines, as well as being built on the dry playas today. Periods of soil formation and valley incision with presumed more effective vegetation cover have alternated with periods of mass wasting and aggradation, presumably during nonpluvial times, in agreement with Huntington's principle (p. 266). However, the correlation of massive alluviation on Death Valley fans with pluvial times and entrenchment during nonpluvial times contradicts Huntington's principle. It is possible that the same pluvial conditions that stabilized land surfaces and reduced the erosion rate in semiarid regions caused more erosion in the drainage basins in the mountains that surround the hyperarid Death Valley, with resulting fan aggradation in the basin (Fig. 12–7). Several lesser cycles of arroyo entrenchment and alluviation have occurred within postpluvial time, so an exact correlation with the pluvial–nonpluvial cycle is not likely.

An unsolved problem of morphogenesis in the southwestern United States concerns the role of pluvial climates in pediment development (p. 339). Was pediment cutting accelerated or inhibited by the pluvial morphogenetic system? Were pediments dissected by more effective precipitation during pluvial times? Alternating pluvial–nonpluvial climates must be significant to pediment formation, although no one has yet been able to demonstrate the significance. An even more challenging hypothesis is that the pediments of the Mohave Desert, California, developed in a semiarid pre-Pliocene climate and have subsequently been exhumed and only slightly modified (Oberlander, 1974).

The Mediterranean Basin.   The Mediterranean climate and morphogenetic system is characterized by summer drought and winter rains. Floods and droughts are both tragically frequent. Mountainous regions rank high in denudation rates (Table 12–1 and Fig. 12–2). From about 40,000 to 20,000 years ago, land around the Mediterranean basin was colder because the water temperatures in the region were lowered 5 to 9°C (Fig. 17–17). Although total rainfall was probably less than now, it was distributed more evenly throughout the year because of the southward shift of the westerly wind storm track. Tree pollen, especially oak, indicates the presence of some trees, although the region was not heavily forested as might be interpreted from a simplified model such as shown in Figure 18–2 (Rognon, 1987). Pluvial lakes expanded despite the decreased total precipitation, perhaps because of the cooler temperatures and better seasonal distribution of rainfall. Shortly after 20,000 years ago, the North African region began to be drier, culminating in maximum aridity about 12,000 years ago. Thereafter, the climate has moderated toward its present state, although human activity for at least 4000 years has almost totally deforested the region.

Mediterranean river valleys are characterized by alluvial terraces into which the present stream channels are incised (Vita-Finzi, 1969). The alluvium is stratigraphically younger than shorelines of the last interglacial. It forms fans and terraces that were graded to a sea level lower than present, contains pollen of plants that indicate cooler temperature, and in places can be traced upstream into periglacial deposits. All these lines of evidence prove that valley aggradation occurred during the last glacial age, probably during the interval of ice-sheet expansion and falling sea level. Intensified frost weathering on highlands and at least seasonally more intense rain have been claimed as causes of the aggradation (Vita-Finzi, 1969, p. 100). Calcareous tufa deposits around springs suggest correlative higher groundwater levels, also the result of more effective precipitation.

About 9000 years ago, the alluvial fans and floodplains of the Mediterranean coast began to be dissected. While the lower reaches of rivers became estuaries because of the late-glacial sea-level rise, their upstream portions were entrenching as postglacial streams established more gentle gradients. Human intervention in the form of dams and soil-conservation systems became a factor in subsequent valley development. Upland erosion was inhibited by check dams, but by about 1000 years ago, a new cycle of floodplain aggradation had begun to bury classical Roman dams and buildings. Finally, for the last five centuries, erosion has again dominated.

The correlation of aggradational phases in Mediterranean valleys with the cooler and pluvial times of the last glaciation and of the "little ice age" of post-Roman times in Europe contradicts Huntington's principle (p. 266). However, if vegetational cover was incomplete in the Mediterranean basin, owing either to climatic or human factors, increased precipitation could cause upland erosion and floodplain aggradation instead of landscape stabilization and valley incision. The last pluvial was approximately correlative with glaciation in northern Europe and local periglacial conditions. It was a time of cooler temperature, decreased total precipitation but more effective precipitation/evaporation ratio, sparse forest cover, alluviation in river valleys, and lower sea level. Clearly, the modern Mediterranean landscape is the result of extremely complex alternating morphogenetic systems, with the further complication of many thousands of years of human impact.

Southern Australia. A series of pluvial lakes formed in the alluvial plains of the semiarid Murray Basin in southeastern Australia. Sediment cores from some of these lakes have been radiocarbon dated to give a pluvial chronology for the last 40,000 years, which is broadly similar throughout the region (Bowler et al., 1976, especially Fig. 6). The pollen record of vegetation shows that the temperature was 8 to 10°C colder during the last pluvial intervals, and precipitation was substantially reduced (Kershaw, 1988, p. 263). Nevertheless, until about 17,000 years ago, the lakes contained more water than now, the effect of a favorable precipitation/evaporation ratio. Many of the lakes dried completely about 17,000 years ago. Those in more arid regions have not held water for more than brief seasonal intervals since.

Those in somewhat more humid regions record weak pluvial periods between 10,000 and 5000 years ago and again during the last 2000 years. The area is at about 35°S, a latitude equivalent to similar lakes in the southwestern United States. Until 17,000 years ago, the correlation with glacial and pluvial events in the northern hemisphere is excellent, strongly supporting theories that in the southern as well as northern middle latitudes, pluvial intervals correlate with glaciations. For the last 17,000 years, however, the Australian lakes record brief pluvial periods that do not correlate with the southwestern United States record. As temperature rose in Holocene time, the United States lakes dried up, but the Australian lakes recovered again between 10,000 and 5000 years ago. An unexplained increase in Australian precipitation must be invoked, possibly related to the rising sea-surface temperatures and changes in atmospheric circulation. Bowler et al. (1976) maintained that the pluvial intervals were due primarily to reduced evaporation, but Dury (1973) gave hydraulic-geometry arguments for at least 50 percent greater precipitation during times of higher lake levels.

Major changes in fluvial morphogenesis in southeastern Australia are correlated with pluvial–nonpluvial conditions. In the last 40,000 years, three cycles of fluvial morphogenesis occurred on the Murray River and its tributaries. Each began with pluvial high discharge and intense aggradation by large rivers and ended with soil formation and smaller channel networks. The oldest cycle is poorly dated. The second of the three pluvial events is dated at between 30,000 and 25,000 years ago. The youngest of the events dates from 25,000 or 20,000 years ago to about 16,000 or 13,000 years ago (Bowler et al., 1976, p. 381; Bowler and Wasson, 1983, p. 187). During similar cyclic pluvials, the Murrumbidgee River channels may have carried a discharge five times that of the modern river (Schumm, 1968, p. 27). Meander wavelengths, meander radii, and channel widths of the ancestral fluvial systems are much larger than the equivalent dimensions of modern streams. Bowler and Wasson (1983, p. 190) suggested that the very wide channels and meander belts that characterized many Australian rivers during the interval between 30,000 and 15,000 years ago may have been due to a high groundwater table inherited from the preceding humid interval. High groundwater pore pressure in stream banks would promote bank failure, rapid meander formation, and wide channels. The inference is that although the pluvial lakes dried up about 17,000 years ago groundwater continued to affect fluvial geomorphic processes for another 2000 years. The modern cycle of channel development has continued for the last 13,000 years, when discharge and channel dimensions have been roughly equivalent to those of the modern rivers. A decrease in discharge of three to fivefold since 16,000 years ago can be inferred. Clearly, the pluvial morphogenetic interval for the southeastern Australian lakes and rivers is correlative with the last glaciation in the northern hemisphere, although postpluvial conditions in Australia began 6000 years before postglacial time (as defined in Europe). Subsequently, the pluvial conditions have recurred without obvious northern hemisphere correlatives. Earlier pluvial stages, beyond the 40,000-year range of radiocarbon dating, are also recognized in the Australian sites.

Southeastern Brazil. Alternating climatic morphogenesis, in which pluvials correlate with interglacial or Holocene conditions in higher latitudes, is well illustrated by an explanatory description of the landscape in southeastern Brazil (Bigarella et al., 1965). The landscape of the presently humid, forested, southeastern coast of the country has steplike valley cross profiles of nonstructural benches alternating with steep slopes. Traced into more arid northeastern Brazil, each bench becomes part of a regional pediment or "pediplain" separated from the next lower pediment by an escarpment.

Three cycles of alternating semiarid sediment transport and humid valley incision are hypothesized (Fig. 18–10). Within each, minor fluctuations toward the opposite extreme are recognized. During the semiarid half of each cycle, pediments were cut across uplands, and coarse alluvium was transported down high-gradient braided alluvial channels to the sea, which was lower than present, thus establishing the basic correlation of nonpluvial with glacial conditions. During the humid half of each cycle, pediments and alluvial surfaces alike were forested and deeply weathered. The surface stabilization and higher stream discharge allowed rivers to entrench their upper valleys, even though their mouths were graded to a higher sea level. Minor semiarid episodes during humid intervals promoted scarp retreat and valley widening, and minor humid episodes during semiarid intervals caused minor valley entrenchment.

Because coarse alluvium and steep pediments are graded to lower sea level, and because humid valley incision is graded to a sea level comparable to today's, the basic pluvial = interglacial correlation is recognized in the Brazilian alternating morphogenetic explanation. Further proof has been provided by a radiocarbon date of about 14,000 years from organic debris in coarse terrace alluvium on one of the rivers (Pflug, 1969). In the Rio Dôce valley, after an earlier stage of deep erosion, alluvium was aggraded to so high a level that tributary streams were ponded against it as lakes. At that time, about 14,000 years ago, the uplands were providing large amounts of feldspar-rich, mechanically weathered sand and gravel. Subsequently, the river has again entrenched the alluvium as deep as 20 m, but the dammed lakes in the lower tributaries are still held about 20 m above the stream by the alluvium. The entire region is under tropical rainforest today.

East Africa. Supposed evidence of higher level lakes in the rift valleys of eastern Africa (Fig. 3–17) led to the early concept of "pluvials" that were intuitively correlated with the glacial sequence of the Alps. No dating methods or means of correlation were then available. Subsequently, some of the "lake beds" proved to be volcanic tephra, others are alluvium, and none can be proved to be related to glaciation. All of the names given to the alleged pluvials have been abandoned. Radiometric dates on both lacustrine sediments and volcanic rocks definitely show major climatic changes, but the supposed correlation with glacial events is disproved (Butzer et al., 1972; Street and Grove, 1976, 1979) (Fig. 18–11). Key eastern African lakes were at high levels sometime prior to 21,000 years ago but were at low levels and high salinity until about 10,000 years ago, nearly at the end of the last glaciation. Then, for 2000 years, they were at maximum levels, and several of them overflowed into adjacent tectonic basins. Some of them also have had several moderately high levels during the Holocene, as with the Australian lakes. In the eastern African lakes, the major pluvial interval was neither glacial nor interglacial but rather at the transition from glacial to interglacial conditions (Street and Grove, 1979). An explanation may be that during full-glacial time about 18,000 years ago the western Indian Ocean was no colder than it is today (Fig. 17–17), and precipitation on the adjacent eastern African highlands was comparable to today's. However, during the maximum northern hemisphere seasonality contrast in solar radiation about 9000 years ago (Fig. 18–4) the seasonal monsoon circulation over the Indian Ocean was intensified, and more abundant seasonal moisture reached the adjacent eastern African highlands as well as other regions peripheral to the northern Indian Ocean (COHMAP members, 1988). Nile crocodiles and hippopotamuses expanded their range into lakes and swamps in the present Sahara; shallow lakes were widespread between the great longitudinal dunes of the presently hyperarid Rub' al Khali desert in Arabia (McClure, 1976).

India. As a final example of low-latitude, alternating pluvial and interpluvial systems, the few available studies from India can be summarized. River valleys in the Deccan region of northwestern peninsular India record successive Quater-

Figure 18-10. Hypothesis of slope evolution in southeastern Brazil under alternating semiarid and humid conditions. I–III: Semiarid pediments lowered by slight changes in local base level and minor climatic changes. IV: Entrenchment in a humid climate. V: Valley widening and alluviation by minor climatic changes within a humid phase. VI: Renewed pedimentation under a semiarid climate. VII–IX: Repetition of stages III–V. (Bigarella et al., 1965, Figs. 6, 7, and 8.)

Figure 18-11. History of lake levels in eastern and central Africa. All lakes presently lack outlets. (Butzer et al., 1972, Fig. 4; © 1972, American Association for the Advancement of Science.)

nary episodes of (1) deep erosion; (2) deposition of "older alluvium" tens of meters in thickness, with the development of alluvial plains several kilometers broad; (3) entrenchment and terrace cutting into the older alluvium; and (4) deposition of Holocene silty alluvium. The older alluvium has lenses of cemented gravel in a silt or sand matrix, which suggests deposition as alluvial fans or channel and point-bar deposits. It contains artifacts that formerly were assigned to the Middle Pleistocene, predating the last glaciation, but the age of the artifacts is now doubted, and the older alluvium is probably 30,000 to 17,000 years old (Kale and Rajaguru, 1987). The coarse older alluvium seems to correlate with aggradation during the full-glacial weak monsoon system and aridity as noted in the preceding paragraph. The later phase of aggradation by fine-grained overbank sediment and the subsequent early Holocene intrenchment are correlated with the more abundant wet monsoons of 9000 years ago and younger. As in Brazil, aridity and aggradation in India correlate with lower sea level on the adjacent coast, and Holocene stream incision correlates with postglacial high sea level. The full glacial arid phase in India could have been accentuated by the fall of sea level, which widened the coastal plain of northwestern India by 200 km and further reduced the effective precipitation on the Deccan uplands.

In the Thar desert, just north of the Deccan Plateau, an episode of Pleistocene dune building choked former river valleys. The arid phase ended abruptly about 10,000 years ago, when lakes formed behind the dunes. The lakes have gradually shrunk during the Holocene but were fresh until about 3000 years ago. Now they are saline (Singh et al., 1972; Singh et al., 1974). Clearly, the Thar desert was drier during at least the late stages of the last glacial age and then abruptly more humid. The chronology is similar to that of the eastern African rift lakes, Arabia, and the Deccan Plateau. Other evidence of dune growth and soil formation in western India also supports the theory that glacial ages were nonpluvial in the Indian subcontinent (Verstappen, 1970).

SUMMARY

The examples that have been presented generally support the hypotheses of Quaternary climatic change approximately as shown by Figures 18–2, 18–3, and 18–4. There is some indication that rapid geomorphic change occurred during the transition from glacial to interglacial or pluvial to nonpluvial rather than during the peak of either extreme, especially in the tropics. As noted at the beginning of this chapter, the time scale of major Quaternary climatic cycles is on the order of 100,000 years, with lesser oscillations on the order of 20,000 years each. These intervals are long enough to put distinctive morphogenetic imprints on landscapes but far too short to shape any region totally. Two possible exceptions are the humid equatorial tropics and the continental interior desert regions, which may have felt the least impact of Quaternary alternating morphogenetic systems. Perhaps this is one reason why the oldest of the earth's landscapes remain in the arid parts of Australia and Africa. The humid tropics, even though buffered from Quaternary climatic change by an excess of heat and moisture, could not escape the geomorphic impact of sea-level fluctuations. Arid regions with internal drainage were unaffected even by this otherwise worldwide morphogenetic factor.

REFERENCES

BENSON, L., and THOMPSON, R. S., 1987, The physical record of lakes in the Great Basin, *in* Ruddiman, W. F., and Wright, H. E., Jr., North America and adjacent oceans during the last deglaciation: Geol. Soc. America, The geology of North America, v. K-3, pp. 241–60.

BERGER, A., 1978, Long-term variation of caloric insolation resulting from the earth's orbital elements: Quaternary Res., v. 9, pp. 139–67.

BIGARELLA, J. J., MOUSINHO, M. R., and DA SILVA, J. X., 1965, Processes and environments of the Brazilian Quaternary: Universidade de Paraná, Curitiba, Brasil, 69 pp.

BLUEMLE, J. P., 1972, Pleistocene drainage development in North Dakota: Geol. Soc. American Bull., v. 83, pp. 2189–94.

BOWLER, J. M., HOPE, G. S., JENNINGS, J. N., SINGH, G., and WALKER, D., 1976, Late Quaternary climates of Australia and New Guinea: Quaternary Res., v. 6, pp. 359–94.

BOWLER, J. M., and WASSON, R. J., 1983, Glacial age environments of inland Australia, *in* Vogel, J. C., ed., Late Cenozoic paleoclimates of the southern hemisphere: A.A. Balkema, Rotterdam, The Netherlands, pp. 183–208.

BÜDEL, J. K., 1957, The ice age in the tropics: Universitas (Quarterly English Language Ed.), v. 1, pp. 183–91.

BUTZER, K. W., ISAAC, G. L., RICHARDSON, J. L., and WASHBOURN-KAMAU, C., 1972, Radiocarbon dating of East African lake levels: Science, v. 175, pp. 1069–76.

COHMAP MEMBERS, 1988, Climatic changes of the last 18,000 years: Observations and model simulations: Science, v. 241, pp. 1043–52.

DOHRENWEND, J. C., 1987, Basin and Range, *in* Graf, W. L., ed., Geomorphic systems of North America: Geol. Soc. America Centennial Spec., Vol. 2, pp. 303–42.

DORN, R. I., DeNIRO, M. J., and AJIE, H. O., 1987, Isotopic evidence for climatic influence on alluvial-fan development in Death Valley, California: Geology, v. 15, pp. 108–10.

DORN, R. I., JULL, A. J. T., DONAHUE, D. J., LINICK, T. W., and TOOLIN, L. J., 1989, Accelerator mass spectrometry radiocarbon dating of rock varnish: Geol. Soc. America Bull., v. 101, pp. 1363–72.

DURY, G. H., 1973, Paleohydrologic implications of some pluvial lakes in northwestern New South Wales, Australia: Geol. Soc. America Bull., v. 84, pp. 3663–76.

ESCHMAN, D. F., 1985, Summary of the Quaternary history of Michigan, Ohio, and Indiana: Jour. Geol. Education, v. 33, pp. 161–67.

FLINT, R. F., 1971, Glacial and Quaternary geology: John Wiley & Sons, Inc., New York, 892 pp.

FLOHN, H., 1953, Studien über die atmosphärische Zirkulation in der letzten Eiszeit: Erdkunde, v. 7, pp. 266–75.

FRAKES, L. A., and KEMP, E. M., 1972, Influence of continental positions on early Tertiary climates: Nature, v. 240, pp. 97–100.

GALLOWAY, R. W., HOPE, G. S., LÖFFLER, E., and PETERSON, J. A., 1973, Late Quaternary glaciation and periglacial phenomena in Australia and New Guinea, *in* van Zinderen Bakker, E. M., ed., Paleoecology of Africa and the surrounding islands and Antarctica: A.A. Balkema, Cape Town, South Africa, v. 8, pp. 125–38.

GUILCHER, A., 1976, Les côtes à rias de Corée et leur évolution morphologique: Ann. Géographie, v. 85, pp. 641–71.

VAN DER HAMMEN, T., WIJMSTRA, T. A., and ZAGWIJN, W. H., 1971, The floral record of the late Cenozoic of Europe, *in* Turekian, K. K., ed., Late Cenozoic glacial ages: Yale Univ. Press, New Haven, Conn., pp. 391–424.

HOPE, G. S., and PETERSON, J. A., 1975, Glaciation and vegetation in the high New Guinea mountains: New Zealand Royal Soc. Bull. 13, pp. 155–62.

————, 1976, Paleoenvironments, *in* Hope, G. S., Peterson, J. A., Radok, U., and Allison, I., eds., The equatorial glaciers of New Guinea: A.A. Balkema, Rotterdam, The Netherlands, pp. 173–205.

JACOBSON, G. L., JR., WEBB, T., III, and GRIMM, E. C., 1987, Patterns and rates of vegetation change during the deglaciation of eastern North America, *in* Ruddiman, W. F., and Wright, H. E., Jr., eds., North America and adjacent oceans during the last deglaciation: Geol. Soc. America, The geology of North America, v. K-3, pp. 277–88.

JOHNSON, W. H., 1985, Stratigraphy and correlation of the glacial deposits of the Lake Michigan lobe prior to 14 ka BP: Quaternary Sci. Rev., v. 5, pp. 17–22.

KALE, V. S., and RAJAGURU, S. N., 1987, Late Quaternary alluvial history of the northwestern Deccan upland region: Nature, v. 325, pp. 612–14.

KEMPTON, J. P., 1983, Stratigraphy of the sediments within the Mahomet Bedrock Valley in Illinois [abs.]: Geol. Soc. America Abstracts with Programs, v. 15, p. 610.

KENNETT, J. P., HOUTZ, R. E., ANDREWS, P. B., EDWARDS, A. R., GOSTIN, V. A., HAJOS, M., HAMPTON, M. A., JENKINS, D. G., MARGOLIS, S. V., OVENSHINE, A. T., and PERCH-NIELSON, K., 1974, Development of the Circum-Antarctic Current: Science, v. 186, pp. 144–47.

KERSHAW, A. P., 1988, Australasia, *in* Huntley, B., and Webb, T., III, eds., Vegetation history: Kluwer Academic Publishers, Dordrecht, The Netherlands, pp. 237–306.

LÖFFLER, E., 1972, Pleistocene glaciation in Papua and New Guinea: Zeitschr. für Geomorph., Supp. v. 13, pp. 32–58.

————, 1977, Geomorphology of Papua New Guinea: Commonwealth Scientific and Industrial Research Organization, Canberra, 195 pp.

MAXWELL, J. A., and DAVIS, M. B., 1972, Pollen evidence of Pleistocene and Holocene vegetation on the Allegheny Plateau, Maryland: Quaternary Res., v. 2, pp. 506–30.

MCCLURE, H. A., 1976, Radiocarbon chronology of late Quaternary lakes in the Arabian desert: Nature, v. 263, pp. 755–56.

MIFFLIN, M. D., and WHEAT, M. M., 1979, Pluvial lakes and estimated pluvial climates of Nevada: Nevada Bur. Mines, Geology, Bull. 94, 57 pp.

OBERLANDER, T. M., 1974, Landscape inheritance and the pediment problem in the Mohave Desert of southern California: Am. Jour. Sci., v. 274, pp. 849–75.

OVIATT, C. G., and NASH, W. P., 1989, Late Pleistocene basaltic ash and volcanic eruptions in the Bonneville basin, Utah: Geol. Soc. America Bull., v. 101, p. 292–303.

PFLUG, R., 1969, Quaternary lakes of eastern Brazil: Photogrammetria, v. 24, pp. 29–35.

PORTER, S. C., 1964, Composite Pleistocene snow line of Olympic Mountains and Cascade Range, Washington: Geol. Soc. America Bull., v. 75, pp. 477–82.

ROGNON, P., 1987, Late Quaternary climatic reconstruction for the Maghreb (North Africa): Palaeogeography, Palaeoclimatology, Palaeoecology, v. 58, pp. 11–34.

RUDDIMAN, W. F., and McINTYRE, A., 1973, Time-transgressive deglacial retreat of polar waters from the North Atlantic: Quaternary Res., v. 3, pp. 117–30.

SCHUMM, S. A., 1968, River adjustment to altered hydrologic regimen—Murrumbidgee River and paleochannels, Australia: U.S. Geol. Survey Prof. Paper 598, 65 pp.

SHACKLETON, N. J., and OPDYKE, N. D., 1973, Oxygen isotope and paleomagnetic stratigraphy of equatorial Pacific core V28-238: Oxygen isotope temperatures and ice volumes on a 10^5 year and 10^6 year scale: Quaternary Res., v. 3, pp. 39–55.

———, 1976, Oxygen-isotope and paleomagnetic stratigraphy of Pacific core V28-239 late Pliocene to latest Pleistocene, in Cline, R. M., and Hays, J. D., eds., Investigations of late Quaternary paleoceanography and paleoclimatology: Geol. Soc. America Mem. 145, pp. 449–64.

SINGH, G., JOSHI, R. D., CHOPRA, S. K., and SINGH, A. B., 1974, Late Quaternary history of vegetation and climate of the Rajasthan Desert, India: Royal Soc. London Philos. Trans., ser. B, v. 267, pp. 467–501.

SINGH, G., JOSHI, R. D., and SINGH, A. B., 1972, Stratigraphic and radiocarbon evidence for the age and development of three salt lake deposits in Rajasthan, India: Quaternary Res., v. 2, pp. 496–505.

SMITH, G. I., and STREET-PERROTT, F. A., 1983, Pluvial lakes of the western United States, in Porter, S. C., ed., Late Quaternary environments of the United States, v. 1, The late Pleistocene: Univ. Minnesota Press, Minneapolis, pp. 190–212.

SPAULDING, W. G., LEOPOLD, E. B., and VAN DEVENDER, T. R., 1983, Late Wisconsin paleoecology of the American southwest, in Porter, S. C., ed., Late Quaternary environments of the United States, v. 1, The late Pleistocene: Univ. Minnesota Press, Minneapolis, pp. 259–93.

STREET, F. A., and GROVE, A. T., 1976, Environmental and climatic implications of late Quaternary lake-level fluctuations in Africa: Nature, v. 261, pp. 385–89.

———, 1979, Global maps of lake-level fluctuations since 30,000 yr B.P.: Quaternary Res., v. 12, pp. 83–118.

SWADLEY, W C, 1980, New evidence supporting Nebraskan age for origin of Ohio River in north-central Kentucky: U.S. Geol. Survey Prof. Paper 1126-H, 7 pp.

TELLER, J. T., 1973, Preglacial (Teays) and early glacial drainage in the Cincinnati area, Ohio, Kentucky, and Indiana: Geol. Soc. America Bull., v. 84, pp. 3677–88.

TIGHT, W. G., 1903, Drainage modifications in southeastern Ohio and adjacent parts of West Virginia and Kentucky: U.S. Geol. Survey Prof. Paper 13, 111 pp.

TWIDALE, C. R., 1983, Australian laterites and silcretes: Ages and significance: Rev. Geol. Dynam. Geog. Phys., v. 24, pp. 35–45.

VALLADAS, H., GENESTE, J. M., JORON, J. L., and CHADELLE, J. P., 1986, Thermoluminescence dating of Le Moustier (Dordogne, France): Nature, v. 322, pp. 452–54.

VERSTAPPEN, H. T., 1964, Geomorphology of the Star Mountains: Nova Guinea, Geology, v. 5, pp. 101–58.

———, 1970, Eolian geomorphology of the Thar desert and paleoclimates: Zeitschr. für Geomorph., Supp. v. 10, pp. 104–20.

VITA-FINZI, C., 1969, The Mediterranean valleys: Geological changes in historical times: Cambridge Univ. Press, Cambridge, England, 140 pp.

WATERS, M. R., 1989, Late Quaternary lacustrine history and paleoclimatic significance of pluvial Lake Cochise, southeastern Arizona: Quaternary Res., v. 32, pp. 1–11.

WEST, R. G., 1968, Pleistocene geology and biology, with especial reference to the British Isles: John Wiley & Sons, Inc., New York, 377 pp.

PART IV

COASTAL GEOMORPHOLOGY

The fluid that covers 71 percent of the earth's rocky crust is seawater, not air. Weathering, mass wasting, and erosion proceed in very different ways under water than under air. Deposition dominates over erosion. The deeper sea floor is mantled with a thin layer of sediment that settles out of the overlying water mass, vaguely analogous to the loess mantle that covers large areas of subaerial landscape. The fine sediment, both biogenic and terrigenous, that covers most of the sea floor is in places overwhelmed by a thickness of many thousands of meters of coarser sediment derived from subaerial erosion. Most of the coarse sediment stays near the continental margins, but some is carried by subaqueous mass wasting into the deepest ocean basins.

The systematic geomorphology of coasts involves a localized set of processes and landforms encountered neither in the subaerial nor in the deeper submarine landscape. Energy is fed into the coastal geomorphic system primarily by surface ocean waves and is expended over an area that is narrow in height and width relative to its great length. If the relative level of land and sea had not changed, all coastal landforms would be in a zone about 10 m above and below mean sea level and a few kilometers inland and seaward of the average shoreline position. Both tectonic movements of the land and eustatic (p. 435) sea-level changes have occurred, but even so, coastal landforms have a narrow distribution.

A peculiar and important aspect of coastal geomorphology is that the narrow vertical operating range of coastal processes ensures that small relative changes of land and sea level will produce relict coastal landforms above or below those that are now forming. Considering the frequency and magnitude of Quaternary sea-level fluctuations (Chap. 17), it is hardly surprising that modern coastal morphology is highly complex. Relict and active landforms are intimately mixed and frequently confused.

Subaerial processes interfere with or contribute to coastal geomorphic change as well.

The coastal ocean, including estuaries, is a zone of enormous biologic productivity. In warm tropical oceans, organisms build major shallow-water or coastal landforms. Entire island archipelagos are emerged coral reefs, uplifted by tectonic movements or relict from times of higher Quaternary sea levels. There is no subaerial analog for these huge biogenic constructional landforms.

Climatic morphogenesis is not a major topic in coastal geomorphology. Certainly coral reefs and mangrove swamps are tropical, hurricanes and typhoons are low-latitude phenomena, and ice-pushed beach ridges are high-latitude landforms. However, long-period surface waves transport energy across entire oceans with minimal loss, diffusing climatic morphogenetic gradients. The degree of exposure to regional surface wave energy is more significant to coastal geomorphology than the local climate.

Within the narrow, almost linear coastal zone, a complex range of landforms are eroded and built, organisms evolve special forms of adaptation, and human activity is focused. By United Nations estimates, two-thirds of the world's people live in the 10 percent of land defined as the coastal zone. Food resources, commerce, and recreation are critically dependent on the dynamic landforms of coasts. Coastal scenery has been the subject of great artistic expression and massive engineering projects.

A major current concern is that the present trend of rising sea level and climate warming could cause a collapse of the West Antarctic Ice Sheet over the next few centuries (p. 409). The resulting glacial-eustatic rise of sea level would be about 6 m, enough to displace a large fraction of the human population and flood most of the world's major cities. It would be futile to build sea walls and dikes against such a global catastrophe; cities would have to be rebuilt and people would be forced to migrate. The political and social consequences of such an event are awesome, but must be evaluated realistically.

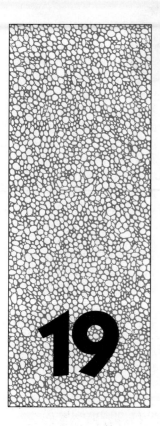

19 Shore-Zone Processes and Landforms

The **shoreline** is the line of demarcation between land and water. It fluctuates from moment to moment, influenced by waves and tides. The *shore zone,* or simply **shore,** is the zone affected by wave action. It is conveniently subdivided into (1) the **offshore,** the shallow bottom seaward of the breaking waves; (2) the **nearshore,** between the breaker zone and low-tide level; (3) the **foreshore,** which extends from low-tide level to the limit of high-tide, storm-wave effects; and (4) the **backshore,** from the limit of recent or frequent storm waves landward to the base of a cliff, dune, or vegetated beach ridge. These definitions are not precise: "Offshore" oil is sought on the continental shelves to a depth of 200 m or more. In nautical use, the offshore zone is a place of hazardous navigation in contrast to the "open" sea. Lakes and ponds as well as oceans have shores, but the more general word **coast** applies only to the geographic region adjacent to an oceanic shore.

ENERGETICS OF SHORE-ZONE PROCESSES

Wind Waves

Most of the energy expended in the shore zone is transmitted there by wind-generated waves on the water surface (Fig. 5–2). It is easy to visualize how a shearing stress by wind over water produces mass transport in the surface-water layer, but why the water surface should assume a wave form is not simple to explain. One theory emphasizes the inherent gustiness of wind; another theory explains waves as a result of slight pressure differences over small ripples. The obvious phenomenon of waves being kicked up by a wind is in fact an elusive and complex process (Kinsman, 1965).

In an area of strong winds and storminess over the ocean, the water surface is thrown into a confused mass of waves that intersect in peaks and troughs. Smaller waves and ripples run up the backs of larger waves. Mariners have for centuries

469

called such an area of wave generation a **sea.** When "a sea is running," it is better for ships to avoid the area. Waves as high as 34 m have been reported, but accurate observation of such waves is understandably difficult. The height of waves generated by wind depends on wind *speed,* the *duration* of wind from one direction, and the *fetch,* or length of water surface over which the wind blows. For example, a near gale (wind speeds to 17 m/s) must blow for 6 to 12 hours over a fetch of 40 to 140 km to produce waves 50 to 75 percent as high as the theoretical maximum height of 6 m. Tables of predicted wave heights for various combinations of wind speed, duration, and fetch are available. The role of fetch in wave generation explains why even large lakes never have wave heights comparable to those measured in the open ocean.

Radiating outward from the generating area of a sea, waves with the longest wavelengths advance most rapidly. Steep waves, with large height/length ratios, decay rapidly, but long, low waves radiate for thousands of kilometers across oceans with little energy loss (Fig. 19–1). The regular pattern of smooth, rounded waves that characterizes the surface of the ocean during fair weather is called **swell.** Swell is usually composed of many wave trains of different wavelengths, moving outward from more than one generating area. From the air, swell patterns look like a grid of intersecting lines. Swell approaches a shore as a complex of wave trains that combine to produce alternately a succession of higher waves and a succession of lower waves.

Swell waves transmit energy outward from a stormy area in two forms: potential and kinetic. The height of a wave determines the potential energy of position above still-water level. The motion of individual water particles as a wave passes is a measure of the kinetic energy of the wave (Fig. 19–2). The potential energy of the waves moves forward with the wave form, but the kinetic energy of each moving water particle is expended in the nearly circular orbit of the particle. The orbital diameters of water particles beneath a wave decrease rapidly with depth in a geometric progression related to the wavelength (Gross, 1990, p. 196). The orbital diameter is halved for each increase in depth equal to one-ninth of the wavelength. Thus, at a depth equal to the wavelength, the orbital diameter is only about 1/512, or 0.2 percent, of the surface diameter. This rapid decrease in water motion with depth explains why a submarine can ride out the most severe storm by submerging 50 m or more. A wave is assumed to be moving in deep water when the water depth exceeds

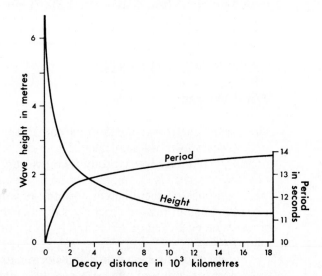

Figure 19-1. Effect of swell decay on the wave period and height of a wave initially 6.1 m high with a period of 10 seconds. (Davies, 1980, Fig. 16.)

one-half the wavelength. At that depth, the drag of the wave on the bottom is negligible. Swell waves are not deflected by the Coriolis effect (p. 110) but move across deep water in a direction perpendicular to their crests, along great-circle tracks (Komar, 1976, p. 99; Davies, 1980, p. 33). This tendency, plus the persistence of swell waves, means that energy for coastal erosion can be generated by storms in the opposite hemisphere as much as 15,000 km away.

Breakers and Surf

As deep-water swell approaches a shore, the form of the waves and the method of water and energy transport change. The orbital velocity u of a water particle in a surface wave is

$$u = \pi H/T$$

where H is the orbital diameter (wave height) and T is the wave period, the time required for the particle to move through a complete orbit as the wave moves forward one full wavelength. This particle velocity decreases rapidly with depth, as noted in the preceding paragraph. The shoaling sea floor distorts the orbital motion beneath the waves from a circle to an ellipse, then to a to-and-fro linear motion. If the orbital velocity at the bottom exceeds the threshold of erosion of the bottom sediment, which is at least 15 cm/s (Fig. 9–10a), sediment on the ocean floor is moved back and forth by the waves and absorbs energy from the moving water. At this depth, the conversion of wave energy to geomorphic work begins.

The depth at which waves begin to stir bottom sediment is determined by wave height, wavelength, and period, plus the grain size and sorting of the bottom sediment. Although very long-period swell waves may stir water and bottom sediment to great depths, the rapid decay of the orbital velocity with depth ensures that ordinary waves will not significantly transport or erode bottom sediment at a depth greater than about 10 m. This depth, which is called **wave base,** or **surf base** (Dietz, 1963), is a convenient reference level, although we should keep in mind that the work of waves does not abruptly cease at a certain depth.

When a swell wave begins to "feel" a shoaling bottom and become a geomorphic agent, the wave velocity is decreased by bottom friction. The wavelength is correspondingly shortened, the wave height increases, and the wave form steepens (Horikawa, 1988, Part I). As the crest of each wave approaches, the water moves rapidly forward as well as upward. Ultimately, the forward motion of the water mass in the wave equals the decreasing forward motion of the wave front. The wave assumes a steep face, and then its crest collapses forward into the trough. The deep-water swell becomes **surf,** or lines of **breakers** (Fig. 19–16). Breakers form in the surf zone where the bottom depth is about one-third greater than the breaker

Figure 19-2.   Motion of surface water particles during the passage of a wave. The orbital diameter decreases rapidly with depth beneath the wave.

height. On gently sloping coasts, the first line of breakers may be as much as a kilometer offshore, and the incoming water may reform into lower waves that break again at successively shallower depths. Wave energy is expended throughout the zone of bottom friction.

Swell waves can be *reflected* by a vertical seawall or steep cliff with almost no transfer of energy from the waves to the structure, because the mass of water in each wave is not moving forward significantly more than the diameter of the particle orbit. However, when a breaking wave hits a cliff or seawall, thousands of tons of water are in motion against the structure. The greatest pressures are exerted by breaking waves that curl over at their crest and trap air between the wave face and a steep wall or cliff so that the air is compressed; shock pressures of 608 kPa (12,700 lb/ft²) have been recorded against seawalls. The duration of such great pressure is less than one-hundredth of a second, but large blocks of rock can be pried loose and moved by the repeated assault of the waves.

The wave power expended against the coral reef of Bikini Atoll was carefully calculated by Munk and Sargent (1954). For a typical average breaker height of 7 ft (2.1 m), the waves' total power against the reef is 3.73×10^5 kW (500,000 horsepower). On the sector most exposed to the trade winds, the power reaches 20 kW/m (8 horsepower per linear foot) of reef front. Many coasts have even more energetic environments. Attempts have been made to harness wave energy for electric power generation, but the oscillating nature of the energy is difficult to convert to a useful form.

As a deep-sea swell crosses an irregular shoaling coastal bottom, wave *refraction* occurs, analogous to the bending of light rays on passing through a lens. The segment of each wave in the shallowest water is slowed so that in plan view the wave crest becomes concave forward. The effect is best illustrated by constructing a series of orthogonals perpendicular to wave crests (Fig. 19–3). A shallow submarine ridge thus focuses wave energy against its flanks, whereas a submarine valley or basin causes wave orthogonals to diverge and wave crests to be attenuated. Because of the postglacial rise of sea level, most offshore submarine ridges and valleys are continuations of the present subaerial topography, so that modern headlands are usually fronted by submarine ridges, and bays by submarine valleys. The effect is to cause wave energy to converge on headlands and diverge into bays.

Wave refraction is the basis for two important generalizations about the evolution of coasts. First, an irregular initial shoreline becomes more regular as initial seaward protrusions caused by sloping ridges tend to be eroded faster than adjacent shoreline segments in coves or bays at the heads of submarine valleys. Second, wave refraction generates currents flowing along the shore from headlands where the breakers are focused and the water level is higher to adjacent coves where waves are attenuated and the water level is lower. These wave-generated **longshore** or **littoral currents,** along with tidal and other currents, transport the sediment eroded from headlands into adjacent coves where beaches are built (Fig. 19–3). The simplification of a shoreline by refracted waves thus involves the filling of bays as well as the erosion of headlands. Even on straight coasts such as those lined by sandy barriers (p. 491), submerged obstructions, tidal inlets and other features break the coastal currents into *littoral cells,* within which sand is transported from available sources in the shore zone or nearshore bottom to a variety of sediment traps, where it is deposited (p. 488). The power expended in the coastal system by wind-driven waves is large enough (Fig. 5–2) that even if only 10 percent of it were converted into longshore currents, those currents could transport 58,000 times more sediment along coasts than is supplied to the world's coast by all subaerial erosion (Inman and Brush, 1973, p. 26). As with the subaerial landscape, net erosion is the most likely coastal condition, with deposition occurring only in special, sheltered locations.

Figure 19-3. Wave refraction over an irregular shoaling bottom. Orthogonals are drawn perpendicular to the wave crests, equally spaced along the swell so that segments 1 to 4 have equal amounts of energy. Note the convergence of energy on the headland and divergence into the coves.

Waves are *diffracted* when a segment of wave crest passes between headlands or artificial constrictions into open water such as a river mouth or bay. The littoral currents generated within bays and harbors by waves erode and transport shore sediment until the bay shoreline "fits" the refraction and diffraction patterns closely. In theory, shorelines shaped by refracted and diffracted waves should assume gentle curvatures parallel to the wave fronts of the dominant swell. In fact, headlands are commonly composed of more resistant rock than the shores of adjacent coves, and even though wave attack is concentrated on the headlands, the more resistant rocks continue to form structurally controlled seaward projections as the entire shoreline retreats (Fig. 19–17). Straight shorelines are characteristic only of coasts on terranes of uniform erodibility.

Storm Surges and Seismic Sea Waves

As a cyclonic (low atmospheric pressure) storm approaches a coast, a series of processes combine to endanger coastal installations and promote rapid geomorphic change. A drop of 100 Pa (1 mb) in atmospheric pressure causes the sea surface to rise 1 cm (1 bar ≃ 10 m seawater) by the "inverted barometer" effect. Hurricanes or typhoons frequently have central atmospheric pressure 100 mb below normal, so the general rise of the sea surface under the storm may approach 1 m. Even more important is the shear stress of high, persistent winds on the sea surface, actually driving masses of water forward and raising mean sea level several meters along several hundred kilometers of coast. This "wind setup" combined with the "inverted barometer" effect produces a **storm surge,** with shore-zone water depths several times as great as the highest normal tides. Waves, by far the most effective product of such storms, are even more effective because they move shoreward through water several meters deeper than normal and break against land that is poorly shaped to absorb wave energy (Hayes, 1967).

A *seismic sea wave,* often called by the Japanese name **tsunami,** is entirely distinct from wind waves. Sudden motion of the sea floor during an earthquake (Fig. 2–2) creates a solitary wave, or at most two or three waves, of very low amplitude that moves rapidly over the deep ocean. Successive arrival times prove that a tsunami moves with a velocity of 700 to 800 km/h, with a wavelength of 200 to 300 km, a height of about 2 m, and a period of about 20 minutes. Because of their extremely long wavelength and low height, tsunamis cannot be detected as they pass through the swell patterns of the open ocean, science fiction stories notwithstanding. As they cross a shoaling bottom, the wavelength shortens, and the wave height greatly increases to 15 m or more, with disastrous consequences when it breaks. The height may increase even more if the wave is funneled into a narrowing bay or estuary. Tsunamis are not rare in the Pacific Ocean; their recurrence interval for Hawaii is about 25 to 50 years.

Tides

Tides are a unique form of energy for geomorphic change. Power is dissipated by the tidal drag of the moon and sun on the rotating earth at a rate of about 3×10^9 kW (Fig. 5–2). The complex periods and heights of tides are driven by lunar and solar gravity but are affected by the size and depth of the various ocean basins, the shape of their shorelines, and the latitudes of the basins, among many other variables (Neumann and Pierson, 1966, pp. 298–325). Seasonal winds and atmospheric pressure patterns also influence the tides. Each ocean, gulf, and sea has its own tidal pattern. The energy of tides is primarily dissipated by shallow seas such as the Bering Sea, the Sea of Okhotsk, the Timor Sea, the Patagonian shelf, and the Hudson Straits. These five shallow seas alone account for nearly one-third of the total energy conversion (Hendershott, 1973, p. 76). We can justifiably conclude that the coastal and submarine geomorphology of shallow seas are strongly shaped by tidal energy.

Tidal ranges are high only where the tides enter a semienclosed sea or gulf that has a shape and depth to produce a natural period of oscillation of water that is in resonance with some period of the tide. The greatest tidal range on earth occurs in the Bay of Fundy, Canada, where sea level may vary 16 m between high and low water. The Bay of Fundy has a natural period of oscillation of slightly over 6 hours, almost in phase with the ideal semidiurnal tide (Garrett, 1972).

In addition to their primary geomorphic role of raising and lowering the level of wave attack against the land, tides also generate strong currents. Tidal currents compete with waves and river currents in shaping deltas (Fig. 10–7). The most rapid currents develop in narrow channels that connect basins with different tidal periods. Hell Gate in the East River at New York City is one such place. The high tide in Long Island Sound takes almost 3 hours to move westward through the Sound. In that time, the tide in the less restricted lower New York Harbor has crested and is halfway down to low water again. Because of the difference in level, the water in the

Sound *ebbs* (flows out) southward through Hell Gate at speeds of 9 km/h (1 km/h = 28 cm/s). About 6 hours later, the tide in the west end of the Sound is low, but the water in New York Harbor is nearing high-tide level, and the current *floods* (flows in) northward into the Sound almost as fast as it previously ebbed. The name of the strait is an obvious reference to the violent and changeable tidal currents.

Where a tide enters a large funnel-shaped river mouth, it begins to drag the bottom like any long wave and steepen its forward slope. In extreme examples, the river mouth develops a **tidal bore,** a single wave that races upstream at 15 to 30 km/h as a vertical wall of water as much as 3 m high. Tidal currents and bores are obviously capable of extensive erosion and transportation of sediment (Ludwick, 1974). The current velocities just cited can erode and transport any sediment from compact clay to boulders (Fig. 9–10). If abrasive tools are available, deep channels can be eroded in bedrock as well.

Wind

Wind is the key link in the conversion of solar energy to wave energy (Fig. 5–2). In the shore zone, wind also acts directly as a geomorphic agent. Low tides expose extensive areas of wave-transported sand for several hours twice each day. Onshore winds dry the sand and blow it into dunes or transport it inland beyond the reach of waves. On some sandy shores, wind accounts for the major loss of sand from the shore system. If wave energy and sediment sources are adequate, the sand is replaced from offshore or alongshore. If not, beaches can be stripped of sand by wind.

The limited travel distance of saltation-transported sand grains (p. 354) keeps most of the windblown sand within the coastal region. Sand dunes are an inherent component of sandy coasts. They are usually of the transverse or parabolic type (Figs. 14–10 and 14–11), because sand supply is renewed by waves, winds are irregular or seasonal, and vegetation restricts dune expansion. Some desert coasts are backed by fields of barchan dunes migrating inland from the beach (Fig. 14–14).

Shore-Zone Weathering and Mass Wasting

On oceanic shores, repeated wetting and drying by saline water is an effective weathering agent (Trenhaile, 1987, Chap. 2). In addition, a remarkable variety of shore-zone plants and animals bore, scrape, and dissolve rocks, providing biological energy for shore-zone processes (Trenhaile, 1987, Chap. 4). At our present stage of knowledge, it is not practical to attempt a detailed classification of shore-zone biologic, chemical, and mechanical weathering processes, because the three are complexly interrelated.

Intertidal life zones and weathering processes are strongly affected by the length of the alternating emergent and submergent intervals. As each zone has a diagnostic flora and fauna, so it has an associated set of weathering processes (Stephenson and Stephenson, 1972). The spray zone, just above normal high tides, is a zone of especially active chemical weathering, with measured surface lowering of 0.625 mm/yr in southern England (Mottershead, 1989). Lower and higher zones, which remain either wet or dry for longer intervals, are less affected. On some rocky shores, the lower intertidal zone is dominated by algal or molluscan encrusting deposits, whereas on other shores, burrowing organisms such as sea urchins abrade craters into the rock. The net effect is a set of shore-zone landforms with very pronounced vertical zonation.

Because the recent postglacial rise of sea level has put the shoreline against a variety of former subaerial landscapes, many coasts include cliffs that drop abruptly to the shore. Wave erosion at the base of a coastal cliff maintains its steep slope, but above the reach of the waves, the cliff evolves by the same processes of mass wasting that are found on any inland cliff (Chap. 8). If wave erosion is weakened or eliminated, for instance by beach deposition, the cliff may become fossil and evolve to a steep grassy or wooded slope. If wave erosion remains active, the cliff may provide the abrasive tools to erode the shore platform (p. 478) at its base, or it may provide so much debris that it buries the shore platform and forces the shoreline to advance into the sea. A complex equilbrium is maintained between the processes of subaerial mass wasting and the processes of coastal erosion and deposition, as described in the following paragraphs.

EROSIONAL SHORE-ZONE LANDFORMS

Sea level has risen close to its present level only within the last 6000 years with the melting of the large ice sheets of the Wisconsin glaciation. Considerable uncertainty remains about the last 6000 years, during which sea level rose a few meters above its present level and then subsided on some coasts, was very nearly stable on others, and has continued a slow rise to the present level on still others. The primary cause for the great variety of coastal Holocene sea-level histories is the complex interaction of the postglacial eustatic rise of sea level of about 120 m between 15,000 and 6000 years ago (p. 435) and the isostatic response of the coastal lithosphere (Clark and Lingle, 1979). Some coasts in formerly glaciated regions have isostatically risen more than the postglacial sea level rise and show flights of emerged shorelines (Fig. 17–13). Other coasts have isostatically subsided under the load of the water that was returned to the oceans by melting ice. Probably no two coastal regions have identical Holocene sea-level histories (Bloom, 1983). The details are complicated by tectonic movements, which have deformed many shorelines (pp. 24 and 500).

There are scarcely any coasts of resistant rock that show significant Holocene erosional shore-zone landforms. Those that do have inherited the basic landforms from interglacial ages, the latest of which was about 120,000 years ago, when sea level was a few meters higher than now (p. 437). If an inherited shore platform or wave-cut cliff base has been reoccupied by the sea for even a few thousand years, it is likely to appear deceptively well developed (Fig. 19–4). Only careful search and good fortune can lead to the discovery of uneroded remnants of sedimentary cover on ancient platforms by which their great age can be proved.

The Holocene is a particularly complex time to study shore-zone landforms because so many are relict or exhumed. However, enough weak-rock coasts have modern erosional forms, and enough relict forms are preserved, to give us a good idea of what shore-zone erosion could do if enough time were available (Cotton, 1955).

Shore Platforms

Where the postglacial rise of sea level has created a shoreline on a former hill slope, shore-zone processes cut a cliff and bench. The cliff is controlled by mass wasting and structural factors. The eroded bench or platform, approximately intertidal, is given the nongenetic descriptive name **shore platform.**

Processes of Shore-Platform Erosion. The hydraulic pressure and turbulence of breaking waves can easily erode weathered and joint-bounded fresh bedrock (Fig. 19–5). The initial stage of shore-platform erosion is probably the direct removal of rock by waves, which by analogy with river and glacial work is called *wave plucking* or *quarrying.* Even though wave plucking is restricted to a narrow vertical zone, high cliffs can result if sliderock is exported from the talus at the cliff base. **Sea caves** or a **notch** may be cut at the base of a cliff.

Figure 19-4. The chalk cliffs at Sangat, France, are fronted by a prominent shore platform. However, a Paleolithic site on the platform near the low-water line proves that the surface was cut during the last interglacial or earlier, was a periglacial hillside terrace during the last glacial, and has only recently been reoccupied by the sea.

Figure 19-5. Betty Moody's Cave, Star Island, New Hampshire. Waves have quarried a columnar basalt dike from between granite walls.

Figure 19-6. Abrasion ramp cut across dipping strata, Bay of Fundy coast near Joggins, Nova Scotia.

As plucking deepens a shore-zone notch or erodes back a sea cliff, the shore platform widens. It is in the surf zone, and sediment from wave plucking and cliff retreat is dragged back and forth across the platform to shape an **abrasion ramp** (Fig. 19–6). Wave abrasion requires constant removal of plucked or mass-wasted sediment from the abrasion ramp lest the surface is buried. Thus a seaward *slope of transportation* develops on an eroding surface thinly veneered with rock debris in active transport. The abrasion ramp extends seaward until the debris is either abraded to fine sediment, dissolved, or settles into water too deep for further wave movement. From careful profiles across modern abrasion ramps and emerged coastal terraces in California, Bradley (1958) concluded that their typical seaward gradient is about 1° (about 2 percent). On a tideless shore, such a gradient would extend below the 10-m depth of ordinary wave erosion on a bench no wider than 500 m. If a tide range of 5 m is assumed, an abrasion ramp with a 1° gradient could still be no wider than 800 m. Shore platforms wider than that cannot develop by abrasion at a fixed sea level, although they could form during progressive submergence.

Abrasion platforms can form rapidly on weak rocks. A subhorizontal shore platform up to 200 m in width was cut in 600 years across erodible mudstones on the east coast of New Zealand during slow tectonic subsidence. Then it was abruptly uplifted, probably during an earthquake about 2200 years ago. The abrupt emergence preserved the fossiliferous cover strata on the platform. Radiocarbon dates on mollusk shells in the cover strata prove that the platform was youngest at the base of the fossil cliff at its landward edge, and had progressively widened at a rate of 0.2 to 0.4 m/yr (Hull, 1987).

Older geomorphic literature proposed continental-scale marine planation as an alternative to subaerial peneplanation. However, the necessity for a seaward slope on an abrasion ramp ensures that even a small island could not be planed off to sea level by abrasion alone. Wide shore platforms, or those with gentle gradients, must be cut by processes other than abrasion (Bradley and Griggs, 1976).

Water-level weathering (Wentworth, 1938) seems to be the dominant process that widens and flattens shore platforms on noncarbonate rocks. Intertidal pools and salt spray on abrasion ramps or in wave-plucked notches promote weathering in the zone just above permanent saturation. Wave erosion during high tides carries

(a)   (b)

Figure 19-7. "Tesselated pavement," an intertidal platform shaped by water-level weathering on horizontal, jointed sandstone south of Hobart, Tasmania. Blocks have either raised rims (a), or raised centers (b).

off the weathered detritus (Fig. 19–7). On exposed platforms, all stages can be seen from initial ponding in joint-bounded basins to coalescing flat-floored pools separated by serrate ridges. Bedding planes and joints are frequently etched in high relief.

Shore platforms are developed by water-level weathering at various heights relative to tide level, depending on structural factors such as permeability and fractures and also on wave energy, tidal range, and climate. On Hawaiian coasts exposed to vigorous surf, platforms develop as much as 7 m above sea level by weathering at the edges of spray-filled pools, although better-developed platforms are at 3 to 4 m (Wentworth, 1938, p. 20). These high platforms on headlands converge landward into bays, where, in the absence of vigorous surf and salt spray, they are only about 1 m above sea level. Such a range in contemporary water-level weathering platforms warns of the danger in attempting to establish the heights of ancient sea levels from the evidence of relict shore platforms and abrasion ramps.

Water-level weathering puts the finishing touches on surfaces prepared by wave abrasion and plucking, especially on volcanic rocks and weakly indurated pyroclastic sediments. The abundance and high perfection of shore platforms around the islands of the peripheral archipelagos of the Pacific Ocean are in part related to the abundance of Cenozoic volcanic rocks in those regions.

On carbonate rocks in the shore zone, solution is so common that a special class of **solution platforms** is distinguished (Wentworth, 1939). These are the most complex shore platforms to describe, because their origin involves both chemical and biologic solution and deposition. Solution platforms are generally intertidal. Their surfaces are flat and smooth but not abraded (Fig. 19–8). Small pools alternate with low domes and ridges, often formed by living carbonate-secreting algae. At the landward edge of a solution platform may be a horizontal *solution notch* as deep as 3 m. Because seawater is generally saturated with calcium bicarbonate, it has been argued that a solution platform must form at the level of freshwater saturation, or water table. However, good solution platforms are also common on desert coral islands that have no fresh groundwater. Diurnal changes in the pH of intertidal pools by photosynthesis and temperature change offers one explanation for alternating solution and deposition by seawater (Revelle and Emery, 1957).

Figure 19-8.   Solution platform almost truncating residual karst blocks of calcareous eolianite near Robe, South Australia.

Even more effective but complex is the role of biochemical weathering and erosion in developing solution platforms and notches on carbonate rocks. Many intertidal plants, especially algae, penetrate their substrate with microscopic strands of organic tissue (p. 151). *Lithophagic* (rock-destroying) animals graze on the limestone surface and abrade or dissolve the perforated rock surface to obtain the algae. The net chemical and mechanical effect is rapid erosion, as much as 1 mm/yr by repeated measurements on the exposed parts of pins driven into rock (Hodgkin, 1964). An estimated 154 g/yr of rock was abraded from a square meter area in Barbados by one species of lithophagic gastropod (McLean, 1967, 1972). Subtidal erosion rates of 2 to 10 mm/yr by marine boring organisms have been estimated for calcareous mudstones in California (Warme and Marshall, 1969).

Considering the complex nature of the processes that shape them, solution platforms may be misnamed. However, the complexity of intertidal carbonate geochemistry is so great and so poorly understood that "solution" is an adequate word for the present. On coral reef coasts, the geomorphology is even more complicated, because solution platforms of the sort described here grade seaward into constructional reef flats and shoreward into abrasion ramps. In the low tidal range common on tropical coral islands, the entire complex may form a flat surface more than a kilometer in width with a relief of only 1 to 2 m.

Theoretical Equilibrium Form of Shore Platforms.   If we assume a maximum depth of wave abrasion of about 10 m and certain wave energy parameters, we can deduce an equilibrium profile for an abrasion platform that can migrate shoreward but retain the same form. Zenkovich (1967, pp. 151–53) deduced that the stable profile of an abrasion ramp would be one on which the "effective" wave, or the wave form that combines optimum frequency with optimum abrasive energy, would not break. The profile should be convex upward, with the minimum slope at the shoreline and steepening seaward. The deduced profile should be longer and more gently sloping for larger effective waves. Some shore platforms in the Black Sea may approach the ideal form, but only because they are eroded on limestone by abrasive siliceous beach sediment. The lack of other examples testifies to the general disequilibrium of the shore zone in relation to present sea level. Computer models give useful predictions of the shapes of late Quaternary shore platforms that might have evolved during repeated glacial-eustatic sea-level fluctuations (Trenhaile, 1989).

Most modern shore platforms are probably inherited from relict shore platforms that were slightly higher. Old abrasion ramps or other shore platforms, eroded during the last interglacial age when sea level was a few meters above present level, form ideal surfaces on which water-level weathering and solution can operate. Because there is no definite tidal level for the modern forms, it is not possible to assert by altitude alone whether a present-day shore platform is totally relict, an inherited form modified by present processes, or entirely modern. Only where shore platforms are being exhumed from beneath cover deposits of the last glacial age (Fig. 19–4), or where they grade into constructional reefs that can be radiometrically dated as of last interglacial age, can their relict character be demonstrated.

Coastal Cliffs

The cliffs that rise from the back of shore platforms are commonly called *wave-cut* cliffs, but as with any subaerial escarpment, mass wasting is the dominant process of cliff retreat. Wave quarrying at the cliff base keeps it fresh at the appropriate angle of repose. With stable sea level, abrasion ramps eventually widen until wave attack becomes infrequent on the cliff face, when it becomes a vegetated, subdued hillside. The many bold, cliffed coasts of the world are another reminder of the essential disequilibrium between coastal landforms and present sea level.

On some cliffed coasts, the cliffs are so steep that wave-plucked debris falls into water too deep to develop a wave-cut notch or abrasion ramp (Fig. 19–9). These steep coasts are especially common on volcanic terranes, fault scarps, and glaciated coasts. The cliffs are said to *plunge* into deep water (Cotton, 1952). A striking feature of plunging cliffs is that swell waves do not break against them but are reflected back out to sea (p. 472). If rock is strong enough to form a steep cliff,

Figure 19-9. Plunging cliffs on the drowned margins of a basalt dome, Banks Peninsula, Canterbury, New Zealand. (Photo: D. L. Homer, New Zealand Geological Survey.)

and if that cliff happens to rise from water more than about 10 m deep, it is subject to only minor wave attack. Only when a submarine talus forms at the base of a cliff to create breakers can the full energy of waves be impressed on the cliff.

Laboratory wave-tank experiments using a weak sand and cement mixture to simulate a sea cliff and an oscillating paddle to create waves 7 to 9 cm in height have proved useful in documenting cliff and abrasion platform evolution (Sunamura, 1975, 1977). With the original cliff face sloping at 75°, experiments were run with (1) standing waves in "deep" water (20 cm), (2) breaking waves at the cliff face, and (3) broken waves that had re-formed after breaking in shallower water at some distance in front of the cliff face. In the first experiment, deep-water standing waves, approximately swell waves, were simply reflected from the artificial cliff with negligible erosion. The water rose and fell against the cliff face and the wave train was reflected back through the incoming waves. Waves breaking at the model cliff face

(a)

(b)

Figure 19-10. (a) Cliff erosion and beach development by broken waves in a wave tank. Time in hours. (b) Temporal variation of eroded volume in the experiment illustrated above. See text for discussion. (From Sunamura, 1977, Figs. 1 and 2.)

were the most effective erosive agents, cutting a notch as deep as 40 cm in about 60 hours. With broken waves, the erosional effects were similar but only about half as fast. With both breaking and broken waves, the erosion rate was not uniform (Fig. 19–10). Initial slow erosion freed sand grains from the weak cement, which then became abrasional tools for an interval of accelerated erosion. However, as additional sand accumulated in front of notch, wave energy was dissipated in the sand and the rate of notch deepening decreased (Sunamura, 1977). The offshore depositional profile seemed to approach a convex-upward equilibrium shape as erosion progressed. In the laboratory experiments, only a basal notch was cut at the model cliff base. In real sea cliffs, jointing and other structures would cause the entire cliff to retreat.

Structural Control of Shore-Zone Landforms

Factors of rock strength, lithology, and solubility have been repeatedly mentioned as contributors to shore-zone erosional morphology. As in the subaerial environment, some rocks are more resistant to weathering and erosion than others. Resistant rocks on coasts tend to protrude seaward as well as to maintain higher elevations. Even though wave refraction focuses more energy onto headlands of resistant rock, they continue to form headlands as erosion progresses. Wave refraction around a protruding headland can be so extreme that only a narrow neck of land is left, connecting a peninsula to the mainland. In suitable structures, especially flat-lying sedimentary rocks, wave erosion may quarry through the neck, creating a **natural arch**. With further erosion, arches collapse leaving isolated pinnacles or **sea stacks** standing on the shore platform (Fig. 19–11).

Shore platforms on coastal terraces of nearly flat-lying sedimentary rocks are especially difficult to interpret. It was noted that water-level weathering takes advantage of preexisting benches. In horizontal strata, structural benches readily form and may hold enough water to promote water-level weathering at several different

Figure 19-11. The Twelve Apostles, sea stacks eroded from flat-lying Miocene limestone, Port Campbell National Park, Victoria, Australia. (Photo: J. F. Bliney, contributed by B. G. Thom.)

intertidal and supertidal levels simultaneously. Even more commonly, slight structural dips are eliminated by water-level weathering to form horizontal platforms that truncate structure but are nevertheless basically controlled by structure, not by shore-zone processes.

CONSTRUCTIONAL SHORE-ZONE LANDFORMS

Organic Reefs

A reef is by mariners' tradition an obstruction or menace to navigation. By definition, then, **reefs** are shallow-water submarine landforms. The term is usually applied to solid, rocky structures rather than to sand bars. Most reefs are constructed by marine organisms, although some are structurally controlled, shallow, submerged ledges such as hogbacks, dikes, or lava flows.

Ecology of Organic Reefs. Throughout Phanerozoic time, since organisms with hard parts evolved, reef building has been a repeated and persistent form of marine growth, especially for colonial organisms. In shallow oceans of the Cenozoic Era, the Scleractinian colonial corals and associated calcareous algae are the most important builders of organic reefs, although mollusks, sponges, and a variety of other organisms build reefs in special situations. Reefs are built up in shallow water to take advantage of sunlight for photosynthesis, to shed or stay above smothering detrital mud, and to provide a large surface area for continued growth. Waves breaking over and refracting around reefs create currents that sweep nutrients and oxygenated water over the sedentary colonial organisms. Reefs are usually well adapted to local wave and tidal energy so that they are nourished by the energy expenditure rather than being destroyed. Larvae have a free-swimming stage before they settle and attach to the bottom, so reef-builders disperse widely.

Coral and coralline algae reefs are widespread in tropical waters (Guilcher, 1988). They require shallow (less than 70 m but with their best growth in less than 10 m) and clear water for photosynthesis. Reef growth is best in water that does not cool below about 18°C during winter months. Salinity should be close to normal. Fresh water from rivers or torrential rains is especially damaging. A hard substrate is necessary for a colony to become established, but small patch reefs can grow on soft sediments and gradually fuse to form a suitable base. Loose sand is poor for coral growth, as is mud. Despite all these restrictions, the coastal area of coral reefs is very large (Fig. 19–12). One of the most fundamental subdivisions of coasts could

Figure 19-12. Distribution of Holocene coral reefs. Some form only thin veneers on older structures. Shaded region includes the areas of most prolific reef growth and almost all the atolls. (Davies, 1980, Fig. 45.)

be between those that have corals and all others (Davies, 1980, p. 5). Organic reefs other than coral are trivial landforms and are subsequently ignored.

Coral Reef Morphology. A coral reef is an enormous mass of limestone that is organically built. The seaward edge of the reef, especially on a windward coast, is where wave-resistant growth forms thrive in the most nutrient-rich environment and provide coralline detritus in the largest quantities. Some detritus breaks off and forms a **fore-reef talus** on the ocean side of the reef. Other, often finer, detritus is swept into the sheltered region behind the reef. In the surf zone on the extreme seaward rim of many Indo-Pacific reefs, calcareous algae build smooth, rounded mounds or rims. The algal limestone is more dense than coral limestone and is a vital contributor to reef stability. More delicate coral species grow in the lee of the seaward reef, adding more detrital sediment. Either a sandy lagoon floor or an intertidal **reef flat** may extend from the lee of the reef (Fig. 19–13).

Coral reefs are generally limited in upward growth to mean low-tide level. The extreme productivity of the reef provides sediment for many subaerial shore-zone landforms, however. Large storms and tsunamis tear away tons of reef-front coralline debris and hurl it over the algal rim onto the reef flat behind (Maragos et al., 1973). Most reef flats are dotted with large erratic blocks of storm-tossed coral rock, often deeply notched by solution. On the reef flat, islands of coralline sand and gravel may build up high enough to support a fresh groundwater lens and vegetation. The carbonate sand and gravel is readily recemented, especially if exposed to rain and fresh groundwater. Major storms can remove any uncemented part of a coral-reef island, but if a mass of debris is not disturbed for a few decades or centuries, it becomes relatively indurated and resistant to all but the worst hurricanes or typhoons. Cementation is generally close to the level of mean high tide in the zone of saturation. Because the tide range in tropical oceans is generally less than 2 m, low platforms of cemented reef rubble are exposed on most eroding shores at about the same level as high-tide solution platforms (Fig. 19–14). Both kinds of high-tide platforms have been frequently mistaken for emerged reef flats by which a Holocene fall of sea level has been widely inferred.

Some coral islands become the nesting grounds for sea birds. The phosphate-rich guano reacts with limestone to form phosphate rock, much sought for fertilizer

Figure 19-13. Fringing reef with a low algal rim in the surf zone, Tongatapu, Kingdom of Tonga. Note solution notches on talus blocks and cliffs.

Figure 19-14. High-tide platform of recemented coral debris about 1.5 m above modern reef flat, Turtle Islands, northern Queensland, Australia.

manufacture. Phosphatic islands have a rugged karst surface of lapies and karren (Fig. 7–2) under a thin cover of weathered coral sand and gravel.

Classification of Coral Reefs.   The traditional mariners' terms for reefs included **fringing reef,** which is attached to and extended seaward from a coast, and **barrier reef,** which is separated from the coast by a lagoon. Later, the word **atoll,** from the Maldive Islands, was adopted into the scientific literature for an annular reef that encloses a central lagoon (Fig. 19–15). Darwin (1837) conceived of these

Figure 19-15.   Ari Atoll, Maldive Islands. An atoll 90 km long made of numerous smaller, elongate atolls on its rim (faros). Land areas (in black) are very small.

three reef forms as a genetic sequence by which a fringing reef around a submerging central island would become progressively a barrier reef and then an atoll. Darwin's deduction required subsidence of the ocean floor over vast regions (Fig. 1–5) and sparked a spirited scientific controversy in the late nineteenth century. A series of deep boreholes on atolls have subsequently demonstrated that they are underlain by 1000 m or more of shallow-water reef limestone on basaltic basement, generally validating Darwin's theory for the origin of atolls in the deep ocean basins.

Darwin's theory does not explain **shelf atolls,** the similar forms on shallow submerged continental shelves. Fairbridge (1950) suggested that shelf atolls could form by crescentic reefs extending their "horns" downwind until they closed into a ring. A simple genetic classification of coral reefs is not possible, because most of the shape is probably inherited from alternating periods of growth and subaerial erosion during Quaternary sea-level oscillations. A simple descriptive classification is given by Table 19–1.

TABLE 19–1
Descriptive Scheme for Classifying Organic Reefs

Adjacent to coasts or separated by a shallow channel				Fringing reefs
Separated from coast by deep channel				Barrier reefs
			large	Platform reefs
	without central lagoon		small	Patch reefs
Forming islands			very small	Coral pinnacles
	with central lagoon	deep lagoon		Oceanic atolls
		shallow lagoon	large	Shelf atolls
			small	Faros[a]

Source: Davies (1980), p. 70.

[a] Small, elongate atolls with narrow lagoons along the rim of large reefs (Fig. 19–15). Their origin may involve multiple periods of growth separated by subaerial karst development (MacNeil, 1954).

Inheritance and Reef Morphology. Coral reefs are very strictly controlled by sea-level datum. Emergence or submergence of even a few meters completely displaces all the depositional and erosional zones of the reef and creates relict forms. Quaternary sea-level fluctuations (Figs. 17–16 and 18–1) have left a particularly complex record on coral reefs (Hopley, 1982; Guilcher, 1988). When sea level was low, reefs were exposed as flat-topped or hill-rimmed islands that were subject to karst solution and peripheral wave erosion or reef growth. Drowned karst topography (p. 158) is well shown on the floors of some atoll lagoons (Shepard, 1970). It is hard to avoid the conclusion that most reef morphology is inherited, in part from constructional forms that date from the last interglacial when sea level was several meters higher than now, and in part from destructional karst landforms that formed during glacial-age low sea levels (Bloom, 1974; Purdy, 1974; Strecker et al., 1987). On a longer time scale, the Great Barrier Reef of eastern Australia, which extends almost 2000 km from 9 to 24° S latitude, began to grow in its northern part between 16 and 25 million years ago. The central part began to grow 10 to 15 million years ago, and the southern part only a few million years ago (Davies et al., 1987). The progression relates to the northward movement of the Indo-Australian lithospheric plate, with northeastern Australia moving progressively into the tropics throughout the late Cenozoic era (p. 436). Until sophisticated radiometric dating techniques became available, there was no way to evaluate, or often even to detect, the glacial-age unconformities in coral reef stratigraphy. Now, uplifted reefs of known age can be used to measure rates of tectonic uplift (Figs. 20–1 to 20–4), absolute rates of denudation, and the history of Quaternary sea-level oscillations (Figs. 2–3 and 17–16).

Sandy Shores

Beaches. If some wave energy were not absorbed in the work of moving sediment along coasts, far more erosional work would be accomplished. The sediment in motion along a shore is the **beach.** Its role is analogous to that of floodplain alluvium in river valleys; it absorbs energy and moves during times of storm and thereby stabilizes the rate of wave-energy conversion into geomorphic work. A continuous beach along a coast is as good a criterion of the graded condition as is a continuous floodplain on the floor of a river valley or a continuous sheet of sediment on a semiarid pediment.

The postglacial rise of sea level has submerged a variety of sediments, including glacial drift, sand dunes, older marine deposits, colluvium, and alluvium. As waves move shoreward over the loose sediments, silt and clay are stirred into suspension and carried long distances, usually coming to rest either on tidal mud flats or in water too deep for further wave action. On a shoaling bottom, sand sizes are the first to be stirred by waves (Fig. 9–10), but the sand is moved by saltation, not in suspension, so it stays within the zone of wave action. A **null line,** or *null point* on a bottom profile, can be defined, seaward of which sand is either not moved or moved seaward. Landward of the null line, there is a net shoreward movement of sand by waves (Komar, 1976, pp. 310–14). The rate of landward sand movement is a function of grain size and wave energy. Nearshore sand becomes very well sorted as it is moved shoreward, permitting the sandy bottom to respond to small changes in wave energy. Gravel beaches are also common but move only at much higher energy levels.

Although the null line and the beginning of net onshore sand transport are well seaward of the breaker zone, breakers cause the most dramatic sediment transport. Sand is thrown into suspension by each breaking wave, and before it can settle to the bottom again, it can be moved many meters shoreward or laterally by turbulent currents. In cross profile (Fig. 19–16), two zones of erosion are seen; the most important zone is under the breakers, but another zone of intense sand erosion is in the **swash** zone, where small waves run up the beach and collide with backwash. Between these two eroding zones, when the waves are approaching at an angle, one vector of the water motion is parallel to the shoreline and a powerful **littoral current** is generated. When waves break nearly parallel to the shore, only weak littoral currents are produced, and sand movement in the surf zone is nearly perpendicular to the beach. With a breaker angle and height that generate a littoral current of 30 cm/s, sand moves along the shore but with an outward component toward the breaker line. At littoral current velocities of 80 to 100 cm/s, a strong movement of sand parallel to the shore occurs, like a shallow alluvial river with one bank on the beach and the other in the breaker zone (Ingle, 1966). Within the surf zone, a ribbon of sand 100 m or more in width and 5 to 10 cm in thickness moves at a rate of 1000 m³/day or more (Das, 1971). Under such dynamic conditions, sandy beaches can adjust to changing wave and tidal energy within hours, and the nearshore profile is constantly regrading to changing conditions. New sand is added to the littoral system from offshore, from river mouths, and from eroding sea cliffs. Sand is lost from the system by wind transport inland, by tidal-current deposition in sheltered lagoons, by *rip currents* into deep water, and rarely by being trapped in the heads of submarine canyons that lead it out to deep-sea basins (Inman and Brush, 1973). Most sandy shores can be divided into distinct *littoral cells* ranging in length from about 1 to 100 km, each defined by a sand source, a transportational segment, and a sand sink where sediment is removed from the system. Sources can be erodible sediments offshore or alongshore; sinks can be tidal inlets into lagoons, strong currents that carry the sand offshore beyond the depth of wave transport or wind that blows sand off the beach into inland dunes.

Figure 19-16. Major dynamic zones in the nearshore zone of a sandy coast. Two regions of maximum sand suspension and transport are shaded. (Ingle, 1966, Fig. 116.)

	Oscillatory Waves	Wave Collapse	Waves of Translation (bores): Longshore Currents: Seaward Return Flow; Rip Currents	Collision	Swash; Backwash	Wind
WATER MOTION						
DYNAMIC ZONE	Offshore	Breaker	Surf	Transition	Swash	Berm Crest
PROFILE			Mean lower low water			
SEDIMENT SIZE TRENDS	← Coarser	Coarsest Grains → Coarser ←		Bi-modal Lag deposit	← Coarser →	Wind-winnowed lag deposit
PREDOMINANT ACTION	Accretion	Erosion	Transportation	Erosion	Accretion and Erosion	
SORTING	← Better	Poor	Mixed	Poor	← Better →	
ENERGY	← Increase	High	← Gradient →	High	↑ ↓	

Wave refraction causes mobile beaches to change their orientation readily until the plan view of the shoreline conforms closely to the shape of the refracted wave fronts. On coasts with alternating headlands and beaches and a dominant swell that approaches obliquely, beaches assume a "half-heart," or **zetaform,** curvature (Fig. 19–17). The outline of these headland-bay beaches has been shown to closely approximate a logarithmic spiral (Yasso, 1965; Davies, 1980). Each zeta-

Figure 19-17. Zetaform coast of northern New South Wales, Australia. Longest beaches face the dominant swell from the southeast and trend obliquely to a line connecting the headlands. (Photo courtesy B. G. Thom.)

form beach segment seems to be offset relative to its neighbors with the straight segment of beach rotated from the mean trend of the coast toward the dominant swell direction. Grain size, sorting, and beach profiles all change systematically along each zetaform beach. Each is nearly a closed littoral cell of sand supply and movement, although some sand probably drifts around the downwave headland to nourish the next beach. The pattern of offset zetaform beaches dominates high-altitude photographs and satellite images of many coasts (Bloom, 1986) and is easily recognizable on coastal maps and charts.

Barriers. On many coasts, the postglacial submergence was across plains of low relief and abundant sediment supply. During the later stages of the submergence, the marine transgression created growing ridges of sand that eventually become too massive to be moved farther shoreward. The sea flooded the area behind them to form **lagoons,** but the massive sand ridges extend for hundreds of kilometers along many coasts as **barriers,** only rarely tied to headlands (Hoyt, 1967, 1968). Barriers and lagoons are said to extend along 13 percent of the world's coasts (Zenkovich, 1967, pp. 288, 390). Approximately 17 percent of the coastline of the United States is along barriers and lagoons (Cronin, 1986, p. 33).

The origin of barriers was one of the durable controversies of geomorphology. As long as the submarine offshore topography was poorly known, and shore-zone sediment transport unstudied, barriers were thought to represent submarine sand bars that had grown very large and then emerged by tectonic uplift or a drop of sea level. Some authorities argued that wave action actually built barriers above sea level, aided by high storm levels and wind transport. Others maintained that barriers were built along shallow coasts by littoral drift and were not driven landward but grew in place.

The identification of barriers with coastal emergence has now been thoroughly disproved. Numerous boreholes in barriers and their associated lagoons prove that many were built in the last few thousand years, during and after the final phase of postglacial submergence (Kraft and Chrzastowski, 1985, Fig. 9–11). Most of them rest on older foundations, perhaps dating from the last interglacial interval (Otvos, 1972). As with any beach, barriers are supplied with sand both from offshore and by littoral drifting. Many seem to have grown more rapidly prior to a few thousand years ago and are now eroding (Thom, 1984). Sand supply was more abundant as the sea steadily transgressed new terrain, whereas with the nearly stable sea level today, sand is being lost into dunes, lagoons, and the offshore zone and is not being replaced.

Barriers are large constructional landforms as much as 1 km wide and 100 m high to the crests of the highest dunes. Unless underlain by a relict landform, the high relief is eolian. Sand moving onshore or alongshore is blown off beaches at low tide. If a plentiful supply of sand is available, it is replaced, and the barrier grows. If sand supply is inadequate, the barrier is eroded or driven landward by washover and loss of sand into inlets during major storms (Fig. 19–18). Extensive reclamation and waterfront stabilization projects actually endanger barriers, because unless sand is free to move with changing wave conditions, erosion results (Dolan et al., 1973).

Although landward migration by overwash and landward sediment movement into inlets is a long-range continuing process, on the time scale of centuries it seems to be erratic. Major coastal storms every few decades have major impacts on barriers. Barriers may grow narrower by erosion of their exposed beach face as well as on their lagoon side, until they are so narrow that they become unstable and begin to migrate (Leatherman, 1983). They do not necessarily maintain a constant mass of sand as they retreat, either. During the historical submergence trend that typically amounts to a few millimeters per year (Fig. 2–4), barriers on the southeast-

Figure 19-18. Core Banks, Cape Lookout National Seashore, North Carolina. A barrier in its natural state, broad and low, subject to washover during storms. (Photo: P. J. Godfrey, U.S. National Park Service.)

ern Atlantic coast of the United States are retreating 1 to 6 m/yr, but in an erratic and unpredictable way (Dolan et al., 1980; Leatherman et al., 1982). It is a startling fact that in the face of the ever-increasing cost of real estate destruction on barrier islands, government agencies permit continued development on them (Dean et al., 1988).

Most barriers are segmented at intervals by **tidal inlets** that allow tidal currents to ebb and flow into the lagoons and marshes behind the barriers and allow river runoff to escape. Inlets are systematically spaced to provide adequate tidal circulation. They are deep channels, often cutting through the entire cross section of the barrier and into the substrate. As a consequence, if a tidal inlet is forced by littoral sand movement to migrate downcurrent, it completely removes the barrier on the eroding side and builds a new barrier on the depositional side. Historical records show that when inlets migrate too far in this way, a new inlet breaks through the barrier somewhere upcurrent and in turn migrates in the direction of sand transport (Fisher, 1968). On most barrier coasts, the tidal inlets are depositional traps rather than sources for nearshore sediment. On coasts that consist of segmented barrier islands, some sand moves laterally across the mouths of tidal inlets, but some is permanently lost into the inlets. Unless this sand is replaced by headland erosion and river deposition upcurrent or by onshore transport by waves, the barrier will migrate landward over its lagoon or marsh, or grow thinner, or both. Modern coastal conservation techniques include calculations of the sand budgets for each section of a barrier coast. To correct negative budgets, additional sand is provided at the upcurrent end of each littoral cell. The quantities, measured in thousands of cubic meters per day (Das, 1971; Komar, 1976, p. 218), can only be provided by dredging sand from the ocean floor well to seaward of the null line. Fortunately, the postglacial rise of sea level seems to have frequently generated incipient barriers that were subsequently overtopped and bypassed and are now relict submarine sand ridges. If these ridges are truly relict and not part of an as-yet-unknown modern shelf sediment-transport system, the sand in them can safely be used to restore and maintain modern barriers and other eroding coasts.

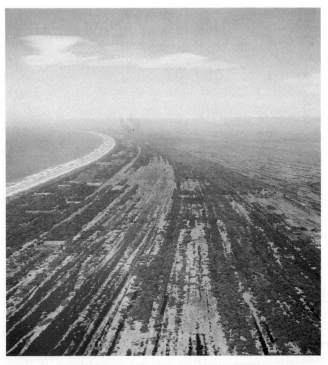

Figure 19-19. Multiple beach ridges on a prograding coastal plain, Nayarit, Mexico. (Photo: F. B. Phleger, courtesy J. R. Curray.)

Other Landforms of Sandy Shores. The ready readjustment of beaches and barriers to wave, wind, and tidal energy gives rise to many dynamic landforms. Terminology is extensive and complex but can be found in many books on coastal geomorphology (Zenkovich, 1967; Komar, 1976; Davies, 1980; Pethick, 1984). Many words, such as **spit** (a sand ridge attached at one end but free on the other) and **bar** (a submerged or intertidal sand ridge), are old mariners' terms. **Tombolo,** a beach that ties an island to the mainland or another island, is of Italian origin. The beach crest, just above high-tide wave action, is called a **berm** by analogy with the curved cross profile of a well-graded roadway. Dunes, blowouts, and other eolian forms (Chap. 14) are common on beaches. On prograded coasts, multiple beach ridges numbering into the hundreds (Fig. 19–19) record the seaward migration of the shoreline by deposition, or **progradation** (Curray et al., 1969).

Muddy Shores

Intertidal Mud Flats. Clay-size and silt-size particles settle slowly from suspension and are carried by littoral current as suspended load. But in seawater, unlike in rivers, fine sediment *flocculates* into loose, large aggregates and settles out of suspension in quiet water as poorly sorted mud. Mud in suspension that is carried out beyond the null line is lost to the deep ocean floor, but much of the mud stirred up in the shore zone, or carried there by rivers, moves landward into estuaries, bays, lagoons, and other shallow, quiet water. Tidal ebb and flow, which might be thought to be comparable to a river that symmetrically reverses direction every 6 hours, actually cause a net shoreward transfer of mud. During the slack-water intervals when each tide reverses, mud begins to settle. Mud on the shoreward, shallow edge of the flood tide reaches the bottom and coheres, resisting current erosion on the ebb, or outward, tide. However, at the end of the ebb cycle, mud that moves seaward and settles an equal distance has not yet reached the bottom, so it is carried shoreward again by the next flood tide (van Straaten and Kuenen, 1958; Postma,

1967). This tendency to move mud into shallow, quiet water, while waves stir and winnow all the mud out of the surf zone, explains why beaches of clean, well sorted sand are so commonly backed by intertidal mud flats, tidal marshes, or muddy lagoons.

Accretion on mud flats is not only a mechanical process but a biological one as well. Many animals burrow in the mud and feed by filtering suspended organic debris out of the overlying water during tidal submersion. The waste products, including mud, are secreted as fecal pellets that are hydraulically similar to sand grains except that they tend to stick together. These pellets of agglutinated mud account for a significant proportion of the total mud-flat accumulation.

Tidal Marshes. When mud flats aggrade until they are exposed for about one-half of the tidal cycle, they can be colonized by **halophytic** (salt-tolerant) plants. Algae are the first colonizers, but several genera of grasses, sedges, and trees are uniquely adapted to tolerate immersion in salt water. Once vegetation is established on the mud flats, suspended mud is more effectively filtered out of the tidal waters, and **tidal marshes** quickly accrete to about mean high tide (Frey and Basan, 1985). An equilibrium is reached when the reduced amount of tidal flooding decreases the rate of accretion. The marsh buildup is supplemented by organic accumulation but countered by decay, compaction, and fire.

Tidal marshes are as rigorously controlled by high-tide flooding as coral reefs are by low-tide exposure. For that reason, studies of tidal marsh stratigraphy have been useful to document the history of the postglacial rise of sea level. That history is described more fully in Chapter 20. Marshes, once built to high-tide level, cannot be built higher by marine processes. However, plant succession can permit upland plants to build freshwater peat seaward over the tidal marshes and convert them to new, although very low, coastal plains. The process is slow and on most coasts has been countered by a continuing slow submergence.

Among the most remarkable landforms of the world are the tropical **mangrove swamps** (Fig. 19–20). A few species of trees from several families have convergently

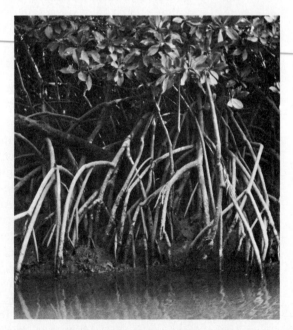

Figure 19-20. Prop roots of *Rhizophora* mangroves in the intertidal mud of Three Mile Creek, near Townsville, northern Queensland, Australia. (Photo: A. P. Spenceley.)

evolved to survive intertidal saltwater immersion (Lugo and Snedaker, 1974). The adaptations involve specialized root structures and buttressed trunks to support trees in soft mud, pneumatophores to permit oxygen to reach the roots buried in strongly anoxic mud, and viviparous seedlings that remain attached to the parent tree until they have developed leaves and a protoroot. When the seedlings drop, they float right-side up until they wash aground on a suitable bottom and take root.

Mangroves are tropical plants with a distribution almost coincident with coral reefs (Fig. 19–12). Tropical deltaic coasts are vast mangrove swamps with meandering distributaries often completely bridged by mangroves. Huge areas of delta coasts in Africa, India, Malaysia, Indonesia, New Guinea, and Australia, and on the Caribbean coasts of Central and South America are mangrove swamps. Whether mangroves expand onto exposed mud flats and initiate deposition or simply exploit previous mud deposits (Bird, 1972), they add great regions of semiflooded landscapes to tropical coasts. Progradation rates of 50 to 100 m/yr in recent centuries have been suggested for the Sumatra coast on the Malacca Straits.

REFERENCES

BIRD, E. C. F., 1972, Mangroves and coastal morphology in Cairns Bay, North Queensland: Jour. Tropical Geog., v. 35, pp. 11–16.

BLOOM, A. L., 1974, Geomorphology of reef complex, *in* Laporte, L. F., ed., Reefs in time and space: Soc. Econ. Paleontologists and Mineralogists Spec. Pub. 18, pp. 1–8.

———, 1983, Sea level and coastal changes, *in* Wright, H. E., Jr., ed., Late-Quaternary environments of the United States, v. 2, The Holocene: Univ. Minnesota Press, Minneapolis, pp. 42–51.

———, 1986, Coastal landforms, *in* Short, N. M., and Blair, R. W., Jr., eds., Geomorphology from space: NASA Spec. Pub. Sp-486, Washington, D.C., pp. 353–406.

BRADLEY, W. C., 1958, Submarine abrasion and wave-cut platforms: Geol. Soc. America Bull., v. 69, pp. 967–74.

———, and GRIGGS, G. B., 1976, Form, genesis, and deformation of central California wave-cut platforms: Geol. Soc. America Bull., v. 87, pp. 433–49.

CLARK, J. A., and LINGLE, C. S., 1979, Predicted relative sea-level changes (18,000 years B. P. to present) caused by late-glacial retreat of the Antarctic ice sheet: Quaternary Res., v. 11, pp. 279–88.

COTTON, C. A., 1952, Criteria for the classification of coasts: Internat. Geog. Cong., 17th, Washington 1952, Proc., pp. 315–19 [1957] (reprinted 1974 *in* Bold coasts: A.H. & A.W. Reed, Wellington, New Zealand, pp. 118–25).

———, 1955, Theory of secular marine planation: Am. Jour. Sci., v. 253, pp. 580–89 (reprinted 1974 *in* Bold coasts: A.H. & A.W. Reed, Wellington, New Zealand, pp. 164–74).

CRONIN, T. M., 1986, Ostracods from late Quaternary deposits, south Texas coastal complex, *in* Shideler, G. L., ed., Stratigraphic studies of a late Quaternary barrier-type coastal complex, Mustang Island–Corpus Christi Bay area, south Texas Gulf Coast: U.S. Geol. Survey Prof. Paper 1328, p. 33–63.

CURRAY, J. R., EMMEL, F. J., and CRAMPTON, P. J., 1969, Holocene history of a strand plain lagoonal coast, Nayarit, Mexico: Internat. Sympos. Coastal Lagoons, Mexico City, 1967, Proc., pp. 63–100.

DARWIN, C., 1837, On certain areas of elevation and subsidence in the Pacific and Indian Oceans, as deduced from the study of coral formations: Geol. Soc. London Proc., v. 2, pp. 552–54.

DAS, M. M., 1971, Longshore sediment transport rates: A compilation of data: U.S. Army Coastal Engineering Res. Center, Misc. Paper 1-71, 81 pp.

DAVIES, J. L., 1980, Geographical variation in coastal development, 2nd ed.: Longman, Inc., New York, 212 pp.

DAVIES, P. J., SYMONDS, P. A., FEARY, D. A., and PIGRAM, C. J., 1987, Horizontal plate motion: A key allocyclic factor in the evolution of the Great Barrier Reef: Science, v. 238, pp. 1697–1700.

DEAN, R. G., PILKEY, O. H., JR., and HOUSTON, J. R., 1988, Eroding shorelines impose costly choices: Geotimes, v. 33, no. 5, pp. 9–14.

DIETZ, R. S., 1963, Wave-base, marine profile of equilibrium, and wave-built terraces: A critical appraisal: Geol. Soc. America Bull., v. 74, pp. 971–90.

DOLAN, R., GODFREY, P. J., and ODUM, W. E., 1973, Man's impact on the barrier islands of North Carolina: Am. Scientist, v. 61, pp. 152–62.

DOLAN, R., LINS, H., and STEWART, J., 1980, Geographical analysis of Fenwich Island, Maryland, a middle Atlantic coastal barrier island: U.S. Geol. Survey Prof. Paper 1177A, 24 pp.

FAIRBRIDGE, R. W., 1950, Recent and Pleistocene coral reefs of Australia: Jour. Geology, v. 58, pp. 330–401.

FISHER, J. J., 1968, Barrier island formation: Discussion: Geol. Soc. America Bull., v. 79, pp. 1421–26.

FREY, R. W., and BASAN, P. B., 1985, Coastal salt marches, *in* Davis, R. A., Jr., ed., Coastal sedimentary environments, Second revised, expanded edition: Springer-Verlag New York Inc., New York, pp. 225–301.

GARRETT, C. J. R., 1972, Tidal resonance in the Bay of Fundy and Gulf of Maine: Nature, v. 238, pp. 441–43.

GROSS, M. G., 1990, Oceanography: A view of the earth, 5th ed.: Prentice-Hall, Englewood Cliffs, New Jersey, 441 pp.

GUILCHER, A., 1988, Coral reef geomorphology: John Wiley & Sons Ltd., Chichester, England, 228 pp.

HAYES, M. O., 1967, Hurricanes as geological agents: Case studies of hurricanes Carla, 1961, and Cindy, 1963: Texas Bureau of Econ. Geology, Rept. of Invest. no. 61, 56 pp.

HENDERSHOTT, M. C., 1973, Ocean tides: EOS (Am. Geophys. Un., Trans.), v. 54, pp. 76–86.

HODGKIN, E. P., 1964, Rate of erosion of intertidal limestone: Zeitschr. für Geomorph., v. 8, pp. 385–92.

HOPLEY, D., 1982, The geomorphology of the Great Barrier Reef: Quaternary development of coral reefs: John Wiley & Sons., Inc., New York, 453 pp.

HORIKAWA, K., ed., 1988, Nearshore dynamics and coastal processes: Univ. Tokyo Press, Tokyo, 522 pp.

HOYT, J. H., 1967, Barrier island formation: Geol. Soc. America Bull., v. 78, pp. 1125–36.

———, 1968, Barrier island formation: Reply: Geol. Soc. America Bull., v. 79, pp. 1427–30.

HULL, A. G., 1987, A late Holocene marine terrace on the Kidnappers coast, North Island, New Zealand: Quaternary Res., v. 28, pp. 183–95.

INGLE, J. C., 1966, The movement of beach sand: Advances in sedimentology, v. 5, Elsevier Publishing Co., New York, 221 pp.

INMAN, D. L., and BRUSH, B. M., 1973, The coastal challenge: Science, v. 181, pp. 20–32.

KINSMAN, B., 1965, Wind waves: Their generation and propagation on the ocean surface: Prentice-Hall, Inc., Englewood Cliffs, N.J., 676 pp.

KOMAR, P. D., 1976, Beach processes and sedimentation: Prentice-Hall, Inc., Englewood Cliffs, N.J., 429 pp.

KRAFT, J. C., and CHRZASTOWSKI, M. J., 1985, Coastal stratigraphic sequences, *in* Davis, R. A., Jr., ed., Coastal sedimentary environments, Second revised, expanded edition: Springer-Verlag New York Inc., New York, pp. 625–63.

LEATHERMAN, S. P., 1983, Barrier dynamics and landward migration with Holocene sea-level rise: Nature, v. 301, pp. 415–18.

————, RICE T. E., and GOLDSMITH, V., 1982, Virginia barrier island configuration: A reappraisal: Science, v. 215, pp. 285–87.

LUDWICK, J. C., 1974, Tidal currents and zig-zag sand shoals in a wide estuary entrance: Geol. Soc. America Bull., v. 85, pp. 717–26.

LUGO, A. E., and SNEDAKER, S. C., 1974, Ecology of mangroves: Ann. Rev. Ecology and Systematics, v. 5, pp. 39–64.

MACNEIL, F. S., 1954, Shape of atolls: An inheritance from subaerial erosion forms: Am. Jour. Sci., v. 252, pp. 402–27.

MARAGOS, J. E., BAINES, G. B. K., and BEVERIDGE, P. J., 1973, Tropical cyclone Bebe creates a new land formation on Funafuti Atoll: Science, v. 181, pp. 1161–64.

MCLEAN, R. F., 1967, Measurements of beachrock erosion by some tropical marine gastropods: Bull. Marine Sci., v. 17, pp. 551–61.

————, 1972, Nomenclature for rock-destroying organisms: Nature, v. 240, p. 490.

MOTTERSHEAD, D. N., 1989, Rates and patterns of bedrock denudation by coastal salt spray weathering: A seven-year record: Earth Surface Processes and Landforms, v. 14, pp. 383–98.

MUNK, W. H., and SARGENT, M. C., 1954, Adjustment of Bikini Atoll to ocean waves: U.S. Geol. Survey Prof. Paper 260-C, pp. 274–80.

NEUMANN, G., and PIERSON, W. J., JR., 1966, Principles of physical oceanography: Prentice-Hall, Inc., Englewood Cliffs, N.J., 545 pp.

OTVOS, E. G., 1972, Mississippi Gulf Coast Pleistocene beach barriers and the age problem of the Atlantic–Gulf Coast "Pamlico"–"Ingleside" beach ridge system: Southeastern Geology, v. 14, pp. 241–50.

PETHICK, J., 1984, An introduction to coastal geomorphology: Edward Arnold (Publishers) Ltd., London, 260 pp.

POSTMA, H., 1967, Sediment transport and sedimentation in the estuarine environment, *in* Lauff, G. H., ed., Estuaries: Am. Assoc. Adv. Sci. Pub. no. 83, pp. 158–79.

PURDY, E. G., 1974, Reef configurations: Cause and effect, *in* Laporte, L. F., ed., Reefs in time and space: Soc. Econ. Paleontologists and Mineralogists Spec. Pub. 18, pp. 9–76.

REVELLE, R., and EMERY, K. O., 1957, Chemical erosion of beach rock and exposed reef rock: U.S. Geol. Survey Prof. Paper 260-T, pp. 699–709.

SHEPARD, F. P., 1970, Lagoonal topography of Caroline and Marshall Islands: Geol. Soc. America Bull., v. 81, pp. 1905–14.

STEPHENSON, T. A., and STEPHENSON, S., 1972, Life between tidemarks on rocky shores: W.H. Freeman and Company, San Francisco, 425 pp.

VAN STRAATEN, L. M. J. U., and KUENEN, P. H., 1958, Tidal action as a cause of clay accumulation: Jour. Sed. Petrol., v. 28, pp. 406–13.

STRECKER, M. R., BLOOM, A. L., and LECOLLE, J., 1987, Time span for karst development on Quaternary coral limestones: Santo Island, Vanuatu, *in* Godard, A., and Rapp, A., eds., Processes and measurement of erosion: Editions du Centre National de la Recherche Scientifique, Paris, pp. 369–86.

SUNAMURA, T., 1975, Laboratory study of wave-cut platform formation: Jour. Geology, v. 83, pp. 389–97.

————, 1977, Relationship between wave-induced cliff erosion and erosive force of waves: Jour. Geology, v. 85, pp. 613–18.

THOM, B. G., 1984, Sand barriers of eastern Australia: Gippsland—a case study, *in* Thom, B. G., ed., Coastal geomorphology in Australia: Academic Press Australia, North Ryde, N.S.W., Australia, pp. 233–61.

TRENHAILE, A. S., 1987, The geomorphology of rocky coasts: Oxford Univ. Press, Oxford, England, 384 pp.

————, 1989, Sea level oscillations and the development of rock coasts, *in* Lakhan, V. C., and Trenhaile, A. S., eds., Applications in coastal modeling: Elsevier Oceanography Series, 49, Elsevier Science Publishers B. V., New York, pp. 271–95.

WARME, J. E., and MARSHALL, N. F., 1969, Marine borers in calcareous terrigenous rocks of the Pacific coast: Am. Zoologist, v. 9, pp. 765–74.

WENTWORTH, C. K., 1938, Marine bench-forming processes: Water-level weathering: Jour. Geomorphology, v. 1, pp. 6–32.

———, 1939, Marine bench-forming processes: II, solution benching: Jour. Geomophology, v. 2, pp. 3–25.

YASSO, W. E., 1965, Plan geometry of headland-bay beaches: Jour. Geology, v. 73, pp. 702–14.

ZENKOVICH, V. P., 1967, Processes of coastal development (transl. by D. G. Fry): Wiley-Interscience, New York, 738 pp.

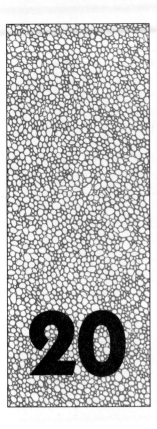

20 Explanatory Description of Coasts

Among the earliest historical observations of geomorphic change were reports of coastal erosion and deposition. Long before scholars could accept ideas that subaerial landscapes evolve, they readily noted the retreat of sea cliffs, progressive shoaling of harbors, and forward growth of deltas. Attempts to develop rational classifications and systematic descriptions of coastal landscapes continued in parallel with similar attempts to describe and classify subaerial landscapes. However, even in the modern era, coasts are generally treated as a separate topic of geomorphology, in acknowledgment of the fundamentally different energy transfer mechanisms for coastal change and the narrow vertical range of shore processes.

Systematic coastal geomorphology has paralleled other kinds of systematic geomorphology in recognizing the fundamental roles of structure, process, and time in coastal landscape evolution. However, the three components of geomorphic explanatory description are not applied to coastal studies in quite the same way as to subaerial landscapes. Too much of the coastal landscape is relict, inherited either from earlier shore processes now inoperative or from subaerial or submarine processes. No generally accepted classification of coasts has emerged from two centuries of geomorphic research. The role of structure, process, and time in coastal evolution can be recognized, however. In this chapter, these three component parts of explanatory description are reviewed in succession to illustrate both the similarities and contrasts of coastal and subaerial geomorphic systems.

STRUCTURAL FACTORS OF COASTAL DESCRIPTION

Tectonic Coasts

The narrow vertical range of shore processes ensures that any tectonic movement that affects the relative level of the land and sea will be recorded by relict or modified coastal landforms. However, the postglacial eustatic rise of sea level of about

499

120 m has drowned all but the most strongly uplifted coasts, so that many recent tectonic movements are totally masked by late-glacial drowning. The distinction between coasts that have been deformed or displaced by tectonic movements and those that have been passively drowned by the postglacial rise of sea level was made the basis of a coastal classification by Cotton (1952). He recognized the following categories of coasts:

1. *Coasts of stable regions* These have all been affected by recent submergence:
   a. Coasts dominated by features produced by recent submergence.
   b. Coasts dominated by features inherited from earlier episodes of emergence.
   c. Miscellaneous coasts (fiord, volcanic, etc.).
2. *Coasts of mobile regions* These have been affected by uplift or depression of the land as well as by recent submergence.
   a. Coasts on which the effects of marine submergence have not been counteracted by uplift of the land.
   b. Coasts on which recent uplift of the land has led to emergence.
   c. Fold and fault coasts.
   d. Miscellaneous coasts (fiord, volcanic, etc.)

Cotton also attempted to classify mobile coasts by the style and orientation of the tectonic movement (Table 20–1). He deduced a series of initial coasts that result from tectonic motion parallel to or transverse to the shoreline, plus some constructional and erosional forms, and then deduced a series of possible subsequent move-

TABLE 20–1

Classification of Initial Coasts

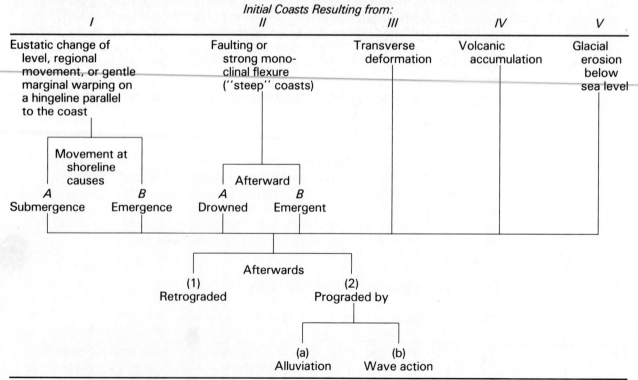

Source: Cotton (1942).

ments, erosion, or deposition. Of interest is the distinction between coasts that are deformed parallel to the shoreline and those that are deformed transverse to it. In the case of tilting parallel to the shore, the coast might either emerge or submerge, depending on the location of the tilt axis. In the case of transverse tilting, the coast might consist of alternating emerged and submerged segments. In New Zealand, where Cotton tested his classification, as well as elsewhere on the rim of the Pacific Basin and in the Mediterranean and Caribbean Seas, a descriptive classification based on the nature of recent or continuing tectonic movements has great usefulness. Nevertheless, in all the examples cited by Cotton, tectonic movement has only modified, but not negated, the effects of glacially controlled eustatic sea-level fluctuations.

As noted in Chapter 2, tectonic movements in coastal regions can be categorized as slow or rapid based on their comparision with the amplitude and rates of glacially controlled sea-level oscillations. But uplift in excess of even the most rapid measured short-term coastal uplift rates, approaching 10 mm/yr, would have had to continue for 10,000 years from about 15,000 to 5000 years ago in order to have kept up with the 120 m or more of rapidly rising sea level at the close of the Wisconsin glaciation (Fig. 17–16). No coasts are known where sediments or landforms of the last full-glacial interval are now above sea level, and it is unlikely that any will be found.

Figures 20–1 to 20–4 illustrate a conceptual framework for interpreting tectonic movements and sea-level changes during the last 500,000 years (Bloom, 1980). Each figure shows on its ordinate the present altitude of some hypothetical shore-zone landforms (such as coral-reef crests) that developed during interglacial intervals of the last 500,000 years. Tectonic uplift or subsidence at a constant rate is assumed for each figure. The basic sea-level curve used in each figure is from Figure 18–1, redrawn with a vertical scale of sea-level change on the assumption that the typical deviation of the $^{18}O/^{16}O$ isotopic ratio in foraminifera between glacial and interglacial conditions is equal to 100 m of sea-level equivalent. According to the chronology of Figure 18–1, sea level was near its Holocene level about 125,000 years ago (isotope stage 5e, the highest part of stage 5 in Fig. 18–1) and at numerous other times earlier in the Pleistocene. Coral-reef dating programs have demonstrated that sea level was 5 to 10 m higher relative to many tropical coasts about 125,000 years ago. A value of +6 m was assumed in drawing Figures 20–1 to 20–4.

Figure 20–1 illustrates the predicted terrace sequence on a coral coast that is rising at a slow but constant tectonic rate of 0.1 mm/yr. A mid-Holocene reef on such a coast might now be slightly above present sea level, but would be within the zone of storm waves and would be difficult to interpret. However, a reef that built to a sea level about 6 m higher than present during the last interglacial would now be emerged 18 to 20 m, and a succession of older and higher terraces to an altitude of 50 m might date back to 400,000 or 500,000 years ago. In most tropical environments, weathering will have degraded such old reef limestone to such a degree that the original terrace surfaces will be lost. As illustrated by Figure 20–1, a few interstadial reef terraces younger than 125,000 years might be above present sea level, as on Barbados (Mesolella et al., 1969), but they might be underlain by reef rock more than 200,000 years old. The Quaternary uplift of Barbados has been in the range of 0.1 to 0.3 mm/yr, which makes it an excellent example of slow uplift on a coral coast.

By contrast to Figure 20–1, Figure 20–2 illustrates the predicted stratigraphic and morphologic succession on a slowly subsiding coast. The assumed subsidence rate of 0.1 mm/yr is two to three times faster than the typical "cooling curve" subsidence rates for oceanic lithosphere more than about 10 million years old (Fig. 1–5), on which the subsiding volcanic foundations of many atolls are to be found. Figure 20–2 illustrates that at such slow but geologically probable rates of subsidence of

TIME—Kiloans B.P.

Figure 20-1. Predicted terrace sequence on a coast rising at a long-term rate of 0.1 mm/yr. Reefs or abrasion platforms are presumed to develop during intervals of tangency between uplift and oscillating sea level. Holocene and last-interglacial terraces may be on antecedent surfaces of much greater age. Sea level is derived from the oxygen isotope curve of Shackleton and Opdyke (1973, Fig. 9). (Bloom, 1980, Fig. 2.)

Time (thousands of years ago)

Subsidence: (0.1mm/yr)

Figure 20-2. Predicted stratigraphic succession of reef limestones on a subsiding atoll or continental margin. Subsidence at a rate of only 0.1 mm/yr is sufficient to drown or bury all interglacial and interstadial deposits except reefs built to a sea level higher than present, 125,000 years ago. (Bloom, 1980, Fig. 3.)

atoll foundations, reef limestones formed at intervals of high sea level during the last 500,000 years would now be found to depths of 50 m in atoll boreholes. The reef of last interglacial age, even if it grew 6 m above present sea level, would be encountered below sea level or under about 6 m of Holocene reef limestone. In general, this is what is found in shallow boreholes into modern reefs (p. 487).

A similar sequence might be expected on a subsiding continental margin in the coral-reef zone. Florida in the United States might have a sequence analogous to the model shown in Figure 20–2, although the subsidence rate of Florida is probably less than 0.1 mm/yr. The Key Largo limestone is dated as 95,000 to 140,000 years old (Broecker and Thurber, 1965) and at various localities throughout southern Florida and the Florida Keys it is slightly above or slightly below present sea level. As a meter or more of surface lowering could have been caused by weathering during times of glacially lowered sea level, a subsidence rate cannot be confidently calculated for Florida from the few dates now available.

Figures 20–3 and 20–4 illustrate the predicted sequence of Quaternary reef terraces on strongly rising tropical coasts, such as those of New Guinea (Fig. 2–3) (Chappell, 1974; Bloom et al., 1974). Many South Pacific islands along tectonic island arcs have emerged Holocene terraces that demonstrate tectonic uplift of 1 to 2 mm/yr. Unless erosion has removed the Pleistocene rock, such emerged Holocene reef terraces are very likely to abut on their inland edges against older and higher terraces. It has been frequently supposed that each such terrace represents a sudden coseismic uplift event; many tectonic analyses of island-arc terraces have erron-

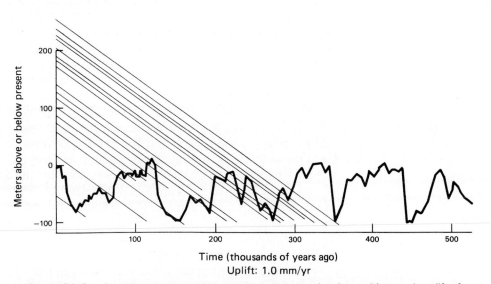

Time (thousands of years ago)
Uplift: 1.0 mm/yr

Figure 20-3. Predicted terrace sequence on a coast experiencing rapid tectonic uplift of 1.0 mm/yr. Some interstadial reef crests appear in the terrace sequence, and many interstadial and interglacial reefs, including Holocene forms, may be veneers on older platforms produced during times of lower sea levels. (Bloom, 1980, Fig. 4.)

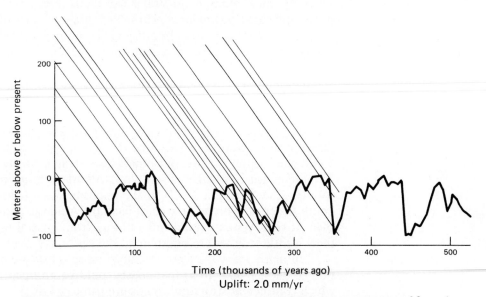

Time (thousands of years ago)
Uplift: 2.0 mm/yr

Figure 20-4. Predicted terrace sequence on a coast rising at a constant rate of 2 mm/yr. Compare with Figure 20-3. At uplift of 2 mm/yr or more, many interstadial reefs form discrete hillside terraces. The clustering of terraces due to sea-level history between 200,000 and 300,000 years ago is pronounced. (Bloom, 1980, Fig. 5.)

eously attempted to interpret the terrace sequences as the record of large earthquakes. However, the terraces are not evidence for intermittent uplift; on the contrary, the assumption of a constant uplift rate superimposed on an oscillating glacial eustatic sea level guarantees that all such terraces will appear in crudely clustered sets, progressively younger at lower altitudes, as shown by Fig. 20–3 and 20–4.

An uplift rate of about 1 mm/yr results in about 100 m of uplift during each major glacial eustatic cycle of about 100,000 years. That amount of tectonic uplift is approximately equal to the range of glacially controlled sea-level oscillations on the same time scale. This fact has important implications for deductions about coral-reef development on rising tectonic blocks such as are common on the frontal arcs of island-arc systems. It is possible that reef growth on a rapidly rising tectonic block will be initiated during full glacial times, when low sea level first exposes the rising submarine block to shallow-water reef growth (Taylor and Bloom, 1977). Once established, reefs may grow upward fast enough to maintain their upper surfaces near sea level even during the rapid deglacial rises of sea level at the end of major glacial cycles. Note the near tangency of some rising sea level trends with the assumed linear uplift rates on Figure 20–4, especially in the time interval of 200,000 to 300,000 years ago. The result should be massive reef terraces now 200 to 400 m above sea level. Such massive terraced landforms, of reef limestones too altered for radiometric dating, are distinctive features on many high islands in the tropics. Chappel and Shackleton (1986) extrapolated the dated reef terrace sequence on the Huon Peninsula of Papua New Guinea back in time and higher on the terraced mountainside to a reasonable, though unverified, 240,000 years. On Barbados, even older terraces have been dated by the uranium–helium method to about 700,000 years B.P., although the slow uplift of Barbados has raised the oldest terraces only a maximum of 300 m above present sea level (Bender et al., 1979).

A more fundamental tectonic classification of coasts has grown out of the plate tectonic theory (Inman and Nordstrom, 1971). As the surface of earth is a series of lithospheric plates on which continents and ocean basins are of only minor significance, we should be able to recognize only four tectonic categories of coasts: (1) Those at the edge of active diverging plates, (2) those at zones of plate convergence, (3) those along major transform faults (Fig. 3–10), and (4) those on continents embedded within large plates, far from any zones of convergence, spreading, or translatory plate motion. In category (1) could be placed block-faulted coasts and rift coasts such as the Red Sea coasts of Africa and Arabia. Category (2) includes island-arc systems, such as Indonesia, Japan, and other parts of the western Pacific, and mountainous continental coasts such as the west side of South America and the north side of the Mediterranean Sea (pp. 29–31). In category (3) are steep transform-fault coasts such as southern California, Venezuela, and the north side of New Guinea. In category (4) are most of the Atlantic Ocean coasts, as well as the Indian Ocean coasts of India, Africa, and Australia. Some of these coasts have shallow, broad continental shelves, but others have steep escarpments and narrow shelves. The difference is whether rivers flow toward or away from the particular coast (Audley-Charles et al., 1977). If the opposite side of the continent is high, as in South America, a broad, low coast faces the Atlantic, and rivers have built a broad shelf. If river systems drain inland away from coastal escarpments, as in Africa, there is little sediment to build broad shelves except at major deltas (p. 261).

In many ways, the tectonic categories of coasts are reminiscent of the suggestion of Suess (1904, p. 5) that coasts could be grouped into an Atlantic type, in which continental structural trends are truncated at the coast (actually at the submerged continental shelf margin), and a Pacific type, in which island arcs or mountain ranges parallel the coast (Bloom, 1986). The Atlantic coasts are primarily "trailing edges" of continents, whereas the Pacific coasts are "leading edges," either converging or transform-fault coasts.

The tectonic component of coastal classification includes such issues as the direction and rate of past and present vertical crustal movement, the amount of volcanic activity at the coast or offshore, and the trend of deformational axes relative to the coast.

Structurally Controlled Coasts

As in the case of subaerial landforms (Chaps. 3 and 11), we must distinguish between coasts that have a history of continuing tectonic movements and those that have inherited zones of unequal rock strength from ancient depositional and tectonic events, now inactive. The latter are *structurally controlled* in the sense that inherited structural factors rather than contemporary movements determine the plan view and profile of the coastal region.

Much of the structural control exhibited by coasts is inherited from subaerial landscape evolution. Rocks resistant to subaerial erosion form regions of high relief, and where the sea intersects these regions, cliffed coasts result. Weak-rock lowlands are drowned by the sea to form low coastal plains, perhaps with low islands offshore. Alluvial plains are very likely to develop barriers, because the initial gentle gradients and abundant sediment supply promote constructional shore-zone landforms.

Bold Coasts. In coastal regions of resistant rocks, marine erosion and deposition are quick to exploit structural weaknesses. Joints and faults are the loci for active wave plucking or quarrying. Dikes, especially with columnar joints, are especially subject to selective wave erosion and form deep clefts when they are quarried from between walls of more resistant rocks (Fig. 19–5). Pocket beaches, often of gravel rather than sand, accumulate in coves along shattered zones in otherwise massive rock (Fig. 20–5). In plan view, many coasts on metamorphic terrains show intersecting linear patterns of fiords or estuaries, marking former glacial or fluvial erosion along structural weaknesses (Figs. 11–7 and 20–6). Glacial erosion has been especially effective in etching out structural weaknesses in crystalline rocks prior to their drowning by the postglacial rise of sea level.

Where metamorphic or folded sedimentary rocks of contrasting erodibility have been drowned, the typical valley-and-ridge subaerial landscape takes on an even more impressive aspect. Two variants are known. If the drowned fold belt is parallel to the coast, the landscape is called the **Dalmatian type,** from the coastal region of Dalmatia, Yugoslavia (Baulig, 1930; Cotton, 1956). The drowned valleys parallel the coast but are cross-connected by channels. The regional landscape is best appreciated from map study (Fig. 20–7) or satellite images (Bloom, 1986, Plate C-8). If the structurally controlled valleys are perpendicular to the regional coastline, a typical **ria** coast develops (Fig. 11–7). Along much of the coasts of western Europe and northeastern North America, Paleozoic fold belts truncated by the opening of the Atlantic basin form deeply indented ria coasts.

Fault-line scarps and monoclines also form bold coasts. As with subaerial forms, some care must be taken to distinguish tectonic coastal scarps from those eroded along old structural trends. On some steep coasts, such as the southeastern Australia coast south of Sydney and the west side of the Indian peninsula, eroded escarpments rise abruptly from the back of narrow coastal plains. No structural control for the scarps is apparent, although they may have retreated many kilometers from initial structural-controlled features that are now submerged.

Low Coasts. Piedmont alluvial plains (p. 264), deltas, and pediments form substantial portions of the coasts of the world. In fact, "coastal plain" has become synonymous with a variety of weakly indurated sedimentary aprons in which strati-

Figure 20-5. Broad Cove beach, Appledore Island, Maine. Granite abrasion ramp in foreground has only a single layer of boulders and cobbles overlying it.

Figure 20-6. Hardanger fiord, Norway. The structural grain of the coastal metamorphic rocks is clearly shown beneath a thin snow cover. Fault lines and joints also influence the dissection. (NASA ERTS E-1299-10211.)

fication broadly parallels the surface slope and dips gently seaward. Many such coasts are prograding today, with new sedimentary layers accumulating in the off-shore zone. Little structural control is shown by these low coasts, exemplified by the Gulf coast of the United States. On some low coasts, emerged and eroded resistant layers form low cuestas on the plain, but these rarely affect coastal form. Barriers and lagoons are characteristic.

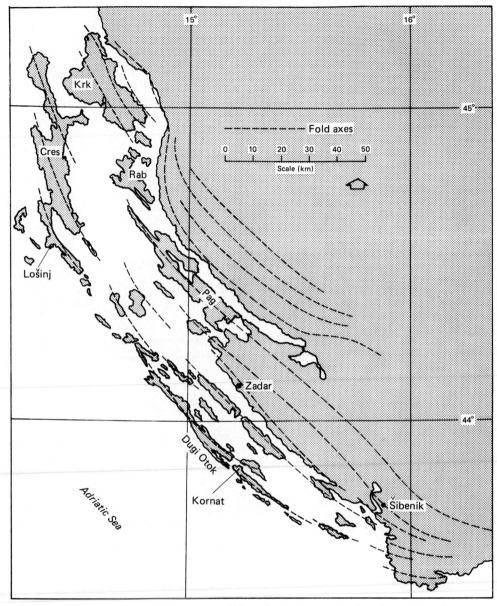

Figure 20-7. The Dalmatian coast of Yugoslavia, the type locality of an embayed coast formed by submergence of a fold belt that parallels the coast. Compare with Figure 11-7.

Low coasts have been strongly affected by Quaternary sea-level fluctuations. Because the weak sediments are easily eroded, rivers entrenched deep valleys across coastal plains during times of low sea levels. Subsequently, the valleys have drowned to form estuaries. On the south and east coasts of the United States, several episodes of drowning and estuarine sedimentation have alternated with sea levels higher than present relative to the land when beach ridges and wave-cut cliffs formed as high as 50 m above present sea level. The ages and even the number of the submergence episodes are widely debated subjects (Cooke, 1930; Flint, 1971, pp. 334–39; Wehmiller and Belknap, 1982; Cronin, 1988). It is not known whether the southeastern coasts of the United States are now emerging or submerging, although the coastal plain is made primarily of emerged shallow marine Tertiary sedimentary rocks and has been emerging during Cenozoic time. The episodes of alternating emergence and submergence that built the shoreline features inland and eroded the estuaries below present sea level were at least in part glacially controlled. Some large estuaries, such as the Chesapeake and Delaware Bays, are obviously the drowned lower segments of very large river valleys, but the age of their erosion may date from one or more of the early Quaternary glaciations.

A few low coasts are anomalous in that they are on resistant rocks. The **strandflat** of Norway is the classic example (Fig. 20–8). It is a low coastal region of crystalline rocks crossed by deep fiords but consisting mainly of low islands and shallow drowned terrain of low relief. The strandflat has been thought to be down-faulted, wave-abraded, ice-eroded, or a cryoplanation (p. 383) surface. Its origin is still not known (Holtedahl, 1958; Gjessing, 1967).

Figure 20-8.   The Norwegian strandflat in the Brønnøysund area (65°N). View is north from an isolated hill 260 m high toward a dissected plateau 500 m high. The intervening low terrain is cut on a variety of metamorphic rocks and has a total submerged and emerged relief of less than 60 m except for fiords that cross it. (Photo: Just Gjessing.)

PROCESS FACTORS IN COASTAL DESCRIPTION

Sea-Level Fluctuations

Worldwide, or *eustatic,* changes of sea level have been the major factor in the development of Quaternary coastal landscapes (pp. 435 and 476). The magnitude of the glacier-controlled sea-level fluctuations far exceeded the 10-meter vertical range of modern shore processes, and their frequency exceeded the rate of development of erosional equilibrium landforms. Depositional coasts, especially coral reefs and sandy barriers, have been able to develop equilibrium landforms even during the brief times, perhaps lasting only 10,000 years, of interglacial sea-level stability. However, the reefs and sand barriers that together form such a large part of our present coastal landscape are not the result of Holocene events alone: Almost without exception, the modern forms have been shown to be thin veneers resting on ancient but genetically similar foundations dating from the last interglacial or even earlier (Figs. 20–1 to 20–4).

Either by eustatic fluctuation or by tectonic movements, some coasts have *emerged* from the sea, and others have *submerged.* These useful terms refer only to the relative movement of land and sea, or the algebraic sum of all vertical movements at the shoreline. Emergence is by geomorphic convention a net positive movement of the land; submergence is considered a net negative movement. By contrast, stratigraphers, especially those who study marine strata, prefer the opposite connotation of submergence as a positive movement (a marine *transgression*) and emergence as a negative, or *regressive,* phenomenon.

The dichotomy of submergence and emergence has been the basis for most coastal classifications since it was formally proposed by Gulliver (1899) and rigidly codified by Johnson (1919). The supplemental categories of *neutral coasts* and *compound coasts* were added to cover coasts that were the result of progradation, as by a delta, with neither emergence or submergence a significant factor, or coasts that show evidence of both emergence and submergence. Although extremely influential, the descriptive system of submerged, emerged, neutral, and compound coasts has lost much of its utility with the growing knowledge of multiple Quaternary sea-level fluctuations. With few exceptions, all coasts must be classified as compound in this classification.

Valentin (1952, 1970) devised an ingenious coordinate system (Fig. 20–9) that gave submergence and emergence equivalent status with shoreline erosion (retrogradation) and progradation in determining whether a coast has *advanced* seaward or *retreated* landward. The addition of a time axis to this scheme enables the history of a coast to be graphically portrayed (Fig. 20–10) and suggests additional categories for classifying coasts, although the modification is intended primarily for explanatory description rather than classification (Bloom, 1965).

Wave Energy; Coastal Exposure

The rate of energy expenditure at a shoreline has some potential as a basis for classifying coastal landscapes (Tanner, 1960; Davies, 1980). *High-energy coasts* are characterized by bold erosional forms or massive deposits of rather coarse sediment. *Low-energy coasts,* by contrast, are of low relief, often with salt marshes or mangrove swamps. Terrestrial processes, such as deltaic progradation, dominate over shoreline processes.

Shoreline energy is not a random effect of exposure or local coastal configuration. Certain regions, such as coasts in latitudes 45 to 60°, and east-facing coasts in the low-latitude trade-wind regions, are generally subject to higher than average storm-induced waves (Fig. 20–11). Equatorial coasts and subarctic coasts generally have low tides and low waves, although the causes for the low energy differ. Subtropical east-facing coasts are especially subjected to cyclonic storms (hurricanes or typhoons) that occur with disastrous regularity every year or two.

Figure 20-9. Principal genetic types of coastal configuration. Coasts can be categorized as "emerged and prograded," "submerged and prograded," and so on. A balance of vertical and horizontal components can also keep a coast at quasi-equilibrium. (Valentin, 1970.)

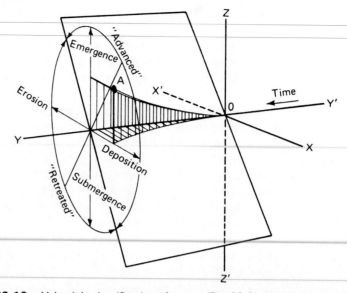

Figure 20-10. Valentin's classification of coasts (Fig. 20-9) simplified and with a time axis added. The history of a coast can be graphed on these axes. For instance, point A is shown as the result of emergence and deposition through time. (Bloom, 1965.)

The geometric factor of fetch, or exposure, the climatic factors of wind speed, duration, and storm frequency, and geologic factors of offshore depth and sediment supply can be combined to determine the rate at which energy is fed into a coastal segment. Regional patterns of energy distribution can be combined with structural factors of regional coastal rock type to deduce sets of possible landforms. The energy approach is especially useful for explanatory description of modern shore-zone landforms. The broader coastal landscape is likely to show additional relict forms of tectonic or eustatic origins or climatic change.

Climatic Factors

In addition to their fundamental role in determining wave energy input, climatic variables offer other possibilities for explanatory description of coasts. Glaciated coasts, with their distinctive assemblage of glacial erosional and depositional landforms such as fiords, outwash plains, and drumlin fields, are very distinct from non-glaciated coasts (Fitzgerald and Rosen, 1987). On almost any world map or globe, the deep embayments of fiords poleward of 40 to 50° latitude on west-facing coasts indicate the former extent of tidewater glaciers. The pattern is obvious in both hemispheres, in New Zealand, Chile, Scandinavia, the northern British Isles, and the coast in the states of Washington and Alaska and Canadian British Columbia. Certain shore processes, ice-push by grounded drift ice on high-latitude coasts for instance, are climate-determined. The new adjective "glaciel" was proposed by L. E. Hamelin in 1959 to describe the geomorphic work of floating ice (Dionne, 1974). Coral-reef and mangrove coasts are uniquely tropical. Coasts in the humid tropics are generally lacking in dune fields, because tropical vegetation quickly stabilizes new beach sand during progradation (Bird and Hopley, 1969). Arid coasts have salt flats, prominent barrier systems built of the abundant sand-size sediment provided from flash floods, and spectacular coastal dunes. A climatic morphogenetic system (p. 327) could be erected for coastal landforms, as for subaerial forms.

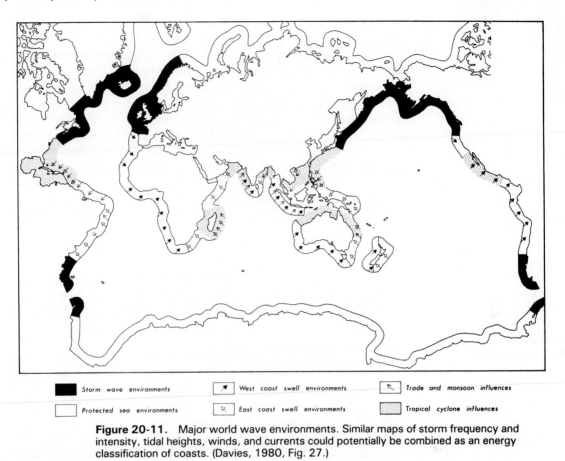

Figure 20-11. Major world wave environments. Similar maps of storm frequency and intensity, tidal heights, winds, and currents could potentially be combined as an energy classification of coasts. (Davies, 1980, Fig. 27.)

Legend:

- ■ Storm wave environments
- □ Protected sea environments
- ↗ West coast swell environments
- ↖ East coast swell environments
- ↖ Trade and monsoon influences
- ░ Tropical cyclone influences

TIME AS A FACTOR IN COASTAL GEOMORPHOLOGY

The Concept of Maturity

To geomorphologists of the late nineteenth and early twentieth centuries, initial structures and active processes interacted in a deduced series of stages, analogous to the youth, maturity, and old age of organisms. Time was a subjective variable, marked by the attainment of certain diagnostic landforms rather than by the passage of years or millenniums of real time. It was inevitable that the concept of "stages" in the "life cycle" of subaerial landscapes would be applied to coastal landscapes as well. However, the scheme was never fully adopted for explanatory description of coasts. W. M. Davis, the great champion of the geomorphic cycle concept, cautioned (1905, p. 159):

> One significant peculiarity of the development of the shoreline [as compared to the development of the subaerial landscape] is its immediate recognition of changes of level or interruptions. . . . This contrast leads to the inference that the product of long-continued work of the shore-line forces on a fixed level or on a uniformly changing level is less likely to the found than the product of long-continued work on an inland region, where a series of small and frequent interruptions (elevations and depressions) might hardly make themselves felt.

Despite this early recognition of the vital importance of emergence and submergence on coasts, subsequent geomorphologists continued to deduce sequential stages in the erosional evolution of coasts on the assumption of sea-level stability. As an abstract logical exercise, no fault can be found with the deductions. The difficulty came in trying to fit real coastal landscapes into the deduced schemes. By analogy with fluvial landscapes, maturity of a coast meant a graded condition, both in plan view and in cross section perpendicular to the shoreline. Cliffs were to be cut back to alignment with prograded beaches until sediment could move freely alongshore. The nearshore profile was to be "graded," but insufficient coastal work had been done to define that condition. To D. W. Johnson (1925) and others of his time, the youthful stage of coastal erosion implied irregularities in both plan and profile. "Maturity" was reached only when the entire shoreline had been eroded back beyond the heads of any initial bays. The analogy was to the complete removal of waterfalls in the "mature" river profile, but time required for headlands to be eroded and a continuous shore platform to develop greatly exceeds the time intervals of sea-level stability. For that reason, few examples of "mature" coasts on resistant rocks could be cited (Johnson, 1919, 1925). Although widely hailed as a triumph of deductive geomorphic reasoning, the "cycles of erosion" on submerged, emerged, neutral, and compound coasts were never very useful in explanatory description.

Primary and Secondary Coasts

As an alternative to the deductive scheme of submerged and emerged coasts with subsequent temporal stages of erosion, Shepard (1937, 1973) subdivided his dominant class of submerged coasts into *primary* and *secondary* categories. Primary coasts are those that are submerged but otherwise unmodified. Secondary coasts have been significantly modified by shoreline erosion or deposition after submergence. No sequential development of secondary forms was deduced. Neither was any provision made for inherited forms, despite the fact that many, perhaps most, coasts have relict landforms.

It is easy to criticize Shepard's subdivision on the grounds that the present degree of marine influence on coastal configuration is dominated by structure and wave energy rather than by a temporal sequence. To a first approximation, the latest eustatic rise of sea level affected all coasts simultaneously, and rather than passing sequentially from the primary to the secondary categories, some coasts have continuously responded to marine influence, either because of weak structures or high energy input, while others have retained the imprint of tectonic forms or subaerial processes and have passively drowned. More rarely, coasts have emerged because of strong tectonic uplift. However, Shepard's dichotomy was not necessarily meant to imply an evolutionary change but rather to provide convenient "pigeonholes" for coastal classification. Attempts to compare it with the earlier sequential scheme distort its intended purpose.

Tempo of Erosion

Rather than treating time as a direct variable in coastal classification, either as measured in years or as in stages of sequential changes, the rate, or *tempo,* of coastal landscape evolution can be used (Cotton, 1952). Some coasts have evolved at a rapid tempo in postglacial time; others have been negligibly modified by marine processes. In a sense, coasts with a rapid tempo of evolution are similar to Shepard's secondary coasts, and slowly changing coasts are the equivalent of Shepard's primary class. There are important differences, however. If steep cliffs were cut by the sea at a lower level, they might now "plunge" (Fig. 19–9) and reflect rather than re-fract incoming waves. Thus plunging cliffs might presently change only at a very slow tempo, even though they were shaped by marine erosion at a lower sea level.

The concept of tempo has been applied primarily to erosional coasts but could be extended to prograded coasts as well. Deltas and coral reefs have extended some coasts seaward despite postglacial submergence. A rapid tempo of progradation can offset the effects of submergence (Fig. 20–9) just as a rapid tempo of erosion can offset the effects of tectonic emergence. Tempo is not a very promising base for a geomorphic classification of coasts, because rapid tempo might be caused by high wave energy, or weak structures, or changing levels. Too many variables could have combined to produce the same observed results. The concept of tempo in coastal evolution is important in applied research on coasts, however. Coasts undergoing rapid change, either of erosion or progradation, present many more problems for human utilization (Komar, 1983).

Inheritance of Coastal Landforms

The area covered by each of the successive Pleistocene ice sheets was remarkably similar, differing by less than 10 percent. If the thicknesses of the ice sheets were also similar, as glacier-flow theory predicts (Chap. 16), then sea level has probably oscillated through a range of 120 ± 60 m during the later Cenozoic Era (Figs. 17–16, 18–1, and 20–1 to 20–4), repeatedly reoccupying shoreline features cut or built much earlier. Most modern coral reefs seem to be veneers of Holocene limestone about 10 m thick, deposited during the last 8000 years or less on foundations that date from the last interglacial. We cannot say how much of other modern coastal landscapes is relict, because we have almost no dating techniques that are suitable for nonreef coasts. Many broad shore platforms, barriers, and reefs that appear to be in equilibrium with modern sea level and wave energy may be only slightly modi-fied inherited forms (Fig. 19–4). Perhaps the unsatisfactory status of coastal explan-atory description is due to the fact that we have not yet learned to separate active from relict forms.

REFERENCES

AUDLEY-CHARLES, M. G., CURRAY, J. R., and EVANS, G., 1977, Location of major deltas: Geology, v. 5, pp. 341–44.

BAULIG, H., 1930, Le littoral Dalmate: Ann. Géogr., no. 219, pp. 305–10.

BENDER, M. L., FAIRBANKS, R. G., TAYLOR, F. W., MATTHEWS, R. K., GODDARD, J. T., and BROECKER, W. S., 1979, Uranium-series dating to the Pleistocene reef tracts of Bar-bados, West Indies: Geol. Soc. America Bull., v. 90, pp. 577–94.

BIRD, E. C. F., and HOPLEY, D., 1969, Geomorphological features on a humid tropical sec-tor of the Australian coast: Australian Geog. Studies, v. 7, pp. 89–108.

BLOOM, A. L., 1965, The explanatory description of coasts: Zeitschr. für Geomorph., v. 9, pp. 422–36.

———, 1980, Late Quaternary sea level change on South Pacific coasts: A study in tectonic diversity, in Mörner, N.-A., ed., Earth rheology, isostasy and eustasy: John Wiley & Sons Ltd., Chichester, England, pp. 505–15.

———, 1986, Coastal landforms, in Short, N. M., and Blair, R. W., Jr., eds., Geomorphology from space: NASA Spec. Pub. SP-486, Washington, D.C., pp. 353–406.

———, BROECKER, W. S., CHAPPELL, J. M. A., MATTHEWS, R. K., and MESOLELLA, K. J.,

1974, Quaternary sea level fluctuations on a tectonic coast: New ^{230}Th/^{234}U dates from the Huon Peninsula, New Guinea: Quaternary Res., v. 4, pp. 185–205.

BROECKER, W. S., and THURBER, D. L., 1965, Uranium-series dating of corals and oolites from Bahaman and Florida Key limestones: Science, v. 149, pp. 58–60.

CHAPPELL, J., 1974, Geology of coral terraces, Huon Peninsula, New Guinea: A study of Quaternary tectonic movements and sea-level changes: Geol. Soc. America Bull., v. 85, pp. 553–70.

———, and SHACKLETON, N. J., 1986, Oxygen isotopes and sea level: Nature, v. 324, pp. 137–40.

COOKE, C. W., 1930, Correlation of coastal terraces: Jour. Geology, v. 38, pp. 577–89.

COTTON, C. A., 1942, Shorelines of transverse deformation: Jour. Geomorph., v. 5, pp. 45–58 (reprinted 1974 *in* Bold coasts: A.H. & A.W. Reed, Wellington, New Zealand, pp. 34–45).

———, 1952, Criteria for the classification of coasts: Internat. Geog. Cong., 17th, Washington 1952, Proc., pp. 315–19 [1957] (reprinted 1974 *in* Bold coasts: A.H. & A.W. Reed, Wellington, New Zealand, pp. 118–25).

———, 1956, Rias *sensu stricto* and *sensu lato:* Geog. Jour., v. 122, pp. 360–64 (reprinted 1974 *in* Bold coasts: A.H. & A.W. Reed, Wellington, New Zealand, pp. 187–94).

CRONIN, T. M., 1988, Evolution of marine climates of the U.S. Atlantic coast during the past four million years: Phil. Trans. Roy. Soc. London, ser. B, v. 318, pp. 661–78.

DAVIES, J. L., 1980, Geographical variation in coastal development, 2nd ed.: Longman, Inc., New York, 212 pp.

DAVIS, W. M., 1905, Complications of the geographical cycle: Internat. Geog. Cong., 8th, Washington 1904, Repts., pp. 150–63 (reprinted 1954 *in* Geographical essays: Dover Publications, Inc., New York, pp. 279–95).

DIONNE, J. C., 1974, Bibliographie annotée sur les aspects géologiques du glaciel (annotated bibliography on the geological aspects of drift ice): Can. Centre Rech. For. Laurentides, Ste-Foy, Québec, Rapp. Inf. LAU-X-9, 122 pp.

FITZGERALD, D. M., and ROSEN, P. S., eds., 1987, Glaciated coasts: Academic Press, Inc., San Diego, California, 364 pp.

FLINT, R. F., 1971, Glacial and Quaternary geology: John Wiley & Sons, Inc., New York, 892 pp.

GJESSING, J., 1967, Norway's paleic surface: Norsk Geografsk Tidsskrift, v. 21, pp. 69–132.

GULLIVER, F. P., 1899, Shoreline topography: Am. Acad. Arts and Sciences Proc., v. 34, pp. 151–258.

HOLTEDAHL, H., 1958, Some remarks on geomorphology of continental shelves off Norway, Labrador, and southeast Alaska: Jour. Geology, v. 66, pp. 461–71.

INMAN, D. L., and NORDSTROM, C. E., 1971, On the tectonic and morphologic classification of coasts: Jour. Geology, v. 79, pp. 1–21.

JOHNSON, D. W., 1919, Shore processes and shoreline development: John Wiley & Sons, Inc., New York, 584 pp.

———, 1925, New England–Acadian shoreline: John Wiley & Sons, Inc., New York, 608 pp.

KOMAR, P. D., ed., 1983, CRC handbook of coastal processes and erosion: CRC Press, Inc., Boca Raton, Florida, 305 pp.

MESOLELLA, K. J., MATTHEWS, R. K., BROECKER, W. S., and THURBER, D. L., 1969, Astronomical theory of climatic change: Barbados data: Jour. Geology, v. 77, pp. 250–74.

SHACKLETON, N. J., and OPDYKE, N. D., 1973, Oxygen isotope and paleomagnetic stratigraphy of equatorial Pacific core V28-238: Oxygen isotope temperatures and ice volumes on a 10^5 year and 10^6 year time scale: Quaternary Res., v. 3, pp. 39–55.

SHEPARD, F. P., 1937, Revised classification of marine shorelines: Jour. Geology, v. 45, pp. 602–24.

———, 1973, Submarine geology, 3rd ed.: Harper & Row, Publishers, New York, 517 pp.

SUESS, E., 1904, The face of the earth, v. 1 (transl. by H. B. C. Sollas): Clarendon Press, Oxford, 604 pp.

TANNER, W. F., 1960, Florida coastal classification: Gulf Coast Assoc. Geol. Soc., v. 10, pp. 259–66.

TAYLOR, F. W., JR., and BLOOM, A. L., 1977, Coral reefs on tectonic blocks, Tonga island arc: Third Internat. Coral Reef Symposium, Miami, Proc., v. 2, pp. 275–81.

VALENTIN, H., 1952, Die Küsten der Erde: Petermanns Geogr. Mitt. Ergänzungsheft 246, Justus Perthes, Gotha, 118 pp.

————, 1970, Principles and problems of a handbook on regional coastal geomorphology of the world: Paper read at the Symposium of the IGU Commission on Coastal Geomorphology, Moscow, 10 pp.

WEHMILLER, J. F., and BELKNAP, D. F., 1982, Amino acid age estimates, Quaternary Atlantic coastal plain: Comparisons with U-series dates, biostratigraphy, and paleomagnetic control: Quaternary Res., v. 18, pp. 311–59.

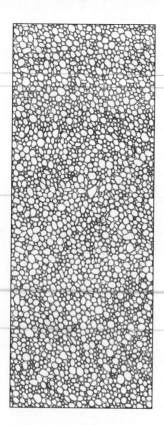

Author Index

(For multi-authored works, first and second authors only)

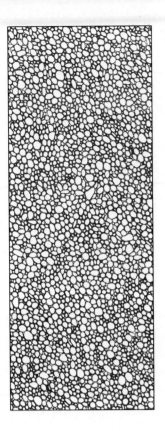

Subject Index